The Elementary Part
of a Treatise on
the Dynamics of a
System of Rigid Bodies

Edward John Routh

CAMBRIDGE
UNIVERSITY PRESS

CAMBRIDGE UNIVERSITY PRESS

Cambridge, New York, Melbourne, Madrid, Cape Town,
Singapore, São Paolo, Delhi, Mexico City

Published in the United States of America by Cambridge University Press, New York

www.cambridge.org
Information on this title: www.cambridge.org/9781108050319

© in this compilation Cambridge University Press 2013

This edition first published 1891
This digitally printed version 2013

ISBN 978-1-108-05031-9 Paperback

CAMBRIDGE LIBRARY COLLECTION

Books of enduring scholarly value

Mathematics

From its pre-historic roots in simple counting to the algorithms powering modern desktop computers, from the genius of Archimedes to the genius of Einstein, advances in mathematical understanding and numerical techniques have been directly responsible for creating the modern world as we know it. This series will provide a library of the most influential publications and writers on mathematics in its broadest sense. As such, it will show not only the deep roots from which modern science and technology have grown, but also the astonishing breadth of application of mathematical techniques in the humanities and social sciences, and in everyday life.

The Elementary Part of a Treatise on the Dynamics of a System of Rigid Bodies

As senior wrangler in 1854, Edward John Routh (1831–1907) was the man who beat James Clerk Maxwell in the Cambridge mathematics tripos. He went on to become a highly successful coach in mathematics at Cambridge, producing a total of twenty-seven senior wranglers during his career – an unrivalled achievement. In addition to his considerable teaching commitments, Routh was also a very able and productive researcher who contributed to the foundations of control theory and to the modern treatment of mechanics. First published in one volume in 1860, this textbook helped disseminate Routh's investigations into stability. This revised fifth edition was published in two volumes between 1891 and 1892. The first part establishes the principles of dynamics, providing formulae and examples throughout. While the growth of modern physics and mathematics may have forced out the problem-based mechanics of Routh's textbooks from the undergraduate syllabus, the utility and importance of his work is undiminished.

Cambridge University Press has long been a pioneer in the reissuing of out-of-print titles from its own backlist, producing digital reprints of books that are still sought after by scholars and students but could not be reprinted economically using traditional technology. The Cambridge Library Collection extends this activity to a wider range of books which are still of importance to researchers and professionals, either for the source material they contain, or as landmarks in the history of their academic discipline.

Drawing from the world-renowned collections in the Cambridge University Library and other partner libraries, and guided by the advice of experts in each subject area, Cambridge University Press is using state-of-the-art scanning machines in its own Printing House to capture the content of each book selected for inclusion. The files are processed to give a consistently clear, crisp image, and the books finished to the high quality standard for which the Press is recognised around the world. The latest print-on-demand technology ensures that the books will remain available indefinitely, and that orders for single or multiple copies can quickly be supplied.

The Cambridge Library Collection brings back to life books of enduring scholarly value (including out-of-copyright works originally issued by other publishers) across a wide range of disciplines in the humanities and social sciences and in science and technology.

THE ELEMENTARY PART

OF A TREATISE ON THE

DYNAMICS OF A SYSTEM OF RIGID BODIES

BEING PART I. OF A TREATISE ON THE WHOLE SUBJECT.

THE ELEMENTARY PART

OF A TREATISE ON THE

DYNAMICS OF A SYSTEM OF RIGID BODIES.

BEING PART I. OF A TREATISE ON THE WHOLE SUBJECT.

With numerous Examples.

BY

EDWARD JOHN ROUTH, Sc.D., LL.D., F.R.S., &c.

HON. FELLOW OF PETERHOUSE, CAMBRIDGE ;

FELLOW OF THE SENATE OF THE UNIVERSITY OF LONDON.

FIFTH EDITION, REVISED AND ENLARGED.

London:

MACMILLAN AND CO.

AND NEW YORK.

1891

First Edition, 1860. *Second Edition*, 1868.
Third Edition, 1877. *Fourth Edition*, 1882. *Fifth Edition*, 1891.

PREFACE.

In this edition many improvements have been made. Though some new matter has been added, most of the changes are in the form of additional explanations and generalizations of theorems already given. In some cases also the proofs have been simplified. The numbering of the articles is the same as in the last edition, except in some cases where the additions were such as to require a rearrangement.

When the last edition was printed, the book was divided into two parts in order to render it less bulky. This division has been retained. All the elements of the subject, together with some methods intended for the more advanced student, are placed in the first volume. In the second part the higher applications are given. In order that the plan of the book may be understood, a short summary of the subjects treated of in the second volume has been added to the table of contents.

As in the former editions, each chapter has been made as far as possible complete in itself, so that all that relates to any one part of the subject may be found in the same place. This arrangement is convenient for those who are already acquainted with dynamics, as it enables them to direct their attention to those parts in which they may feel most interested. It also enables the student to select his own order of reading. The student who is just beginning dynamics may not wish to be delayed by a long chapter of preliminary analysis before he enters on the real subject of the book. He may therefore begin with D'Alembert's Principle and read only those parts of chapter I. to which reference is made. Others may wish to pass on as soon as possible to the great principles of Angular Momentum and Vis Viva. Though a different order may be found advisable for some readers, I have ventured to indicate a list of Articles to which those who are beginning dynamics should first turn their attention.

It will be observed that a chapter has been devoted to the discussion of Motion in Two Dimensions. This course has been adopted because it seemed expedient to separate the difficulties of dynamics from those of solid geometry.

A slight historical notice of each result has been attempted whenever it could be briefly given. Such additions, if not carried too far, add greatly to the interest of the subject. But the success of the attempt is far from complete. In the earlier history there was the guidance of Montucla, and further on there was Prof. Cayley's Report to the British Association. With the help of these the task became comparatively easy; but in some other portions the number of memoirs which have been written is so vast that anything but a slight notice is impossible. A useful theorem is many times discovered, and probably each time with some variations. It is thus often difficult to ascertain who is the first author. It has therefore been found necessary to correct some of the references given in the former editions, and to add references where there were none before. It has not however been thought necessary to refer to the author's own additions to the subject except when they had already been printed elsewhere.

Throughout each chapter there will be found numerous examples, many being very easy, while others are intended for the more advanced student. In order to obtain as great a variety of problems as possible, a further collection has been added at the end of each chapter, taken from the Examination Papers which have been set in the University and in the Colleges. As these problems have been constructed by many different examiners, I hope that this selection will enable the student to acquire facility in solving all kinds of dynamical problems.

In constructing the examples my first care has been to follow closely the principle which each is intended to illustrate. But such instruments or applications of principles have been sought for as have been found useful in practice. Whenever some useful instrument has been found, which did not require so lengthy a description as to unfit it for an illustration, it has been preferred as an example to a merely curious and artificial construction.

In the former editions differential coefficients with regard to the time have been represented by accents in the chapters after the seventh. However unsuitable such a notation may be when several independent variables are used in the same investigation, it has some advantages in such a subject as dynamics, where the differentiations are nearly always taken with regard to the time. It was not used in the earlier chapters because it was thought that it would add to the initial difficulties of the subject those of an unaccustomed notation. But now that the representation of differential coefficients by dots has been used in several standard books both on elementary and on advanced mechanics, this reason has lost much of its force. Dots and accents have therefore been used throughout this edition whenever a shortened notation has appeared to be desirable. One objection to the use of this notation is that the meaning of the symbol may be changed by a slight error in the number of the dots or accents. As this might increase the difficulties of the subject to a beginner, the use of dots in the earlier chapters has been restricted chiefly to the working of examples, and care has been taken that the results should be clearly stated.

I cannot conclude without expressing how much I am indebted to Mr J. M. Dodds of Peterhouse for his assistance in correcting the proof sheets. I hope that the work, having had the advantage of his revision, will be found clear of serious errors.

EDWARD J. ROUTH.

PETERHOUSE,
December 8, 1890.

CONTENTS.

CHAPTER I.

ON MOMENTS OF INERTIA.

CHAPTER II.

D'ALEMBERT'S PRINCIPLE, &C.

CHAPTER III.

MOTION ABOUT A FIXED AXIS.

CHAPTER IV.

MOTION IN TWO DIMENSIONS.

CHAPTER V.

MOTION IN THREE DIMENSIONS.

CHAPTER VI.

ON MOMENTUM.

CHAPTER VII.

VIS VIVA.

CHAPTER VIII.

LAGRANGE'S EQUATIONS.

CHAPTER IX.

SMALL OSCILLATIONS.

CHAPTER X.

ON SOME SPECIAL PROBLEMS.

The following subjects will be treated of in the second volume.

Theory of moving axes, Clairaut's theorem, and motion relative to the earth.

Theory of small oscillations with several degrees of freedom both about a position of equilibrium and about a state of steady motion.

Motion of a body about a fixed point under no forces.

Motion of a body under any forces.

Theory of free and forced oscillations.

Methods of Isolation and of Multipliers.

Applications of the calculus of finite differences.

Applications of the calculus of variations.

Precession and Nutation.

Motion of a string or chain.

Motion of a membrane.

The student, to whom the subject is entirely new, is advised to read *first* the following articles: Chap. I. 1—25, 33—36, 47—52. Chap. II. 66—87. Chap. III. 88—93, 98—104, 110, 112—118. Chap. IV. 130—164, 168—175, 179—186, 199. Chap. V. 214—245, 248—256, 261—269. Chap. VI. 282—285, 288—295, 299—309. Chap. VII. 332—374. Chap. VIII. 395—409. Chap. IX. 432—463, 467—476. Chap. X. 483, 488—499.

ERRATUM.

Page 190, line 25, *for* $\left(\dfrac{gt^2}{2} - a\theta \right)$ *read* $\left(\dfrac{gt^2}{2} - a\theta \right)^2$.

CHAPTER I.

1. IN the subsequent pages of this work it will be found that certain integrals continually recur. It is therefore convenient to collect these into a preliminary chapter for reference. Though their bearing on dynamics may not be obvious beforehand, yet the student may be assured that it is as useful to be able to write down moments of inertia with facility as it is to be able to quote the centres of gravity of the elementary bodies.

In addition however to these necessary propositions there are many others which are useful as giving a more complete view of the arrangement of the axes of inertia in a body. These also have been included in this chapter though they are not of the same importance as the former.

2. All the integrals used in dynamics as well as those used in statics and some other branches of mixed mathematics are included in the one form

$$\iiint x^\alpha y^\beta z^\gamma \, dx \, dy \, dz,$$

where (α, β, γ) have particular values. In statics two of these three exponents are usually zero, and the third is either unity or zero, according as we wish to find the numerator or denominator of a co-ordinate of the centre of gravity. In dynamics of the three exponents one is zero, and the sum of the other two is usually equal to 2. The integral in all its generality has not yet been fully discussed, probably because only certain cases have any real utility. In the case in which the body considered is a homogeneous ellipsoid the value of the general integral has been found in gamma functions by Lejeune Dirichlet in Vol. IV. of Liouville's journal. His results were afterwards extended by Liouville in the same volume to the case of a heterogeneous ellipsoid in which the strata of uniform density are similar ellipsoids.

In this treatise, it is intended chiefly to restrict ourselves to the consideration of moments and products of inertia, as being the only cases of the integral which are useful in dynamics.

3. **Definitions.** If the mass of every particle of a material system be multiplied by the square of its distance from a straight line, the sum of the products so formed is called the *moment of inertia* of the system about that line.

If M be the mass of a system and k be such a quantity that Mk^2 is its moment of inertia about a given straight line, then k is called the *radius of gyration* of the system about that line.

The term "moment of inertia" was introduced by Euler, and has now come into general use wherever Rigid Dynamics is studied. It will be convenient for us to use the following additional terms.

If the mass of every particle of a material system be multiplied by the square of its distance from a given plane or from a given point, the sum of the products so formed is called the moment of inertia of the system with reference to that plane or that point.

If two straight lines Ox, Oy be taken as axes, and if the mass of every particle of the system be multiplied by its *two* co-ordinates x, y, the sum of the products so formed is called the *product of inertia* of the system about those two axes.

This might, perhaps more conveniently, be called the product of inertia of the system with reference to the two co-ordinate planes xz, yz.

The term *moment of inertia with regard to a plane* seems to have been first used by M. Binet in the *Journal Polytechnique*, 1813.

4. Let a body be referred to any rectangular axes Ox, Oy, Oz meeting in a point O, and let x, y, z be the co-ordinates of any particle m, then according to these definitions the moments of inertia about the axes of x, y, z respectively will be

$$A = \Sigma m \left(y^2 + z^2\right), \quad B = \Sigma m \left(z^2 + x^2\right), \quad C = \Sigma m \left(x^2 + y^2\right).$$

The moments of inertia with regard to the planes yz, zx, xy, respectively, will be

$$A' = \Sigma m x^2, \quad B' = \Sigma m y^2, \quad C' = \Sigma m z^2.$$

The products of inertia with regard to the axes yz, zx, xy, will be

$$D = \Sigma m yz, \quad E = \Sigma m zx, \quad F = \Sigma m xy.$$

Lastly, the moment of inertia with regard to the origin will be

$$H = \Sigma m \left(x^2 + y^2 + z^2\right) = \Sigma m r^2,$$

where r is the distance of the particle m from the origin.

5. **Elementary Propositions.** The following propositions may be established without difficulty, and will serve as illustrations of the preceding definitions.

(1) The three moments of inertia A, B, C about three rectangular axes are such that the sum of any two of them is greater than the third.

(2) The sum of the moments of inertia about any three rectangular axes meeting at a given point is always the same; and is equal to twice the moment of inertia with respect to that point.

For $A + B + C = 2\Sigma m\,(x^2 + y^2 + z^2) = 2\Sigma mr^2$, and is therefore independent of the directions of the axes.

(3) The sum of the moments of inertia of a system with reference to any plane through a given point and its normal at that point is constant and equal to the moment of inertia of the system with reference to that point.

Take the given point as origin and the plane as the plane of xy, then $C' + C = \Sigma mr^2$, which is independent of the directions of the axes.

Hence we infer that

$$A' = \tfrac{1}{2}(B + C - A), \quad B' = \tfrac{1}{2}(C + A - B), \quad \text{and} \quad C' = \tfrac{1}{2}(A + B - C).$$

(4) Any product of inertia as D cannot numerically be so great as $\tfrac{1}{2}A$.

(5) If A, B, F be the moments and product of inertia of a lamina about two rectangular axes in its plane, then AB is greater than F^2.

If t be any quantity we have $At^2 + 2Ft + B = \Sigma m\,(yt + x)^2 = $ a positive quantity. Hence the roots of the quadratic $At^2 + 2Ft + B = 0$ are imaginary, and therefore $AB > F^2$.

(6) Prove that for any body

$$(A + B - C)(B + C - A) > 4E^2,$$
$$(A + B - C)(B + C - A)(C + A - B) > 8DEF.$$

(7) The moment of inertia of the surface of a sphere of radius a and mass M about any diameter is $M\tfrac{2}{3}a^2$.

Since every element is equally distant from the centre its moment of inertia about the centre is Ma^2. Hence by (2) the result follows.

(8) The moment of inertia of the surface of a hemisphere of radius a and mass M about a diameter is $M\tfrac{2}{3}a^2$.

This follows immediately from (7) by completing the sphere.

6. It is clear that the process of finding moments and products of inertia is merely that of integration. We may illustrate this by the following example.

To find the moment of inertia of a uniform triangular plate about an axis in its plane passing through one angular point.

Let ABC be the triangle, Ay the axis about which the moment is required. Draw Ax perpendicular to Ay and produce

BC to meet Ay in D. The given triangle ABC may be regarded as the difference of the triangles ABD, ACD. Let us then first

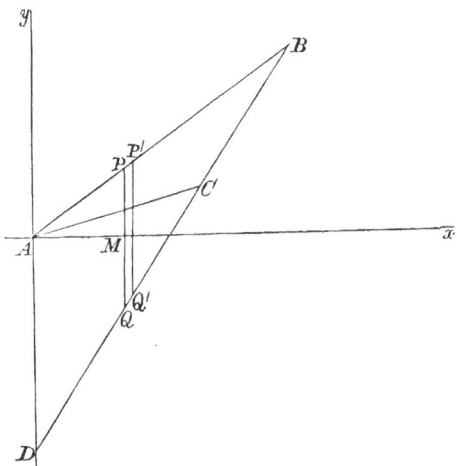

find the moment of inertia of ABD. Let $PQP'Q'$ be an elementary area whose sides PQ, $P'Q'$ are parallel to the base AD, and let PQ cut Ax in M. Let β be the distance of the angular point B from the axis Ay, $AM = x$ and $AD = l$.

Then the elementary area $PQP'Q'$ is clearly $l \dfrac{\beta - x}{\beta} dx$, and its moment of inertia about Ay is $\mu l \dfrac{\beta - x}{\beta} dx \cdot x^2$, where μ is the mass per unit of area. Hence the moment of inertia of the triangle ABD

$$= \mu \int_0^\beta l \left(1 - \frac{x}{\beta} \right) x^2 dx = \tfrac{1}{12} \mu l \beta^3.$$

Similarly if γ be the distance of the angular point C from the axis Ay, the moment of inertia of the triangle ACD is $\tfrac{1}{12} \mu l \gamma^3$. Hence the moment of inertia of the given triangle ABC is $\tfrac{1}{12} \mu l (\beta^3 - \gamma^3)$. Now $\tfrac{1}{2} l \beta$ and $\tfrac{1}{2} l \gamma$ are the areas of the triangles ABD, ACD. Hence if M be the mass of the triangle ABC, the moment of inertia of the triangle about the axis Ay is

$$\tfrac{1}{6} M (\beta^2 + \beta\gamma + \gamma^2).$$

Ex. If each element of the mass of the triangle be multiplied by the nth power of its distance from the straight line through the angle A, then it may be proved in the same way that the sum of the products is

$$\frac{2M}{(n+1)(n+2)} \frac{\beta^{n+1} - \gamma^{n+1}}{\beta - \gamma}.$$

7.   *When the body is a lamina the moment of inertia about an axis perpendicular to its plane is equal to the sum of the moments of inertia about any two rectangular axes in its plane drawn from the point where the former axis meets the plane.*

For let the axis of z be taken normal to the plane, then, if A, B, C be the moments of inertia about the axes, we have,

$$A = \Sigma my^2, \quad B = \Sigma mx^2, \quad C = \Sigma m\,(x^2 + y^2),$$

and therefore $C = A + B.$

We may apply this theorem to the case of the triangle. Let β', γ', be the distances of the points B, C from the axis Ax. Then the moment of inertia of the triangle about a normal to the plane of the triangle through the point A is

$$= \tfrac{1}{6} M \,(\beta^2 + \beta\gamma + \gamma^2 + \beta'^2 + \beta'\gamma' + \gamma'^2).$$

Ex.   Prove that the moment of inertia of the perimeter of a circle of radius a and mass M about any diameter is $\tfrac{1}{2} Ma^2$.

Since every element is equally distant from the axis of the circle, the moment of inertia about that axis is Ma^2. The result follows at once.

8.   **Reference Table.** The following moments of inertia occur so frequently that they have been collected together for reference. The reader is advised to commit to memory the following table:

The moment of inertia of

(1)   A rectangle whose sides are $2a$ and $2b$

about an axis through its centre in its plane perpendicular to the side $2a$ $\Big\} = \text{mass}\ \dfrac{a^2}{3}$,

about an axis through its centre perpendicular to its plane $\Big\} = \text{mass}\ \dfrac{a^2 + b^2}{3}$.

(2)   An ellipse semi-axes a and b

about the major axis $a = \text{mass}\ \dfrac{b^2}{4}$,

about the minor axis $b = \text{mass}\ \dfrac{a^2}{4}$,

about an axis perpendicular to its plane through the centre $\Big\} = \text{mass}\ \dfrac{a^2 + b^2}{4}$.

In the particular case of a circle of radius a, the moment of inertia about a diameter $= \text{mass}\ \dfrac{a^2}{4}$, and that about a perpendicular to its plane through the centre $= \text{mass}\ \dfrac{a^2}{2}$.

(3)   An ellipsoid semi-axes a, b, c

about the axis $a = \text{mass} \dfrac{b^2 + c^2}{5}$.

In the particular case of a sphere of radius a the moment of inertia about a diameter $= \text{mass } \dfrac{2}{5} a^2$.

(4)   A right solid whose sides are $2a$, $2b$, $2c$

$$\left.\begin{array}{l}\text{about an axis through its centre perpendicular}\\ \text{to the plane containing the sides } b \text{ and } c\end{array}\right\} = \text{mass} \dfrac{b^2 + c^2}{3}.$$

These results may be all included in one rule, which the author has long used as an assistance to the memory.

$$\left.\begin{array}{l}\text{Moment of inertia}\\ \text{about an axis}\\ \text{of symmetry}\end{array}\right\} = \text{mass} \dfrac{\text{(sum of squares of perpendicular semi-axes)}}{3,\ 4 \text{ or } 5}.$$

The denominator is to be 3, 4 or 5, according as the body is rectangular, elliptical or ellipsoidal.

Thus, if we require the moment of inertia of a circle of radius a about a diameter, we notice that the perpendicular semi-axes in its plane is the radius a, and that the semi-axis perpendicular to its plane is zero, the moment of inertia required is therefore $M\dfrac{a^2}{4}$, if M be the mass.   If we require the moment about a perpendicular to its plane through the centre, we notice that the perpendicular semi-axes are each equal to a and the moment required is therefore $M\dfrac{a^2 + a^2}{4} = M\dfrac{a^2}{2}$.

9.   As the process for determining these moments of inertia is very nearly the same for all these cases, it will be sufficient to consider only two instances.

To determine the moment of inertia of an ellipse about the minor axis.

Let the equation to the ellipse be $y = \dfrac{b}{a}\sqrt{a^2 - x^2}$.   Take any elementary area PQ parallel to the axis of y, then clearly the moment of inertia is

$$4\mu \int_0^a x^2 y\, dx = 4\mu \frac{b}{a} \int_0^a x^2 \sqrt{a^2 - x^2}\, dx,$$

where μ is the mass of a unit of area.

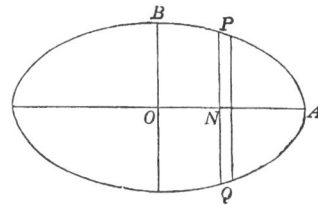

To integrate this, put $x = a \sin \phi$, then the integral becomes

$$a^4 \int_0^{\frac{\pi}{2}} \cos^2 \phi \sin^2 \phi \, d\phi = a^4 \int_0^{\frac{\pi}{2}} \frac{1 - \cos 4\phi}{8} \, d\phi = \frac{\pi a^4}{16} \, ;$$

$$\therefore \text{ the moment of inertia} = \mu \pi a b \frac{a^2}{4} = \text{mass } \frac{a^2}{4} \, .$$

In the same way we may show that the product of inertia of an elliptic quadrant about its axes $= \text{mass } \dfrac{ab}{2\pi}$.

To determine the moment of inertia of an ellipsoid about a principal diameter.

Let the equation to the ellipsoid be $\dfrac{x^2}{a^2} + \dfrac{y^2}{b^2} + \dfrac{z^2}{c^2} = 1$.   Take any elementary area PNQ parallel to the plane of yz. Its area is evidently $\pi PN \cdot QN$. Now PN is the

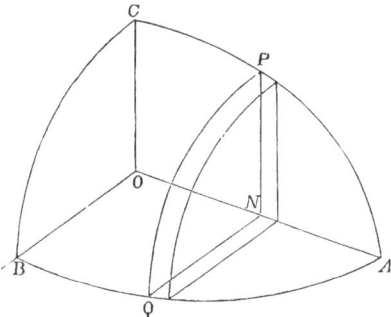

value of z when $y = 0$, and QN the value of y when $z = 0$, as obtained from the equation to the ellipsoid; $\therefore PN = \dfrac{c}{a} \sqrt{a^2 - x^2}, \quad QN = \dfrac{b}{a} \sqrt{a^2 - x^2};$

$$\therefore \text{ the area of the element} = \frac{\pi bc}{a^2} (a^2 - x^2).$$

Let μ be the mass of the unit of volume, then the whole moment of inertia

$$= \mu \int_{-a}^{a} \frac{\pi bc}{a^2} (a^2 - x^2) \frac{PN^2 + QN^2}{4} \, dx$$

$$= \mu \frac{\pi}{4} \frac{bc}{a^2} \int_{-a}^{a} (a^2 - x^2) \frac{b^2 + c^2}{a^2} (a^2 - x^2) \, dx$$

$$= \mu \frac{4}{3} \pi abc \frac{b^2 + c^2}{5} = \text{mass } \frac{b^2 + c^2}{5} \, .$$

In the same way we may show that the product of inertia of the octant of an ellipsoid about the axes of $(x, y) = \text{mass } \dfrac{2ab}{5\pi}$.

Ex. 1.   The moment of inertia of an arc of a circle whose radius is a and which subtends an angle 2α at the centre

(a)   about an axis through its centre perpendicular to its plane $= Ma^2$,

(b)   about an axis through its middle point perpendicular to its plane

$$= 2M \left(1 - \frac{\sin \alpha}{\alpha} \right) a^2,$$

(c) about the diameter which bisects the arc $= M \left(1 - \dfrac{\sin 2\alpha}{2\alpha}\right) \dfrac{a^2}{2}$.

Ex. 2. The moment of inertia of the part of the area of a parabola cut off by any ordinate at a distance x from the vertex is $\frac{3}{7} M x^2$ about the tangent at the vertex, and $\frac{1}{5} M y^2$ about the principal diameter, where y is the ordinate corresponding to x.

Ex. 3. The moment of inertia of the area of the lemniscate $r^2 = a^2 \cos 2\theta$ about a line through the origin in its plane and perpendicular to its axis is $M \dfrac{3\pi + 8}{48} a^2$.

Ex. 4. A lamina is bounded by four rectangular hyperbolas, two of them have the axes of co-ordinates for asymptotes, and the other two have the axes for principal diameters. Prove that the sum of the moments of inertia of the lamina about the co-ordinate axes is $\frac{1}{4}(a^2 - a'^2)(\beta^2 - \beta'^2)$, where aa', $\beta\beta'$ are the semi-major axes of the hyperbolas.

Take the equations $xy = u$, $x^2 - y^2 = v$, then the two moments of inertia are $A = \iint x^2 J\,du\,dv$ and $B = \iint y^2 J\,du\,dv$, where $1/J$ is the Jacobian of (u, v) with regard to (x, y). This gives at once $A + B = \frac{1}{2} \iint du\,dv$, where the limits are clearly $u = \frac{1}{2} a^2$ to $\frac{1}{2} a'^2$, $v = \beta^2$ to $v = \beta'^2$.

Ex. 5. A lamina is bounded on two sides by two similar ellipses, the ratio of the axes in each being m, and on the other two sides by two similar hyperbolas, the ratio of the axes in each being n. These four curves have their principal diameters along the co-ordinate axes. Prove that the product of inertia about the co-ordinate axes is $\dfrac{(a^2 - a'^2)(\beta^2 - \beta'^2)}{4(m^2 + n^2)}$, where aa', $\beta\beta'$ are the semi-major axes of the curves.

Ex. 6. If $d\sigma$ be an element of the surface of a sphere referred to any rectangular axes meeting at the centre, prove that $\displaystyle\int x^{2n}\,d\sigma = \dfrac{4\pi}{2n+1}\, r^{2n+2}$, where r is the radius of the sphere and n is integral.

Ex. 7. Taking the same axes as in the last example, prove that

$$\int x^{2f} y^{2g} z^{2h} d\sigma = \frac{4\pi}{2n+1}\, r^{2n+2} \frac{L(f)\,L(g)\,L(h)}{L(n)},$$

where $n = f + g + h$ and $L(f)$ stands for the quotient of the product of all the natural numbers up to $2f$ by the product of the same numbers up to f, both included.

To prove this, we notice that by the last example we have

$$\int (\lambda x + \mu y + \nu z)^{2n}\, d\sigma = (\lambda^2 + \mu^2 + \nu^2)^n \frac{4\pi r^{2n+2}}{2n+1}.$$

Expand both sides and equate the coefficients of $\lambda^{2f} \mu^{2g} \nu^{2h}$.

If we multiply the result by $D\,dr$ we have the value of the integral for any homogeneous shell of density D and thickness dr. Regarding D as a function of r, and integrating with regard to r, we can find the value of the integral for any heterogeneous sphere in which the strata of equal density are concentric spheres.

Ex. 8. If $d\sigma$ be an element of the surface of an ellipsoid referred to its principal diameters, and if p be the perpendicular from the centre on the tangent plane, prove

$$\int x^{2f} y^{2g} z^{2h} p\,d\sigma = \frac{4\pi}{2n+1} \frac{L(f)\,L(g)\,L(h)}{L(n)}\, a^{2f+1} b^{2g+1} c^{2h+1},$$

where a, b, c are the semi-axes and the rest of the notation is the same as before.

This result follows at once from the corresponding one for a spherical shell by the *method of projections.*

Ex. 9. Show that the volume V, the surface S, and the moment of inertia I with regard to the plane perpendicular to the co-ordinate x_1, of the sphere in space of n dimensions, whose equation is $x_1{}^2 + x_2{}^2 + \dots + x_n{}^2 = r^2$, are given by

$$V = r^n (\Gamma\tfrac{1}{2})^n / \Gamma(\tfrac{1}{2}n + 1), \qquad S = \frac{n}{r} V, \qquad I = V \frac{r^2}{n+2}.$$

These results follow easily from Dirichlet's theorem. See also Art. 5 (2).

10. **Method of Differentiation.** Many moments of inertia may be deduced from those given in Art. 8 by the method of differentiation. Thus the moment of inertia of a solid ellipsoid of uniform density ρ about the axis of a is known to be $\dfrac{4}{3} \pi abc\rho \dfrac{b^2 + c^2}{5}$. Let the ellipsoid increase indefinitely little in size, then the moment of inertia of the enclosed shell is

$$d \left\{ \frac{4}{3} \pi abc\rho \, \frac{b^2 + c^2}{5} \right\}.$$

This differentiation can be effected as soon as the law according to which the ellipsoid alters is given. Suppose the bounding ellipsoids to be similar, and let the ratio of the axes in each be given by $b = pa$, $c = qa$. Then

$$\text{moment of inertia of solid ellipsoid} = \frac{4}{3} \pi \rho pq \, \frac{p^2 + q^2}{5} \, a^5 ;$$

$$\therefore \ \text{moment of inertia of shell} = \frac{4}{3} \pi \rho pq \, (p^2 + q^2) \, a^4 da.$$

In the same way the mass of solid ellipsoid $= \dfrac{4}{3} \pi \rho pq a^3$;

$$\therefore \ \text{mass of shell} = 4\pi \rho pq a^2 da.$$

Hence the moment of inertia of an indefinitely thin ellipsoidal shell of mass M bounded by similar ellipsoids is $M \dfrac{b^2 + c^2}{3}$.

By reference to Art. 8, it will be seen that this is the same as the moment of inertia of the circumscribing right solid of equal mass. These two bodies therefore have equal moments of inertia about their axes of symmetry at the centre of gravity.

11. The moments of inertia of a heterogeneous body whose boundary is a surface of uniform density may sometimes be found by the method of differentiation. Suppose the moment of inertia of a homogeneous body of density D, bounded by any surface of uniform density, to be known. Let this when expressed in terms of some parameter a be $\phi(a) D$. Then the moment of inertia of a stratum of density D will be $\phi'(a) D da$. Replacing D by the variable density ρ, the moment of inertia required will be $\int \rho \phi'(a) da$.

Ex. 1. Show that the moment of inertia of a heterogeneous ellipsoid about the major axis, the strata of uniform density being similar concentric ellipsoids, and the density along the major axis varying as the distance from the centre, is $\frac{2}{9} M (b^2 + c^2)$.

Ex. 2. The moment of inertia of a heterogeneous ellipse about the minor axis, the strata of uniform density being confocal ellipses and the density along the minor axis varying as the distance from the centre, is $\frac{3M}{20} \frac{4a^5 + c^5 - 5a^3c^2}{2a^3 + c^3 - 3ac^2}$.

Other methods of finding moments of inertia.

12. The moments of inertia given in the table in Art. 8 are only a few of those in continual use. The moments of inertia of an ellipse, for example, about its principal axes are there given, but we shall also frequently want its moments of inertia about other axes. It is of course possible to find these in each separate case by integration. But this is a tedious process, and it may be often avoided by the use of the two following propositions.

The moments of inertia of a body about certain axes through its centre of gravity, which we may take as axes of reference, are regarded as given in the table. In order to find the moment of inertia of *that body* about any other axis we shall investigate,

(1) A method of comparing the required moment of inertia with that about a parallel axis through the centre of gravity. This is the theorem of parallel axes.

(2) A method of determining the moment of inertia about this parallel axis in terms of the given moments of inertia about the axes of reference. This is the theorem of the six constants of a body.

13. **Theorem of Parallel Axes.** *Given the moments and products of inertia about all axes through the centre of gravity of a body, to deduce the moments and products about all other parallel axes.*

The moment of inertia of a body or system of bodies about any axis is equal to the moment of inertia about a parallel axis through the centre of gravity plus the moment of inertia of the whole mass collected at the centre of gravity about the original axis.

The product of inertia about any two axes is equal to the product of inertia about two parallel axes through the centre of gravity plus the product of inertia of the whole mass collected at the centre of gravity about the original axes.

Firstly, take the axis about which the moment of inertia is required as the axis of z. Let m be the mass of any particle of the body, which generally will be any small element. Let x, y, z be the co-ordinates of m, $\bar{x}$, $\bar{y}$, $\bar{z}$ those of the centre of gravity G of the whole system of bodies, x', y', z' those of m referred to a system of parallel axes through the centre of gravity.

Then since $\dfrac{\Sigma mx'}{\Sigma m}$, $\dfrac{\Sigma my'}{\Sigma m}$, $\dfrac{\Sigma mz'}{\Sigma m}$ are the co-ordinates of the centre of gravity of the system referred to the centre of gravity as the origin, it follows that $\Sigma mx' = 0$, $\Sigma my' = 0$, $\Sigma mz' = 0$.

The moment of inertia of the system about the axis of z is

$$= \Sigma m \, (x^2 + y^2),$$
$$= \Sigma m \, \{(\bar{x} + x')^2 + (\bar{y} + y')^2\},$$
$$= \Sigma m \, (\bar{x}^2 + \bar{y}^2) + \Sigma m \, (x'^2 + y'^2) + 2\bar{x} \, . \, \Sigma mx' + 2\bar{y} \, . \, \Sigma my'.$$

Now $\Sigma m \, (\bar{x}^2 + \bar{y}^2)$ is the moment of inertia of a mass Σm collected at the centre of gravity, and $\Sigma m \, (x'^2 + y'^2)$ is the moment of inertia of the system about an axis through G, also $\Sigma mx' = 0$, $\Sigma my' = 0$; whence the proposition is proved.

Secondly, take the axes of x, y as the axes about which the product of inertia is required. The product required is

$$= \Sigma m \, xy = \Sigma m \, (\bar{x} + x') \, (\bar{y} + y'),$$
$$= \bar{x}\bar{y} \, . \, \Sigma m + \Sigma mx'y' + \bar{x}\Sigma my' + \bar{y}\Sigma mx'$$
$$= \bar{x}\bar{y}\Sigma m + \Sigma mx'y'.$$

Now $\bar{x}\bar{y} \, . \, \Sigma m$ is the product of inertia of a mass Σm collected at G and $\Sigma mx'y'$ is the product of the whole system about axes through G; whence the proposition is proved.

Let there be two parallel axes A and B at distances a and b from the centre of gravity of the body. Then, if M be the mass of the material system,

$$\left. \begin{matrix} \text{moment of inertia} \\ \text{about } A \end{matrix} \right\} - Ma^2 = \left\{ \begin{matrix} \text{moment of inertia} \\ \text{about } B \end{matrix} \right. - Mb^2.$$

Hence when the moment of inertia of a body about one axis is known, that about any other parallel axis may be found. It is obvious that a similar proposition holds with regard to the products of inertia.

14. The preceding proposition may be generalized as follows. Let any system be in motion, and let x, y, z be the co-ordinates at time t of any particle of mass m, then $\dfrac{dx}{dt}$, $\dfrac{dy}{dt}$, $\dfrac{dz}{dt}$ are the velocities, and $\dfrac{d^2x}{dt^2}$, $\dfrac{d^2y}{dt^2}$, $\dfrac{d^2z}{dt^2}$ the accelerations of the particle

resolved parallel to the axes.   Suppose

$$V = \Sigma m\phi \left(x, \frac{dx}{dt}, \frac{d^2x}{dt^2}, y, \frac{dy}{dt}, \frac{d^2y}{dt^2}, z, \frac{dz}{dt}, \frac{d^2z}{dt^2} \right)$$

to be a given function depending on the structure and motion of the system, the summation extending throughout the system. Also let ϕ be an algebraic function of the first or second order. Thus ϕ may consist of such terms as

$$ax^2 + bx\frac{dy}{dt} + c\left(\frac{dz}{dt}\right)^2 + eyz + fx + \ldots\ldots$$

where a, b, c, &c. are some constants.   Then the following general principle will hold.

The value of V *for any system of co-ordinates is equal to the value of* V *obtained for a parallel system of co-ordinates with the centre of gravity for origin plus the value of* V *for the whole mass collected at the centre of gravity with reference to the first system of co-ordinates.*

For let $\bar{x}$, $\bar{y}$, $\bar{z}$ be the co-ordinates of the centre of gravity, and let $x = \bar{x} + x'$, &c. $\therefore \frac{dx}{dt} = \frac{d\bar{x}}{dt} + \frac{dx'}{dt}$, &c.

Now since ϕ is an algebraic function of the second order of $x, \frac{dx}{dt}, \frac{d^2x}{dt^2}$; y, &c. it is evident that on making the above substitution and expanding, the process of squaring &c. will lead to three sets of terms, those containing only $\bar{x}, \frac{d\bar{x}}{dt}, \frac{d^2\bar{x}}{dt^2}$, &c., those containing the products of $\bar{x}$, x' &c., and lastly those containing only $x', \frac{dx'}{dt}$, &c.   The first of these will on the whole make up $\phi\left(\bar{x}, \frac{d\bar{x}}{dt}, \text{&c.}\right)$, and the last $\phi\left(x', \frac{dx'}{dt}, \text{&c.}\right)$.

Hence $V = \Sigma m\phi\left(\bar{x}, \frac{d\bar{x}}{dt} \ldots\right) + \Sigma m\phi\left(x', \frac{dx'}{dt} + \ldots\right)$

$$+ \Sigma m\left(A\bar{x}\frac{dx'}{dt} + B\frac{d\bar{x}}{dt}x' + C\bar{x}\frac{dy'}{dt} + \ldots\right),$$

where A, B, C, &c. are some constants.

Now the term $\Sigma m\left(\bar{x}\frac{dx'}{dt}\right)$ is the same as $\bar{x}\Sigma m\frac{dx'}{dt}$, and this vanishes.   For since $\Sigma mx' = 0$, it follows that $\Sigma m\frac{dx'}{dt} = 0$.   Similarly all the other terms in the second line vanish.

Hence the value of V is reduced to two terms. But the first of these is the value of V for the whole mass collected at the centre of gravity, and the second of these the value of V for the whole system referred to the centre of gravity as origin. Hence the proposition is proved.

The proposition would obviously be true if $\dfrac{d^3x}{dt^3}$, $\dfrac{d^3y}{dt^3}$, $\dfrac{d^3z}{dt^3}$, or any higher differential coefficients were also present in the function V.

15. **Theorem of the six constants of a body.** *Given the moments and products of inertia about three straight lines at right angles meeting in a point, to deduce the moments and products of inertia about all other axes meeting in that point.*

Take these three straight lines as the axes of co-ordinates. Let A, B, C be the moments of inertia about the axes of x, y, z; D, E, F the products of inertia about the axes of yz, zx, xy. Let α, β, γ be the direction-cosines of any straight line through the origin, then the moment of inertia I of the body about that line will be given by the equation

$$I = A\alpha^2 + B\beta^2 + C\gamma^2 - 2D\beta\gamma - 2E\gamma\alpha - 2F\alpha\beta.$$

Let P be any point of the body at which a mass m is situated, and let x, y, z be the co-ordinates of P. Let ON be the line whose direction-cosines are α, β, γ, draw PN perpendicular to ON.

Since ON is the projection of OP, it is clearly

$$= x\alpha + y\beta + z\gamma,$$

also $OP^2 = x^2 + y^2 + z^2$, and $1 = \alpha^2 + \beta^2 + \gamma^2$.

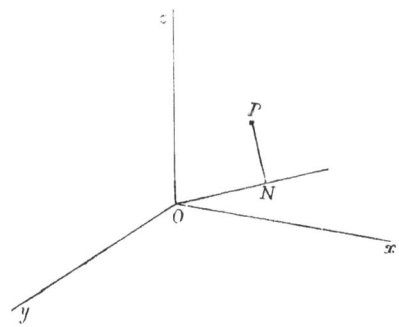

The moment of inertia I about $ON = \Sigma m PN^2$

$$= \Sigma m \left\{x^2 + y^2 + z^2 - (\alpha x + \beta y + \gamma z)^2\right\}$$
$$= \Sigma m \left\{(x^2 + y^2 + z^2)(\alpha^2 + \beta^2 + \gamma^2) - (\alpha x + \beta y + \gamma z)^2\right\}$$
$$= \Sigma m \,(y^2 + z^2)\,\alpha^2 + \Sigma m \,(z^2 + x^2)\beta^2 + \Sigma m \,(x^2 + y^2)\,\gamma^2$$
$$\qquad - 2\Sigma myz \,.\, \beta\gamma - 2\Sigma mzx \,.\, \gamma\alpha - 2\Sigma mxy \,.\, \alpha\beta$$
$$= A\alpha^2 + B\beta^2 + C\gamma^2 - 2D\beta\gamma - 2E\gamma\alpha - 2F\alpha\beta.$$

It may be shown in exactly the same manner that if $A'B'C'$ be the moments of inertia with regard to the *planes yz, zx, xy*, then the moment of inertia with regard to the *plane* whose direction-cosines are α, β, γ is

$$I' = A'\alpha^2 + B'\beta^2 + C'\gamma^2 + 2D\beta\gamma + 2E\gamma\alpha + 2F\alpha\beta.$$

It should be remarked that this formula differs from that giving the moment about a straight line in the signs of the three last terms.

16.   When three straight lines at right angles and meeting in a given point are such that if they be taken as axes of co-ordinates the products Σmxy, Σmyz, Σmzx all vanish, these are said to be *Principal Axes* at the given point.

The three planes which pass each through two principal axes are called the *Principal Planes* at the given point.

The moments of inertia about the principal axes at any point are called the *Principal moments of inertia* at that point.

The fundamental formula in Art. 15 may be much simplified if the axes of co-ordinates can be chosen so as to be principal axes at the origin.   In this case the expression takes the simple form

$$I = A\alpha^2 + B\beta^2 + C\gamma^2.$$

A method will presently be given by which we can always find these axes, but in some simpler cases we may determine their position by inspection.   Let the body be symmetrical about the plane of xy.   Then for every element m on one side of the plane whose co-ordinates are (x, y, z) there is another element of equal mass on the other side whose co-ordinates are $(x, y, -z)$. Hence for such a body $\Sigma mxz = 0$ and $\Sigma myz = 0$.   If the body be a lamina in the plane of xy, then the z of every element is zero, and we have again $\Sigma mxz = 0$, $\Sigma myz = 0$.

Recurring to the table in Art. 8, we see that in every case the axes, about which the moments of inertia are given, are principal axes.   Thus in the case of the ellipsoid, the three principal sections are all planes of symmetry, and therefore, by what has just been said, the principal diameters are principal axes of inertia.   In applying the fundamental formula of Art. 15 to any body mentioned in the table, we may therefore always use the modified form given in this article.

17.   Let us now consider how the two important propositions of Arts. 13 and 15 are to be applied in practice.

Ex. 1.   Suppose we want the moment of inertia of an elliptic area of mass M and semiaxes a and b about a diameter making an angle θ with the major axis.   The moments of inertia about the axes of a and b respectively are $\frac{1}{4}Mb^2$ and $\frac{1}{4}Ma^2$. Then by Art. 16 the moment of inertia about the diameter is

$$\tfrac{1}{4} Mb^2 \cos^2 \theta + \tfrac{1}{4} Ma^2 \sin^2 \theta.$$

If r be the length of the diameter this is known from the equation to the ellipse to be the same as $\dfrac{M}{4} \dfrac{a^2 b^2}{r^2}$, which is a very convenient form in practice.

Ex. 2.　Suppose we want the moment of inertia of the same ellipse about a tangent. Let p be the perpendicular from the centre on the tangent, then by Art. 13, the required moment is equal to the moment of inertia about a parallel axis through the centre together with $Mp^2 = \dfrac{M}{4} \dfrac{a^2 b^2}{r^2} + Mp^2 = \dfrac{5M}{4} p^2$, since $pr = ab$.

Ex. 3.　As an example of a different kind, let us find the moment of inertia of an ellipsoid of mass M and semiaxes (a, b, c) with regard to a diametral *plane* whose direction-cosines referred to the principal planes are (a, β, γ). By Art. 8, the moments of inertia with regard to the principal axes are $\tfrac{1}{5} M (b^2 + c^2)$, $\tfrac{1}{5} M (c^2 + a^2)$, $\tfrac{1}{5} M (a^2 + b^2)$. Hence by Art. 5, the moments of inertia with regard to the principal *planes* are $\tfrac{1}{5} Ma^2$, $\tfrac{1}{5} Mb^2$, $\tfrac{1}{5} Mc^2$. Hence the required moment of inertia is $\tfrac{1}{5} M (a^2 a^2 + b^2 \beta^2 + c^2 \gamma^2)$. If p be the perpendicular on the parallel tangent plane, we know by solid geometry that this is the same as $M \dfrac{p^2}{5}$.

Ex. 4.　The moment of inertia of a rectangle whose sides are $2a$, $2b$ about a diagonal is $\dfrac{2M}{3} \dfrac{a^2 b^2}{a^2 + b^2}$.

Ex. 5.　If k_1, k_2 be the radii of gyration of an elliptic lamina about two conjugate diameters, then $\dfrac{1}{k_1^2} + \dfrac{1}{k_2^2} = 4 \left(\dfrac{1}{a^2} + \dfrac{1}{b^2} \right)$.

Ex. 6.　The sum of the moments of inertia of an elliptic area about any two tangents at right angles is always the same.

Ex. 7.　If M be the mass of a right cone, a its altitude and b the radius of the base, then the moment of inertia about the axis is $M \dfrac{3}{10} b^2$; that about a straight line through the vertex perpendicular to the axis is $M \dfrac{3}{5} \left(a^2 + \dfrac{1}{4} b^2 \right)$, that about a slant side $M \dfrac{3b^2}{20} \dfrac{6a^2 + b^2}{a^2 + b^2}$; that about a perpendicular to the axis through the centre of gravity is $M \dfrac{3}{80} (a^2 + 4b^2)$.

Ex. 8.　If a be the altitude of a right cylinder, b the radius of the base, then the moment of inertia about the axis is $\tfrac{1}{2} Mb^2$ and that about a straight line through the centre of gravity perpendicular to the axis is $\tfrac{1}{4} M (\tfrac{1}{3} a^2 + b^2)$.

Ex. 9.　The moment of inertia of a body of mass M about a straight line whose equation is $\dfrac{x - f}{l} = \dfrac{y - g}{m} = \dfrac{z - h}{n}$ referred to any rectangular axes meeting at the centre of gravity is

$$Al^2 + Bm^2 + Cn^2 - 2Dmn - 2Enl - 2Flm + M \{ f^2 + g^2 + h^2 - (fl + gm + hn)^2 \},$$

where (l, m, n) are the direction-cosines of the straight line.

Ex. 10.　The moment of inertia of an elliptic disc whose equation is

$$ax^2 + 2bxy + cy^2 + 2dx + 2ey + 1 = 0,$$

about a diameter parallel to the axis of x, is $\dfrac{M}{4} \cdot \dfrac{-Ha}{(ac - b^2)^2}$, where M is the mass and H is the determinant $ac - b^2 + 2bed - ae^2 - cd^2$, usually called the discriminant.

Ex. 11. The moment of inertia of the elliptic disc whose equation in areal co-ordinates is $\phi(xyz) = 0$ about a diameter parallel to the side a is

$$- M \left(\frac{\Delta}{a}\right)^2 \frac{H}{2K^2} \left(\frac{d}{dy} - \frac{d}{dz}\right)^2 \phi,$$

where Δ is the area, H the discriminant and K the bordered discriminant.

18. Method of transformation of axes. The method used in Art. 15 to find the moment of inertia about the straight line ON is really equivalent to a change of co-ordinate axes in which this straight line is taken as a new axis, say, of ξ, those of η and ζ not being required. We may now generalize this into a method which is often of great practical use.

Let us suppose that $\phi(\xi\eta\zeta)$ is any quadric function, say

$$\phi = L_1 \xi^2 + L_2 \eta^2 + L_3 \zeta^2 + 2K_1 \eta\zeta + 2K_2 \zeta\xi + 2K_3 \xi\eta,$$

and that it is required to find $\Sigma m\phi(\xi\eta\zeta)$ the summation extending throughout any body.

Select some convenient set of axes which we may call x, y, z having the same origin such that the *six constants of the body*, viz. Σmx^2, Σmy^2, Σmz^2, Σmxy, Σmyz, Σmzx are all known or can be easily found. Let the direction-cosines of these axes be given by the diagram in the margin.

	ξ	η	ζ
x	α	β	γ
y	α'	β'	γ'
z	α''	β''	γ''

We then have $\xi = \alpha x + \alpha' y + \alpha'' z$, $\eta = \beta x + \beta' y + \beta'' z$, $\zeta = \gamma x + \gamma' y + \gamma'' z$. Substituting these values and expanding we obtain an expression for $\Sigma m\phi(\xi\eta\zeta)$ in terms of the six known constants of the body.

The result may appear at first sight to be rather complicated, but if the new axes be properly chosen it reduces in most cases to a few terms. Thus if the *axes of xyz are principal axes* the terms Σmxy, Σmyz, Σmzx are all zero. Supposing this choice to be made, the formula reduces to the convenient form

$$\Sigma m\phi(\xi\eta\zeta) = \phi(\alpha\beta\gamma) \Sigma mx^2 + \phi(\alpha'\beta'\gamma') \Sigma my^2 + \phi(\alpha''\beta''\gamma'') \Sigma mz^2.$$

In using this formula, the coefficient of Σmx^2 is obtained by substituting for $(\xi\eta\zeta)$ in $\phi(\xi\eta\zeta)$ the direction-cosines of the *new axis of x*, i.e. the cosines in the row of the diagram marked x. The coefficient of Σmy^2 may be obtained by substituting the direction-cosines of the new axis of y, i.e. the cosines in the row marked y, and so on.

If it be required to change the origin of co-ordinates also, this may be done by an application of the theorem in Art. 14.

Ex. 1. The co-ordinates of the centre of an elliptic area are (fgh) and the direction-cosines of its axes are $(\alpha\beta\gamma)(\alpha'\beta'\gamma')$, prove that

$$\Sigma m\zeta^2 = M\left(h^2 + \tfrac{1}{4}a^2\gamma^2 + \tfrac{1}{4}b^2\gamma'^2\right).$$

Ex. 2. Let Ox, Oy, Oz be the principal axes at the origin, prove that the product of inertia $F' = \Sigma m\xi\eta$ about two rectangular axes $O\xi$, $O\eta$ whose directions are $(\alpha\alpha'\alpha'')(\beta\beta'\beta'')$ is given by either of the formulae

$$\Sigma m\xi\eta = \alpha\beta\Sigma mx^2 + \alpha'\beta'\Sigma my^2 + \alpha''\beta''\Sigma mz^2$$
$$= -\alpha\beta A - \alpha'\beta'B - \alpha''\beta''C.$$

The second result follows from the first since $\alpha\beta + \alpha'\beta' + \alpha''\beta'' = 0$.

Ex. 3. Let $(\gamma\gamma'\gamma'')$ be the direction-cosines of a fixed axis $O\zeta$. Then as $O\xi$, $O\eta$ turn round $O\zeta$, prove that $D'^2 + E'^2$ and $A'B' - F'^2$ are both constant where A', B', C', D', E', F' are the moments and products of inertia of the body referred to these moving axes.

For by Ex. 2, $-D' = A\beta\gamma + B\beta'\gamma' + C\beta''\gamma''$, $-E' = A\alpha\gamma + Ba'\gamma' + Ca''\gamma''$;

$$\therefore \ D'^2 + E'^2 = A^2\gamma^2(a^2 + \beta^2) + 2AB\gamma\gamma'(aa' + \beta\beta') + \&c.,$$

since $a^2 + \beta^2 = 1 - \gamma^2 = \gamma'^2 + \gamma''^2$ and $aa' + \beta\beta' = -\gamma\gamma'$ we have

$$D'^2 + E'^2 = (A - B)^2(\gamma\gamma')^2 + (B - C)^2(\gamma'\gamma'')^2 + (C - A)^2(\gamma''\gamma)^2.$$

Similarly $A'B' - F'^2 = BC\gamma^2 + CA\gamma'^2 + AB\gamma''^2.$

The Ellipsoids of Inertia.

19.   The expression which has been found in Art. 15 for the moment of inertia I about a straight line whose direction-cosines are (α, β, γ),

$$I = A\alpha^2 + B\beta^2 + C\gamma^2 - 2D\beta\gamma - 2E\gamma\alpha - 2F\alpha\beta,$$

admits of a very useful geometrical interpretation.

Let a radius vector OQ move in any manner about the given point O, and be of such length that the moment of inertia about OQ may be *proportional* to the inverse square of the length. Then if R represent the length of the radius vector whose direction-cosines are (α, β, γ), we have $I = \dfrac{M\epsilon^4}{R^2}$, where ϵ is some constant introduced to keep the dimensions correct, and M is the mass. Hence the polar equation to the locus of Q is

$$\frac{M\epsilon^4}{R^2} = A\alpha^2 + B\beta^2 + C\gamma^2 - 2D\beta\gamma - 2E\gamma\alpha - 2F\alpha\beta.$$

Transforming to Cartesian co-ordinates, we have

$$M\epsilon^4 = AX^2 + BY^2 + CZ^2 - 2DYZ - 2EZX - 2FXY,$$

which is the equation to a quadric. Thus to every point O of a material body there is a corresponding quadric which possesses the property that the moment of inertia about any radius vector is represented by the inverse square of that radius vector. The convenience of this construction is, that the relations which exist between the moments of inertia about straight lines meeting at any given point may be discovered by help of the known properties of a quadric.

Since a moment of inertia is essentially positive, being by definition the sum of a number of squares, it is clear that every radius vector R must be real. Hence the quadric is always an ellipsoid. It is called the *momental ellipsoid*, and was first used by Cauchy, *Exercises de Math.* Vol. II.

So much has been written on the ellipsoids of inertia that it is difficult to determine what is really due to each of the various authors. The reader will find much information on these points in Prof. Cayley's report to the British Association on the *Special problems of Dynamics*, 1862.

20. The Invariants. The momental ellipsoid is defined by a *geometrical* property, viz. that any radius vector is equal to some constant divided by the square root of the moment of inertia about that radius vector. Hence whatever co-ordinate axes are taken, we must always arrive at the same ellipsoid. If therefore the momental ellipsoid be referred to any set of rectangular axes, the coefficients of X^2, Y^2, Z^2, $-2YZ$, $-2ZX$, $-2XY$ in its equation will still represent the moments and products of inertia about these axes.

Since the discriminating cubic determines the lengths of the axes of the ellipsoid, it follows that its coefficients are unaltered by a transformation of axes. But these coefficients are

$$A + B + C,$$
$$AB + BC + CA - D^2 - E^2 - F^2,$$
$$ABC - 2DEF - AD^2 - BE^2 - CF^2.$$

Hence for all rectangular axes having the same origin, these are invariable and all greater than zero.

21. It should be noticed that the constant ϵ is arbitrary, though when once chosen it cannot be altered. Thus we have a series of similarly and similarly situated ellipsoids, any one of which may be used as a momental ellipsoid.

When the body is a plane lamina, a section of the ellipsoid corresponding to any point in the lamina by the plane of the lamina, is called a *momental ellipse* at that point.

22. If principal axes at any point O of a body be taken as axes of co-ordinates, the equation to the momental ellipsoid takes the simple form $AX^2 + BY^2 + CZ^2 = M\epsilon^4$, where M is the mass and ϵ^4 any constant. Let us now apply this to some simple cases.

Ex. 1. To find the momental ellipsoid at the centre of a material elliptic disc. Taking the same notation as before, we have $A = \frac{1}{4}Mb^2$, $B = \frac{1}{4}Ma^2$, $C = \frac{1}{4}M(a^2 + b^2)$. Hence the ellipsoid is $\frac{1}{4}Mb^2 X^2 + \frac{1}{4}Ma^2 Y^2 + \frac{1}{4}M(a^2 + b^2) Z^2 = M\epsilon^4$. Since ϵ is any constant, this may be written

$$\frac{X^2}{a^2} + \frac{Y^2}{b^2} + \left(\frac{1}{a^2} + \frac{1}{b^2}\right) Z^2 = \epsilon'.$$

When $Z = 0$, this becomes an ellipse similar to the boundary of given disc. Hence we infer that the momental ellipse at the centre of an elliptic area is any similar and similarly situated ellipse. This also follows from Art. 17, Ex. 1.

Ex. 2. To find the momental ellipsoid at any point O of a material straight rod AB of mass M and length $2a$. Let the straight line OAB be the axis of x, O the origin, G the middle point of AB, $OG = c$. If the material line can be regarded as indefinitely thin, $A = 0$, $B = M\left(\frac{1}{3}a^2 + c^2\right) = C$, hence the momental ellipsoid is $Y^2 + Z^2 = \epsilon'^2$, where ϵ' is any constant. The momental ellipsoid is therefore an elongated spheroid, which becomes a right cylinder having the straight line for axis, when the rod becomes indefinitely thin.

Ex. 3. The momental ellipsoid at the centre of a material ellipsoid is

$$(b^2 + c^2)X^2 + (c^2 + a^2)Y^2 + (a^2 + b^2)Z^2 = \epsilon^4,$$

where ϵ is any constant. It should be noticed that the longest and shortest axes of the momental ellipsoid coincide in direction with the longest and shortest axes respectively of the material ellipsoid.

23. **Elementary Properties of Principal Axes.** By a consideration of some simple properties of ellipsoids, the following propositions are evident:

I. *Of the moments of inertia of a body about axes meeting at a given point, the moment of inertia about one of the principal axes is greatest and about another least.*

For, in the momental ellipsoid, the moment of inertia about a radius vector from the centre is least when that radius vector is greatest and *vice versâ*. And it is evident that the greatest and least radii vectores are two of the principal diameters.

It follows by Art. 5 that of the moments of inertia with regard to all *planes* passing through a given point, that with regard to one principal plane is greatest and with regard to another is least.

II. *If the three principal moments at any point O are equal to each other, the ellipsoid becomes a sphere.* Every diameter is then a principal diameter, and the radii vectores are all equal. Hence *every straight line through O is a principal axis at O, and the moments of inertia about them are all equal.*

For example, the perpendiculars from the centre of gravity of a cube on the three faces are principal axes; for, the body being referred to them as axes, we clearly have $\Sigma mxy = 0$, $\Sigma myz = 0$, $\Sigma mzx = 0$. Also the three moments of inertia about them are by symmetry equal. Hence every axis through the centre of gravity of a cube is a principal axis, and the moments of inertia about them are all equal.

Next suppose the body to be a regular solid. Consider two planes drawn through the centre of gravity each parallel to a face of the solid. The relations of these two planes to the solid are in all respects the same. Hence also the momental ellipsoid at the centre of gravity must be similarly situated with regard to each of these planes, and the same is true for planes parallel to all the faces. Hence the ellipsoid must be a sphere and the moment of inertia will be the same about every axis.

Ex. 1.   Three equal particles A, B, C are placed at the corners of an equilateral triangle ; prove that the momental ellipse at their centre of gravity G is a circle.

By symmetry the diameters GA, GB, GC of the momental ellipse at G must be equal.   The ellipse is therefore a circle.

Ex. 2.   Four equal particles are placed at the corners of a tetrahedron.   If the momental ellipsoid at their centre of gravity is a sphere prove that the tetrahedron is regular.

Ex. 3.   Any point O in a body being given and any plane drawn through it, prove that two straight lines at right angles can be drawn in this plane through O such that the product of inertia about them is zero.

These are the axes of the section of the momental ellipsoid at the point O formed by the given plane.

24.   *At every point of a material system there are always three principal axes at right angles to each other.*

Construct the momental ellipsoid at the given point.   Then it has been shown that the products of inertia about the axes are half the coefficients of $-XY$. $-YZ$, $-ZX$ in the equation to the momental ellipsoid referred to these straight lines as axes of co-ordinates.   Now if an ellipsoid be referred to its principal diameters as axes, these coefficients vanish.   Hence the principal diameters of the ellipsoid are the principal axes of the system. But every ellipsoid has at least three principal diameters, hence every material system has at least three principal axes.

25. Ex. 1.   The principal axes at the centre of gravity being the axes of reference, prove that the momental ellipsoid at the point (p, q, r) is

$$\left(\frac{A}{M}+q^2+r^2\right)X^2+\left(\frac{B}{M}+r^2+p^2\right)Y^2+\left(\frac{C}{M}+p^2+q^2\right)Z^2-2qr\,YZ-2rp\,ZX-2pq\,XY=\epsilon^4,$$

when referred to its centre as origin.

Ex. 2.   Show that the cubic equation to find the three principal moments of inertia at any point (p, q, r) may be written in the form of a determinant

$$\begin{vmatrix} \dfrac{I-A}{M}-q^2-r^2 & pq & rp \\[2mm] pq & \dfrac{I-B}{M}-r^2-p^2 & qr \\[2mm] rp & qr & \dfrac{I-C}{M}-p^2-q^2 \end{vmatrix}=0.$$

If (l, m, n) be proportional to the direction-cosines of the axes corresponding to any one of the values of I, their values may be found from the equations

$$\left.\begin{aligned} \{I-(A+Mq^2+Mr^2)\}\,l+Mpqm+Mrpn=0, \\ Mpql+\{I-(B+Mr^2+Mp^2)\}\,m+Mqrn=0, \\ Mrpl+Mqrm+\{I-(C+Mp^2+Mq^2)\}\,n=0. \end{aligned}\right\}$$

Ex. 3.   If $S=0$ be the equation to the momental ellipsoid at the centre of gravity O referred to any rectangular axes written in the form given in Art. 19, then the momental ellipsoid at the point P whose co-ordinates are (p, q, r) is

$$S+M\,(p^2+q^2+r^2)\,(X^2+Y^2+Z^2)-M\,(pX+qY+rZ)^2=0.$$

Hence show (1) that the conjugate planes of the straight line OP in the momental ellipsoids at O and P are parallel and (2) that the sections perpendicular to OP have their axes parallel.

26. **Ellipsoid of Gyration.** The reciprocal surface of the momental ellipsoid is another ellipsoid, which has also been employed to represent, geometrically, the positions of the principal axes and the moment of inertia about any line.

We shall require the following elementary proposition. The reciprocal surface of the ellipsoid $\dfrac{x^2}{a^2} + \dfrac{y^2}{b^2} + \dfrac{z^2}{c^2} = 1$ is the ellipsoid $a^2x^2 + b^2y^2 + c^2z^2 = \epsilon^4$.

Let ON be the perpendicular from the origin O on the tangent plane at any point P of the first ellipsoid, and let l, m, n be the direction-cosines of ON, then $ON^2 = a^2l^2 + b^2m^2 + c^2n^2$. Produce ON to Q so that $OQ = \epsilon^2/ON$, then Q is a point on the reciprocal surface. Let $OQ = R$; $\therefore$ $\epsilon^4 = (a^2l^2 + b^2m^2 + c^2n^2) R^2$. Changing this to rectangular co-ordinates, we get $\epsilon^4 = a^2x^2 + b^2y^2 + c^2z^2$.

To each point of a material body there corresponds a series of similar momental ellipsoids. If we reciprocate these we get another series of similar ellipsoids coaxial with the first, and such that the moments of inertia of the body about the perpendiculars on the tangent planes to any one ellipsoid are proportional to the squares of those perpendiculars. It is, however, convenient to call *that particular ellipsoid the ellipsoid of gyration which makes the moment of inertia about a perpendicular on a tangent plane equal to the product of the mass into the square of that perpendicular.* If M be the mass of the body and A, B, C the principal moments, the equation to the ellipsoid of gyration is

$$\frac{X^2}{A} + \frac{Y^2}{B} + \frac{Z^2}{C} = \frac{1}{M}.$$

It is clear that the constant on the right-hand side must be $1/M$, for when Y and Z are put equal to zero, MX^2 must by definition be A.

27. Conversely, the series of momental ellipsoids at any point of a body may be regarded as the reciprocals, with different constants, of the ellipsoid of gyration at that point. They are all of an opposite shape to the ellipsoid of gyration, having their longest axes in the direction of the shortest axis and their shortest axes in the direction of the longest axis of the ellipsoid of gyration. The momental ellipsoids however resemble the general shape of the body more nearly than the ellipsoid of gyration. They are protuberant where the body is protuberant and compressed where the body is compressed. The exact reverse of this is the case in the ellipsoid of gyration. See Art. 22, Ex. 3.

28. Ex. 1. To find the ellipsoid of gyration at the centre of a material elliptic disc. Taking the values of A, B, C given in Art. 22, Ex. 1, we see that the ellipsoid of gyration is $\dfrac{X^2}{b^2} + \dfrac{Y^2}{a^2} + \dfrac{Z^2}{a^2 + b^2} = \dfrac{1}{4}$.

Ex. 2. The ellipsoid of gyration at any point O of a material rod AB is $\dfrac{X^2}{0} + \dfrac{Y^2}{\frac{1}{3}a^2 + c^2} + \dfrac{Z^2}{\frac{1}{3}a^2 + c^2} = 1$, taking the same notation as in Art. 22, Ex. 2. It is thus a very flat spheroid which, when the rod is indefinitely thin, becomes a circular area, whose centre is at O, whose radius is $\sqrt{\frac{1}{3}a^2 + c^2}$ and whose plane is perpendicular to the rod.

Ex. 3. It may be shown that the general equation to the ellipsoid of gyration referred to any set of rectangular axes meeting at the given point of the body is

$$\begin{vmatrix} A & -F & -E & MX \\ -F & B & -D & MY \\ -E & -D & C & MZ \\ MX & MY & MZ & M \end{vmatrix} = 0,$$

or, when expanded,

$$(BC - D^2)\,X^2 + (CA - E^2)\,Y^2 + (AB - F^2)\,Z^2 + 2\,(AD + EF)\,YZ + 2\,(BE + FD)\,ZX$$
$$+ 2\,(CF + DE)\,XY = \frac{1}{M}\,(ABC - AD^2 - BE^2 - CF^2 - 2DEF).$$

The right-hand side, when multiplied by M, is the discriminant obtained by leaving out the last row and the last column, and the coefficients of X^2, Y^2, Z^2, $2ZX$, $2XY$, $2YZ$ are the minors of this discriminant.

29.   The use of the ellipsoid whose equation referred to the principal axes at the centre of gravity is

$$\frac{X^2}{\Sigma mx^2} + \frac{Y^2}{\Sigma my^2} + \frac{Z^2}{\Sigma mz^2} = \frac{5}{M},$$

has been suggested by Legendre in his *Fonctions Elliptiques*. This ellipsoid is to be regarded as a homogeneous solid of such density that its mass is equal to that of the body. By Art. 8, Ex. 3, it possesses the property that its moments of inertia with regard to its principal axes, and therefore by Art. 15 its moments of inertia with regard to all planes and axes, are the same as those of the body. We may call this ellipsoid the *equimomental* ellipsoid or *Legendre's* ellipsoid.

Ex. If a plane move so that the moment of inertia with regard to it is always proportional to the square of the perpendicular from the centre of gravity on the plane, then this plane envelopes an ellipsoid similar to Legendre's ellipsoid.

30.   There is another ellipsoid which is sometimes used. By Art. 15 the moment of inertia with reference to a plane whose direction-cosines are (a, β, γ) is

$$I' = \Sigma mx^2 \cdot a^2 + \Sigma my^2 \cdot \beta^2 + \Sigma mz^2 \cdot \gamma^2 + 2\Sigma myz \cdot \beta\gamma + 2\Sigma mzx \cdot \gamma a + 2\Sigma mxy \cdot a\beta.$$

Hence, as in Art. 19, we may construct the ellipsoid

$$\Sigma mx^2 \cdot X^2 + \Sigma my^2 \cdot Y^2 + \Sigma mz^2 \cdot Z^2 + 2\Sigma myz \cdot YZ + 2\Sigma mzx \cdot ZX + 2\Sigma mxy \cdot XY = M\epsilon^4.$$

Then the moment of inertia with regard to any plane through the centre of the ellipsoid is represented by the inverse square of the radius vector perpendicular to that plane.

If we compare the equation of the momental ellipsoid with that of this ellipsoid, we see that one may be obtained from the other by subtracting the same quantity from each of the coefficients of X^2, Y^2, Z^2. Hence the two ellipsoids have their circular sections coincident in direction.

This ellipsoid may also be used to find the moments of inertia about any straight line through the origin. For we may deduce from Art. 15 that the moment of inertia about any radius vector is represented by the difference between the inverse square of that radius vector and the sum of the inverse squares of the semi-axes. This ellipsoid is a reciprocal of Legendre's ellipsoid. All these ellipsoids have their principal diameters coincident in direction, and any one of them may be used to determine the directions of the principal axes at any point.

31. When the body considered is a lamina, the section of the ellipsoid of gyration at any point of the lamina by the plane of the lamina is called the *ellipse of gyration*. If the plane of the lamina be the plane of xy, we have $\Sigma mz^2 = 0$. The section of the fourth ellipsoid is then clearly the same as an ellipse of gyration at the point. If any momental ellipse be turned round its centre through a right angle it evidently becomes similar and similarly situated to the ellipse of gyration. Thus, in the case of a lamina, any one of these ellipses may be easily changed into the others.

32. **Equimomental Cone.** *A straight line passes through a fixed point* O *and moves about it in such a manner that the moment of inertia about the line is always the same and equal to a given quantity* I. *To find the equation to the cone generated by the straight line.*

Let the principal axes at O be taken as the axes of co-ordinates, and let (α, β, γ) be the direction-cosines of the straight line in any position. Then by Art. 16 we have $A\alpha^2 + B\beta^2 + C\gamma^2 = I$.

Hence the equation to the locus is

$$(A - I)\,\alpha^2 + (B - I)\,\beta^2 + (C - I)\,\gamma^2 = 0,$$

or, transforming to Cartesian co-ordinates,

$$(A - I)\,x^2 + (B - I)\,y^2 + (C - I)\,z^2 = 0.$$

It appears from this equation that the principal diameters of the cone are the principal axes of the body at the given point.

The given quantity I must be less than the greatest and greater than the least of the moments A, B, C. Let A, B, C be arranged in descending order of magnitude; then if I be less than B, the cone has its concavity turned towards the axis C, if I be greater than B the concavity is turned towards the axis A, if $I = B$ the cone becomes two planes which are coincident with the central circular sections of the momental ellipsoid at the point O.

The geometrical peculiarity of this cone is that its circular sections in all cases are coincident in direction with the circular sections of the momental ellipsoid at the vertex.

This cone is called an *equimomental cone* at the point at which its vertex is situated.

On Equimomental Bodies.

33. Two bodies or systems of bodies are said to be equi-momental when their moments of inertia about all straight lines are equal each to each.

34. If two systems have the same centre of gravity, the same mass, the same principal axes and principal moments at the centre of gravity, it follows from the two fundamental propositions of Arts. 13 and 15 that their moments of inertia about all straight lines are equal, each to each.

That the converse theorem is also true may be shown thus. We know by Art. 13 that of all straight lines having a given direction in a body, that straight line has the least moment of inertia which passes through the centre of gravity. It is clear that these least moments of inertia cannot be equal in two bodies for *all* directions unless they have a common centre of gravity. Of all straight lines through the centre of gravity those which have the greatest and least moments of inertia are two of the principal axes, hence these and therefore also the third principal axis must be coincident in direction if the two bodies are equi-momental. The principal moments of inertia must then be equal, because all the moments are equal. Lastly, by Art. 13, the two systems cannot have equal moments about two parallel axes, each to each, unless their masses are equal.

It is easy to see that two equimomental systems must have the same momental ellipsoid, and therefore the same principal axes at every point.

35. **Case of a Triangle.** *To find the moments and products of inertia of a triangle about any axes whatever.*

If β and γ be the distances of the angular points B, C, of a triangle ABC from any straight line AX through the angle A, in the plane of the triangle, it is known that the moment of inertia of the triangle about AX is $\dfrac{M}{6}(\beta^2 + \beta\gamma + \gamma^2)$, where M is the mass of the triangle.

Let three equal particles, the mass of each being $\dfrac{M}{3}$, be placed at the middle points of the three sides. Then it is easily seen, that the moment of inertia of the three particles about AX is

$$\frac{M}{3}\left\{\left(\frac{\beta+\gamma}{2}\right)^2 + \left(\frac{\gamma}{2}\right)^2 + \left(\frac{\beta}{2}\right)^2\right\},$$

which is the same as that of the triangle. The three particles, treated as one system, and the triangle have the same centre of gravity. Let this point be called O. Draw any straight line OX' through the common centre of gravity O parallel to AX, then it is evident that the moments of inertia of the two systems about OX' are also equal.

Since this equality exists for all straight lines through O in the plane of the triangle, it will be true for two straight lines OX', OY' at right angles, and therefore also for a straight line OZ' perpendicular to the plane of the triangle.

One of the principal axes at O of the triangle, and of the systems of three particles, is normal to the plane, and therefore the same for the two systems. The principal axes at O in the plane, are those two straight lines about which the moments of inertia are greatest and least, and therefore by what precedes these axes are the same for the two systems. If at any point two systems have the same principal axes and principal moments, they have also the same moments of inertia about all axes through that point, and the same products of inertia about any two straight lines meeting in that point. And if this point be the centre of gravity of both systems, the same thing will also be true for any other point.

If then a particle whose mass is one-third that of the triangle be placed at the middle point of each side, the moment of inertia of the triangle about any straight line, is the same as that of the system of particles, and the product of inertia about any two straight lines meeting one another, is the same as that of the system of particles.

36. The existence of equimomental points is of the greatest utility in finding the moments and products of inertia of a body about any axes. They may also be used for more general integrations. Thus suppose any given body to be equimomental to three particles whose co-ordinates are $(x_1 y_1 z_1) (x_2 y_2 z_2) (x_3 y_3 z_3)$. Since the masses placed at these points may not in all cases be equal, let these masses be respectively $M_1 M_2 M_3$, where of course the sum is equal to the mass of the body. Let $\phi(xyz)$ be any function of x, y, z which does not contain any power higher than the second. Let it be required to find the value of the integral or sum $\Sigma m \phi(xyz)$ taken throughout the body, where m is an element of the mass. The required integral is evidently equal to

$$M_1 \phi(x_1 y_1 z_1) + M_2 \phi(x_2 y_2 z_2) + M_3 \phi(x_3 y_3 z_3).$$

By properly choosing the equivalent points we may use a similar rule when ϕ is any *cubic* or *quartic* function of xyz, but as these cases are not wanted in rigid dynamics we shall merely state a few results a little farther on.

The same body may be equimomental to several systems of points, and some of these sets may be more convenient than the others. In order that a set of equimomental points may be useful it is necessary (1) that the points should be so conveniently placed in the body that their co-ordinates can be easily found with regard to any given axes, (2) that the number of points employed in the set should be as small as possible. Of these two requisites the first is by far the most important.

Equimomental points have another use besides that of shortening integrations which may otherwise be troublesome. It will be presently seen that they have a dynamical importance.

37. *A momental ellipsoid at the centre of gravity of any triangle may be found as follows.*

Let an ellipse be inscribed in the triangle touching two of the sides AB, BC in their middle points F, D. Then, by Carnot's theorem, it touches the third side CA in its middle point E. Since DF is parallel to CA the tangent at E, the straight line joining E to the middle point N of DF passes through the centre, and therefore the centre of the conic is at O the centre of gravity of the triangle.

This conic may be shown to be a momental ellipse of the triangle at O. To prove this, let us find the moment of inertia of the triangle about OE. Let $OE = r$, and let r' be the semi-conjugate diameter, and ω the angle between r and r'. Now $ON = \frac{1}{2}r$, and hence from the equation to the ellipse $FN^2 = \frac{3}{4}r'^2$,

$$\left.\begin{array}{l}\text{therefore moment of} \\ \text{inertia about } OE\end{array}\right\} = \frac{2}{3}M \cdot \frac{3}{4}r'^2 \sin^2\omega, \; = \frac{M}{2} \cdot \frac{\Delta'^2}{\pi^2 r^2};$$

where Δ' is the area of the ellipse, so that the moments of inertia of the system about OE, OF, OD are proportional inversely to OE^2, OF^2, OD^2. If we take a momental ellipse of the right dimensions, it will cut the inscribed conic in E, F, and D, and therefore also at the opposite ends of the diameters through these points. But two conics cannot cut each other in six points unless they are identical. Hence this conic is a momental ellipse at O of the triangle.

A normal at O to the plane of the triangle is a principal axis of the triangle (Art. 16). Hence a momental ellipsoid of the triangle has the inscribed conic for one principal section. If $2a$ and $2b$ be the lengths of the axes of this conic, $2c$ that of the axis of the ellipsoid which is perpendicular to the plane of the lamina, we have, by Arts. 7 and 19, $1/c^2 = 1/a^2 + 1/b^2$.

If the triangle be an equilateral triangle, the momental ellipsoid becomes a spheroid, and every axis through the centre of gravity in the plane of the triangle is a principal axis.

Since any similar and similarly situated ellipse is also a momental ellipse, we may take the ellipse circumscribing the triangle, and having its centre at the centre of gravity, as the momental ellipse of the triangle.

38. **Ex. 1.** A momental ellipse at an angular point of a triangular area touches the opposite side at its middle point and bisects the adjacent sides.

Ex. 2. The principal radii of gyration at the centre of gravity of a triangle are the roots of the equation

$$x^4 - \frac{a^2 + b^2 + c^2}{36} x^2 + \frac{\Delta^2}{108} = 0,$$

where Δ is the area of the triangle.

Ex. 3. The direction of the principal axes at the centre of gravity O of a triangle may be constructed thus. Draw at the middle point D of any side BC lengths $DH = \dfrac{6k^2}{p}$, $DH' = \dfrac{6k'^2}{p}$ along the perpendicular, where p is the perpendicular from A on BC and k, k' are the principal radii of gyration found by the last example. Then OH, OH' are the directions of the principal axes at O, whose moments of inertia are respectively Mk^2 and Mk'^2.

Ex. 4. The directions of the principal axes and the principal moments at the centre of gravity may also be determined thus. Draw at the middle point D of any side BC a perpendicular $DK = BC/2\sqrt{3}$. Describe a circle on OK as diameter and join D to the middle point of OK by a line cutting the circle in R and S, then OR, OS are the directions of the principal axes, and the moments of inertia about them are respectively $M\dfrac{DS^2}{2}$, and $M\dfrac{DR^2}{2}$.

Ex. 5. Let four particles each one-sixth of the mass of the area of a parallelogram be placed at the middle points of the sides and a fifth particle one-third of the same mass at the centre of gravity, *then these five particles and the area of the parallelogram are equimomental systems.*

Ex. 6. Let particles each equal to one-twelfth of the mass of a quadrilateral area be placed at each corner and let a fifth particle of negative mass but also one-twelfth be placed at the intersection of the diagonals. Then the centre of gravity of the quadrilateral area is the centre of gravity of these five particles. Let a sixth particle equal to three-quarters of the mass of the quadrilateral be placed at the centre of gravity thus found. Prove that *these six particles are equimomental to the quadrilateral area.*

Ex. 7. Let particles each equal to one quarter of the mass of an elliptic area be placed at the middle points of the chords joining the extremities of any pair of conjugate diameters. Prove that *these four particles are equimomental to the elliptic area.*

Ex. 8. Any sphere of radius a and mass M is equimomental to a system of four particles each of mass $\dfrac{3M}{20}\left(\dfrac{a}{r}\right)^2$ placed so that their distances from the centre make equal angles with each other and are each equal to r, and a fifth particle equal to the remainder of the mass of the sphere placed at the centre.

39. **Case of a Tetrahedron.** *To find the moments and products of inertia of a tetrahedron about any axes whatever, i.e. to find a system of equimomental particles.*

Let $ABCD$ be the tetrahedron. Through one angular point D draw any plane and let it be taken as the plane of xy. Let D

be the area of the base ABC, α, β, γ the distances of its angular points from the plane of xy, and p the length of the perpendicular from D on the base ABC.

Let PQR be any section parallel to the base ABC and of thickness du, where u is the perpendicular from D on PQR. The moment of inertia of the triangle PQR with respect to the *plane* of xy is the same as that of three equal particles, each one-third its mass, placed at the middle points of its sides. The volume of the element $PQR = \dfrac{u^2}{p^2} D du$. The ordinates of the middle points of the sides AB, BC, CA are respectively $\dfrac{\alpha+\beta}{2}$, $\dfrac{\beta+\gamma}{2}$, $\dfrac{\gamma+\alpha}{2}$. Hence, by similar triangles, the ordinates of the middle points of PQ, QR, RP are $\dfrac{\alpha+\beta}{2}\dfrac{u}{p}$, $\dfrac{\beta+\gamma}{2}\dfrac{u}{p}$, $\dfrac{\gamma+\alpha}{2}\dfrac{u}{p}$.

The moment of inertia of the triangle PQR with regard to the plane xy is therefore

$$\frac{1}{3}\frac{u^2}{p^2} D du \left\{ \left(\frac{\beta+\gamma}{2}\frac{u}{p}\right)^2 + \left(\frac{\gamma+\alpha}{2}\frac{u}{p}\right)^2 + \left(\frac{\alpha+\beta}{2}\frac{u}{p}\right)^2 \right\}.$$

Integrating from $u=0$ to $u=p$, we have the moment of inertia of the tetrahedron with regard to the plane xy

$$= \frac{V}{10}\left\{ \alpha^2 + \beta^2 + \gamma^2 + \beta\gamma + \gamma\alpha + \alpha\beta \right\},$$

where V is the volume.

If particles each one-twentieth of the mass of the tetrahedron were placed at each of the angular points and the rest of the mass, viz. four-fifths, were collected at the centre of gravity, the moment of inertia of these five particles with regard to the plane of xy would be

$$= V\frac{4}{5}\left(\frac{\alpha+\beta+\gamma}{4}\right)^2 + \frac{V}{20}\alpha^2 + \frac{V}{20}\beta^2 + \frac{V}{20}\gamma^2,$$

which is the same as that of the tetrahedron.

The centre of gravity of these five particles is the centre of gravity of the tetrahedron, and together they make up the mass of the tetrahedron. Hence, by Art. 13, the moments of inertia of the two systems with regard to any plane through the centre of gravity are the same, and by the same article this equality will exist for all planes whatever. It follows, by Art. 5, that the moments of inertia about any straight line are also equal. *The two systems are therefore equimomental*.

---

* This result was proposed as a problem in the Mathematical Tripos between the dates of the publication of the preceding and following results, thus anticipating the author by a short time.

40. Theory of Projections. If the distance of every point in a given figure in space from some fixed plane be increased in a fixed ratio, the figure thus altered is called the *projection* of the given figure. By projecting a figure from three planes at right angles as base planes in succession, the figure may be often much simplified. Thus an ellipsoid can always be projected into a sphere, and any tetrahedron into a regular tetrahedron.

It is clear that if the base plane from which the figure is projected be moved parallel to itself into a position distant D from its former position, no change of form is produced in the projected figure. If n be the fixed ratio of projection the projected figure has merely been moved through a space nD perpendicular to the base plane. We may therefore suppose the base plane to pass through any given point which may be convenient.

41. *If two bodies are equimomental, their projections are also equimomental.*

Let the origin be the common centre of gravity, then the two bodies are such that $\Sigma m = \Sigma m'$; $\Sigma mx = 0$, $\Sigma m'x' = 0$, &c., $\Sigma mx^2 = \Sigma m'x'^2$, $\Sigma myz = \Sigma m'y'z'$, &c., unaccented letters referring to one body and accented letters to the other. Let both the bodies be projected from the plane of xy in the fixed ratio $1:n$. Then any point whose co-ordinates are (x, y, z) is transferred to (x, y, nz) and (x', y', z') to (x', y', nz'). Also the elements of mass m, m' become nm and nm'. It is evident that the above equalities are not affected by these changes, and that therefore the projected bodies are equimomental.

The projection of a momental ellipse of a plane area is a momental ellipse of the projection.

Let the figure be projected from the axis of x as base line, so that any point (x, y) is transferred to (x, y') where $y' = ny$, and any element of area m becomes m' where $m' = nm$. Then

$$\Sigma mx^2 = \frac{1}{n}\Sigma m'x^2, \quad \Sigma mxy = \frac{1}{n^2}\Sigma m'xy', \quad \Sigma my^2 = \frac{1}{n^3}\Sigma m'y'^2.$$

The momental ellipses of the primitive and the projection are

$$\Sigma my^2 X^2 - 2\Sigma mxy\, XY + \Sigma mx^2 Y^2 = M\epsilon^4,$$
$$\Sigma m'y'^2 X'^2 - 2\Sigma m'xy'\, X'Y' + \Sigma m'x^2 Y'^2 = M'\epsilon'^4.$$

To project the former we put $X' = X$, $Y' = nY$. Its equation becomes identical with the latter by virtue of the above equalities when we put $\epsilon'^4 = \epsilon^4 n^2$.

42. Ex. 1. A momental ellipse of the area of a square at its centre of gravity is easily seen to be the inscribed circle. By projecting this figure first with one side as base line, and secondly with a diagonal as base, the square becomes successively a rectangle and a parallelogram. Hence one momental ellipse at the centre of gravity of a parallelogram is the inscribed conic touching the sides at their middle points.

Ex. 2. By projecting an equilateral triangle into any triangle, we may infer the results of some of the previous articles, but the method will be best explained by its application to a tetrahedron.

Ex. 3. Since any ellipsoid may be obtained by projecting a sphere, we infer by Art. 38, Ex. 8, that any solid ellipsoid of mass M is equimomental to a system of four particles each of mass $\dfrac{3M}{20}\dfrac{1}{n^2}$ placed on a similar ellipsoid whose linear dimensions are n times as great as those of the material ellipsoid, so that the eccentric lines of the particles make equal angles with each other, and a fifth particle equal to the remainder of the mass of the ellipsoid placed at the centre of gravity.

If this material ellipsoid be the Legendre's ellipsoid of any given body, we see that any body whatever is equimomental to a system of five particles placed as above described on an ellipsoid similar to the Legendre's ellipsoid of the body.

Ex. 4. Show that a solid oblique cone on an elliptic base is equimomental to a system of three particles each one-tenth of the mass of the cone placed on the circumference of the base so that the differences of their eccentric angles are equal, a fourth particle equal to three-tenths of the cone placed at the middle point of the straight line joining the vertex to the centre of gravity of the base, and a fifth particle to make up the mass of the cone placed at the centre of gravity of the volume.

43. *To find an ellipsoid equimomental to any tetrahedron.*

The moments of inertia of a regular tetrahedron with regard to all *planes* through the centre of gravity O are equal by Art. 23. If r be the radius of the inscribed sphere, the moment with regard to a plane parallel to one face is easily seen by Art. 39 to be $M\dfrac{3r^2}{5}$. If then we describe a sphere of radius $\rho = \sqrt{3}r$, with its centre at the centre of gravity, and its mass equal to that of the tetrahedron, this sphere and the tetrahedron will be equimomental. Since the centre of gravity of any face projects into the centre of gravity of the projected face, we infer that the ellipsoid to which any tetrahedron is equimomental is similar and similarly situated to that inscribed in the tetrahedron and touching each face in its centre of gravity, but has its linear dimensions greater in the ratio $1 : \sqrt{3}$. It may also be easily seen that the sphere whose radius is $\rho = \sqrt{3}r$, touches each edge of the regular tetrahedron at its middle point. Hence we infer that the ellipsoid equimomental to any tetrahedron touches each edge at its middle point and has its centre at the centre of gravity of the volume.

These results may also be deduced from Art. 25, Ex. 2, without the use of projections.

Ex. 1. If E^2 be the sum of the squares of the edges of a tetrahedron, F^2 the sum of the squares of the areas of the faces and V the volume, show that the semi-axes of the ellipsoid inscribed in the tetrahedron, touching each face in the centre of gravity and having its centre at the centre of gravity of the tetrahedron, are the roots of

$$\rho^6 - \frac{E^2}{2^4 \cdot 3}\rho^4 + \frac{F^2}{2^4 \cdot 3^2}\rho^2 - \frac{V^2}{2^6 \cdot 3} = 0,$$

and that, if the roots be $\pm\rho_1$, $\pm\rho_2$, $\pm\rho_3$, the moments of inertia with regard to the principal *planes* of the tetrahedron are $M\dfrac{3\rho_1{}^2}{5}$, $M\dfrac{3\rho_2{}^2}{5}$, $M\dfrac{3\rho_3{}^2}{5}$.

Ex. 2. If a perpendicular EP be drawn at the centre of gravity E of any face $=4\rho^2/p$, where p is the perpendicular from the opposite corner of the tetrahedron on that face, then P is a point on the principal plane corresponding to the root ρ of the cubic.

44. *Four particles of equal mass can always be found which are equimomental to any given solid body.*

Let O be the centre of gravity of the body, Ox, Oy, Oz, the principal axes at O. Let the moments of inertia with regard to the co-ordinate *planes* be Ma^2, $M\beta^2$, and $M\gamma^2$. By Art. 34, the mass of each particle must be $\frac{1}{4}M$. Let $(x_1 y_1 z_1)$ &c. $(x_4 y_4 z_4)$ be the required co-ordinates of these four points. Then these twelve co-ordinates must satisfy the nine equations

$$\Sigma x^2=4a^2,\ \ \Sigma y^2=4\beta^2,\ \ \Sigma z^2=4\gamma^2,\ \ \Sigma xy=0,\ \ \Sigma yz=0,\ \ \Sigma zx=0,\ \ \Sigma x=0,\ \ \Sigma y=0,\ \ \Sigma z=0.$$

Now if we write $x_1=a\xi_1$, $x_2=a\xi_2$ &c. $y_1=\beta\eta_1$, $y_2=\beta\eta_2$ &c. $z_1=\gamma\zeta_1$ &c. we have nine equations to find the twelve co-ordinates $(\xi_1 \eta_1 \zeta_1)$ &c. $(\xi_4 \eta_4 \zeta_4)$ which differ from those just written down only in having a^2, β^2, γ^2 each replaced by unity. These modified equations express that the momental ellipsoid at O of the four particles must be a sphere. The equations are therefore satisfied if the four points, whose co-ordinates are represented by the *Greek letters*, are the corners of a regular tetrahedron. (See also Art. 23, Ex. 2.) This tetrahedron may be regarded as inscribed in a sphere whose radius is $\sqrt{3}$. If we project this sphere into an ellipsoid whose semi-axes are a β γ the regular tetrahedron will be deformed into an oblique tetrahedron. The corners of this oblique tetrahedron are the required equimomental points.

In the same way we may prove that three particles of equal mass can always be found which are equimomental to any plane area. If Ma^2, $M\beta^2$, and zero are the moments of inertia of the area about the principal *planes* at the centre of gravity, the result is that these particles must lie on the ellipse $\beta^2x^2+a^2y^2=2a^2\beta^2$. It also follows that, if one of these points, as D, be taken anywhere on this ellipse, the other two points, E and F, are at the opposite extremities of that chord which is bisected in some point N by the produced radius DO so that $ON=\frac{1}{2}OD$.

45. **Moments with higher powers.** These moments are not often wanted in dynamics though useful in other subjects. It will therefore be sufficient to state some general results, the demonstrations of which are left to the reader.

Let $d\sigma$ be any elementary area or volume as the case may be. Let z be its ordinate referred to any plane of xy. Our object is to find the value of the integral $\int z^n d\sigma$ for a triangle, quadrilateral, tetrahedron, &c.

Let the co-ordinates of the corners of the body considered be $(x_1 y_1 z_1)$, $(x_2 y_2 z_2)$, &c. Let $H_n(z_1 z_2$ &c.$)$ represent the *arithmetic mean* of the different homogeneous products of $z_1 z_2$, &c. of n dimensions, for example $H_3(z_1 z_2) = \frac{1}{4}(z_1{}^3 + z_1{}^2 z_2 + z_1 z_2{}^2 + z_2{}^3)$.

Then for a triangle of area Δ,

$$\int z^n d\sigma = \Delta H_n (z_1 z_2 z_3).$$

For a quadrilateral of area Δ,

$$\int z^n d\sigma = \Delta \left\{ H_n (z_1 z_2 z_3 z_4) - z' H_{n-1} (z_1 z_2 z_3 z_4) \right\},$$

where z' is the ordinate of the intersection of the diagonals.

For a tetrahedron of volume V,

$$\int z^n d\sigma = V H_n (z_1 z_2 z_3 z_4).$$

For two tetrahedra joined together, whose united volume is V,

$$\int z^n d\sigma = V \left\{ H_n (z_1 \ldots z_5) - z' H_{n-1} (z_1 \ldots z_5) \right\},$$

where z' is the ordinate of the point of intersection of the common base with the straight line joining the two vertices.

We notice that, except for the factor Δ or V representing the area or volume, these four expressions are functions of the ordinates only of the corners and are not functions of the differences of the abscissæ.

When the index n is not a positive integer these expressions take more complicated forms. For these we refer the reader to a paper by the author in the *Quarterly Journal of Mathematics*, No. 83, 1886.

When the value of $\int z^n d\sigma$ is known that of $n \int x z^{n-1} d\sigma$ can be found by performing the operation $x_1 \dfrac{d}{dz_1} + x_2 \dfrac{d}{dz_2} + \ldots$ on the former result. The value of $n(n-1) \int x^2 z^{n-2} d\sigma$ can be found by repeating the operation and so on.

Lastly, it may be shown that when two bodies are such that the values of $\int z^n d\sigma$ are equal, each to each, for all planes of xy these bodies are equimomental.

Ex. 1. If $\phi (xyz)$ be a function not higher than the third degree the value of $\int \phi d\sigma$ for any triangle can be found by using seven equivalent or equimomental points. We collect *one-twentieth* of the mass of the area at each corner, *two-fifteenths* at the middle point of each side, and *the rest, viz. nine-twentieths*, at the centre of gravity.

Ex. 2. If $\phi (xyz)$ be not higher than the third degree the value of $\int \phi d\sigma$ for a tetrahedron can be represented by eight equivalent points. We collect *nine-fortieths* of the volume at the centre of gravity of each face and *one-fortieth* at each corner.

Other examples may be found in the paper already referred to.

46.   **Theory of Inversion.**   *To explain how the theory of inversion can be applied to find moments of inertia.*

Let a radius vector drawn from some fixed origin O to any point P of a figure be produced to P', where the rectangle $OP \cdot OP' = \kappa^2$, κ being some given quantity. Then as P travels all over the given figure, P' traces out another which is called the inverse of the given figure.

Let (x, y, z) be the co-ordinates of P, (x', y', z') those of P'; r, r' the radii vectores, dv, dv' corresponding polar elements of volume; ρ, ρ', dm, dm' their respective densities and masses.   Let $d\omega$ be the solid angle subtended at O by either dv or dv'.   Then

$$dv' = r'^2 d\omega dr' = \left(\frac{\kappa}{r}\right)^6 r^2 d\omega dr = \left(\frac{\kappa}{r}\right)^6 dv,$$

and since $\dfrac{x'}{r'} = \dfrac{x}{r}$ we have $x'^2 dv' = \left(\dfrac{\kappa}{r}\right)^{10} x^2 dv$.   Now $dm = \rho dv$, $dm' = \rho' dv'$.   If then we take $\dfrac{\rho'}{\rho} = \left(\dfrac{r}{\kappa}\right)^{10}$ we have $\Sigma x'^2 dm' = \Sigma x^2 dm$, with similar equalities in the case of all the other moments and products of inertia.

When the body is an area or an arc the ratio of dv' to dv is different.   We have in these cases respectively $\dfrac{dv'}{dv} = \left(\dfrac{\kappa}{r}\right)^4$ or $\left(\dfrac{\kappa}{r}\right)^2$.   Similar results however follow which may be all summed up in the following theorem.

THEOR. I.   *Let any body be changed into another by inversion with regard to any point* O.   *If the densities at corresponding points be denoted by* ρ, ρ' *and their distances from* O *by* r, r'; *let* $\rho' = \rho\left(\dfrac{\kappa}{r'}\right)^n$.   *Then these two bodies have the same moments of inertia with regard to all straight lines through* O.   Here $n = 10$, 8 or 6 according as the body is a *volume*, an *area* or an *arc*.

It also follows that the two bodies have the same principal axes at the point O, and the same ellipsoids of gyration.

We may also obtain the following theorem by the use of Sir W. Thomson's method of finding the potentials of attracting bodies by Inversion.

THEOR. II.   *Let any body be changed into another body by inversion with regard to any point* O.   *If the densities at corresponding points* P, P' *be denoted by* ρ, ρ', *and their distances from* O *by* r, r', *let* $\rho' = \rho\left(\dfrac{\kappa}{r'}\right)^n$.   *Then the moment of inertia of the second body with regard to any point* C' *is equal to that of the first body with regard to the corresponding point* C *multiplied by either of the equal quantities* $\left(\dfrac{\kappa}{OC}\right)^2$ *or* $\dfrac{OC'}{OC}$.   Here $n = 8$, 6 or 4 according as the body is a *volume*, *area* or *arc*.

To prove this, consider the case in which the body is a volume.   By similar triangles $CP \cdot r' = C'P' \cdot OC$.   Hence proceeding as before, we find

$$\rho dv \, (CP)^2 \left(\frac{\kappa}{OC}\right)^2 = \rho' dv' \, (C'P')^2.$$

This being true for every element the theorem follows at once.

Ex.   The density of a solid sphere varies inversely as the tenth power of the distance from an external point O.   Prove that its moment of inertia about any straight line through O is the same as if the sphere were homogeneous and its density equal to that of the heterogeneous sphere at a point where the tangent from O meets the sphere.   Prove that if the density had varied inversely as the sixth power of the distance from O, the masses of the two spheres would have been equal.   What is the condition that they should have a common centre of gravity?   Math. Tripos.

47. **Centre of Pressure.**   The theory of equimomental particles is of considerable use in finding the centre of pressure of any area vertically immersed in a homogeneous fluid under the action of gravity.   It may be proved from hydrostatical principles

that if the axis of x be in the effective surface, and the axis of y vertically downwards, the co-ordinates of the centre of pressure are

$$X = \frac{\text{Product of inertia about the axes}}{\text{moment of area about } Ox},$$

$$Y = \frac{\text{Moment of inertia about } Ox}{\text{moment of area about } Ox}.$$

We see therefore that two equimomental areas have the same centre of pressure.

Let the given area be equimomental to particles whose masses are m_1, m_2 &c. and let (x_1, y_1), (x_2, y_2), &c. be the co-ordinates of these particles. Then

$$X = \frac{\Sigma m x y}{\Sigma m y}, \qquad Y = \frac{\Sigma m y^2}{\Sigma m y}.$$

But these are the formulæ to find the centre of gravity of particles whose masses are proportional to $m_1 y_1$, $m_2 y_2$ &c. having the same co-ordinates as before. Hence this rule,

If any area be equimomental to a series of particles, the centre of pressure of the area is the centre of gravity of the same particles with their masses increased in the ratio of their depths.

For example the centre of pressure of a triangle wholly immersed is the centre of gravity of three weights placed at the middle points of the sides and each proportional to the depth of the point at which it is placed.

Ex. 1. If p, q, r be the depths of the corners of a triangular area wholly immersed in a fluid, prove that the areal co-ordinates of its centre of pressure referred to the sides of the triangle itself are $\frac{1}{4}(1 + p/s)$, $\frac{1}{4}(1 + q/s)$, $\frac{1}{4}(1 + r/s)$, where $s = p + q + r$.

This may be proved by replacing the triangle by three weights situated at the middle points of the sides proportional to their depths, and taking moments about the sides in succession to find their centre of gravity.

Ex. 2. Let any vertical area be referred to Cartesian rectangular axes Ox, Oy, with the origin at the centre of gravity. Let the depth of the centre of gravity be h, and let the intersection of the area with the surface of the fluid make an angle θ with the axis of x, and let this intersection in the standard case cut the positive side of the axis of y. Let A, B and F be the moments and product of inertia of the area about the axes. Then by taking moments about Ox, Oy we see that the co-ordinates of the centre of pressure are

$$X = \frac{B \sin \theta - F \cos \theta}{ha}, \qquad Y = \frac{F \sin \theta - A \cos \theta}{ha},$$

where a is the area.

Ex. 3. If the area turn round its centre of gravity in its own plane the locus of its centre of pressure *in the area* is an ellipse and *in space* is a circle. The ellipse has its principal diameters coincident in direction with the principal axes of the area at the centre of gravity. The circle has its centre in the vertical through the centre of gravity.

Ex. 4. In a heterogeneous fluid the pressure at any point P referred to a unit of area is given by $p = a + bz^n$ where z is the ordinate of P. Prove that the ordinate of the centre of pressure of any triangular area wholly immersed at any inclination to the horizon is $\dfrac{aH_1 + bH_{n+1}}{aH_0 + bH_n}$ where H has the meaning given in Art. 45.

Ex. 5. In rotating fluids the pressure at any point P is given by $p = a + bz + cr^2$ where r is the distance of P from the axis of z. Show that the pressure on any part of the area of the containing vessel is given by

(1) whole pressure $= \int (a + bz + cr^2)\, d\sigma = (a + b\bar{z})\, \sigma + c\sigma k^2$

where $\bar{z}$ is the ordinate of the centre of gravity of the area σ, and σk^2 its moment of inertia about the axis of z.

(2) Vertical pressure $= \iint (a + bz + cr^2)\, dx\, dy = aP + bV + cPk'^2$

where P is the projection of σ on the plane of xy, V the volume between σ and its projection and Pk'^2 the moment of inertia of the projection P about the axis of z.

It is evident that in all these cases the values of the integrals can in general be written down by the rules given in this chapter; so that actual integrations are for the most part unnecessary.

On the positions of the Principal Axes of a system.

48. PROP. *A straight line being given it is required to find at what point in its length it is a principal axis of the system, and if any such point exist to find the other two principal axes at that point.*

This point may be conveniently called the principal point of the straight line.

Take the straight line as axis of z, and any point O in it as origin. Let C be the point at which it is a principal axis, and let Cx', Cy' be the other two principal axes.

Let $CO = h$, $\theta =$ angle between Cx' and Ox. Then

$$\left. \begin{aligned} x' &= x\cos\theta + y\sin\theta \\ y' &= -x\sin\theta + y\cos\theta \\ z' &= z - h \end{aligned} \right\}$$

Hence

$$\left. \begin{aligned} \Sigma mx'z' &= \cos\theta\,\Sigma mxz + \sin\theta\,\Sigma myz \\ &\quad - h\,(\cos\theta\,\Sigma mx + \sin\theta\,\Sigma my) \end{aligned} \right\} = 0 \dots\dots\dots(1),$$

$$\left. \begin{aligned} \Sigma my'z' &= -\sin\theta\,\Sigma mxz + \cos\theta\,\Sigma myz \\ &\quad - h\,(-\sin\theta\,\Sigma mx + \cos\theta\,\Sigma my) \end{aligned} \right\} = 0 \dots\dots\dots(2),$$

$$\Sigma mx'y' = \Sigma m\,(y^2 - x^2)\,\frac{\sin 2\theta}{2} + \Sigma mxy\cos 2\theta = 0 \dots\dots\dots(3).$$

3—2

The last equation shows that

$$\tan 2\theta = \frac{2\Sigma mxy}{\Sigma m\,(x^2 - y^2)} \quad\dotsb\dotsb\dotsb\dotsb\dotsb(4)$$

$$= \frac{2F}{B - A},$$

according to the previous notation.

The equations (1) and (2) must be satisfied by the same value of h. Eliminating h we get $\Sigma mxz\,\Sigma my = \Sigma myz\,\Sigma mx$ as the condition that the axis of z should be a principal axis at some point in its length. Substituting in (1) we have

$$h = \frac{\Sigma myz}{\Sigma my} = \frac{\Sigma mxz}{\Sigma mx} \quad\dotsb\dotsb\dotsb\dotsb\dotsb(5).$$

The equation (5) expresses the condition that the axis of z should be a principal axis at some point in its length; and the value of h gives the position of this point. The positions of the other two principal axes may then be found by equation (4).

If $\Sigma mxz = 0$ and $\Sigma myz = 0$, the equations (1) and (2) are both satisfied by $h = 0$. These are therefore the sufficient and necessary conditions that the axis of z should be a principal axis at the origin.

If the system be a plane lamina and the axis of z be a normal to the plane at any point, we have $z = 0$. Hence the conditions $\Sigma mxz = 0$ and $\Sigma myz = 0$ are satisfied. Therefore one of the principal axes at any point of a plane lamina is a normal to the plane at that point.

In the case of a surface of revolution bounded by planes perpendicular to the axis, the axis is a principal axis at any point of its length.

Again, equation (4) enables us, when one principal axis is given, to find the other two. If $\theta = \alpha$ be the first value of θ, all the others are included in $\theta = \alpha + \frac{1}{2}n\pi$; hence all these values give only the same axes over again.

49. Since (4) does not contain h, it appears that if the axis of z be a principal axis at more than one point, the principal axes at those points are parallel. Again, in that case (5) must be satisfied by more than one value of h. But, since h enters only in the first power, this cannot be unless

$$\Sigma mx = 0, \qquad \Sigma my = 0,$$
$$\Sigma mxz = 0, \qquad \Sigma myz = 0\,;$$

so that the axis must pass through the centre of gravity and be a principal axis at the origin, and therefore (since the origin is arbitrary) a principal axis at every point in its length.

If the principal axes at the centre of gravity be taken as the axes of x, y, z, (1) and (2) are satisfied for all values of h. Hence, if a straight line be a principal axis at the centre of gravity, it is a principal axis at every point in its length.

50. Let the system be projected on a plane perpendicular to the given straight line, so that the ratios of the elements of mass to each other are unaltered. The given straight line, which has been taken as the axis of z, cuts this plane in O, and will be a principal axis of the projection at O, because, the projected system being a plane lamina, the conditions $\Sigma mxz = 0$, $\Sigma myz = 0$ are both satisfied. Since z does not appear in equation (4), it follows that, if the given straight line be a principal axis at some point C in its length, the other two principal axes at C will be parallel to the principal axes of the projected system at O. These last may often be conveniently found by the next proposition.

51. Ex. 1.   The principal axes of a right-angled triangle at the right angle are, one perpendicular to the plane and two others inclined to its sides at the angles $\frac{1}{2} \tan^{-1} \frac{ab}{a^2 - b^2}$, where a and b are the sides of the triangle adjacent to the right angle.

We have $\tan 2\theta = \dfrac{2F}{B - A}$, Art. 48, and by Art. 35, $A = M \dfrac{a^2}{6}$, $B = M \dfrac{b^2}{6}$, $F = M \dfrac{ab}{12}$.

Ex. 2.   The principal axes of a quadrant of an ellipse at the centre are, one perpendicular to the plane and two others inclined to the principal diameters at the angles $\frac{1}{2} \tan^{-1} \frac{4}{\pi} \frac{ab}{a^2 - b^2}$, where a and b are the semi-axes of the ellipse.

Ex. 3.   The principal axes of a cube at any point P are, the straight line joining P to O the centre of gravity of the cube, and any two straight lines at P perpendicular to PO, and perpendicular to each other.

Ex. 4.   Prove that the locus of a point P at which one of the principal axes is parallel to a given straight line is a rectangular-hyperbola in the plane of which the centre of gravity of the body lies, and one of whose asymptotes is parallel to the given straight line. But if the given straight line be parallel to one of the principal axes at the centre of gravity, the locus of P is that principal axis or the perpendicular principal plane.

Take the origin at the centre of gravity, and one axis of co-ordinates parallel to the given straight line.

Ex. 5.   An edge of a tetrahedron will be a principal axis at some point in its length only when it is perpendicular to the opposite edge. [Jullien.]

Conversely, if this condition be satisfied, the edge will be a principal axis at a point C, such that $OC = \frac{2}{3} ON$, where N is the middle point of the edge and O is the foot of the perpendicular distance between it and the opposite edge.

Ex. 6.   The axes Ox, Oy are so placed that the product of inertia F or Σmxy is zero. If A and B are the moments of inertia about these axes, prove that the product of inertia about two perpendicular axes Ox', Oy' in the plane xy is

$$F' = \tfrac{1}{2} (A - B) \sin 2\theta$$

where θ is the angle xOx' measured in the positive direction from Ox.

52. Foci of Inertia. *Given the positions of the principal axes* Ox, Oy, Oz *at the centre of gravity* O, *and the moments of inertia about them, to find the positions of the principal axes at any point* P *in the plane of* xy, *and the moments of inertia about those axes.*

Let the mass of the body be M, and let A, B be the moments of inertia about the axes Ox, Oy, of which we shall suppose A the greater. Take two points S and H in the axis of greatest moment, one on each side of the origin so that

$$OS = OH = \sqrt{\frac{A-B}{M}}.$$

These points may be called the foci of inertia for that principal plane.

Because these points are in one of the principal axes at the centre of gravity, the principal axes at S and H are parallel to the axes of co-ordinates, and the moments of inertia about those in the plane of xy are respectively A and $B + M \cdot OS^2 = A$. These being equal, any straight line through S or H in the plane of xy is a principal axis at that point, and the moment of inertia about it is equal to A.

If P be any point in the plane of xy, then one of the principal axes at P will be perpendicular to the plane xy. For, if p, q be the co-ordinates of P, the conditions that this line should be a principal axis are

$$\Sigma m\,(x-p)\,z = 0\}$$
$$\Sigma m\,(y-q)\,z = 0\}\ ,$$

which are obviously satisfied, because the centre of gravity is the origin, and the principal axes the axes, of co-ordinates.

The other two principal axes may be found thus. If two straight lines meeting at a point P be such that the moments of inertia about them are equal, then, provided they are in a principal plane, the principal axes at P bisect the angles between these two straight lines. For, if with centre P we describe the momental

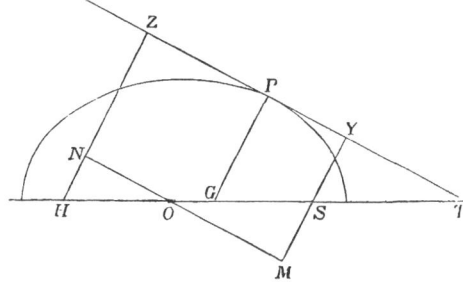

ellipse, the axes of this ellipse bisect the angles between any two equal radii vectores.

Join SP and HP; the moments of inertia about SP, HP are each equal to A. Hence, if PG and PT are the internal and external bisectors of the angle SPH, PG, PT are the principal axes at P. *If therefore with* S *and* H *as foci we describe any ellipse or hyperbola, the tangent and normal at any point are the principal axes at that point.*

53. Take any straight line MN through the origin, making an angle θ with the axis of x. Draw SM, HN perpendiculars on MN. The moment of inertia about MN is

$$= A \cos^2\theta + B \sin^2\theta$$
$$= A - (A - B) \sin^2\theta$$
$$= A - M . (OS \sin \theta)^2$$
$$= A - M . SM^2.$$

Through P draw PT parallel to MN, and let SY and HZ be the perpendiculars from S and H on it. The moment of inertia about PT is then

$$= \text{moment about } MN + M . MY^2$$
$$= A + M (MY - SM)(MY + SM)$$
$$= A + M . SY . HZ.$$

In the same way it may be proved that the moment of inertia about a line PG passing between H and S is *less* than A by the mass into the product of the perpendiculars from S and H on PG.

If therefore with S *and* H *as foci we describe any ellipse or hyperbola, the moment of inertia about any tangent to either of these curves is constant.*

It follows from this that the moments of inertia about the principal axes at P are equal to $B + M \left(\dfrac{SP \pm HP}{2} \right)^2$.

For if a and b be the axes of the ellipse we have $a^2 - b^2 = OS^2$ $= \dfrac{A - B}{M}$, and hence

$$A + M . SY . HZ = A + Mb^2 = B + Ma^2 = B + M \left(\frac{SP + HP}{2} \right)^2,$$

and the hyperbola may be treated in a similar manner.

54. This reasoning may be extended to points lying in any given plane passing through the centre of gravity O of the body. Let Ox, Oy be the axes in the given plane such that the product of inertia about them is zero (Art. 23). Construct the points S and H as before, so that OS^2 and OH^2 are each equal to the

difference of the moments of inertia about Ox and Oy divided by the mass. Draw Sy' a parallel through S to the axis of y, the product of inertia about Sx, Sy' is equal to that about Ox, Oy together with the product of inertia of the whole mass collected at O. Both these are zero, hence the section of the momental ellipsoid at S is a circle, and the moment of inertia about every straight line through S in the plane xOy is the same and equal to that about Ox. We can then show that the moments of inertia about PH and PS are equal; so that PG, PT, the internal and external bisectors of the angle SPH, are the principal diameters of the section of the momental ellipsoid at P by the given plane. And it also follows that the moments of inertia about the tangents to a conic whose foci are S and H are the same.

55. Ex. 1. To find the foci of inertia of an elliptic area. The moments of inertia about the major and minor axes are $\frac{1}{4}Mb^2$ and $\frac{1}{4}Ma^2$. Hence the minor axis is the axis of greatest moment. The foci of inertia therefore lie in the *minor* axis at a distance from the centre $=\frac{1}{2}\sqrt{a^2-b^2}$, i.e. half the distance of the geometrical foci from the centre.

Ex. 2. Two particles each of mass m are placed at the extremities of the minor axis of an elliptic area of mass M. Prove that the principal axes at any point of the circumference of the ellipse will be the tangent and normal to the ellipse, provided that $\dfrac{m}{M} = \dfrac{5}{8}\dfrac{e^2}{1-2e^2}$.

Ex. 3. At the points which have been called foci of inertia *two* of the principal moments are equal. Show that it is not in general true that a point exists such that the moments of inertia about all axes through it are the same, and find the conditions that there may be such a point. Such points when they exist in a solid body may be called the spherical points of inertia of that solid.

Refer the body to the principal axes at the centre of gravity. Let P be the point required, (x, y, z) its co-ordinates. Since the momental ellipsoid at P is to be a sphere, the products of inertia about all rectangular axes meeting at P are zero. Hence, by Art. 13, $xy=0$, $yz=0$, $zx=0$. It follows that two of the three x, y, z must be zero, so that the point must be on one of the principal axes at the centre of gravity. Let this be called the axis of z. Since the moments of inertia about three axes at P parallel to the co-ordinate axes are $A + Mz^2$, $B + Mz^2$ and C, we see that these cannot be equal unless $A = B$ and each is less than C. There are then two points on the axis of unequal moment which are equimomental for all axes. [Poisson and Binet.]

Ex. 4. The spherical points of a hemispherical surface are the centre and a point on the surface. Find also the spherical points of a solid hemisphere.

By Art. 5, Ex. 8 the moments of inertia about every axis through the centre are the same. Hence the centre is one spherical point. Since the centre of gravity bisects the distance between the points the position of the other follows at once.

56. **Arrangement of Principal axes.** *Given the positions of the principal axes at the centre of gravity* O *and the moments*

of inertia about them, to find the positions of the principal axes* *and the principal moments at any other point* P.

Let the body be referred to its principal axes at the centre of gravity O, let A, B, C be its principal moments, the mass of the body being taken as unity. Construct a quadric-confocal with the ellipsoid of gyration, and let the squares of its semi-axes be $a^2 = A + \lambda$, $b^2 = B + \lambda$, $c^2 = C + \lambda$. Let us find the moment of inertia with regard to any tangent plane.

Let (α, β, γ) be the direction angles of the perpendicular to any tangent plane. The moment of inertia, with regard to a parallel plane through O, is

$$\tfrac{1}{2}(A + B + C) - (A \cos^2\alpha + B \cos^2\beta + C \cos^2\gamma).$$

The moment of inertia, with regard to the tangent plane, is found by adding the square of the perpendicular distance between the planes, viz.

$$(A + \lambda) \cos^2\alpha + (B + \lambda) \cos^2\beta + (C + \lambda) \cos^2\gamma.$$

We get $\left.\begin{array}{l}\text{moment of inertia with} \\ \text{regard to a tangent plane}\end{array}\right\} = \tfrac{1}{2}(A + B + C) + \lambda$

$$= \tfrac{1}{2}(B + C - A) + a^2.$$

Thus the moments of inertia with regard to all tangent planes to any one quadric confocal with the ellipsoid of gyration are the same.

These planes are all principal planes at the point of contact. For draw any plane through the point of contact P, then in the case in which the confocal is an ellipsoid, the tangent plane parallel to this plane is more remote from the origin than this plane. Therefore, the moment of inertia with regard to any plane through P is less than the moment of inertia with regard to a tangent plane to the confocal ellipsoid through P. That is, the tangent plane to the ellipsoid is the principal plane of greatest moment. In the same way the tangent plane to the confocal hyperboloid of two sheets through P is the principal plane of least moment. It follows that the tangent plane to the confocal hyperboloid of one sheet is the principal plane of mean moment.

Through a given point P, three confocals can be drawn, and the normals to these confocals are the principal axes at P. By Art. 5, Ex. 3, the principal axis of *least* moment is normal to the confocal *ellipsoid* and that of greatest moment normal to the confocal hyperboloid of two sheets.

---

* Some of the following theorems were given by Sir William Thomson and Mr Townsend, in two articles which appeared at the same time in the *Mathematical Journal*, 1846. Their demonstrations are different from those given in this treatise.

57. The moment of inertia with regard to the *point* P is, by Art. 14, $\frac{1}{2}(A + B + C) + OP^2$. Hence, by Art. 5, Ex. 3, the moments of inertia about the normals to the three confocals through P whose parameters are $\lambda_1, \lambda_2, \lambda_3$ are respectively

$$OP^2 - \lambda_1, \quad OP^2 - \lambda_2, \quad OP^2 - \lambda_3.$$

58. If we describe any other confocal and draw a tangent cone to it whose vertex is P, the axes of this cone are known to be the normals to the three confocals through P. This gives another construction for the principal axes at P.

If the confocal diminish without limit, until it becomes a focal conic, we see that the principal diameters of the system at P are the principal diameters of a cone whose vertex is P and base a focal conic of the ellipsoid of gyration at the centre of gravity.

59. If we wish to use only one quadric, we may consider the confocal ellipsoid through P. We know* that the normals to

---

* These propositions are to be found in books on solid geometry, they may also be proved as follows.

Let the confocal ellipsoid pass near P and approach it indefinitely. The base of the enveloping cone is ultimately the Indicatrix ; and as the cone becomes ultimately a tangent plane, one of its axes is ultimately a perpendicular to the plane of the Indicatrix. Now in any cone two of its axes are parallel to the principal diameters of any section perpendicular to the third axis. Hence the axes of the enveloping cone are the normal to the surface and parallels to the principal diameters of the Indicatrix. But all parallel sections of an ellipsoid are similar and similarly situated, hence the principal diameters of the Indicatrix are parallel to the principal diameters of the diametral section parallel to the tangent plane at P.

To find the principal moments, we may reason as follows. Let a tangent plane to the ellipsoid be drawn perpendicular to any radius vector OQ of the diametral

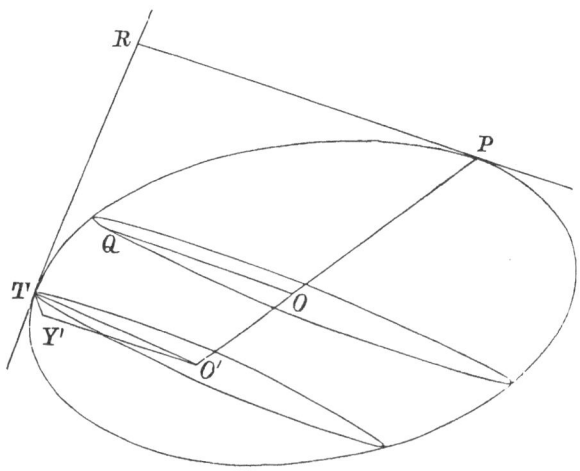

the other two confocals are tangents to the lines of curvature on the ellipsoid, and are also parallel to the principal diameters of the diametral section made by a plane parallel to the tangent plane at P. And if D_1, D_2 be these principal semi-diameters, we know that

$$\lambda_2 = \lambda_1 - D_1^2, \qquad \lambda_3 = \lambda_1 - D_2^2.$$

Hence, if through any point P we describe the quadric

$$\frac{x^2}{A+\lambda} + \frac{y^2}{B+\lambda} + \frac{z^2}{C+\lambda} = 1,$$

the axes of co-ordinates being the principal axes at the centre of gravity, then the principal axes at P are the normal to this quadric, and parallels to the axes of the diametral section made by a plane parallel to the tangent plane at P. And if these axes be $2D_1$ and $2D_2$, the principal moments at P are

$$OP^2 - \lambda, \quad OP^2 - \lambda + D_1^2, \quad OP^2 - \lambda + D_2^2.$$

Ex. If two bodies have the same centre of gravity, the same principal axes at the centre of gravity and the differences of their principal moments equal, each to each, then these bodies have the same principal axes at all points.

60. Condition that a line should be a principal axis.
The axes of co-ordinates being the principal axes at the centre of gravity it is required to express the condition that any given straight line may be a principal axis at some point in its length and to find that point.

Let the equations to the given straight line be

$$\frac{x-f}{l} = \frac{y-g}{m} = \frac{z-h}{n} \quad \dots\dots\dots\dots\dots(1),$$

section of OP, then the point of contact T, OQ and OP will lie in one plane when OQ is an axis of the section. For draw through T a section parallel to the diametral section, and let O' be its centre, and let $O'Y'$ be a perpendicular from O' on the tangent plane which touches at T. Then OQ, $O'Y'$ and OP are in one plane. Now consider the section whose centre is O'; $O'Y'$ is the perpendicular on the tangent to an ellipse whose point of contact is T. Hence $O'Y'$, $O'T$ do not coincide unless $O'Y'$ be the direction of the axis of the ellipse. But this section is similar to the diametral section to which it was drawn parallel. Hence OQ is an axis of the diametral section.

Let PR be a straight line drawn through P parallel to OQ to meet in R the tangent plane which touches in T. Then RP, RT are two tangents at right angles to the ellipse PQT. Hence

$OR^2 = $ sum of the squares of the semi-axes of the ellipse $= OP^2 + OQ^2$

because OP, OQ are conjugate diameters.

The moment of inertia about PR, a perpendicular to a tangent plane, has been proved above to be $OR^2 - \lambda$, hence the moment of inertia about a parallel through P to the axis OQ is $OP^2 + OQ^2 - \lambda$.

then it must be a normal to some quadric

$$\frac{x^2}{A+\lambda} + \frac{y^2}{B+\lambda} + \frac{z^2}{C+\lambda} = 1 \quad \ldots\ldots\ldots\ldots\ldots(2)$$

at the point at which the straight line is a principal axis.

Hence comparing the equation of the normal to (2) with (1), we have

$$\frac{x}{A+\lambda} = \mu l, \quad \frac{y}{B+\lambda} = \mu m, \quad \frac{z}{C+\lambda} = \mu n \ldots\ldots\ldots\ldots(3).$$

These six equations must be satisfied by the same values of x, y, z, λ and μ. Substituting for x, y, z from (3) in (1), we get

$$A\mu - \frac{f}{l} = B\mu - \frac{g}{m} = C\mu - \frac{h}{n}.$$

Equating the values of μ given by these equations we have

$$\frac{\dfrac{f}{l} - \dfrac{g}{m}}{A - B} = \frac{\dfrac{g}{m} - \dfrac{h}{n}}{B - C} = \frac{\dfrac{h}{n} - \dfrac{f}{l}}{C - A} \quad \ldots\ldots\ldots\ldots\ldots\ldots (4).$$

This clearly amounts to only one equation, and is the required condition that the straight line should be a principal axis at some point in its length.

Substituting for x, y, z from (3) in (2), we have

$$\lambda (l^2 + m^2 + n^2) = \frac{1}{\mu^2} - (Al^2 + Bm^2 + Cn^2),$$

which gives one value only to λ. The values of λ and μ having been found, equations (3) will determine x, y, z the co-ordinates of the point at which the straight line is a principal axis.

The geometrical meaning of this condition may be found by the following considerations, which were given by Mr Townsend in the *Mathematical Journal*. The normal and tangent plane at every point of a quadric will meet any principal plane in a point and a straight line, which are pole and polar with regard to the focal conic in that plane. Hence, to find whether any assumed straight line is a principal axis or not, draw any plane perpendicular to the straight line and produce both the straight line and the plane to meet any principal plane at the centre of gravity. If the line of intersection of the plane be parallel to the polar line of the point of intersection of the straight line with respect to the focal conic, the straight line will be a principal axis, if otherwise it will not be so. And the point at which it is a principal axis may be found by drawing a plane through the polar line perpendicular to the straight line. The point of intersection is the required point.

The analytical condition (4) exactly expresses the fact that the polar line is parallel to the intersection of the plane.

61. **Ex. 1.** Show that the straight line $a(x-a)=b(y-b)=c(z-c)$ is at some point in its length a principal axis of an ellipsoid whose semi-axes are abc.

Ex. 2. Show that any straight line drawn on a lamina is a principal axis of that lamina at some point. Where is this point if the straight line pass through the centre of gravity?

Ex. 3. Given a plane $\dfrac{x}{f}+\dfrac{y}{g}+\dfrac{z}{h}-1=0$, there is always some point in it at which it is a principal plane. Also this point is its intersection with the straight line $fx-A=gy-B=hz-C$.

Ex. 4. Let two points P, Q be so situated that a principal axis at P intersects a principal axis at Q. Then if two planes be drawn at P and Q perpendicular to these principal axes, their intersection will be a principal axis at the point where it is cut by the plane containing the principal axes at P and Q. [Mr Townsend.]

For let the principal axes at P, Q meet any principal plane at the centre of gravity in p, q, and let the perpendicular planes cut the same principal plane in LN, MN. Also let the perpendicular planes intersect each other in RN. Then RN is perpendicular to the plane containing the points P, Q, p, q. Also since the polars of p and q are LN, MN, it follows that pq is the polar of the point N. Hence the straight line RN satisfies the criterion of the last Article.

Ex. 5. If P be any point in a principal plane at the centre of gravity, then every axis which passes through P, and is a principal axis at some point, lies in one of two perpendicular planes. One of these planes is the principal plane at the centre of gravity, and the other is a plane perpendicular to the polar line of P with regard to the focal conic. Also the locus of all the points Q at which QP is a principal axis is a circle passing through P and having its centre in the principal plane. [Mr Townsend.]

Ex. 6. The edge of regression of the developable-surface which is the envelope of the normal planes of any line of curvature drawn on a confocal quadric is a curve such that all its tangents are principal axes at some point in each.

62. Locus of equal Moments. *To find the locus of the points at which two principal moments of inertia are equal to each other.*

The principal moments at any point P are

$$I_1 = OP^2 - \lambda, \quad I_2 = OP^2 - \lambda + D_1^2, \quad I_3 = OP^2 - \lambda + D_2^2.$$

If we equate I_1 and I_2 we have $D_1 = 0$, and the point P must lie on the elliptic focal conic of the ellipsoid of gyration.

If we equate I_2 and I_3 we have $D_1 = D_2$, so that P is an umbilicus of any ellipsoid confocal with the ellipsoid of gyration. The locus of these umbilici is the hyperbolic focal conic.

In the first of these cases we have $\lambda = -C$, and D_2 is the semi-diameter of the focal conic conjugate to OP. Hence $D_2^2 + OP^2 =$ sum of squares of semi-axes $= A - C + B - C$. The three principal moments are therefore $I_1 = I_2 = OP^2 + C$, $I_3 = A + B - C$, and the axis of unequal moment is a tangent to the focal conic.

The second case may be treated in the same way by using a confocal hyperboloid, we therefore have $I_2 = I_3 = OP^2 + B$, $I_1 = A + C - B$, and the axis of unequal moment is a tangent to the focal conic.

These results follow also by combining Arts. 57 and 58. The cone which envelopes the ellipsoid of gyration and has its vertex at P must by these articles be a right cone if two principal moments at P are equal. But we know from solid geometry that this only happens when the vertex lies on a focal conic, and the unequal axis is then a tangent to that conic.

63. *To find the curves on any confocal quadric at which a principal moment of inertia is equal to a given quantity* I.

Firstly. The moment of inertia about a normal to a confocal quadric is $OP^2 - \lambda$. If this be constant, we have OP constant, and therefore the required curve is the intersection of that quadric with a concentric sphere. Such a curve is a sphero-conic.

Secondly. Let us consider those points at which the moment of inertia about a tangent is constant.

Construct any two confocals whose semi-major axes are a and a'. Draw any two tangent planes to these which cut each other at right angles. The moment of inertia about their intersection is the sum of the moments of inertia with regard to the two planes, and is therefore

$$B + C - A + a^2 + a'^2.$$

Thus the moments of inertia about the intersections of perpendicular tangent planes to the same confocals are equal to each other.

Let a, a', a'' be the semi-major axes of the three confocals which meet at any point P, then since confocals cut at right angles the moment of inertia about a tangent to the intersection of the confocals a', a'' is

$$I_1 = B + C - A + a'^2 + a''^2.$$

The intersection of these two confocals is a line of curvature on either. *Hence the moments of inertia about the tangents to any line of curvature are equal to one another; and these tangents are principal axes at the point of contact.*

On the quadric a draw a tangent PT making angles ϕ and $\frac{1}{2}\pi - \phi$ with the tangents to the lines of curvature at the point of contact P. If I_2, I_3 be the moments about the tangents to these lines of curvature, the moment of inertia about the tangent PT

$$= I_2 \cos^2\phi + I_3 \sin^2\phi$$
$$= B + C - A + (a''^2 + a^2) \cos^2\phi + (a^2 + a'^2) \sin^2\phi.$$

But, along a geodesic on the quadric a, $a'^2 \sin^2\phi + a''^2 \cos^2\phi$ is constant. *Hence the moments of inertia about the tangents to any geodesic on the quadric are equal to each other.*

64. **Ex. 1.** If a straight line touch any two confocals whose semi-major axes are a, a', the moment of inertia about it is $B + C - A + a^2 + a'^2$.

Ex. 2. When a body is referred to its principal axes at the centre of gravity, show how to find the co-ordinates of the point P at which the three principal moments are equal to the three given quantities I_1, I_2, I_3. [Jullien's Problem.]

The elliptic co-ordinates of P are evidently $a^2 = \frac{1}{2}(I_2 + I_3 - I_1 - B - C + A)$ &c.; and the co-ordinates (x, y, z) may then be found by Dr Salmon's formulæ,

$$x^2 = \frac{a^2 a'^2 a''^2}{(A - B)(A - C)} \text{ &c.}$$

Ex. 3. Let two planes at right angles touch two confocals whose semi-major axes are a, a'; and let a, a' be the values of a, a' for confocals touching the intersection of the planes; then $a^2 + a'^2 = \mathrm{a}^2 + \mathrm{a}'^2$, and the product of inertia with regard to the two planes is $(a^2 a'^2 - \mathrm{a}^2 \mathrm{a}'^2)^{\frac{1}{2}}$.

65. **Equimomental Surface.** The locus of all those points at which one of the principal moments of inertia of the body is equal to a given quantity is called an *equimomental surface*.

To find the equation to such a surface we have only to put I_1 constant, this gives $\lambda = r^2 - I$. Substituting in the equation to the subsidiary quadric, the equation to the surface becomes

$$\frac{x^2}{x^2 + y^2 + z^2 + A - I} + \frac{y^2}{x^2 + y^2 + z^2 + B - I} + \frac{z^2}{x^2 + y^2 + z^2 + C - I} = 1.$$

Through any point P on an equimomental surface describe a confocal quadric such that the principal axis is a tangent to a line of curvature on the quadric. By Art. 63, one of the intersections of the equimomental surface and this quadric is the line of curvature. Hence the principal axis at P about which the moment of inertia is I is a tangent to the equimomental surface.

Again, construct the confocal quadric through P such that the principal axis is a normal at P, then one of the intersections of the momental surface and this quadric is the sphero-conic through P. The normal to the quadric, being the principal axis, has just been shown to be a tangent to the surface. Hence the tangent plane to the equimomental surface is the plane which contains the normal to the quadric and the tangent to the sphero-conic.

To draw a perpendicular from the centre O on this tangent plane we may follow Euclid's rule. Take PP' a tangent to the sphero-conic, drop a perpendicular from O on PP', this is the radius vector OP, because PP' is a tangent to the sphere. At P in the tangent plane draw a perpendicular to PP', this is the normal PQ to the quadric. From O drop a perpendicular OQ on this normal, then OQ is a normal to the tangent plane. Hence this construction:

If P *be any point on an equimomental surface whose para-meter is* I, *and* OQ *a perpendicular from the centre on the tangent plane, then* PQ *is the principal axis at* P *about which the moment of inertia is* I.

The equimomental surface becomes Fresnel's wave surface when I is greater than the greatest principal moment of inertia at the centre of gravity. The general form of the surface is too well known to need a minute discussion here. It consists of two sheets, which become a concentric sphere and a spheroid when two of the principal moments at the centre of gravity are equal. When the principal moments are unequal, there are two singularities in the surface.

(1) The two sheets meet at a point P in the plane of the greatest and least moments. At P there is a tangent cone to the surface. Draw any tangent plane to this cone, and let OQ be a perpendicular from the centre of gravity O on this tangent plane. Then PQ is a principal axis at P. Thus there are an infinite number of principal axes at P because an infinite number of tangent planes can be drawn to the cone. But at any given point there cannot be more than three principal axes unless two of the principal axes be equal, and then the locus of the principal axes is a plane. Hence the point P is situated on a focal conic, and the locus of all the lines PQ is a normal plane to the conic. The point Q lies on a sphere whose diameter is OP, hence the locus of Q is a circle.

(2) The two sheets have a common tangent plane which touches the surface along a curve. This curve is a circle whose plane is perpendicular to the plane of greatest and least moments. Let OP' be a perpendicular from O on the plane of the circle, then P' is a point on the circle. If R be any other point on the circle the principal axis at R is RP'. Thus there is a circular ring of points, at each of which the principal axis passes through the same point, and the moments of inertia about these principal axes are all equal.

The equation to the equimomental surface may also be used for the purpose of finding the three principal moments at any point whose co-ordinates (x, y, z) are given. If we clear the equation of fractions, we have to determine I a cubic whose roots are the three principal moments.

Thus let it be required to find the locus of all those points at which any symmetrical function of the three principal moments is equal to a given quantity. We may express this symmetrical function in terms of the coefficients of the cubic by the usual rules, and the equation to the locus is found.

Ex. 1. If an equimomental surface cut a quadric confocal with the ellipsoid of gyration at the centre of gravity, then the intersections are a sphero-conic and a line of curvature. But, if the quadric be an ellipsoid, these cannot be both real.

For if the surface cut the ellipsoid in both, let P be a point on the line of curvature, and P' a point on the sphero-conic, then by Art. 59, $OP^2 + D_1^2 = OP'^2$, which is less than $A + \lambda$. But $OP^2 + D_1^2 + D_2^2 = A + B + C + 3\lambda$, therefore $D_2^2 > B + C + 2\lambda$, which is $> A + 2\lambda$. Hence $D_2 >$ the greatest radius vector of the ellipsoid, which is impossible.

Ex. 2. Find the locus of all those points in a body at which

(1) the sum of the principal moments is equal to a given quantity I,

(2) the sum of the products of the principal moments taken two and two together is equal to I^2,

(3) the product of the principal moments is equal to I^3.

The results are

(1) a sphere whose radius is $\sqrt{\dfrac{I - (A + B + C)}{2M}}$, Art. 13,

(2) the surface
$$\left.\begin{array}{l}(x^2 + y^2 + z^2)^2 + (A + B + C)(x^2 + y^2 + z^2) \\ + Ax^2 + By^2 + Cz^2 + AB + BC + CA\end{array}\right\} = I^2, \text{ Art. 65,}$$

(3) the surface $A'B'C' - A'y^2z^2 - B'z^2x^2 - C'x^2y^2 - 2x^2y^2z^2 = I^3$,

where $A' = A + y^2 + z^2$, with similar expressions for B', C'.

CHAPTER II.

D'ALEMBERT'S PRINCIPLE, &c.

66. THE principles, by which the motion of a single particle under the action of given forces can be determined, will be found discussed in any treatise on dynamics of a particle. These principles are called the three laws of motion. It is shown that if (x, y, z) be the co-ordinates of the particle at any time t referred to three rectangular axes fixed in space, m its mass, X, Y, Z the forces resolved parallel to the axes, the motion may be found by solving the simultaneous equations,

$$m\frac{d^2x}{dt^2} = X, \quad m\frac{d^2y}{dt^2} = Y, \quad m\frac{d^2z}{dt^2} = Z.$$

If we regard a rigid body as a collection of material particles connected by invariable relations, we may write down the equations of the several particles in accordance with the principles just stated. The forces on each particle are however no longer known, some of them being due to the mutual actions of the particles.

We assume (1) that the action between two particles is along the line which joins them, (2) that the action and reaction between any two are equal and opposite. Suppose there are n particles, then there will be $3n$ equations, and, as shown in any treatise on statics, $3n - 6$ unknown reactions. To find the motion it will be necessary to eliminate these unknown quantities. We shall thus obtain six resulting equations, and these will be shown, a little further on, to be sufficient to determine the motion of the body.

When there are several rigid bodies which mutually act and react on each other the problem becomes still more complicated. But it is unnecessary for us to consider in detail either this or the preceding case, for D'Alembert has proposed a method by which all the necessary equations may be obtained without writing down the equations of motion of the several particles, and without making any assumption as to the nature of the mutual actions except the following, which may be regarded as a natural consequence of the laws of motion :

The internal actions and reactions of any system of rigid bodies in motion are in equilibrium amongst themselves.

67.   *To explain D'Alembert's principle.*

In the application of this principle it will be convenient to use the term *effective force*, which may be defined as follows.

When a particle is moving as part of a rigid body, it is acted on by the external impressed forces and also by the molecular reactions of the other particles. If we consider this particle to be separated from the rest of the body, and all these forces removed, there is some one force which, under the same initial conditions, would make it move in the same way as before. This force is called the effective force on the particle. It is evidently the resultant of the impressed and molecular forces on the particle.

Let m be the mass of the particle, (x, y, z) its co-ordinates referred to any fixed rectangular axes at the time t. The accelerations of the particle are $\dfrac{d^2x}{dt^2}$, $\dfrac{d^2y}{dt^2}$ and $\dfrac{d^2z}{dt^2}$. Let f be the resultant of these, then, as explained in dynamics of a particle, the effective force is measured by mf.

Let F be the resultant of the impressed forces, R the resultant of the molecular forces on the particle. Then mf is the resultant of F and R. Hence if mf be reversed, the three F, R and mf are in equilibrium.

We may apply the same reasoning to every particle of each body of the system. We thus have a group of forces similar to R, a group similar to F, and a group similar to mf, the three groups forming a system of forces in equilibrium. Now by D'Alembert's principle the group R will itself form a system of forces in equilibrium. Whence it follows that the group F will be in equilibrium with the group mf. Hence

If forces equal to the effective forces but acting in exactly opposite directions were applied at each point of the system these would be in equilibrium with the impressed forces.

By this principle the solution of a dynamical problem is reduced to that of a problem in statics. The process is as follows. We first choose some quantities by means of which the position of the system in space may be determined. We then express the effective forces on each element in terms of these quantities. These, when reversed, will be in equilibrium with the given impressed forces.   Lastly, the equations of motion for each body may be formed, as is usually done in statics, by resolving in three directions and taking moments about three straight lines.

68.   Before the publication of D'Alembert's principle a vast number of dynamical problems had been solved.   These may be found scattered through the early volumes of the Memoirs of St Petersburg, Berlin and Paris, in the works of John Bernoulli and the *Opuscula* of Euler.   They require for the most part the determination of the motions of several bodies with or without weight which push or pull each other by means of threads or levers to which they are fastened or along which they can glide, and which having a certain impulse given them at first are then left to themselves or are compelled to move in given lines or surfaces.

The postulate of Huyghens, "that if any weights are put in motion by the force of gravity they cannot move so that the centre of gravity of them all shall rise higher than the place from which it descended," was generally one of the principles of the solution : but other principles were always needed in addition to this, and it required the exercise of ingenuity and skill to detect the most suitable in each case.   Such problems were for some time a sort of trial of strength among mathematicians.   The *Traité de dynamique* published by D'Alembert in 1743 put an end to this kind of challenge by supplying a direct and general method of resolving, or at least throwing into equations, any imaginable problem.   The mechanical difficulties were in this way reduced to difficulties of pure mathematics.   See Montucla, Vol. III. page 615, or Whewell's version of the same in his *History of the Inductive Sciences*.

D'Alembert uses the following words :—"Soient A, B, C, &c. les corps qui composent le système, et supposons qu'on leur ait imprimé les mouvemens a, b, c, &c. qu'ils soient forcés, à cause de leur action mutuelle, de changer dans les mouvemens a, b, c, &c.   Il est clair qu'on peut regarder le mouvement a imprimé au corps A comme composé du mouvement a, qu'il a pris, et d'un autre mouvement α ; qu'on peut de même regarder les mouvemens b, c, &c. comme composés des mouvemens b, β ; c, γ ; &c., d'ou il s'ensuit que le mouvement des corps A, B, C, &c. entr'eux auroit été le même, si au lieu de leur donner les impulsions a, b, c, on leur eût donné à-la-fois les doubles impulsions a, α ; b, β ; &c.   Or par la supposition les corps A, B, C, &c. ont pris d'eux-mêmes les mouvemens a, b, c, &c. donc les mouvemens α, β, γ, &c. doivent être tels qu'ils ne dérangent rien dans les mouvemens a, b, c, &c. c'est-à-dire que si les corps n'avoient reçu que les mouvemens α, β, γ, &c. ces mouvemens auroient dû se détruire mutuellement, et le système demeurer en repos.   De là resulte le principe suivant pour trouver le mouvement de plusieurs corps qui agissent les uns sur les autres.   Décomposez les mouvemens a, b, c, &c. imprimés à chaque corps, chacun en deux autres a, α ; b, β ; c, γ ; etc. qui soient tels que si l'on n'eût imprimé aux corps que les mouvemens a, b, c, &c. ils eussent pu conserver les mouvemens sans se nuire réciproquement ; et que si on ne leur eût imprimé que les mouvemens α, β, γ, &c. le système fût demeuré en repos ; il est clair que a, b, c, &c. seront les mouvemens que ces corps prendront en vertu de leur action.   Ce qu'il falloit trouver."

69.   The following remarks on D'Alembert's principle have been supplied by Sir G. Airy :

I have seen some statements of or remarks on this principle which appear to me to be erroneous.   The principle itself is not a new physical principle, nor any addition to existing physical principles; but is a convenient principle of combination of mechanical considerations, which results in a comprehensive process of great elegance.

The tacit idea, which dominates through the investigation, is this:—That every mass of matter in any complex mechanical combination may be conceived as containing in itself two distinct properties:—one that of connexion in itself, of susceptibility to pressure-force, and of connexion with other such masses, but not of inertia nor of impressions of momentum:—the other that of discrete molecules of matter, held in their places by the connexion-frame, susceptible to externally impressed momentum, and possessing inertia. The union produces an imponderable skeleton, carrying ponderable particles of matter.

Now the action of external momentum-forces on any one particle tends to produce a certain momentum-acceleration in that particle, which (generally) is not allowed to produce its full effect. And what prevents it from producing its full effect? It is the pressure of the skeleton-frame, which pressure will be measured by the difference between the impressed momentum-acceleration and the actual momentum-acceleration for the same. Thus every part of the skeleton sustains a pressure-force depending on that difference of momenta. And the whole mechanical system, however complicated, may now be conceived as a system of skeletons, each sustaining pressure-forces, and (by virtue of their combination) each impressing forces on the others.

And what will be the laws of movement resulting from this connexion? The forces are pressure-forces, acting on imponderable skeletons, and they must balance according to the laws of statical equilibrium. For if they did not, there would be instantaneous change from the understood motion, which change would be accompanied with instantaneous change of momentum-acceleration of the mole-cules, that would produce different pressures corresponding to equilibrium. (It is to be remarked that momentum cannot be changed instantaneously, but mo-mentum-acceleration can be changed instantaneously.)

We come thus to the conclusion that, taking for every molecule the dif-ference between the impressed momentum-acceleration and the actual momentum-acceleration, those differences through the entire machine will statically balance. And—combining in one group all the impressed momentum-accelerations, and in another group all the actual momentum accelerations—it is the same as saying that the impressed momentum-accelerations through the entire machine will balance the actual momentum-accelerations through the entire machine. This is the usual expression of D'Alembert's principle.

70.   The ordinary notation for the successive differential co-efficients of a function is very convenient when we are not always using the same independent variable. In a treatise on dynamics the time is usually the independent variable, and it is unnecessary to be continually calling attention to that fact. For this reason it is usual to represent the successive differential coefficients with regard to the time by accents or dots or some other marks placed over the dependent variable. It will be convenient to restrict the dot notation to represent differentiations with regard to the time solely, thus $\dot{x}$ and $\ddot{x}$ will be simply abbreviations for $\dfrac{dx}{dt}$ and $\dfrac{d^2x}{dt^2}$.

Dots will never be used to represent differentiations with regard to any quantity other than the time. When any other abbreviations are used for differential coefficients they will be preceded by an explanation.

This· abbreviated notation is very convenient in working examples or whenever mistakes cannot be produced by an occasional error in the dots. But in stating results to which reference has afterwards to be made, or in which it is important that there should be no misconception as to the meaning, it will be found better to use the more extended notation.

71. Example of D'Alembert's principle. *A light rod* OAB *can turn freely in a vertical plane about a smooth fixed hinge at* O. *Two heavy particles whose masses are* m *and* m' *are attached to the rod at* A *and* B *and oscillate with it. It is required to find the motion.*

The oscillatory motion of a single particle is usually discussed in treatises on elementary dynamics. It is proved that the time of a small oscillation is proportional to the square root of the radius of the circle described. In our problem we have two particles describing circular arcs of different radii in the same time. Each particle must therefore modify the motion of the other. The particle with the shorter radius hastens the motion of the other and is itself retarded by the slower motion of that other. Our object is to find the resulting motion.

By using D'Alembert's principle we are able to change this dynamical problem into an ordinary statical question, which when solved by the rules of statics gives the differential equations of the motion.

Let $OA = a$, $OB = b$, and let the angle the rod OAB makes with the vertical Oz be θ. The particle A describes a circular arc, hence its effective forces are known by elementary dynamics to be $ma\ddot{\theta}$ and $ma\dot{\theta}^2$, the former being directed along a tangent to the circular arc in the direction in which θ increases and the latter along the radius AO inwards. Similarly the effective forces of the particle B are $m'b\ddot{\theta}$ and $m'b\dot{\theta}^2$ along its tangent and radius respectively. The directions of these effective forces are represented in fig. 1 by the double headed arrows, while the single headed

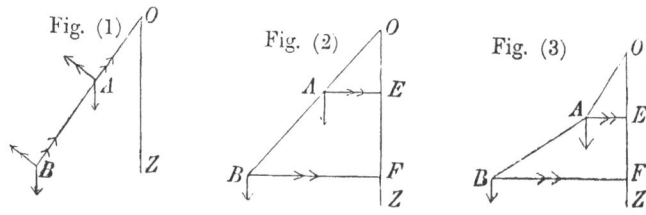

Fig. (1)   Fig. (2)   Fig. (3)

arrows indicate the directions of the weights mg and $m'g$ of the particles.

By D'Alembert's principle the four effective forces when reversed are in equilibrium with the weights of the particles. To avoid introducing the unknown reaction at O and those between the particles and the rod, let us take moments for the whole system about O. The forces $ma\dot\theta^2$ and $m'b\dot\theta^2$ being directed along BAO have no moments. The moments of the other two are $ma^2\ddot\theta$ and $m'b^2\ddot\theta$. Reversing these and adding the moments of the weights we have

$$(ma^2 + m'b^2)\,\ddot\theta + (ma + m'b)\,g\,\sin\theta = 0 \ldots\ldots\ldots\ldots(1).$$

This is the differential equation of motion. When it has been solved and the two arbitrary constants determined by any initial conditions we shall have θ expressed as a function of the time. But without entering here into the analytical solution we may shortly obtain the result.

We notice that if we put $m' = 0$ and write l for a, the equation (1) must give the motion of a *single particle* oscillating in a circle of radius l. This motion is therefore given by

$$l\ddot\theta + g\,\sin\theta = 0 \ldots\ldots\ldots\ldots\ldots\ldots(2).$$

This is of the same form as the equation (1). Hence the rod OAB oscillates as if the two particles were joined together into a single particle and placed at a distance $l = \dfrac{ma^2 + m'b^2}{ma + m'b}$ from the hinge O.

As a variation on this problem, let us *find the motion when the rod* OAB *moves round the vertical as a conical pendulum with uniform angular velocity, the angle θ which* OAB *makes with the vertical being constant.*

In this problem also the particles describe circles, but their planes are horizontal and their centres are at E and F as represented in fig. 2. The motion round the vertical being uniform, the effective force of A resolved along the tangent to its path is zero, while the effective force along its radius AE inwards is $ma\sin\theta\dot\phi^2$, ϕ being the angle made by the plane zOA with any fixed plane passing through Oz. Similarly the whole effective force on B is directed along its radius BF and is equal to $m'b\sin\theta\dot\phi^2$.

The directions of these effective forces are represented by the double headed arrows in fig. 2. Reversing these and taking moments as before about O, we have

$$-(ma^2 + mb^2)\sin\theta\cos\theta\,\dot\phi^2 + (ma + m'b)\,g\,\sin\theta = 0.$$

Hence the angular velocity $\dot\phi$ of the plane zOA round the vertical is given by

$$\dot\phi^2 = \frac{(ma + m'b)\, g}{(ma^2 + m'b^2)\cos\theta} \quad \dots\dots\dots\dots\dots(3),$$

except when the rod is vertical.

In this case again the result shows that the motion of the rod OAB round the vertical is the same as if the particles were collected into a single particle and placed at the same distance from O as in the first problem.

In these problems we have followed the rule given in Art. 67. We first express the effective forces by using the results given in treatises on dynamics of a particle. We reverse these effective forces and express by equations the conditions of equilibrium. These equations are the equations of motion.

Ex. 1. If three particles are attached to the rod at different distances from O, find the motion, (1) when the system oscillates in a vertical plane, and (2) when it revolves uniformly round the vertical.

Ex. 2. If the two particles are attached to O by two *strings* OA, AB as shown in fig. 3, and the system revolves round the vertical with a uniform angular velocity $\dot\phi$, show that

$$(m \cdot AE \cdot OE + m' \cdot BF \cdot OF)\,\dot\phi^2 = (m \cdot AE + m' \cdot BF)\, g.$$

72. **General Equations of Motion.** *To apply D'Alembert's principle to obtain the equations of motion of a system of rigid bodies.*

Let (x, y, z) be the co-ordinates of the particle m at the time t referred to any set of rectangular axes fixed in space. Then $\dfrac{d^2x}{dt^2}$, $\dfrac{d^2y}{dt^2}$, and $\dfrac{d^2z}{dt^2}$ will be the accelerations of the particle. Let X, Y, Z be the impressed accelerating forces on the same particle resolved parallel to the axes. By D'Alembert's principle the forces

$$m\left(X - \frac{d^2x}{dt^2}\right), \quad m\left(Y - \frac{d^2y}{dt^2}\right), \quad m\left(Z - \frac{d^2z}{dt^2}\right),$$

together with similar forces on every particle, will be in equilibrium. Hence by the principles of statics we have the equation

$$\Sigma m\, \frac{d^2x}{dt^2} = \Sigma mX,$$

and two similar equations for y and z; these are obtained by resolving parallel to the axes. Also we have

$$\Sigma m\left(y\, \frac{d^2z}{dt^2} - z\, \frac{d^2y}{dt^2}\right) = \Sigma m\,(yZ - zY),$$

and two similar equations for zx and xy; these are obtained by taking moments about the axes.

These equations may be written in the more convenient forms

$$\frac{d}{dt} \Sigma m \frac{dx}{dt} = \Sigma m X$$
$$\frac{d}{dt} \Sigma m \frac{dy}{dt} = \Sigma m Y \quad \Bigg\} \quad \dots\dots\dots\dots\dots(A).$$
$$\frac{d}{dt} \Sigma m \frac{dz}{dt} = \Sigma m Z$$

$$\frac{d}{dt} \Sigma m \left(y \frac{dz}{dt} - z \frac{dy}{dt} \right) = \Sigma m \left(yZ - zY \right)$$
$$\frac{d}{dt} \Sigma m \left(z \frac{dx}{dt} - x \frac{dz}{dt} \right) = \Sigma m \left(zX - xZ \right) \quad \Bigg\} \quad \dots\dots(B).$$
$$\frac{d}{dt} \Sigma m \left(x \frac{dy}{dt} - y \frac{dx}{dt} \right) = \Sigma m \left(xY - yX \right)$$

In a precisely similar manner, by taking the expressions for the accelerations in polar co-ordinates, we should have obtained another but equivalent set of equations of motion.

73. **Co-ordinates of a body.** The equations of motion of Art. 72 are the general equations of motion of any dynamical system. They are, however, extremely inconvenient in their present form. When the system considered is a rigid body and not merely a finite number of separate particles, the Σ's are all definite integrals. There are also an infinite number of x's, y's and z's all connected together by an infinite number of geometrical equations. It will be necessary, as suggested in Art. 67, to find some finite number of quantities which determine the position of the body in space and to express the effective forces in terms of these quantities. These are called the *co-ordinates of the body* *. It is most important in theoretical dynamics to choose the co-ordinates properly. They should be (1) such that a knowledge of them in terms of the time determines the motion of the body in a convenient manner, and (2) such that the dynamical equations when expressed in terms of them may be as little complicated as possible.

Let us first enquire how many co-ordinates are necessary to fix the position of a body.

The position of a body in space is given when we know the co-ordinates of some point in it and the angles which two straight lines fixed in the body make with the axes of co-ordinates. There are three geometrical relations existing between these six angles, so that the position of a body may be made to depend on *six* independent variables, viz. three co-ordinates and three angles. These might be taken as the co-ordinates of the body.

* Sir W. Hamilton uses the phrase "marks of position," but subsequent writers have adopted the term co-ordinates. See Cayley's *Report to the Brit. Assoc.*, 1857.

It is evident that we may express the co-ordinates (x, y, z) of any particle m of a body in terms of the co-ordinates of that body and quantities which are known and remain constant during the motion. First let us suppose the system to consist only of a single body, then if we substitute these expressions for x, y, z in the equations (A) and (B) of Art. 72, we shall have six equations to determine the six co-ordinates of the body in terms of the time. Thus the motion will be found. If the system consist of several bodies, we shall, by considering each separately, have six equations for each body. If there be any unknown reactions between the bodies, these will be included in X, Y, Z. For each reaction there will be a corresponding geometrical relation connecting the motion of the bodies. Thus on the whole we shall have sufficient equations to determine the motion of the system.

When the motion is in two dimensions these six co-ordinates are reduced to *three*. These are the two co-ordinates of the point fixed in the body, and the angle some straight line fixed in the body makes with a straight line fixed in space.

74. Let us next consider how the equations of motion (A) formed by resolution can be simplified by a proper choice of co-ordinates. We must find the resolved part of the momentum and the resolved part of the effective forces of a system in any direction.

Let the given direction be taken as the axis of x. Let (x, y, z) be the co-ordinates of any particle whose mass is m. The resolved part of its momentum in the given direction is $m \dfrac{dx}{dt}$. Hence the resolved part of the momentum of the whole system is $\Sigma m \dfrac{dx}{dt}$. Let $(\bar{x}, \bar{y}, \bar{z})$ be the co-ordinates of the centre of gravity of the system and M the whole mass. Then $M\bar{x} = \Sigma mx$;

$$\therefore M \frac{d\bar{x}}{dt} = \Sigma m \frac{dx}{dt}.$$

Hence the resolved part of the momentum of a system in any direction is equal to the whole mass multiplied into the resolved part of the velocity of the centre of gravity.

That is, the linear momentum of a system is the same as if the whole mass were collected into its centre of gravity.

In the same way, the resolved part of the effective forces of a system in any direction is equal to the whole mass multiplied into the resolved part of the acceleration of the centre of gravity.

It appears from this proposition that it will be convenient to take the co-ordinates of the centre of gravity of each rigid body in the system as three of the co-ordinates of that body. We can then express in a simple form the resolved part of the effective forces in any direction.

75. Lastly, let us consider how the equations of motion (B) formed by taking moments can be simplified by a proper choice of the three remaining co-ordinates. We must find the moment of the momentum and the moment of the effective forces about any straight line.

Let the given straight line be taken as the axis of x, then just as in statics $yZ - zY$ is the moment of a force about the axis of x, so, replacing Y and Z by $\dot{y}$ and $\dot{z}$, the moment of the momentum about the axis of x is

$$\Sigma m \left(y \frac{dz}{dt} - z \frac{dy}{dt} \right).$$

Now this is an expression of the second degree. If, then, we substitute $y = \bar{y} + y'$, $z = \bar{z} + z'$, we get by Art. 14

$$\Sigma m \left(y' \frac{dz'}{dt} - z' \frac{dy'}{dt} \right) + M \left(\bar{y} \frac{d\bar{z}}{dt} - \bar{z} \frac{d\bar{y}}{dt} \right),$$

where M is the mass of the system or body under consideration.

The second term of this expression is the moment about the axis of x of the momentum of a mass M moving with the centre of gravity.

The first term is the moment about a straight line parallel to the axis of x, not of the actual momenta of all the several particles but of their momenta *relatively* to the centre of gravity. In the case of any particular body it therefore depends only on the motion of the body relatively to its centre of gravity. In finding its value we may suppose the centre of gravity reduced to rest by applying to every particle of the system a velocity equal and opposite to that of the centre of gravity. Hence we infer that

The moment of the momentum of a system about any straight line is equal to the moment of the momentum of the whole mass supposed collected at its centre of gravity and moving with it, together with the moment of the momentum of the system relative to its centre of gravity about a straight line drawn parallel to the given straight line through the centre of gravity.

In the same way, this proposition will be also true if for the "momentum" of the system we substitute its "effective force."

By taking the axis Ox through the centre of gravity, we see that the moment of the *relative* momenta about any straight line through the centre of gravity is equal to that of the *actual* momenta.

It appears from this proposition that it will be convenient to refer the angular motion of a body to a system of co-ordinate axes meeting at the centre of gravity. A general expression for the moment of the effective forces about any straight line through the centre of gravity cannot be conveniently investigated at this stage. Different expressions will be found advantageous under different circumstances. There are three cases to which attention should be particularly directed: (1) that of a body turning about an axis fixed in the body and fixed in space; (2) that of motion in two dimensions, and (3) Euler's expression when the body is turning about a fixed point. These will be found at the beginnings of the third and fourth chapters and in the fifth chapter respectively.

76. Let a rigid body be turning about any point O fixed in the body, such as the centre of gravity. Let $O\xi$, $O\eta$, $O\zeta$ be a new set of rectangular axes fixed in the body. Then the ordinary formulæ for transformation of axes give

$$y = l\xi + m\eta + n\zeta, \quad z = \lambda\xi + \mu\eta + \nu\zeta$$

where the direction-cosines (lmn) $(\lambda\mu\nu)$ are functions of the time. We see therefore that the angular momentum

$$\Sigma m(y'\dot{z}' - z'\dot{y}') = A\Sigma m\xi^2 + B\Sigma m\eta^2 + C\Sigma m\xi\eta + \&c.$$

where $A = l\lambda' - \lambda l'$, and B, C &c. are similar functions of the direction-cosines. Now $\Sigma m\xi^2$, $\Sigma m\eta^2$ &c. and also the coefficients A, B, &c. would be the same for any system of particles equimomental to the given body. We therefore infer that the moment of the effective forces of a rigid body about any straight line is the same as that for any equimomental system which moves with the body.

In the same way we may show that the resolved parts of the effective forces are the same. Hence *in calculating the effective forces of a rigid body we may replace it by any convenient equimomental system which is rigidly connected with it.*

77. The quantity $\Sigma m (x\dot{y} - y\dot{x})$ expresses the moment of the momentum about the axis of z. It is called the *angular momentum* of the system about the axis of z. There is another interpretation which can be given to it. If we transform to polar co-ordinates, we have

$$x\dot{y} - y\dot{x} = r^2\dot{\theta}.$$

Now $\frac{1}{2}r^2 d\theta$ is the elementary area described round the origin in the time dt by the projection of the particle on the plane of xy. If twice this polar area be multiplied by the mass of the particle, it is called the *area conserved* by the particle in the time dt round the axis of z. Hence

$$\Sigma m \left(x\frac{dy}{dt} - y\frac{dx}{dt} \right)$$

is called the *area conserved* by the system in a unit of time, or more simply the area conserved.

78. **Three Important Propositions.** Summing up the results of the articles from 72 onwards, we see that we have established three important propositions.

Since any straight line *fixed in space* may be taken as an axis of co-ordinates, the three equations (A) of Art. 72 may be written in the typical form

$$\frac{d}{dt} \begin{pmatrix} \text{Linear Momentum in any} \\ \text{fixed direction} \end{pmatrix} = \begin{pmatrix} \text{Resolved impressed} \\ \text{force} \end{pmatrix}.$$

For the same reason, the three equations (B) of the same article may be written in the typical form

$$\frac{d}{dt} \begin{pmatrix} \text{Angular Momentum about} \\ \text{a fixed straight line} \end{pmatrix} = \begin{pmatrix} \text{Moment of im-} \\ \text{pressed forces} \end{pmatrix}.$$

Thirdly, we see by Art. 74, that the typical expression for the linear momentum may be written

$$\begin{pmatrix} \text{Linear Momentum in} \\ \text{any fixed direction} \end{pmatrix} = \begin{pmatrix} \text{Mass} \times \text{resolved velocity} \\ \text{of centre of gravity} \end{pmatrix}.$$

The corresponding typical expression for the angular momentum is deferred for the present.

79. Independence of Translation and Rotation. We may now enunciate two important propositions, which follow at once from the preceding results. It will, however, be more useful to deduce them from first principles.

(1) *The motion of the centre of gravity of a system acted on by any forces is the same as if all the mass were collected at the centre of gravity and all the forces were applied at that point parallel to their former directions.*

(2) *The motion of a body, acted on by any forces, about its centre of gravity is the same as if the centre of gravity were fixed and the same forces acted on the body.*

Taking any one of the equations (A) we have

$$\Sigma m \frac{d^2x}{dt^2} = \Sigma m X.$$

If $\bar{x}$, $\bar{y}$, $\bar{z}$ be the co-ordinates of the centre of gravity, then $\bar{x}\Sigma m = \Sigma mx$;

$$\therefore \frac{d^2\bar{x}}{dt^2} \Sigma m = \Sigma m X,$$

and the other equations may be treated in a similar manner.

But these are the equations which give the motion of a mass Σm acted on by forces $\Sigma m X$, &c. Hence the first principle is proved.

Taking any one of the equations (B) we have

$$\Sigma m \left(x \frac{d^2y}{dt^2} - y \frac{d^2x}{dt^2} \right) = \Sigma m \left(xY - yX \right).$$

Let $x = \bar{x} + x'$, $y = \bar{y} + y'$, $z = \bar{z} + z'$, then by Art. 14 this equation becomes

$$\Sigma m \left(x' \frac{d^2 y'}{dt^2} - y' \frac{d^2 x'}{dt^2} \right) + \left(\bar{x} \frac{d^2 \bar{y}}{dt^2} - \bar{y} \frac{d^2 \bar{x}}{dt^2} \right) \Sigma m = \Sigma m \, (xY - yX).$$

Now the axes of co-ordinates are quite arbitrary, let them be so chosen that the centre of gravity is passing through the origin at the moment under consideration. Then $\bar{x} = 0$, $\bar{y} = 0$, but $\dfrac{d\bar{x}}{dt}$, $\dfrac{d\bar{y}}{dt}$ are not necessarily zero. The equation then becomes

$$\Sigma m \left(x' \frac{d^2 y'}{dt^2} - y' \frac{d^2 x'}{dt^2} \right) = \Sigma m \, (x'Y - y'X).$$

This equation does not contain the co-ordinates of the centre of gravity and holds at every separate instant of the motion and therefore is always true. But this and the two similar equations obtained from the other two equations of (B) are exactly the equations of moments we should have had if we had regarded the centre of gravity as a fixed point and taken it as the origin of moments.

80.   These two important propositions are called respectively the principles of the conservation of the motions of translation and rotation. The first was given by Newton in the fourth corollary to the third law of motion, and was afterwards generalized by D'Alembert and Montucla. The second is more recent and seems to have been discovered about the same time by Euler, Bernoulli and the Chevalier d'Arcy.

Another name has also been given to these results. Together they constitute the *principle of the independence of the motions of translation and rotation*. The motion of the centre of gravity is the same as if the whole mass were collected at that point, and is therefore quite independent of the rotation. The motion round the centre of gravity is the same as if that point were fixed, and is therefore independent of the motion of that point.

81.   By the first principle the problem of finding the motion of the centre of gravity of a system, however complex the system may be, is reduced to the problem of finding the motion of a single particle. By the second the problem of finding the angular motion of a free body in space is reduced to that of determining the motion of that body about a fixed point.

Example of first principle.   In using the first principle it should be noticed that the impressed forces are to be applied at the centre of gravity *parallel* to their former directions. Thus, if a rigid body be moving under the influence of a central force, the motion of the centre of gravity is *not* generally the same as if the whole mass were collected at the centre of gravity and it were *then* acted on by the same central force. What the principle asserts is, that, if the attraction of the central force on each element of the body be found, the motion of the centre of gravity is the same as if *these* forces were applied at the centre of gravity parallel to their original directions.

If the impressed forces act always parallel to a fixed straight line, or if they tend to fixed centres and vary as the distance from those centres, the magnitude and direction of their resultant are the same whether we suppose the body collected into its centre of gravity or not. But in most cases care must be taken to find the resultant of the impressed forces as they really act on the body before it has been collected into its centre of gravity.

82. **Example of second principle.** Let us next consider an example of the second principle. Suppose the earth to be in rotation about some axis through its centre of gravity and to be acted on by the attractions of the sun and moon. Then we learn, from the second principle, that if the resultant attraction of these bodies pass through the centre of gravity of the earth, the rotation about the axis will not be in any way affected. In whatever way the centre of gravity of the earth may move in space, the axis of rotation will have its direction fixed in space and the angular velocity will be constant. Two important consequences follow immediately from this result. The centre of gravity of the earth is known to describe an orbit round the sun, which is very nearly in one plane, and the changes of the seasons chiefly depend on the inclination of the earth's axis to the plane of motion of the centre of the earth. The permanence of the seasons is therefore established. Secondly, since the angular velocity is constant, it follows that the length of the sidereal day is invariable.

Strictly speaking the resultant attraction due to any particle of the sun and moon does not pass through the centre of gravity of the earth. The reason is that the earth is not a perfect sphere whose strata of equal density are concentric spheres. But since the ellipticities of these strata are all small the motion of rotation of the earth will be but slightly affected. Nevertheless the sun (for instance) will act with unequal forces on those parts of the earth's equator which are nearer to it and those more remote. Thus the sun's attraction will *tend* to turn the earth about an axis lying in the plane of the equator and which is perpendicular to the radius vector of the sun. The general effect of this couple on the rotation of the earth is very remarkable. It will be proved in a later chapter (1) that the period of rotation of the earth is unaltered, (2) that though the direction of the earth's axis is no longer fixed in space, yet the axis still preserves, on the whole, the same inclination to the plane of the earth's motion round the sun. Thus the permanence of the seasons, as far as these causes are concerned, remains unaffected.

83. **General Method of using D'Alembert's principle.** The general problem in dynamics to be solved may be stated thus.

Any number of rigid bodies press both against each other and against fixed points, curves, or surfaces and are acted on by given forces; find their motion.

The mode of using D'Alembert's principle for the solution may be stated thus.

Let x, y, z be the co-ordinates of the centre of gravity of any one of these bodies referred to three rectangular axes fixed in space. Let three other co-ordinates of this body be chosen so that the three moments of the momentum of the body about three rectangular axes fixed in direction and meeting at the centre of gravity may be found conveniently in terms of them. Let h_1, h_2, h_3 be these three moments of the momentum, and let M be the mass. Then the effective forces of the body are equivalent to the three effective forces $M\dfrac{d^2x}{dt^2}$, $M\dfrac{d^2y}{dt^2}$, $M\dfrac{d^2z}{dt^2}$ and the three effective couples $\dfrac{dh_1}{dt}$, $\dfrac{dh_2}{dt}$, $\dfrac{dh_3}{dt}$. The three effective forces act at the centre of gravity parallel to the axes of x, y, z respectively, and the three couples act round the three axes about which the moments of the momentum were taken. The effective forces of all the other bodies of the system may be expressed in a similar manner.

Then all these effective forces and couples being reversed will be in equilibrium with the impressed forces. The equations of equilibrium may be found by resolving in such directions and taking moments about such straight lines as may be convenient. Instead of reversing the effective forces it is usually found more convenient to write the impressed and effective forces on opposite sides of the equations.

Taking the bodies separately we may thus obtain by three resolutions and three moments six equations of motion for each body.

If two rigid bodies press against each other or against a fixed obstacle there may be one or more unknown reactions. But there will also be in general as many equations to express the conditions of contact. The mode of writing down these conditions of contact will be explained in the chapters which follow.

Thus we shall have as many equations as there are co-ordinates and reactions. But sometimes by a judicious choice of the directions in which we resolve, or of the straight lines about which we take moments, we may (exactly as in statics) avoid introducing some of these reactions into the equations. This will reduce the number of equations which have to be formed. We may also sometimes avoid these reactions by resolving or taking moments for two of the bodies as if they formed for an instant one single body.

These differential equations will then have to be solved. The different methods of proceeding will be explained further on. Generally we can find one integral by a method called the principle of *Vis Viva*. A rule will be given to write down this integral without previously forming the equations of motion.

We have here limited ourselves to the method of forming the equation by resolving and taking moments. But we may proceed otherwise. Thus Lagrange has given a method of writing down the equations of motion by which, amongst other advantages, the labour of eliminating the reactions is avoided.

Application of D'Alembert's Principle to impulsive forces.

84.   If a force F act on a particle of mass m always in the same direction, the equation of motion is

$$m \frac{dv}{dt} = F,$$

where v is the velocity of the particle at the time t. Let T be the interval during which the force acts, and let v, v' be the velocities at the beginning and end of that interval. Then

$$m(v' - v) = \int_0^T F dt.$$

Now suppose the force F to increase without limit while the interval T decreases without limit. Then the integral may have a finite limit. Let this limit be P. Then the equation becomes

$$m(v' - v) = P.$$

The velocity in the interval T has increased or decreased from v to v'. Supposing the velocity to have remained finite, let V be its greatest value during this interval. Then the space described is less than VT. But in the limit this vanishes. Hence the particle has not moved during the action of the force F. It has not had time to move, but its velocity has been changed from v to v'.

We may consider that a proper measure has been found for a force when from that measure we can deduce all the effects of the force. In the case of finite forces we have to determine both the change of place and the change in the velocity of the particle. It is therefore necessary to divide the whole time of action into elementary times and determine the effect of the force during each of these. But in the case of infinite forces which act for an indefinitely short time, the change of place is zero, and the change of velocity is the only element to be determined. It is therefore more convenient to collect the whole force expended into one measure. Such a force is called an impulse. It may be defined as the limit of a force which is infinitely great, but acts only during an infinitely short time. There are of course no such forces in nature, but there are forces which are very great, and act only during a very short time. The blow of a hammer is a force of this kind. They may be treated as if they were impulses, and the results will be more or less correct according to the magnitude of the force and the shortness of the time of action. They may also be treated as if they were finite forces, and the small displacement of the body during the short time of action of the force may be found.

The quantity P may be taken as the measure of the force. *An impulsive force is measured by the whole momentum generated by the impulse.*

85. *In determining the effect of an impulse on a body, the effect of all finite forces which act on the body at the same time may be omitted.*

For let a finite force f act on a body at the same time as an impulsive force F. Then as before we have

$$v' - v = \frac{\int_0^T F dt}{m} + \frac{\int_0^T f dt}{m} = \frac{P}{m} + \frac{fT}{m}.$$

But in the limit fT vanishes. Similarly the force f may be omitted in the equation of moments.

86. *To obtain the general equations of motion of a system acted on by any number of impulses at once.*

Let u, v, w, u', v', w' be the velocities of a particle of mass m parallel to the axes just before and just after the action of the impulses. Let X', Y', Z' be the resolved parts of the impulse on m parallel to the axes.

Taking the same notation as before, we have the equation

$$\Sigma m \ddot{x} = \Sigma m X,$$

or integrating $\Sigma m (u' - u) = \Sigma m \int_0^T X dt = \Sigma X'$ ....(1).

Similarly we have the equations

$$\Sigma m (v' - v) = \Sigma Y' \dots\dots\dots\dots\dots\dots(2),$$

$$\Sigma m (w' - w) = \Sigma Z' \dots\dots\dots\dots\dots\dots(3).$$

Again the equation

$$\Sigma m (x\ddot{y} - y\ddot{x}) = \Sigma m (xY - yX)$$

becomes on integration

$$\Sigma m (x\dot{y} - y\dot{x}) = \Sigma m (x \textstyle\int Y dt - y \int X dt),$$

or taken between limits,

$$\Sigma m \left\{ x (v' - v) - y (u' - u) \right\} = \Sigma (x Y' - y X')\ldots\ldots\ldots(4),$$

and the other two equations become

$$\Sigma m \left\{ y (w' - w) - z (v' - v) \right\} = \Sigma (y Z' - z Y')\ldots\ldots\ldots(5),$$

$$\Sigma m \left\{ z (u' - u) - x (w' - w) \right\} = \Sigma (z X' - x Z')\ldots\ldots\ldots(6).$$

In the following investigations it will be found convenient to use accented letters to denote the states of motion after impact which correspond to those denoted by the same letters unaccented before the action of the impulse. Since the changes in direction and magnitude of the velocities of the several particles of the bodies are the only objects of investigation, it will be found convenient to express the equations of motion in terms of these velocities.

87   In applying D'Alembert's Principle to impulsive forces the only change which must be made is in the mode of measuring the effective forces. If $(u, v, w), (u', v', w')$ be the resolved parts of the velocity of any particle just before and just after the impulse, and if m be its mass, the effective forces will be measured by $m (u' - u)$, $m (v' - v)$, and $m (w' - w)$. The quantity mf in Art. 67 is to be regarded as the measure of the impulsive force which, if the particle were separated from the rest of the body, would produce these changes of momentum.

In this case, if we follow the notation of Arts. 74 and 75, the resolved part of the effective force in the direction of the axis of z is the difference of the values of $\Sigma m \dfrac{dz}{dt}$ just before and just after this action of the impulses, and this is the same as the difference of the values of $M \dfrac{d\bar{z}}{dt}$ at the same instants. In the same way the moment of the effective forces about the axis of z will be the difference of the values of

$$\Sigma m \left(x \frac{dy}{dt} - y \frac{dx}{dt} \right)$$

just before and just after the action of the impulses.

We may therefore extend the general proposition of Art. 83 to impulsive forces in the following manner.

Let $(u, v, w), (u', v', w')$ be the velocities of the centre of gravity of any rigid body of mass M just before and just after the action of the impulses resolved parallel to any three fixed rectangular axes. Let $(h_1, h_2, h_3), (h_1', h_2', h_3')$ be the moments of momentum relative to the centre of gravity about three rectangular axes fixed in direction and meeting at the centre of gravity, the moments being taken respectively just before and just after the impulses. Then the effective forces of the body are equivalent to the three effective forces $M(u' - u)$, $M(v' - v)$, $M(w' - w)$, acting at the centre of gravity parallel to the rectangular axes, together with the three effective couples $(h_1' - h_1)$, $(h_2' - h_2)$, $(h_3' - h_3)$ about those axes.

These effective forces and couples being reversed will be in equilibrium with the impressed forces. The equations of equilibrium may then be formed according to the rules of statics.

EXAMPLES.

Ex. 1. Two particles moving in the same plane are projected in parallel but opposite directions with velocities inversely proportional to their masses. Find the motion of their centre of gravity.

Ex. 2. A person is placed on a perfectly smooth table, show how he may get off.

Ex. 3. Explain how a person sitting on a chair is able to move the chair across the room by a series of jerks, without touching the ground with his feet.

Ex. 4. A person is placed at one end of a perfectly rough board which rests on a smooth table. Supposing he walks to the other end of the board, determine how far the board has moved. If he steps off the board, show how to determine its subsequent motion.

Ex. 5. The motion of the centre of gravity of a shell shot from a gun in vacuo is a parabola, and its motion is unaffected by the bursting of the shell.

Ex. 6. A rod revolving uniformly in a horizontal plane round a pivot at its extremity suddenly snaps in two: determine the motion of each part.

Ex. 7. A cube slides down a perfectly smooth inclined plane with four of its edges horizontal. The middle point of the lowest edge comes in contact with a small fixed obstacle and is reduced to rest. Determine whether the cube is also reduced to rest, and show that the resultant impulsive action along the edge will not act along the inclined plane.

Ex. 8. Two persons A and B are situated on a perfectly smooth horizontal plane at a distance a from each other. A throws a ball to B which reaches B after a time t. Show that A will begin to slide along the plane with a velocity $\dfrac{ma}{Mt}$, where M is his own mass and m that of the ball. If the plane had been perfectly rough, explain in general terms the nature of the pressures between A's feet and the plane which would have prevented him from sliding. Would these pressures have had a single resultant?

Ex. 9. A cannon rests on an imperfectly rough horizontal plane and is fired with such a charge that the relative velocity of the ball and cannon at the moment when the ball leaves the cannon is V. If M be the mass of the cannon, m that of the ball, and μ the coefficient of friction, show that the cannon will recoil a distance $\left(\dfrac{mV}{M+m}\right)^2 \dfrac{1}{2\mu g}$ on the plane.

Ex. 10. A spherical cavity of radius a is cut out of a cubical mass so that the centre of gravity of the remaining mass is in the vertical through the centre of the cavity. The cubical mass rests on a perfectly smooth horizontal plane, but the interior of the cavity is perfectly rough. A sphere of mass m, and radius b, rolls down the side of the cavity starting from rest with its centre on a level with the centre of the cavity. Show that when the sphere next comes to rest, the cubical mass will have moved through a space $\dfrac{2m(a-b)}{M+m}$, where M is the mass of the remaining portion of the cube. Would the result be the same if the cavity were smooth or imperfectly rough?

Ex. 11. Two railway engines drawing the same train are connected by a loose chain and come several times in succession into collision with each other; the leading engine being a little top-heavy and the buffers of both rather low. The fore-wheels of the first engine are observed to jump up and down. What dynamical explanation can be given of this rocking motion? At what level should the buffers be placed that it may not occur? *Camb. Trans.* Vol. VII.

Ex. 12. Sir C. Lyell in his account of the earthquake in Calabria in 1783, mentions two obelisks each of which was constructed of three great stones laid one on the top of the other. After the earthquake, the pedestal of each obelisk was found to be in its original place, but the separate stones above were turned partially round and removed several inches from their position without falling. The shock which agitated the building was therefore described as having been horizontal and vorticose. Show that such a displacement would be produced by a simple rectilinear shock, if the resultant blow on each stone did not pass through its centre of gravity. See *Mallet's dynamics of earthquakes.* Milne in his *Earthquakes* 1886, page 196, discusses the latter explanation and refers to some similar cases which occurred in the earthquake at Yokohama in 1880.

CHAPTER III.

88. The Fundamental Theorem. *A rigid body can turn freely about an axis fixed in the body and in space, to find the moment of the effective forces about the axis of rotation.*

Let any plane passing through the axis and fixed in space be taken as a plane of reference, and let θ be the angle which any other plane through the axis and fixed in the body makes with the first plane. Let m be the mass of any element of the body, r its distance from the axis, and let ϕ be the angle made by a plane through the axis and the element m with the plane of reference.

The velocity of the particle m is $r\dot{\phi}$ in a direction perpendicular to the plane containing the axis and the particle. The moment of the momentum of this particle about the axis is clearly $mr^2\dot{\phi}$. Hence the moment of the momenta of all the particles is $\Sigma(mr^2\dot{\phi})$. Since the particles of the body are rigidly connected with each other, it is obvious that $\dot{\phi}$ is the same for every particle, and equal to $\dot{\theta}$. Hence the moment of the momenta of all the particles of the body about the axis is $\Sigma mr^2\dot{\theta}$, *i.e. the moment of inertia of the body about the axis multiplied into the angular velocity.*

The accelerations of the particle m are $r\ddot{\phi}$ and $-r\dot{\phi}^2$ perpendicular to, and along the direction in which r is measured, the moment of the moving forces of m about the axis is $mr^2\ddot{\phi}$, hence the moment of the moving forces of all the particles of the body about the axis is $\Sigma(mr^2\ddot{\phi})$. By the same reasoning as before this is equal to $\Sigma mr^2\ddot{\theta}$, *i.e. the moment of inertia of the body about the axis into the angular acceleration.*

89. *To determine the motion of a body about a fixed axis under the action of any forces.*

By D'Alembert's principle the effective forces when reversed will be in equilibrium with the impressed forces. To avoid introducing the unknown reactions at the axis, let us take moments about the axis.

Firstly, let the forces be impulsive. Let ω, ω' be the angular velocities of the body just before and just after the action of the forces. Then, following the notation of the last article,

$$\omega' \cdot \Sigma mr^2 - \omega \cdot \Sigma mr^2 = L,$$

where L is the moment of the impressed forces about the axis;

$$\therefore \; \omega' - \omega = \frac{\text{moment of forces about axis}}{\text{moment of inertia about axis}}.$$

This equation will determine the change in the angular velocity produced by the action of the forces.

Secondly, let the forces be finite. Then taking moments about the axis, we have

$$\frac{d^2\theta}{dt^2} \cdot \Sigma mr^2 = L;$$

$$\therefore \; \frac{d^2\theta}{dt^2} = \frac{\text{moment of forces about axis}}{\text{moment of inertia about axis}}.$$

This equation when integrated will give the values of θ and $\dot\theta$ at any time. Two undetermined constants will make their appearance in the course of the solution. These are to be determined from the given initial values of θ and $\dot\theta$. Thus the whole motion can be found.

90. It appears from this proposition that the motion of a rigid body about a fixed axis depends on (1) the moment of the forces about that axis and (2) the moment of inertia of the body about the axis. Let Mk^2 be this moment of inertia, so that k is the radius of gyration of the body. Then if the whole mass of the body were collected into a particle and attached to the fixed axis by a rod without inertia, whose length is the radius of gyration k, and if this system be acted on by forces having the same moment as before, and be set in motion with the same initial values of θ and $\dfrac{d\theta}{dt}$, then the whole subsequent angular or gyratory motion of the rod will be the same as that of the body. *We may say briefly, that a body turning about a fixed axis is dynamically given, when we know its mass and radius of gyration.*

91. *Ex. A perfectly rough circular horizontal board is capable of revolving freely round a vertical axis through its centre. A man whose weight is equal to that of the board walks on and round it at the edge: when he has completed the circuit what will be his position in space?*

Let a be the radius of the board, Mk^2 its moment of inertia about the vertical axis. Let ω be the angular velocity of the board, ω' that of the man about the vertical axis at any time. And let F be the action between the feet of the man and the board.

The equation of motion of the board is by Art. 89,
$$Mk^2\dot{\omega} = -Fa \dots\dots\dots\dots\dots\dots\dots\dots\dots(1).$$
The equation of motion of the man is by Art. 79,
$$Ma\dot{\omega}' = F \dots\dots\dots\dots\dots\dots\dots\dots\dots (2).$$
Eliminating F and integrating, we get
$$k^2\omega + a^2\omega' = 0,$$
the constant being zero, because the man and the board start from rest. Let θ, θ' be the angles described by the board and man round the vertical axis. Then $\omega = \dot{\theta}$; $\omega' = \dot{\theta}'$, and $k^2\theta + a^2\theta' = 0$. Hence, when $\theta' - \theta = 2\pi$, we have $\theta' = \dfrac{k^2}{k^2 + a^2} 2\pi$. This gives the angle in space described by the man. If $k^2 = \dfrac{a^2}{2}$ we have $\theta' = \dfrac{2}{3}\pi$.

Let V be the mean relative velocity with which the man walks along the board. Then $\omega' - \omega = \dfrac{V}{a}$; $\therefore \omega = -\dfrac{Va}{k^2 + a^2} = -\dfrac{2}{3}\dfrac{V}{a}$. This gives the mean angular velocity of the board.

On the Pendulum.

92. *A body acted on by gravity only moves about a fixed horizontal axis, to determine the motion.*

Take the vertical plane through the axis as the plane of reference, and the plane through the axis and the centre of gravity as the plane fixed in the body. Then the equation of motion is
$$\frac{d^2\theta}{dt^2} = \frac{\text{moment of forces}}{\text{moment of inertia}} \dots\dots\dots\dots\dots(1)$$
$$= -\frac{Mgh\sin\theta}{M(k^2 + h^2)},$$
where h is the distance of the centre of gravity from the axis and Mk^2 is the moment of inertia of the body about an axis through the centre of gravity parallel to the fixed axis. Hence
$$\frac{d^2\theta}{dt^2} + \frac{gh}{k^2 + h^2}\sin\theta = 0\dots\dots\dots\dots\dots(2).$$

The equation (2) cannot be integrated in finite terms, but if the oscillations be small, we may reject the cubes and higher powers of θ and the equation will become
$$\frac{d^2\theta}{dt^2} + \frac{gh}{k^2 + h^2}\theta = 0.$$

Hence the time of a complete oscillation is $2\pi\sqrt{\dfrac{k^2 + h^2}{gh}}$. If h and k be measured in feet and $g = 32\cdot18$, this formula gives the time in seconds.

The equation of motion of a particle of any mass suspended by a string l is

$$\frac{d^2\theta}{dt^2} + \frac{g}{l} \cdot \sin\theta = 0 \dots\dots\dots\dots\dots(3),$$

which may be deduced from equation (2) by putting $k = 0$ and $h = l$. Hence the angular motions of the string and the body under the same initial conditions will be identical if

$$l = \frac{k^2 + h^2}{h} \quad \dots\dots\dots\dots\dots(4).$$

This length is called the *length of the simple equivalent pendulum*.

Centre of Oscillation *. Through G, the centre of gravity of the body, draw a perpendicular to the axis of revolution cutting it in C. Then C is called the *centre of suspension*. Produce CG to O so that $CO = l$. Then O is called the *centre of oscillation*. *If the whole mass of the body (or indeed any mass) were collected at the centre of oscillation and suspended by a thread to the centre of suspension, its angular motion and time of oscillation would be the same as that of the body under the same initial circumstances.*

The equation (4) may be put under another form. Since $CG = h$ and $OG = l - h$, we have

$$GC \cdot GO = (\text{rad.})^2 \text{ of gyration about } G,$$
$$CG \cdot CO = (\text{rad.})^2 \text{ of gyration about } C,$$
$$OG \cdot OC = (\text{rad.})^2 \text{ of gyration about } O.$$

Any of these equations show that, if O be made the centre of suspension, and the axis be parallel to the axis about which k was

---

* The position of the centre of oscillation of a body was first correctly determined by Huyghens in his *Horologium Oscillatorium* published at Paris in 1673. The most important of the theorems given in the text were discovered by him. As D'Alembert's principle was not known at that time, Huyghens had to discover some principle for himself. The hypothesis was, that when several weights are put in motion by the force of gravity, in whatever manner they act on each other their centre of gravity cannot be made to mount to a height *greater* than that from which it has descended. Huyghens considers that he assumes here only that a heavy body cannot of itself move upwards. The next step in the argument was, that at any instant the velocities of the particles are such that, if they were separated from each other and properly guided, the centre of gravity could be made to mount to a second position *as high* as its first position. For if not, consider the particles to start from their last positions, to describe the same paths reversed, and then again to be joined together into a pendulum; the centre of gravity would rise to its first position; but if this be higher than the second position, the hypothesis would be contradicted. This principle gives the same equation which the modern principle of Vis Viva would give. The rest of his solution is not of much interest.

taken, C will be the centre of oscillation. Thus *the centres of oscillation and suspension are convertible and the times of oscillation about these points are the same.*

If the time of oscillation be given, l is given and the equation (4) will give two values of h. Let these values be h_1, h_2. Let two cylinders be described with that straight line as axis about which the radius of gyration k was taken, and let the radii of these cylinders be h_1, h_2. Then the times of oscillation of the body about all generating lines of these cylinders are the same, and are approximately equal to $2\pi \sqrt{\dfrac{l}{g}}$.

With the same axis describe a third cylinder whose radius is k. Then $l = 2k + \dfrac{(h-k)^2}{h}$, hence l is always greater than $2k$, and decreases continually as h decreases and approaches the value k. Thus the length of the equivalent pendulum continually decreases as the axis of suspension approaches from without to the circumference of this third cylinder. When the axis of suspension is a generating line of the cylinder the length of the equivalent pendulum is $2k$. When the axis of suspension is within the cylinder and approaches the centre of gravity the length of the equivalent pendulum continually increases, and it becomes infinite as the axis passes through the centre of gravity.

The time of oscillation is therefore least when the axis is a generating line of the circular cylinder whose radius is k. But the time about the axis thus found is not an absolute minimum. It is a minimum only for axes drawn parallel to a given straight line in the body. To find the axis about which the time is absolutely a minimum we must find the axis about which k is a minimum. Now it is proved in Art. 23 that the axis through G about which the moment of inertia is least or greatest is one of the principal axes. Hence the axis about which the time of oscillation is a minimum is parallel to that principal axis through G about which the moment of inertia is least. Also if Mk^2 be the moment of inertia about that axis, the axis of suspension is at a distance k from it measured in any direction.

93. Ex. 1. Find the time of the small oscillations of a cube (1) when one side is fixed, (2) when a diagonal of one of its faces is fixed; the axis in both cases being horizontal. If $2a$ be a side of the cube, show that the length of the simple equivalent pendulum is in the first case $\dfrac{4}{3}\sqrt{2}a$, and in the second case $\dfrac{5}{3}a$.

Ex. 2. An elliptic lamina is such that when it swings about one latus rectum as a horizontal axis, the other latus rectum passes through the centre of oscillation, prove that the eccentricity is $\frac{1}{2}$.

Ex. 3. A circular arc oscillates about an axis through its middle point perpendicular to the plane of the arc. Prove that the length of the simple equivalent pendulum is independent of the length of the arc, and is equal to twice the radius.

Ex. 4. The density of a rod varies as the distance from one end, show that the axis perpendicular to it about which the time of oscillation is a minimum intersects the rod at one of the two points whose distance from the centre of gravity is $\frac{1}{6}\sqrt{2}a$, where a is the length of the rod.

Ex. 5. Find what axis in the area of an ellipse must be fixed that the time of a small oscillation may be a minimum. Show that the axis must be parallel to the major axis, and must bisect the semi-minor axis.

Ex. 6. A uniform stick hangs freely by one end, the other end being close to the ground. An angular velocity in a vertical plane is then communicated to the stick, and, when it has risen through an angle of $90°$, the end by which it was hanging is loosed. What must be the initial angular velocity so that on falling to the ground it may pitch in an upright position? Show that the required angular velocity ω is given by $\omega^2 = \frac{g}{2a}\left(3 + \frac{p^2}{p+1}\right)$, where $2p$ may be any odd multiple of π and $2a$ is the length of the rod.

Ex. 7. Two bodies can move freely and independently under the action of gravity about the same horizontal axis; their masses are m, m', and the distances of their centres of gravity from the axis are h, h'. If the lengths of their simple equivalent pendulums be L, L', prove that when they are fastened together in the positions of equilibrium the length of the equivalent pendulum will be $\dfrac{mhL + m'h'L'}{mh + m'h'}$.

The length of this resultant equivalent pendulum lies between L and L' provided h and h' have the same sign.

If a heavy particle m' be attached to a vibrating pendulum it follows that the period is increased or decreased according as the point of attachment is at a greater or less distance from the axis of suspension than the centre of oscillation.

Ex. 8. When it is required to regulate a clock, such as the great Westminster clock, without stopping the pendulum, it is usual to add some small weight to or subtract it from a platform attached to the pendulum. Show that, in order to make a given alteration in the going of the clock by the addition of the least possible weight, the platform must be placed at a distance from the point of suspension equal to half the length of the simple equivalent pendulum. Show also that a slight error in the position of the platform will not affect the weight required to be added.

Ex. 9. A circular table, centre O, is supported by three legs AA', BB', CC' which rest on a perfectly rough horizontal floor, and a heavy particle P is placed on the table. Suddenly one leg CC' gives way, show that the table and the particle will immediately separate if pc be greater than κ^2; where p and c are the distances of P and O respectively from the line AB joining the tops of the legs, and κ is the radius of gyration of the table with the remaining legs about the line $A'B'$ joining the points where the legs rest on the floor.

The condition of separation is that the initial normal acceleration of the point of the table at P should be greater than the normal acceleration of the particle itself.

Ex. 10. A string without weight is placed round a fixed ellipse whose plane is vertical, and the two ends are fastened together. The length of the string is greater than the perimeter of the ellipse. A heavy particle can slide freely on the string and performs small oscillations under the action of gravity. Prove that the simple equivalent pendulum is the radius of curvature of the confocal ellipse passing through the position of equilibrium of the particle.

94. **Effect of change of temperature.** In a clock which is regulated by a pendulum, it is necessary that the time of oscillation should be invariable. As all substances expand or contract with every alteration of temperature, it is clear that the distance of the centre of gravity of the pendulum from the axis and the moment of inertia about that axis will be continually altering. The length of the simple equivalent pendulum does not however depend on either of these elements simply, but on their ratio. If then we can construct a pendulum such that the expansion or contraction of its different parts does not alter this ratio, the time of oscillation will be unaffected by any change of temperature. For an account of the various methods of accomplishing this which have been suggested, we refer the reader to any treatise[*] on clocks. We shall here only notice for the sake of illustration one simple construction, which has been much used. It was invented by George Graham about the year 1715. He gave an account of it in vol. 34 of the *Phil. Trans.* 1726 (printed 1728).

Some heavy fluid, such as mercury, is enclosed in a cast-iron cylindrical jar. Iron is used partly because there is no chemical action between it and the mercury and partly because its coefficient of expansion is not large. An iron rod is screwed into the top of the jar and then suspended in the usual manner from a fixed point. The downward expansion of the iron on any increase of temperature tends to lower the centre of oscillation, but the upward expansion of the mercury tends on the contrary to raise it. It is required to determine the condition that the position of the centre of oscillation may on the whole be unaltered.

Let Mk^2 be the moment of inertia of the iron jar and rod about the axis of suspension, c the distance of their common centre of gravity from that axis. Let l be the length of the pendulum from the point of suspension to the bottom of the jar, a the internal radius of the jar. Let nM be the mass of the mercury, h the height it occupies in the jar.

The moment of inertia of the cylinder of mercury about a straight line through its centre of gravity perpendicular to its axis is by Art. 18, Ex. 8, $nM\left(\dfrac{h^2}{12}+\dfrac{a^2}{4}\right)$

Hence the moment of inertia of the whole body about the axis of suspension is

$$Mn\left\{\frac{h^2}{12}+\frac{a^2}{4}+\left(l-\frac{h}{2}\right)^2\right\}+Mk^2,$$

and the moment of the whole mass collected at its centre of gravity is

$$Mn\left(l-\frac{h}{2}\right)+Mc.$$

---

[*] Reid on *Clocks;* Denison's treatise on *Clocks and Clockmaking* in Weale's Series, 1867; Captain Kater's treatise on *Mechanics* in Lardner's *Cyclopædia*, 1830.

The length L of the simple equivalent pendulum is the ratio of these two, and on reduction we have

$$L = \frac{n\left(\dfrac{h^2}{3} - lh + l^2 + \dfrac{a^2}{4}\right) + k^2}{n\left(l - \dfrac{h}{2}\right) + c} \quad\text{............................(1).}$$

Let the linear expansion of the substance which forms the rod and jar be denoted by α and that of mercury by β for each degree of the thermometer. If the thermometer used be Fahrenheit's, we have $\alpha = \cdot0000065668$, $\beta = \cdot00003336$, according to some experiments of Dulong and Petit. Thus we see that α and β are so small that their squares may be neglected. In calculating the height of the mercury it must be remembered that the jar expands laterally, and thus the relative vertical expansion of the mercury is $3\beta - 2\alpha$, which we shall represent by γ.

If then the temperature of every part be increased t^0, we have a, l, k, c, all increased in the ratio $1 + at : 1$, while h is increased in the ratio $1 + \gamma t : 1$. Since L is to be unaltered, we have

$$\left(\frac{dL}{da} a + \frac{dL}{dl} l + \frac{dL}{dk} k + \frac{dL}{dc} c\right) a + \frac{dL}{dh} h\gamma = 0.$$

But L is a homogeneous function of one dimension, hence

$$\frac{dL}{da} a + \frac{dL}{dl} l + \frac{dL}{dk} k + \frac{dL}{dc} c + \frac{dL}{dh} h = L.$$

The condition becomes therefore by substitution

$$\frac{a}{a - \gamma} = \frac{h}{L} \frac{dL}{dh}.$$

Let A, B be the numerator and denominator of the expression for L given by equation (1). Then taking the logarithmic differential

$$\frac{1}{L} \frac{dL}{dh} = \frac{n\left(\dfrac{2}{3} h - l\right)}{A} + \frac{\dfrac{1}{2} n}{B} = \frac{n}{B}\left(\frac{\dfrac{2}{3} h - l}{L} + \frac{1}{2}\right).$$

Hence the required condition is

$$\frac{a}{3(\beta - a)} = \frac{h}{l - \dfrac{h}{2} + \dfrac{c}{n}}\left(\frac{l - \dfrac{2}{3} h}{L} - \frac{1}{2}\right) \quad\text{..................... (2).}$$

This calculation has more theoretical than practical importance, for the numerical values of α and β depend a good deal on the purity of the metals and on the mode in which they have been worked. The adjustment must therefore be finally made by experiment. If the rate of the clock is found to be affected by a change of temperature it is usual to alter slightly the quantity of mercury in the jar until by trial the adjustment is found to be satisfactory.

In the investigation we have supposed α and β to be absolutely constant, but this is only a very near approximation. Thus a change of $80°$ Fah. would alter β by less than a fiftieth of its value.

When the adjustment is made the compensation is not strictly correct, for the iron jar and mercury have been supposed to be of uniform temperature. Now the different materials of which the pendulum is composed absorb heat at different rates, and therefore while the temperature is changing there will be some slight error in the clock.

The whole length of a seconds pendulum of this construction is about 44 inches, the expansion and contraction of which is corrected by a column of mercury in the jar about 7 inches long. The radius of the jar is usually about one inch. The weight of the mercury is then about 10 to 12 pounds which, added to that of the jar, frame, and rod brings the total weight to about 14 pounds.

Ex. If, as a first approximation, we regard the mercury as the weight, the jar and the rod being only of sufficient mass to hold up the mercury, and if we also suppose h and a to be so much less than L that we may reject the squares of their ratios to L prove that the equation (1) gives $L = l - \frac{1}{2}h$ and that the equation (2) gives $h = \frac{1}{6}L$.

95. **Buoyancy of Air.** Another cause of error in a clock pendulum is the buoyancy of the air. *This produces an upward force acting at the centre of gravity of the volume of the pendulum equal to the weight of the air displaced.* A very slight modification of the fundamental investigation in Art. 92 will enable us to take this into account. Let V be the volume of the pendulum, D the density of the air; h_1, h_2, the distances of the centres of gravity of the mass and volume respectively from the axis of suspension, Mk^2 the moment of inertia of the mass about the axis of suspension. Let us also suppose the pendulum to be symmetrical about a plane through the axis and either centre of gravity.

The equation of motion is then

$$Mk^2\ddot{\theta} = -Mgh_1 \sin\theta + VDgh_2 \sin\theta \ \dots\dots\dots\dots\dots\dots (1).$$

By the same reasoning as before we infer that if l be the length of the equivalent pendulum

$$\frac{k^2}{l} = h_1 - h_2 \frac{VD}{M} \ \dots\dots\dots\dots\dots\dots\dots\dots\dots\dots\dots\dots\dots (2).$$

The density D of the air is continually changing, the changes being indicated by variations in the height of the barometer. Let h be the value of the right-hand side of this equation for any standard density D. Suppose the actual density to be $D + \delta D$ and let $l + \delta l$ be the corresponding length of the seconds pendulum, then we have by differentiation $\dfrac{k^2 \delta l}{l^2} = h_2 \dfrac{V \delta D}{M}$, and therefore $\dfrac{\delta l}{l} = \dfrac{h_2}{h} \dfrac{VD}{M} \dfrac{\delta D}{D}$.

This formula gives in a convenient form the change in the length of the equivalent pendulum due to a change in the density of the air.

96. Ex. 1. If the centres of gravity of the mass and volume were very nearly coincident and the weight of the air displaced were $\frac{1}{7200}$ of the weight of the pendulum, show that a rise of one inch in the barometer would cause an error in the rate of going of the seconds pendulum of nearly one-fifth of a second per day.

This example will enable us to estimate the general effect of a rise of the barometer on the rate of going of an iron pendulum.

Ex. 2. If we affix to the pendulum rod produced upwards a body of the same volume as the pendulum bob but of very small weight, so that the centre of gravity of the volume lies in the axis of suspension, show that the correction for buoyancy vanishes. This method was suggested in 1871 by Sir George Airy, but he remarks that this construction would probably be inconvenient in practice.

Ex. 3. If a barometer were attached to the pendulum show that the rise or fall of the mercury as the density of the air changed could be so arranged as to keep the time of vibration unaltered. This method was suggested first by Dr Robinson of

Armagh in 1831 in the fifth volume of the memoirs of the Astronomical Society, and afterwards by Mr Denison in the *Astronomical Notices* for Jan. 1873. In the *Armagh Places of Stars* published in 1859, Dr Robinson described the difficulties he found in practice before he was satisfied with the working of the clock.

The jar of mercury in Graham's mercurial pendulum might be used as the cistern of the barometer, as Mr Denison remarks.

The theory of the construction is that in differentiating equation (2) we are to suppose k^2, &c. variable and l constant.

Prof. Rankine read a paper to the British Association in 1853 in which he proposed to use a clock with a centrifugal or revolving pendulum, part of which should consist of a siphon barometer. The rising and falling of the barometer would affect the rate of going of the clock so that the *mean* height of the mercurial column during any long period would register itself.

Ex. 4. If the pendulum be supposed to drag a quantity of air with it which bears a constant ratio to the density D of the surrounding air and adds γD to the moment of inertia of the pendulum without increasing the moving power, show that the change produced in the simple equivalent pendulum by a change of density δD is given by $\delta l = \gamma \dfrac{\delta D}{M \bar{h}_1}$.

97. **Moments of Inertia found by experiment.**

In many experimental investigations it is necessary to determine the moment of inertia of the body experimented on about some axis. If the body be of regular shape and be so far homogeneous that the errors of this assumption are of the order to be neglected, we can determine the moment of inertia by calculation. But sometimes this cannot be done. If we can make the body oscillate under gravity about any axis parallel to the given axis placed in a horizontal position, we can determine by equation (4) of Art. 92 the radius of gyration about a parallel axis through the centre of gravity. This requires however that the distances of the centre of gravity from the axes should be very accurately found. Sometimes it is more convenient to attach the body to a pendulum of known mass whose radius of gyration about a fixed horizontal axis has been previously found by observing the time of oscillation. Then by a new determination of the time of oscillation, the moment of inertia of the compound body, and therefore that of the given body, may be found, the masses being known.

If the body be a lamina, we may thus find the radii of gyration about three axes passing through the centre of gravity. By measuring three lengths along these axes inversely proportional to these radii of gyration, we have three points on a momental ellipse at the centre of gravity. The ellipse may then be easily constructed. The directions of its principal diameters are the principal axes, and the reciprocals of their lengths represent on the same scale as before the principal radii of gyration.

If the body be a solid, six observed radii of gyration will determine the principal axes and moments at the centre of gravity. But in most cases some of the circumstances of the particular problem under consideration will simplify the process.

On the length of the Seconds Pendulum.

98. The oscillations of a rigid body may be used to determine the numerical value of the accelerating force of gravity. Let τ be the half time of a small oscillation of a body made in vacuo about a horizontal axis, h the distance of the centre of gravity from the axis, k the radius of gyration about a parallel axis through the centre of gravity. Then we have by Art. 92,

$$k^2 + h^2 = \lambda h\tau^2 \quad\dots\dots\dots\dots\dots\dots\dots(1),$$

where $\lambda = \dfrac{g}{\pi^2}$, so that λ is the length of the simple pendulum whose complete time of oscillation is two seconds.

We might apply this formula to any regular body for which k and h could be found by calculation. Experiments have thus been made with a rectangular bar, drawn as a wire and suspended from one end. In this case $\dfrac{k^2 + h^2}{h}$, which is the length of the simple equivalent pendulum, is easily seen to be two-thirds of the length of the rod. The preceding formula then gives λ or g as soon as the time of oscillation has been observed. By inverting the rod and taking the mean of the results in the two positions any error arising from want of uniformity in density or figure may be partially obviated. It has, however, been found impracticable to obtain a rod sufficiently uniform to give results in accordance with each other.

99. If we make a body oscillate in succession about two parallel axes not at the same distance from the centre of gravity, we get two equations similar to (1), viz.

$$\left.\begin{array}{l} k^2 + h^2 = \lambda h\tau^2 \\ k^2 + h'^2 = \lambda h'\tau'^2 \end{array}\right\} \quad\dots\dots\dots\dots\dots\dots\dots(2).$$

Between these two we may now eliminate k^2, thus

$$\frac{h^2 - h'^2}{\lambda} = h\tau^2 - h'\tau'^2 \quad\dots\dots\dots\dots\dots\dots(3).$$

This equation gives λ. Since k^2 has disappeared, the form and structure of the body is now a matter of no importance. Let a body be constructed with two apertures into which knife edges can be fixed. The apertures may be triangular to prevent slipping. Resting on these knife edges, the body can be made to oscillate through small arcs. The perpendicular distances h, h', of the centre of gravity from the axes must then be measured with great care. The formula will then give λ.

100. In Capt. Kater's method (*Phil. Trans.*, 1818) the body has a sliding weight in the form of a ring which can be moved up and down by means of a screw. The body itself has the form of a bar and the apertures are so placed that the centre of gravity lies between them. The ring weight is then moved until the two times of oscillation are *exactly equal*. The equation (3) then becomes

$$\frac{h + h'}{\lambda} = \tau^2 \dots\dots\dots\dots\dots\dots\dots(4),$$

which determines λ. The advantage of this construction is that the position of the centre of gravity, which is not found without difficulty by experiment, is not required. All we want is $h + h'$, the exact distance between the knife edges. The disadvantage is that the ring weight has to be moved until two times of oscillation, each of which it is difficult to observe, are made equal.

101. The equation (3) can be written in the form

$$\frac{h + h'}{\lambda} = \frac{\tau^2 + \tau'^2}{2} + \tfrac{1}{2}\frac{h + h'}{h - h'}(\tau^2 - \tau'^2).$$

We now see that, if the body be so constructed that the times of oscillation about the two axes of suspension are very nearly equal, $\tau^2 - \tau'^2$ will be small, and therefore it will be sufficient in the last term to substitute for h and h' their *approximate* values. The position of the centre of gravity is of course to be found as accurately as possible, but any small error in its position is of no very great consequence, for such an error is multiplied by the small quantity $\tau^2 - \tau'^2$. The advantage of this construction over Kater's is that the ring weight may be dispensed with and yet the only element which must be measured with extreme accuracy is $h + h'$, the distance between the knife edges.

102. In order to measure the distance between the knife edges, Captain Kater first compared the different standards of length then in use, in terms of each of which he expressed the length of his pendulum. Since then a much more complete comparison of these and other standards has been made under the direction of the Committee appointed for that purpose in 1843. *Phil. Trans.*, 1857.

Having settled his unit of length, Captain Kater proceeded to measure the distance between the knife edges by means of microscopes. Two different methods were used, which however cannot be described here. As an illustration of the extreme care necessary in these measurements, the following fact may be mentioned.

R. D.   6

Though the images of the knife edges were always perfectly sharp and well defined, their distance when seen on a black ground was 000572 of an inch less than when seen on a white ground. This difference appeared to be the same, whatever the relative illumination of the object and ground might be, so long as the difference of character was preserved. Three sets of measurements were taken, two at the beginning of the experiments, and the third after some time. The object of the last set was to ascertain if the knife edges had suffered from use. The mean results of these three differed by less than a ten-thousandth of an inch from each other, the distance to be measured being 39·44085 inches.

103. The time of a single vibration cannot be observed directly, because this would require the fraction of a second of time as shown by the clock to be estimated either by the eye or ear. The difficulty may be overcome by observing the time, say of a thousand vibrations, and thus the error of the time of a single vibration is divided by a thousand. The labour of so much counting may however be avoided by the use of "the method of coincidences." The pendulum is placed in front of a clock pendulum whose time of vibration is slightly different. Certain marks made on the two pendulums are observed by a telescope at the lowest point of their arcs of vibration. The field of view is limited by a diaphragm to a narrow aperture across which the marks are seen to pass. At each succeeding vibration one pendulum follows the other more closely, and at last its mark is completely covered by the other during the passage across the field of view of the telescope. After a few vibrations it appears again preceding the other. In the interval from one disappearance to the next, one pendulum has made, as nearly as possible, one *complete* oscillation more than the other. We have therefore to count the number of vibrations made by either pendulum in the interval. At the beginning of the counting let one pendulum coincide with the other as nearly as we can judge. Suppose that after n half vibrations of the clock pendulum the next coincidence has not quite arrived, but that after $n + 1$ half vibrations the coincidence has passed. If the clock pendulum be the slower of the two, the other must have made $n + 2$ or $n + 3$ half vibrations in the interval. Thus the time of one half vibration of the pendulum lies between the fractions $\dfrac{n}{n + 2}$ and $\dfrac{n + 1}{n + 3}$ of the period of the clock vibration. Taking either of these estimates as the real time of a half vibration of the pendulum the error is less than the fraction $\dfrac{2}{(n + 2)(n + 3)}$ of the time of a half vibration of the clock pendulum. It appears from this that the error varies nearly inversely as the square of the number of vibrations between two

coincidences. In this manner 530 half vibrations of a clock pendulum, each equal to a second, were found to correspond to 532 of Captain Kater's pendulum. The error of this estimate is so small that in twenty-four hours it would accumulate only to about three-fifths of a second. The ratio of the times of vibration of the pendulum and the clock pendulum may thus be calculated with extreme accuracy. The rate of going of the clock must then be found by astronomical means.

The reader should notice the resemblance between this process of comparing two clocks with the use of the *vernier* in comparing lengths. Of course there are differences, because the vernier is applied to space, and we have here to do with time. But the general principle is the same.

104. The time of vibration thus obtained will require several corrections which are called "reductions." For instance, if the oscillation be not so small that we can put $\sin \theta = \theta$ in Art. 92, we must make a reduction to infinitely small arcs. The general method of effecting this will be considered in the chapter on Small Oscillations. Another reduction is necessary if we wish to reduce the result to what it would have been at the level of the sea. The attraction of the intervening land may be allowed for by Dr Young's rule (*Phil. Trans.* 1819). We may thus obtain the force of gravity at the level of the sea, supposing all the land above this level were cut off and the sea constrained to keep its present level. As the level of the sea is altered by the attraction of the land, further corrections are still necessary if we wish to reduce the result to the surface of that spheroid which most nearly represents the earth. See *Camb. Phil. Trans.* Vol. VIII. *On the variation of gravity at the surface of the earth*, by Sir G. Stokes.

Mr Baily gives as the length of the pendulum whose half time of vibration is a mean solar second in the open air in the latitude of London 39·133 inches, and as the length of a similar pendulum vibrating sidereal seconds 38·919 inches.

105. **Correction for Resistance of the Air.** The observations must be made in the air. To correct for this we have to make a reduction to a vacuum. This reduction consists of three parts : (1) The correction for buoyancy, (2) Du Buat's correction for the air dragged along by the pendulum, (3) The resistance of the air.

Let V be the volume of the pendulum which may be found by measuring the dimensions of the body. As the "reduction to a vacuum" is only a correction, any small unavoidable errors in calculating the dimensions will produce an effect only of the second order on the value of λ. Let ρ be the density of the air when the body is oscillating about one knife edge, ρ' the density when oscillating about the other. If the observation be made within an hour or two hours, we may put $\rho = \rho'$. The effect of buoyancy is allowed for by supposing a force $V\rho g$ to act upwards at the centre of gravity of the *volume* of the body. If the body be made as nearly as possible symmetrical about the two knife edges this centre of gravity will be half way between the knife edges, see Art. 95.

Du Buat discovered by experiment that a pendulum drags with it to and fro a certain mass of air which increases the inertia of the body without adding to the moving force of gravity. This result has been confirmed by theory. The mass dragged bears to the mass of air displaced by the body a ratio which depends on the external shape of the body. Let us represent it by $\mu V\rho$. If the body be symmetrical about the knife edges, so that the external shape is the same whichever edge is made the axis of suspension, μ will be the same for each oscillation. Supposing this mass to be collected at the centre of gravity of the volume, we must add to the k^2 of equation (1) in Art. 92, and therefore also in Art. 98, the term $\dfrac{\mu V\rho}{m}\left(\dfrac{h+h'}{2}\right)^2$.

Taking these two corrections the equation (1) of Art. 98 will now become

$$k^2 + h^2 + \frac{\mu V\rho}{m}\left(\frac{h+h'}{2}\right)^2 = \lambda\tau^2\left(h - \frac{V\rho}{m}\frac{h+h'}{2}\right),$$

where m is the mass of the pendulum. Similarly for the oscillation about the other knife edge,

$$k^2 + h'^2 + \frac{\mu V\rho'}{m}\left(\frac{h+h'}{2}\right)^2 = \lambda\tau'^2\left(h' - \frac{V\rho'}{m}\frac{h+h'}{2}\right).$$

We must eliminate k^2 as before. If the observations about the two knife edges succeed each other at a short interval we may put $\rho = \rho'$, and then Du Buat's correction will disappear. This is of course a very great advantage. We then have[*]

$$\frac{h+h'}{\lambda} = \frac{\tau^2+\tau'^2}{2} + \tfrac{1}{2}\frac{h+h'}{h-h'}(\tau^2 - \tau'^2)\left(1 - \frac{V\rho}{m}\right),$$

the last term being very small, because τ and τ' are nearly equal.

The resistance of the air will be some function of the angular velocity θ of the pendulum. Since the angular velocity is very small we may expand this function and take only the first power. Supposing that Maclaurin's theorem does not fail, and that no coefficient of a higher power than the first is very great, this gives a resistance proportional to $\dot{\theta}$. The equation of motion will therefore take the form

$$\ddot{\theta} + n^2\theta = -2f\dot{\theta},$$

where $2\pi/n$ is the time of a complete oscillation in a vacuum, and the term on the right-hand side is that due to the resistance of the air. The discussion of this equation will be found in the chapter on Small Oscillations.

106. **Construction of a pendulum.** In constructing a reversible pendulum to measure the force of gravity, the following are points of importance.

1. The axes of suspension, or knife edges, must not be at the same distance from the centre of gravity of the mass. They should be parallel to each other.

2. The times of oscillation about the two knife edges should be nearly equal.

3. The external form of the body must be symmetrical, and the same about the two axes of suspension.

---

[*] This formula was mentioned to the author as the one used in the late experiments made by Capt. Heaviside to determine the length of the seconds pendulum.

4. The pendulum must be of such a regular shape that the dimensions of all the parts can be readily calculated.

These conditions are satisfied if the pendulum be of rectangular shape with two cylinders placed one at each end. The external forms of these cylinders should be equal and similar, but one solid and the other hollow, and such that the distance between the knife edges is to be as nearly as possible equal to the length of the simple equivalent pendulum found by *calculation*.

5. The pendulum should be made, as far as possible, of one metal, so that as the temperature changes it may be always similar to itself. In this case since the times of oscillations of similar bodies vary as the square root of their linear dimensions, it is easy to reduce the observed time of oscillation to a standard temperature. The knife edges however must be made of some strong substance not likely to be easily injured.

107. Ex. 1. If the knife edges be not perfectly sharp, let r be the *difference* of their radii of curvature; show that

$$\frac{h^2 - h'^2 + (h + h') r}{\lambda} = h\tau^2 - h'\tau'^2$$

very nearly, when the pendulum vibrates in vacuo. It appears that the correction vanishes if the knife edges be only equally sharp. By interchanging the knife edges we have the same equation with the sign of r changed. By making a few observations we may thus determine r. A proposition similar to this has been ascribed to Laplace by Dr Young.

Let ρ, ρ' be the radii of curvature of the knife edges. Then by taking moments about the instantaneous axis we may show (as in Art. 98) that $k^2 + h^2 = \lambda (h + \rho) \tau^2$.

Since ρ is small we may write this in the form $k^2 + h^2 - (k^2 + h^2) \frac{\rho}{h} = \lambda h \tau^2$. The times of vibration τ, τ' are nearly equal, hence by Art. 92 we have $k^2 = hh'$ very nearly. Substituting this value of k in the small terms we get $k^2 + h^2 - (h + h') \rho = \lambda h \tau^2$. There is a similar equation for the pendulum when it vibrates about the other knife edge, which may be obtained from this by interchanging h, h' and τ, τ'. Eliminating k^2 as in Art. 99, and remembering that $r = \rho' - \rho$, we obtain the result to be proved.

Ex. 2. A heavy spherical ball is suspended by a very fine wire successively from two points of support A and B, whose vertical distance b has been carefully measured, thus forming two pendulums. The lowest point of the ball is, on each suspension, made to be as exactly as possible on the same level, which level is approximately at depths a and a' below A and B respectively. If r be the radius of the ball, which is small compared with a or a', and l, l' the lengths of the simple equivalent pendulums, prove that $\dfrac{l - l'}{b} = 1 - \dfrac{2}{5} \dfrac{r^2}{(a - r)(a' - r)}$ very nearly. By counting the number of oscillations performed in a given time by each pendulum, show how to find the ratio of l to l'. Thence show how to find g and point out which lengths must be most carefully measured and which need only be approximately found, so as to render this method effective. This method is mentioned in Grant's *History of Physical Astronomy*, page 155, as having been used by Bessel.

108. **A Standard of Length.** The length of the seconds pendulum has been used as a national standard of length. By an Act of Parliament passed in 1824, it was declared that the distance between the centres of two points in the gold studs in the straight brass rod then in the custody of the clerk of the House of Commons, whereon the words and figures "standard yard, 1760" were engraved, should be the original and genuine standard of length called a yard, the brass being at the temperature of 62° Fahr. And as it was expedient that the said standard yard if injured should be restored to the same length by reference to some invariable natural standard, it was enacted that the new standard yard should be of such length that the pendulum vibrating seconds of mean time in the latitude of London in a vacuum at the level of the sea should be 39·1393 inches.

On Oct. 16, 1834, occurred the fire at the Houses of Parliament, in which the standards were destroyed. The bar of 1760 was recovered, but one of its gold pins bearing a point was melted out and the bar was otherwise injured.

In 1838 a commission was appointed to report to the Government on the course best to be pursued under the peculiar circumstances of the case.

In 1841 the commission reported that they were of opinion that the definition by which the standard yard is declared to be a certain brass rod was the best which it was possible to adopt. With respect to the provision for restoration they did not recommend a reference to the length of the seconds pendulum. "Since the passing of the act of 1824 it has been ascertained that several elements of reduction of the pendulum experiments therein referred to are doubtful or erroneous: thus it was shown by Dr Young, *Phil. Trans.* 1819, that the reduction to the level of the sea was doubtful; by Bessel, *Astron. Nachr.* No. 128, and by Sabine, *Phil. Trans.* 1829, that the reduction for the weight of air was erroneous; by Baily, *Phil. Trans.* 1832, that the specific gravity of the pendulum was erroneously estimated and that the faults of the agate planes introduced some elements of doubt; by Kater, *Phil. Trans.* 1830, and by Baily, *Astron. Soc. Memoirs,* Vol. IX., that very sensible errors were introduced in the operation of comparing the length of the pendulum with Shuckburgh's scale used as a representative of the legal standard. It is evident, therefore, that the course prescribed by the act would not necessarily reproduce the length of the original yard."

The commission stated that there were several measures which had been formerly accurately compared with the original standard yard, and that by the use of these the length of the original yard could be determined without sensible error.

In 1843 another commission was appointed to compare all the existing measures and to construct from them a new Parliamentary standard. Unexpected difficulties occurred in the course of the comparison, which cannot be described here. A full account of the proceedings of the commission will be found in a paper contributed by Sir G. Airy to the Royal Society in 1857.

In France the standard of length is the mètre. This, like our standard yard, was originally defined by reference to a length given in nature. The ten millionth part of the length of a meridian of the earth measured from the pole to the equator was declared to be the legal mètre. But when new and more accurate measurements were subsequently made, it became evident that the length of the legal mètre could not be altered for each improvement in the measure of the earth. Practically therefore the definition of the mètre is a certain length preserved in Paris.

The use of the seconds pendulum as a standard of length assumes that a standard of time has already been obtained. In this case we must have recourse to some natural standard, and the one usually chosen is the time of rotation of the earth on its axis. This is recommended by its simplicity, for the interval between two successive transits of the same star across the meridian is very nearly equal to the time of rotation of the earth. But other natural standards may also be used to check the clock.

For an account of the recommendations made in the two reports (1873 and 1874) by the Units Committee of the British Association, see Prof. Everett's treatise on *Units and Physical constants*.

Oscillation of a Watch Balance.

109. A rod $B'CB$ can turn freely about its centre of gravity C which is fixed, and is acted on by a very fine spiral spring CPB. The spring has one end C fixed in position in such a manner that the tangent at C is also fixed, and has the other end B attached to the rod so that the tangent at B makes a constant angle with the rod. The rod being turned through any angle, it is required to find the time of oscillation. This is the construction used in watches, just as the pendulum is used in clocks, to regulate the motion. In many watches the rod is replaced by a wheel whose centre is C.

Let Cx be the position of the rod when in equilibrium, and let θ be the angle the rod makes with Cx at any time t, Mk^2 the moment of inertia of the rod about C. Let ρ be the radius of curvature at any point P of the spring, ρ_0 the value of ρ when in equilibrium. Let (x, y) be the co-ordinates of P referred to C as origin and Cx as axis of x. Let us consider the forces which act on the rod and the portion BP of the spring. The forces on the rod are X, Y the resolved parts of the action at C parallel to the axes of co-ordinates, and the reversed effective forces which are equivalent to a couple $Mk^2\dfrac{d^2\theta}{dt^2}$. The forces on the spring are, the reversed effective forces which are so small that they may be neglected, and the resultant action across the section of the spring at P. This resultant action is produced by the tensions of the innumerable fibres which make up the spring, and these are equivalent to a force at P and a couple. When an elastic spring

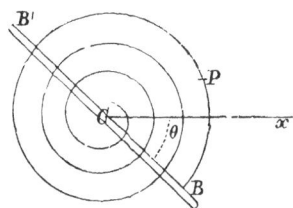

is bent so that its curvature is changed, it is proved both by experiment and theory that this couple is proportional to the change of curvature at P. We may therefore represent it by $E\left(\dfrac{1}{\rho}-\dfrac{1}{\rho_0}\right)$, where E depends only on the material of which the spring is made and on the form of its section.

To avoid introducing the unknown force at P, we take moments about P. This gives

$$Mk^2\frac{d^2\theta}{dt^2}=-E\left(\frac{1}{\rho}-\frac{1}{\rho_0}\right)-Xy+Yx.$$

This equation is true whatever point P may be chosen. Considering the left side constant at any moment and (x, y) variable, this becomes the intrinsic equation to the form of the spring.

Let $BP = s$, multiply this equation by ds and integrate along the whole length l of the spiral spring, we have

$$Mk^2\frac{d^2\theta}{dt^2}l=-E\int\left(\frac{ds}{\rho}-\frac{ds}{\rho_0}\right)+Y\int xds-X\int yds.$$

Now $\dfrac{ds}{\rho}$ is the angle between two consecutive normals, hence $\int\dfrac{ds}{\rho}$ is the angle between the extreme normals. Now at C the normal to the spring is fixed throughout the motion, therefore $\int\left(\dfrac{ds}{\rho}-\dfrac{ds}{\rho_0}\right)$ is the angle between the normals at B in the two positions in which $\theta = \theta$ and $\theta = 0$. But since the normal at B makes a constant angle with the rod, this angle is the angle θ which the rod makes with its position of equilibrium. Also if $\bar{x}$, $\bar{y}$ be the co-ordinates of the centre of gravity of the spring at the time t, we have $\int x ds = \bar{x}l$, $\int y ds = \bar{y}l$. Hence the equation of motion becomes

$$Mk^2 \frac{d^2\theta}{dt^2} = -\frac{E}{l}\theta + Y\bar{x} - X\bar{y}.$$

Let us suppose that in the position of equilibrium there is no pressure on the axis C, then X and Y will, throughout the motion, be small quantities of the order θ. Let us also suppose that the fulcrum C is placed over the centre of gravity of the spring when at rest. Then if the number of spiral turns of the spring be numerous and if each turn be nearly circular, the centre of gravity will never deviate far from C. Thus the terms $Y\bar{x}$ and $X\bar{y}$ are each the product of two small quantities, and are therefore at least of the second order. Neglecting these terms we have

$$Mk^2 \frac{d^2\theta}{dt^2} = -\frac{E}{l}\theta.$$

Hence the time of oscillation is $2\pi\sqrt{\dfrac{Mk^2 l}{E}}$.

It appears that to a first approximation the time of oscillation is independent of the form of the spring in equilibrium, and depends only on its length and on the form of its section.

This brief discussion of the motion of a watch balance is taken from a memoir presented to the Academy of Sciences. The reader is referred to an article in Liouville's *Journal*, 1860, for a further investigation of the conditions necessary for isochronism and for a determination of the best forms for the spring.

When the temperature increases the length l of the balance is increased. For this and other reasons the watch will lose time. The compensation for a change of temperature is now usually effected by altering the moment of inertia of the oscillating body. The circumference of the balance wheel instead of being a complete circle consists of two arcs each less than a semi-circumference. An extremity of each is attached to one extremity of the rod BCB', and each carries a small mass which is attached to it near its free extremity. Each arc is constructed of two thin slips of different metals lying side by side, the outer of which expands more with heat than the inner. As the heat increases, the arcs bend inwards and the moment of inertia of the whole balance is decreased. The proper positions of the masses on the circular arcs are determined by trial and this is usually a troublesome process.

Pressures on the fixed axis.

110.   *A body moves about a fixed axis under the action of any forces, to find the pressures on the axis.*

Firstly. Suppose the body and the forces to be symmetrical about the plane through the centre of gravity perpendicular to the axis. Then it is evident that the pressures on the axis are reducible to a single force at C the centre of suspension.

Let F, G be the actions of the point of support on the body resolved along and perpendicular to CO, where O is the centre of gravity. Let X, Y be the sum of the resolved parts of the impressed forces in the same directions, and L their moment round C. Let $CO = h$ and $\theta =$ angle which CO makes with any straight line fixed in space.

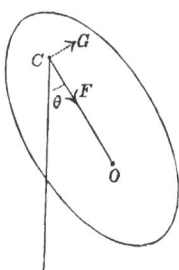

Taking moments about C, we have

$$\frac{d^2\theta}{dt^2} = \frac{L}{M(k^2 + h^2)} \quad\dots\dots\dots\dots\dots\dots\dots(1).$$

The motion of the centre of gravity is the same as if all the forces acted at that point. Since it describes a circle round C, we have, by taking the tangential and normal resolutions,

$$h\frac{d^2\theta}{dt^2} = \frac{Y + G}{M} \quad\dots\dots\dots\dots\dots\dots(2),$$

$$-h\left(\frac{d\theta}{dt}\right)^2 = \frac{X + F}{M} \quad\dots\dots\dots\dots\dots\dots(3).$$

Equation (1) gives the values of $\frac{d^2\theta}{dt^2}$ and $\frac{d\theta}{dt}$, and then the pressures may be found by equations (2) and (3).

If the only force acting on the body be that of gravity, let θ be measured from the vertical. We have

$$X = Mg\cos\theta, \quad Y = -Mg\sin\theta, \quad L = -Mgh\sin\theta;$$

$$\therefore \frac{d^2\theta}{dt^2} = -\frac{gh}{k^2 + h^2}\sin\theta\dots\dots\dots\dots\dots\dots(4).$$

Integrating, we have

$$\left(\frac{d\theta}{dt}\right)^2 = C + \frac{2gh}{k^2 + h^2} \cos\theta \ldots\ldots\ldots\ldots\ldots(5).$$

If the angular velocity of the body be Ω when CO is horizontal, we have $\omega = \Omega$ when $\cos\theta = 0$. We find $C = \Omega^2$. Substituting these values in (2) and (3) we get

$$\left.\begin{array}{l} -\dfrac{F}{M} = \Omega^2 h + g\cos\theta\,\dfrac{k^2 + 3h^2}{k^2 + h^2} \\[2mm] \dfrac{G}{M} = \qquad g\sin\theta\,\dfrac{k^2}{k^2 + h^2} \end{array}\right\} ,$$

where θ is the angle the perpendicular drawn from the centre of gravity of the body on the axis makes with the vertical measured downwards.

It appears from these results that the component of pressure perpendicular to the plane containing the axis and the centre of gravity is independent of the initial conditions. As the body oscillates this component varies as the distance of the centre of gravity from the vertical plane through the axis. On the other hand the component of pressure in the plane containing the axis and the centre of gravity does depend on the initial angular velocity of the body.

If the forces are impulsive, the equations (1), (2), (3) are only slightly altered. Let ω, ω' be the angular velocities of the body just before and just after the action of the impulses. The equations then become

$$\omega' - \omega = \frac{L}{M(k^2 + h^2)}, \quad h\,(\omega' - \omega) = \frac{Y + G}{M}, \quad 0 = X + F,$$

where all the letters have the same meaning as before, except that F, G, X, Y are now impulses instead of finite forces.

Ex. 1. A circular area of weight W can turn freely about a horizontal axis perpendicular to its plane which passes through a point C on its circumference. If it start from rest with the diameter through C vertically above C, show that the resultant pressures on the axis when that diameter is horizontal and vertically below C are respectively $\frac{1}{3}\sqrt{17}W$ and $\frac{11}{3}W$.

Ex. 2. A thin uniform rod, one end of which is attached to a smooth hinge, is allowed to fall from a horizontal position; prove that when the horizontal strain is the greatest possible, the vertical strain on the hinge is to the weight of the rod as 11 : 8.   *Math. Tripos.*

Ex. 3. Let $\dfrac{R}{M}$ be the resultant of $\dfrac{F}{M} + \Omega^2 h$ and $\dfrac{G}{M}$, and let $a = g\,\dfrac{k^2 + 3h^2}{k^2 + h^2}$, $b = g\,\dfrac{k^2}{k^2 + h^2}$. Construct an ellipse with C for centre and axes equal to $2a$ and $2b$ measured along and perpendicular to CO. Let this ellipse be fixed in the body and oscillate with it. Prove that the pressure R varies as the diameter along which it acts. And the direction may be found thus; let the auxiliary circle cut the vertical through O in V, and let the perpendicular from V on CO cut the ellipse in R. Then CR is the direction of the pressure R.

111. In many problems we require the vertical and horizontal components of the pressure, and more particularly the positions of the body in which either of these components changes sign. If, for instance, the body is a wedge supported by its edge on a perfectly rough horizontal table it may be regarded as turning round a horizontal axis. But the table can only exert an upward vertical pressure on the body, if then the vertical pressure changes sign as the body moves, the wedge will leave the table and the whole motion will be different.

Let Q be the vertical pressure on the body, measured upwards when positive, and P the horizontal pressure, measured to the left when the centre of gravity is on the right-hand side of the vertical plane through the axis. For the sake of brevity let $a = g \dfrac{k^2 + 3h^2}{k^2 + h^2}$, and $b = g \dfrac{k^2}{k^2 + h^2}$. Then we find

$$\left. \begin{array}{l} \dfrac{Q}{M} = a \cos^2 \theta + b \sin^2 \theta + \Omega^2 h \cos \theta \\[2mm] \dfrac{P}{M} = (a - b) \sin \theta \cos \theta + \Omega^2 h \sin \theta \end{array} \right\}.$$

As the expression for the vertical pressure is not altered by changing the sign of θ, we need only consider its changes of sign as the centre of gravity moves, say, upwards. Similar changes will occur during the descent of the centre of gravity.

We see also that the vertical pressure is always upwards when the centre of gravity of the body is below the horizontal plane through the axis. Consider then the two extreme positions in which the centre of gravity lies, (1) in the horizontal plane through the axis, and (2) vertically over the axis.

In these two positions Q has opposite signs if $\Omega^2 h > a$ and the intervening vanishing points are given by a quadratic equation. Hence we infer that as the centre of gravity moves from one position to the other, the vertical pressure will vanish *once* and change sign if $\Omega^2 h > a$. If this inequality does not hold the vertical pressure will be directed upwards in both the extreme positions. To determine if it can vanish for any intervening position of the body we must ascertain whether the minimum value of Q is positive or negative. By differentiation we find that Q is least when $\cos \theta = \dfrac{-\Omega^2 h}{2(a - b)}$, and thence by substitution we find the least value of Q to be $M \{b - (a - b) \cos^2 \theta\}$.

If $\Omega^2 h$ be less than $2(a - b)$ and greater than $2\sqrt{b(a - b)}$, this value of $\cos \theta$ will be possible and the minimum value of Q will be negative. The result is that, if both these conditions are satisfied as well as $\Omega^2 h < a$, the vertical pressure will vanish and change sign *twice* as the body moves from one extreme position to the other. But if either condition fail, the vertical pressure will not vanish between the limits. To find the exact positions at which the pressure vanishes, we have to solve the quadratic equation formed by equating Q to zero. We must also remember that the conditions of the question may exclude one or both of these positions. Thus we may show from equation (5) that unless $\Omega^2 h$ exceed $\tfrac{2}{3}(a - b)$, the body cannot go all round. Or again the body may be projected upwards at such an inclination to the vertical that both the roots of the quadratic may be excluded from the arc of oscillation. In such a case the vertical pressure will of course keep one sign during the ascent.

The horizontal pressure vanishes when $\theta = 0$ or π, and when $\cos \theta = \dfrac{-\Omega^2 h}{a - b}$. If this value of $\cos \theta$ be greater than unity, the horizontal pressure vanishes and changes its direction only when the centre of gravity is vertically over or under the axis. If it be numerically less than unity, the pressure vanishes and changes sign in two more positions, which both occur when the centre of gravity is above the axis and at equal heights.

112. *Secondly.* Suppose either the body or the forces not to be symmetrical.

Let the fixed axis be taken as the axis of z with any origin and plane of xz. These we shall afterwards so choose as to simplify our process as much as possible. Let $\bar{x}$, $\bar{y}$, $\bar{z}$ be the co-ordinates of the centre of gravity at the time t. Let ω be the angular velocity of the body, f the angular acceleration, so that $f = \dot{\omega}$.

Now every element m of the body describes a circle about the axis, hence its accelerations along and perpendicular to the radius

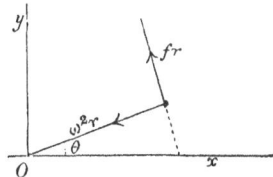

vector r from the axis are $-\omega^2 r$ and fr. Let θ be the angle which r makes with the plane of xz at any time, then from the resolution of forces it is clear that

$$\ddot{x} = -\omega^2 r \cos \theta - fr \sin \theta = -\omega^2 x - fy,$$

and similarly $\ddot{y} = -\omega^2 y + fx$.

These equations may also be obtained by differentiating the equations $x = r \cos \theta$, $y = r \sin \theta$ twice, remembering that r is constant.

Conceive the body to be fixed to the axis at two points, distant a and a' from the origin, and let the reactions of the points on the body resolved parallel to the axes be respectively F, G, H; F', G', H'. The equations of motion of Art. 72 then give

$$\Sigma mX + F + F' = \Sigma m\ddot{x} = \Sigma m \left(-\omega^2 x - fy \right)$$
$$= -\omega^2 M\bar{x} - fM\bar{y} \quad \ldots \ldots \ldots \ldots \ldots \ldots (1),$$

$$\Sigma mY + G + G' = \Sigma m\ddot{y} = \Sigma m \left(-\omega^2 y + fx \right)$$
$$= -\omega^2 M\bar{y} + fM\bar{x} \quad \ldots \ldots \ldots \ldots \ldots \ldots (2),$$

$$\Sigma mZ + H + H' = \Sigma m\ddot{z} = 0 \ldots \ldots \ldots \ldots \ldots \ldots \ldots (3).$$

Taking moments about the axes, we have (Art. 72)

$$\Sigma m\,(yZ - zY) - Ga - G'a' = \Sigma m\,(y\ddot{z} - z\ddot{y}) = - \Sigma m\,(z\ddot{y})$$
$$= \omega^2 \Sigma myz - f\Sigma mxz \ldots\ldots\ldots\ldots\ldots(4):$$

by merely introducing z into the result in (2),

$$\Sigma m\,(zX - xZ) + Fa + F'a' = \Sigma m\,(z\ddot{x} - x\ddot{z})$$
$$= - \omega^2 \Sigma mxz - f\Sigma myz \ldots\ldots\ldots\ldots(5),$$
$$\Sigma m\,(xY - yX) \qquad\qquad = \Sigma m\,(x\ddot{y} - y\ddot{x}) = \Sigma mr^2 \,.\, \dot{\omega}$$
$$= Mk'^2 \,.\, f \ldots\ldots\ldots\ldots\ldots\ldots\ldots\ldots(6).$$

Equation (6) serves to determine f and ω, and equations (1), (2), (4), (5) then determine F, G, F', G'; H and H' are indeterminate, but their sum is given by equation (3).

Looking at these equations, we see that they would be greatly simplified in two cases.

Firstly, if the axis of z be a principal axis at the origin,

$$\Sigma mxz = 0, \quad \Sigma myz = 0,$$

and the calculation of the right-hand sides of equations (4) and (5) would only be so much superfluous labour. Hence, in attempting a problem of this kind, *we should, when possible, so choose the origin that the axis of revolution is a principal axis of the body at that point.*

Secondly, except the determination of f and ω by integrating equation (6), the whole process is merely an algebraic substitution of f and ω in the remaining equations. *Hence our results will still be correct if we choose the plane of xz to contain the centre of gravity at the moment under consideration;* this will make $\bar{y} = 0$, and thus equations (1) and (2) will be simplified.

113. **Impulsive forces.** If the forces which act on the body be impulsive, the equations will require some alterations.

Let u, v, w, u', v', w' be the velocities resolved parallel to the axes of any element m whose co-ordinates are x, y, z. Then $u = - y\omega$, $u' = - y\omega'$, $v = x\omega$, $v' = x\omega'$, and w, w' are both zero. The several equations of Art. 112 will then be replaced by the following:

$$\Sigma X + F + F' = \Sigma m\,(u' - u) = - \Sigma my\,(\omega' - \omega)$$
$$= - M\bar{y}\,(\omega' - \omega)\ldots\ldots\ldots\ldots(1),$$
$$\Sigma Y + G + G' = \Sigma m\,(v' - v) = \Sigma mx\,(\omega' - \omega)$$
$$= M\bar{x}\,(\omega' - \omega)\ldots\ldots\ldots\ldots(2),$$
$$\Sigma Z + H + H' = 0 \ldots\ldots\ldots\ldots\ldots\ldots\ldots(3),$$

$$\Sigma \left(yZ - zY \right) - Ga - G'a' = \Sigma m \left\{ y \left(w' - w \right) - z \left(v' - v \right) \right\}$$
$$= - \Sigma mxz \cdot \left(\omega' - \omega \right) \dots \dots \dots (4),$$
$$\Sigma \left(zX - xZ \right) + Fa + F'a' = \Sigma m \left\{ z \left(u' - u \right) - x \left(w' - w \right) \right\}$$
$$= - \Sigma myz \cdot \left(\omega' - \omega \right) \dots \dots \dots (5),$$
$$\Sigma \left(xY - yX \right) \qquad\qquad = \Sigma m \left(x^2 + y^2 \right) \cdot \left(\omega' - \omega \right) \dots \dots (6).$$

These six equations are sufficient to determine ω', F, F', G, G' and the sum $H + H'$ of the two pressures along the axis.

These equations admit of simplification when the origin can be so chosen that the axis of rotation is a principal axis at that point. In this case the right-hand sides of equations (4) and (5) vanish. Also, if the plane of xz be chosen to pass through the centre of gravity of the body, we have $\bar{y} = 0$, and the right-hand side of equation (1) vanishes.

114. **Analysis of results.** The equations in Art. 112 to find F, G, F', G' are linear; the resolved pressures therefore due to several causes may be found by adding together the resolved pressures due to each separately. It follows that the pressures at the axis are equivalent to two sets of pressures, (1) the statical pressure due to the impressed forces X, Y, Z, &c., and (2) the pressure due to the effective forces $m\ddot{x}$, $m\ddot{y}$, &c.

The resultant statical pressure can be found from the first five equations, their right-hand sides being replaced by zero. These equations are not altered by transferring the impressed forces parallel to themselves to act at points on the axis, provided that we introduce the usual couples. We may then neglect the couple whose axis is Oz, which occurs only in equation (6), and the statical pressure at the axis may be found by compounding the remaining transferred forces. Thus, if the only impressed force on the body is that of gravity, and the axis of suspension is horizontal, the statical pressure on the axis is a vertical force equal to the weight of the body, acting at the foot of the perpendicular drawn from the centre of gravity to the axis of suspension. In the same way if an impulse act on the body perpendicular to the axis, the statical pressure due to it may be found by simply transferring it parallel to itself in a plane perpendicular to the axis to act at a point on that axis.

When the axis of revolution Oz is a principal axis at some point O the pressures due to the effective forces take a simple form. Let F_2, F_2', G_2, G_2' be the parts of the pressures due to these forces alone. Let the plane of xz be chosen to contain the centre of gravity of the body, then $\bar{y} = 0$. Let also the origin be chosen as one of the two points of fixture, then $a = 0$. The equations (1), (2), (4), (5) of Art. 112 then become

$$F_2 + F_2' = - \omega^2 M \bar{x}, \quad G_2 + G_2' = f M \bar{x}, \quad - G_2' a' = 0, \quad F_2' a' = 0.$$

These give $F_2' = 0$, $G_2' = 0$, $F_2 = - \omega^2 M \bar{x}$, $G_2 = f M \bar{x}$.

It follows that the pressures of the axis on the body due to the effective forces are equivalent to a single force, which acts at the principal point O, and is equal to the resultant of the effective forces of the mass of the body collected at its centre of gravity. In the same way the pressure due to the effective forces when the body is acted on by an impulse may be represented by a blow which acts at the principal point O, and whose components parallel to x, y, z are the expressions on the right-hand sides of equations (1), (2), (3) of Art. 113.

Ex. 1. A heavy body can turn freely about a horizontal axis Oz which is a principal axis at O. It starts from rest with the plane GOz through the centre of gravity G horizontal. Show that the pressure due to the effective forces alone makes an angle with the plane GOz whose tangent is half the tangent of the angle which the plane GOz makes with the vertical.

Ex. 2. A quadrant of a circle of radius a can turn freely about a bounding radius as a fixed axis. Show that the pressures on the axis are equivalent to two pressures, one equal to the weight of the lamina acting at a point of the fixed radius distant $4a/3\pi$ from the centre, and the other at a point which divides that radius in the ratio $3 : 5$.

Ex. 3. A lamina can turn freely about an axis Oz in its plane as a fixed axis. It is struck by a blow P at any point A of its area in a direction perpendicular to the lamina. Show that the statical pressure on the axis is equal to a blow P acting at B where AB is a perpendicular to Oz. Show also that the pressure due to the effective forces is equal to a blow $P\bar{x}\xi/k^2$ acting at O in a direction opposite to the blow at B. Here the origin O is the principal point of the axis, $\bar{x}$ and ξ are the distances of the centre of gravity and of A from Oz, Mk^2 is the moment of inertia about Oz. What is the condition that the pressure on the axis should be equivalent to a couple?

Ex. 4. *A door is suspended by two hinges from a fixed axis making an angle a with the vertical. Find the motion and pressures on the hinges.*

Since the fixed axis is evidently a principal axis at the middle point, we shall take this point for origin. Also we shall take the plane of xz so that it contains the centre of gravity of the door at the moment under consideration.

The only force acting on the door is gravity, which may be supposed to act at the centre of gravity. We must first resolve this parallel to the axes. Let ϕ be the angle the plane of the door makes with a vertical plane through the axis of suspension. If we draw a plane zON such that its trace ON on the plane of xOy makes an angle ϕ with the axis of x, this will be the vertical plane through the axis; and if we draw OV in this plane making $zOV = a$, OV will be vertical. Hence the resolved parts of gravity are

$$X = g \sin a \cos \phi, \quad Y = g \sin a \sin \phi, \quad Z = g \cos a.$$

Since the resolved parts of the effective forces are the same as if the whole mass were collected at the centre of gravity, the six equations of motion are

$$Mg \sin \alpha \cos \phi + F + F' = -\omega^2 M\bar{x} \dots\dots\dots\dots\dots(1),$$

$$Mg \sin \alpha \sin \phi + G + G' = f M\bar{x} \dots\dots\dots\dots\dots(2),$$

$$- Mg \cos \alpha + H + H' = 0 \dots\dots\dots\dots\dots\dots(3),$$

$$- Ga + G'a = 0 \dots\dots\dots\dots\dots\dots(4),$$

$$Mg \cos \alpha \bar{x} + Fa - F'a = 0 \dots\dots\dots\dots\dots(5),$$

$$- Mg \sin \alpha \sin \phi . \bar{x} = Mk'^2 . \ddot{\phi} \dots\dots\dots\dots(6).$$

Integrating the last equation, we have

$$C + 2g \sin \alpha \cos \phi \bar{x} = k'^2 \omega^2.$$

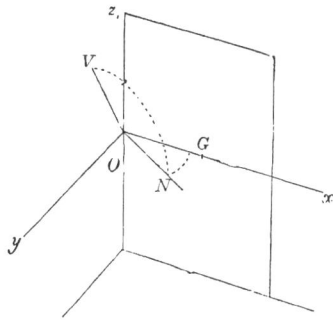

Suppose the door to be initially placed at rest, its plane making an angle β with the vertical plane through the axis; then when $\phi = \beta$, $\omega = 0$; hence

$$k'^2 \omega^2 = 2g\bar{x} \sin \alpha (\cos \phi - \cos \beta)$$

and

$$k'^2 f = - g \sin \alpha \sin \phi . \bar{x}$$

By substitution in the first four equations F, F', G, G' may be found.

115. Dynamical and geometrical similarity.

It should be noticed that the equations of Arts. 112 and 113 do not depend on the form of the body, but only on its moments and products of inertia. We may therefore replace the body by any equimomental body that may be convenient for our purpose.

This consideration will often enable us to reduce the complicated forms of Art. 112 to the simpler ones given in Art. 110. For though the body may not be symmetrical about a plane through its centre of gravity perpendicular to the axis of suspension, yet if the momental ellipsoid at the centre of gravity be symmetrical about this plane we may treat the body as if it were really symmetrical. Such a body may be said to be *dynamically symmetrical*. If at the same time the forces be symmetrical about the same plane, and this will always be the case if the axis of suspension be horizontal and gravity be the only force acting, we know that the pressures on the axis must certainly reduce to a single pressure, which may be found by Art. 110.

116. Ex. 1. A uniform heavy lamina in the form of a sector of a circle is suspended by a horizontal axis parallel to the radius which bisects the arc, and oscillates under the action of gravity. Show that the pressures on the axis are equivalent to a single force, and find its magnitude.

Ex. 2. An equilateral triangle oscillates about any horizontal axis situated in its own plane, show that the pressures are equivalent to a single force and find its magnitude.

117. Permanent axes of Rotation. Let us suppose that any point O of a body under no forces is fixed in space and that it is set in rotation about some axis which we may call Oz. We may enquire what are the necessary conditions that the body should continue to rotate about that axis as if it were fixed in space. When these conditions are satisfied the axis is called a *permanent axis of rotation at the point O*.

To determine these conditions let us suppose some other point A of the axis to be also fixed in space. Then by using the method of Arts. 112 or 113 we may determine the pressures at A which are necessary to fix the axis. If these are zero the attachment at A is unnecessary and may be removed. The body will then continue to rotate about Oz as if it were fixed in space.

Since there are no impressed forces acting on the body, the whole pressure on the axis is that due to the effective forces. If the axis Oz is a principal axis at any point of its length the pressure due to the effective forces will act at that point (Art. 114). Hence the pressure at A cannot be zero unless that point coincide with O. The *conditions are therefore satisfied if the axis of rotation Oz is a principal axis at the fixed point O.*

If the axis Oz is not a principal axis at any point we shall prove that it cannot be a permanent axis of rotation. To prove this we must practically return to the equations (4), (5) and (6) of Art. 112. Let F, G, H, F', G', H' be the pressures at O and A. Then $a = 0$, $a' = OA$. Taking moments about Oz we have $Mk'^2f = 0$; thus the angular velocity of the body about the axis Oz is constant. It easily follows that $\ddot{x} = -\omega^2 x$, $\ddot{y} = -\omega^2 y$, $\ddot{z} = 0$. Taking moments about the axes of x and y we have (Art. 72)

$$- G'a' = \Sigma m\,(y\ddot{z} - z\ddot{y}) = \omega^2\Sigma myz,$$
$$F'a' = \Sigma m\,(z\ddot{x} - x\ddot{z}) = -\omega^2\Sigma mxz.$$

Thus F' and G' cannot be zero unless $\Sigma mxz = 0$ and $\Sigma myz = 0$, i.e. Oz cannot be a permanent axis of rotation unless it is a principal axis at the fixed point O.

The existence of principal axes was first established by Segner in the work *Specimen Theoriæ Turbinum*. His course of investigation is the opposite to that pursued in this treatise. He defines a principal axis to be such that when a body revolves round it the forces arising from the rotation have no tendency to alter the position of the axis. From this dynamical definition he deduces the geometrical properties of these axes. The reader may consult Prof. Cayley's report to the British Association on the special problems of dynamics, 1862, and Bossut, *Histoire des Mathématiques*, Tome II.

118. *A body at rest with one point O fixed in space is acted on by an impulsive couple, it is required to find the initial axis of rotation.*

Let Oz be the initial axis. As before we shall regard this axis as fixed at some other point A, at which the pressures are to be equated to zero. Let L, M, N be the resolved parts of the couple about the axes. The plane of the couple is therefore

$$L\xi + M\eta + N\zeta = 0 \ldots\ldots\ldots\ldots\ldots\ldots\ldots(1).$$

Let u', v', w' be the initial velocities of an element of the body whose co-ordinates are x, y, z, and let ω' be the initial angular velocity of the body. Then $u' = -y\omega'$, $v' = x\omega'$ exactly as in Art. 113. Taking moments about the axes of x, y, z we have

$$L - G'a' = \Sigma m\,(yw' - zv') = -\Sigma mxz\,.\,\omega' \\ M + F'a' = \Sigma m\,(zu' - xw') = -\Sigma myz\,.\,\omega' \\ N = Mk'^2\omega' \Bigg\}.$$

Here F', G' are, as before, resolved parts of the pressure at A, and $OA = a'$. Putting $F' = 0$, $G' = 0$, these equations give the couples which must act on the body to produce rotation about Oz. Substituting the values of L, M, N in (1), the equation to the plane of the couple is

$$-\Sigma mxz\,\xi - \Sigma myz\,\eta + Mk'^2\zeta = 0 \ldots\ldots\ldots\ldots(2).$$

Let the momental ellipsoid at the fixed point O be constructed and let its equation be

$$A\xi^2 + B\eta^2 + C\zeta^2 - 2D\eta\zeta - 2E\zeta\xi - 2F\xi\eta = M\epsilon^4.$$

The diametral plane of the axis of ζ is

$$-E\xi - D\eta + C\zeta = 0 \ldots\ldots\ldots\ldots\ldots\ldots(3).$$

Comparing (2) and (3) we see that the plane of the resultant couple must be the diametral plane of the axis of revolution.

If then a body at rest with one point fixed be acted on by any couple it will begin to rotate about the diametral line of the plane of the couple with regard to the momental ellipsoid at the fixed point.

Thus a body will begin to rotate about a perpendicular to the plane of the couple only when the plane of the couple is parallel to a principal plane of the body at the fixed point.

119. Ex. 1. If a body at rest have one point O fixed and be acted on by any couple whose axis is a radius vector OP of the ellipsoid of gyration at O, the body will begin to turn about a perpendicular from O on the tangent plane at P.

Ex. 2. A solid homogeneous ellipsoid is fixed at its centre, and is acted on by a couple in a plane whose direction-cosines referred to the principal diameters are (l, m, n). Prove that the direction-cosines of the initial axis of rotation are proportional to $\dfrac{l}{b^2 + c^2}$, $\dfrac{m}{c^2 + a^2}$ and $\dfrac{n}{a^2 + b^2}$.

Ex. 3. Any plane section being taken of the momental ellipsoid of a body at a fixed point, the body may be made to rotate about either of the principal diameters of this section by the application of a couple of the proper magnitude whose axis is the other principal diameter.

For assume the body to be turning uniformly about the axis of z. Then the couples which must act on the body to produce this motion are $L = \omega^2 \Sigma myz$, $M = -\omega^2 \Sigma mxz$, $N = 0$. Then by taking the axis of x such that $\Sigma mxz = 0$ we see that the axis of the couple must be the axis of x and the magnitude of the couple will be $L = \omega^2 \Sigma myz$.

Ex. 4. A body having one point O fixed in space is made to rotate about any proposed straight line by the application of the proper couple. The position of the axis of rotation when the magnitude of the couple is a maximum, has been called an axis of *maximum reluctance*. Show that there are six axes of maximum reluctance, two in each principal plane, each two bisecting the angles between the principal axes in the plane in which they are.

Let the axes of reference be the principal axes of the body at the fixed point, let (l, m, n) be the direction-cosines of the axis of rotation, (λ, μ, ν) those of the axis of the couple G. Then by the last question and the second and third examples of Art. 18, we have

$$\frac{\lambda}{(B - C)\, mn} = \frac{\mu}{(C - A)\, nl} = \frac{\nu}{(A - B)\, lm},$$
$$G^2 = (A - B)^2 l^2 m^2 + (B - C)^2 m^2 n^2 + (C - A)^2 n^2 l^2.$$

We have then to make G a maximum by variation of (lmn) subject to the condition $l^2 + m^2 + n^2 = 1$. The positions of these axes were first investigated by Mr Walton in the *Quarterly Journal of Mathematics*, 1865.

The Centre of Percussion.

120. When the fixed axis is given and the body can be so struck that there is no impulsive pressure on the axis, any point in the line of action of the force is called a *centre of percussion*.

When the line of action of the blow is given, the axis about which the body begins to turn is called the *axis of spontaneous rotation*. It obviously coincides with the position of the fixed axis in the first case.

Let us begin by considering the motion in two dimensions. Imagine a lamina at rest and suspended from a point C with the centre of gravity G vertically under C. Let it be struck by a horizontal blow Y which we may suppose to act in the plane of the lamina at some point A in CG produced. Let $CA = a$. Let F and G be the impulsive reactions at the fixed point C. Let ω' be the angular velocity of the body round C just after the blow Y has been given. The equations of motion, exactly as in Art. 110, are therefore

$$\omega' = \frac{Ya}{M(k^2 + h^2)}, \quad h\omega' = \frac{Y + G}{M}, \quad 0 = F.$$

If the pressure G on the fixed point is zero, we have by eliminating Y,
$$k^2 + h^2 = ah.$$

By Art. 92 this shows that A must be the centre of oscillation of the body. *The centre of oscillation is therefore a centre of percussion.*

PROP. *A body is capable of turning freely about a fixed axis. To determine the conditions that there shall be a centre of percussion and to find its position.*

Take the fixed axis as the axis of z, and let the plane of xz pass through the centre of gravity of the body. Let X, Y, Z be the resolved parts of the impulse, and let ξ, η, ζ be the co-ordinates of any point in its line of action. Let Mk'^2 be the moment of inertia of the body about the fixed axis. We have now to find the pressures on the axis, and by equating these to zero we shall discover the conditions for a centre of percussion. The process is virtually the same as that already explained in Art. 113 and again in Art. 117. It seems unnecessary to repeat the steps. Putting $\bar{y} = 0$ and omitting the impulsive pressures on the axis because by hypothesis they are to be equated to zero, the six equations of motion of Art. 113 become

$$X = 0, \quad Y = M\bar{x}(\omega' - \omega), \quad Z = 0 \ldots\ldots\ldots\ldots\ldots\ldots(1).$$

$$\left. \begin{array}{l} \eta Z - \zeta Y = -(\omega' - \omega)\, \Sigma mxz \\ \zeta X - \xi Z = -(\omega' - \omega)\, \Sigma myz \\ \xi Y - \eta X = (\omega' - \omega)\, Mk'^2 \end{array} \right\} \ldots\ldots\ldots\ldots\ldots\ldots(2).$$

From these equations we may deduce the following conditions.

I.   From (1) we see that $X = 0$, $Z = 0$, and therefore the force must act perpendicular to the plane containing the axis and the centre of gravity.

II.   Substituting from (1) in the first two equations of (2) we have $\Sigma myz = 0$ and $\zeta = \dfrac{\Sigma mxz}{M\bar{x}}$. Since the origin may be taken anywhere in the axis of rotation, let it be so chosen that $\Sigma mxz = 0$. Then the axis of z must be a principal axis at the point where a plane passing through the line of action of the blow perpendicular to the axis cuts the axis. Thus there can be no centre of percussion unless the axis be a principal axis at some point in its length.

III.   Substituting from (1) in the last equation of (2) we have $\xi = \dfrac{k'^2}{\bar{x}}$. By Art. 92 this is the equation to determine the centre of oscillation of the body about the fixed axis treated as an axis of suspension. Hence the perpendicular distance between the line of action of the impulse and the fixed axis must be equal to the distance of the centre of oscillation from the axis.

If the fixed axis be parallel to a principal axis at the centre of gravity, the line of action of the blow will pass through the centre of oscillation.

Ex. 1. A circular lamina rests on a smooth horizontal table; how should it be struck that it may begin to turn round a point on its circumference? The line of action of the blow should divide the perpendicular diameter in the ratio 3 : 1.

Ex. 2. A pendulum is constructed of a sphere (radius a, mass M) attached to the end of a thin rod (length b, mass m). Where should it be struck at each oscillation that there may be no impulsive pressures to wear out the point of support? The point is at a distance l from the point of support, where

$$\{M(a+b)+\tfrac{1}{2}mb)\} \, l = M \{\tfrac{2}{5}a^2 + (a+b)^2\} + \tfrac{1}{3}mb^2.$$

The Ballistic Pendulum.

121. It is a matter of considerable importance in the Theory of Gunnery to determine the velocity of a bullet as it issues from the mouth of a gun. By means of it we obtain a complete test of any theory we have reason to form concerning the motion of the bullet in the gun. We may thus find by experiment the separate effects produced by varying the length of the gun, the charge of powder, or the weight of the ball. By determining the velocity of a bullet at different distances from the gun we may discover the laws which govern the resistance of the air.

It was to determine this initial velocity that Robins about 1743 invented the *Ballistic Pendulum*. Before his time but little progress had been made in the true theory of military projectiles. His *New Principles of Gunnery* was soon translated into several languages, and Euler added to his translation of it into German an extensive commentary. The work of Euler was again translated into English in 1784. The experiments of Robins were all conducted with musket balls of about an ounce weight, but they were afterwards continued during several years by Dr Hutton, who used cannon balls of from one to nearly three pounds in weight. These last experiments are still regarded as some of the most trustworthy on smooth-bore guns.

There are two methods of applying the ballistic pendulum, both of which were used by Robins. In the first method, the gun is attached to a very heavy pendulum; when the gun is fired the recoil causes the pendulum to turn round its axis and to oscillate through an arc which can be measured. The velocity of the bullet can be deduced from the magnitude of this arc. In the second method, the bullet is fired into a heavy pendulum. The velocity of the bullet is itself too great to be measured directly, but the angular velocity communicated to the pendulum may be made as small as we please by increasing its bulk. The arc of oscillation being measured, the velocity of the bullet can be found by calculation.

The initial velocity of small bullets may also be determined by the use of some rotational apparatus. Two circular discs of paper are attached perpendicularly to the straight line joining their centres, and are made to rotate about this straight line with a great but known angular velocity. Instead of two discs, a cylinder of paper might be used. The bullet being fired through at least two of the moving surfaces, its velocity can be calculated when the situations of the two small holes made by the bullet have been observed. This was originally an Italian invention, but it was much improved and used by Olinthus Gregory in the early part of this century.

The electric telegraph is now used to determine the instant at which a bullet passes through any one of a number of screens through which it is made to pass. The bullet severs a fine wire stretched across the screen and thus breaks an electric circuit. This causes a record of the time of transit to be made by an instrument expressly prepared for this purpose. By using several screens the velocities of the same bullet at several points of its course may be found.

122. *A rifle is attached in a horizontal position to a large block of wood which can turn freely about a horizontal axis. The rifle being fired, the recoil causes the pendulum to turn round its axis until brought to rest by the action of gravity. A piece of tape is attached to the pendulum, and is drawn out of a reel during the backward motion of the pendulum, and thus serves to measure the amount of the angle of recoil. It is required to find the velocity of the bullet.*

The initial velocity of the bullet is so much greater than that of the pendulum that we may suppose the ball to have left the rifle before the pendulum has sensibly moved from its initial position. The initial momentum of the bullet may be taken as a measure of the impulse communicated to the pendulum.

Let h be the distance of the centre of gravity from the axis of suspension; f the distance from the axis of the rifle to the axis of suspension; c the distance from the axis of suspension to the point of attachment of the tape, m the mass of the bullet; M that of the pendulum and rifle, and n the ratio of M to m; b the chord of the arc of the recoil which is measured by the tape. Let k' be the radius of gyration of the rifle and pendulum about the axis of suspension, v the initial velocity of the bullet.

The explosion of the gunpowder generates equal impulsive actions on the bullet and on the rifle. Since the initial velocity of the bullet is v, this action is measured by mv. The initial angular velocity generated in the pendulum by the impulse is by Art. 89 $\omega = \dfrac{mvf}{Mk'^2}$. The subsequent motion is given (Art. 92) by the

equation
$$\frac{d^2\theta}{dt^2} = -\frac{gh}{k'^2}\sin\theta\,;$$

$$\therefore \left(\frac{d\theta}{dt}\right)^2 = C + \frac{2gh}{k'^2}\cos\theta :$$

when $\theta = 0$ we have $\dfrac{d\theta}{dt} = \omega$, and if α be the angle of recoil, when $\theta = \alpha$, $\dfrac{d\theta}{dt} = 0$. Hence $\omega^2 = \dfrac{2gh}{k^2}(1 - \cos\alpha)$. Eliminating ω we have $v = \dfrac{nk'}{f}\cdot 2\sin\dfrac{\alpha}{2}\sqrt{gh}.$ But the chord of the arc of the recoil is $b = 2c\sin\dfrac{\alpha}{2}.$ Hence the initial velocity of the bullet is

$$v = \frac{nbk'}{cf}\sqrt{gh}.$$

The magnitude of k' may be found experimentally by observing the time of a small oscillation of the pendulum and rifle. If T be a half-time we have $T = \pi\sqrt{\dfrac{k'^2}{gh}}.$ (Art. 97.)

This is the formula given by Poisson in the second volume of his *Mécanique*. The reader will find in the *Philosophical Magazine* for June, 1854, an account of some experiments conducted by Dr S. Haughton from which, by the use of this formula, the initial velocities of rifle bullets were calculated.

123. The formula must however be regarded as only a first approximation, for the recoil of the pendulum when the gun is fired without a ball has been altogether neglected. In Dr Haughton's experiments the charge of powder was comparatively small, and this assumption was nearly correct. But in some of Dr Hutton's experiments, where comparatively large charges of powder were used, the recoil without a ball was found to be very considerable.

To allow for this Dr Hutton, following Mr Robins, assumed that the effect of the charge of powder on the recoil of the gun was the same with as without a ball. If p be the momentum generated by the powder, the whole momentum generated in the pendulum will be $mv + p$ instead of mv. Proceeding as before, we find

$$mv + p = \frac{Mbk'}{cf}\sqrt{gh}.$$

If we now repeat the experiment with an equal charge, and without a ball, we have $p = \dfrac{Mb_0k'}{cf}\sqrt{gh}$, where b_0 is the chord measured by the tape. Subtracting one result from the other, we have

$$v = \frac{M}{m}\frac{(b - b_0)k'}{cf}\sqrt{gh}.$$

Thus Dr Hutton's formula differs from Poisson's in this respect, that the chord of vibration is first found for any charge without a ball and then for an equal charge with a ball: the difference of these chords is regarded as the chord which is due to the recoil of the ball.

When the magnitude of the charge of powder is small, the two methods of using the ballistic pendulum give nearly the same result. With large charges Dr Hutton found that the difference was very considerable, a less velocity being indicated by the method of observing the recoil than by that of firing the ball into the pendulum. He therefore inferred that the effect of the charge of powder on the recoil of the gun was not the same when it was fired without a ball as when it was fired with one.

We may in some measure understand the reason of this discrepancy if we consider separately the effects of the inflamed powder while the ball is in the gun and after it has left the barrel. Supposing, merely as an approximation, that the gas urging the ball forward is of uniform density; its centre of gravity, at the moment when the ball is leaving the gun, will be at the middle point of the barrel and moving relatively to the gun with half the relative velocity of the ball. If μ be the mass of the powder, the angular velocity ω' communicated to the pendulum will be given approximately by $Mk'^2\omega' = \left(m + \dfrac{\mu}{2}\right)vf$. After the ball has left the gun, the inflamed powder escapes from the mouth and continues to exert some pressure tending to increase the recoil. The determination of this motion is a problem in hydrodynamics which has not yet been properly solved and which cannot be discussed here. We may, however, suppose that Robins' principle applies more nearly to this part of the motion than to the whole. If so, the momentum generated by the issuing gas, considered as an impulse, is nearly the same for a given charge and a given gun, whatever the magnitude of the ball may have been.

If p' be the momentum thus generated we have

$$\left(m + \frac{\mu}{2}\right)v + p' = \frac{Mbk'}{cf}\sqrt{gh}.$$

If v_0 and b_0 be the values of v and b when the gun is fired without a ball, we have

$$v - \frac{\mu}{2m}(v_0 - v) = \frac{M}{m}\frac{(b - b_0)k'}{cf}\sqrt{gh}.$$

Since v_0 is greater than v, this equation would show that, for considerable charges, Dr Hutton's formula will give too small a value for v. The value of v_0 is however very imperfectly known.

124. *A gun is placed in front of a heavy pendulum, which can turn freely about a horizontal axis. The ball strikes the pendulum horizontally, penetrates into the wood a short distance, and communicates a momentum to the pendulum. The chord of the arc being measured as before by a piece of tape, find the velocity of the bullet.*

The time, which the bullet takes to penetrate, is so short that we may suppose it completed before the pendulum has sensibly moved from its initial position.

Let i be the distance of the ball from the axis of suspension at the moment when the penetration ceases; let j be the perpendicular distance between the axis and the direction of motion of the bullet; let β be the angle the length j makes with the length represented by i, so that $j = i\cos\beta$. Then if we follow the same notation as before we have at the moment when the impact is concluded

$$mvi\,\cos\beta = (Mk'^2 + mi^2)\,\omega\,;$$

also proceeding as before we may prove

$$(Mk'^2 + mi^2)\,\omega^2 = 2Mgh\,(1 - \cos\alpha) + 2mgi\,\{\cos\beta - \cos(\alpha - \beta)\}.$$

If the gun be placed as nearly as possible opposite the centre of gravity of the pendulum we have $h = j$ nearly, and if the pendulum be rather long β will be very small. Hence, since m is small compared with M, we may as an approximation put $i = h$ and $\beta = 0$ in the terms which contain m as a factor; we thus find

$$v = \frac{M + m}{m}\,\frac{bh}{cj}\,\sqrt{gl},$$

where l is the distance of the centre of oscillation of the pendulum and ball from the axis of suspension.

The inconvenience of this construction as compared with the former is that the balls remain in the pendulum during the time of making one whole set of experiments. The weight, and the positions of the centres of gravity and oscillation, will be changed by the addition of each ball which is lodged in the wood. Even then the changes produced in the pendulum itself by each blow are omitted. A great improvement was made by the French in conducting their experiments at Metz in 1839, and at L'Orient in 1842. Instead of a mass of wood, requiring frequent renewals, as in the English pendulum, a permanent *récepteur* was substituted. This receiver is shaped within as a truncated cone, which is sufficiently long to prevent the shot from passing entirely through the sand with which it is filled. The front is covered with a thin sheet of lead to prevent the sand from being shaken out. This sheet is marked by a horizontal and by a vertical line, the intersection corresponding to the axial line of the cone, so that the actual position of the shot when entering the receiver can be readily determined by these lines.

125. Ex. 1. Show that after each bullet has been fired into a ballistic pendulum constructed on the English plan, h must be increased by $\frac{m}{M}(j - h)$ and l by $\frac{m}{M}(j - l)$ nearly in order to prepare the formula for the next shot.

Ex. 2. Dr Haughton found that, for rifles fired with a constant charge, the initial velocity of the bullet varies as the square root of the mass of the bullet inversely and as the square root of the length of the gun directly. Show from this that the force developed by the explosion of the powder, diminished by the friction of the barrel, is constant as the ball traverses the rifle.

Dr Hutton found that in smooth bores the velocity increases in a ratio somewhat less than the square root of the length of the gun, but greater than the cube root of the length.

Ex. 3. If the velocity of a bullet issuing from the mouth of a gun 30 inches long be 1000 feet per second, show that the time the bullet takes to traverse the gun is about $\frac{1}{200}$ of a second.

Ex. 4. It has been found by experiment that, if a bullet be fired into a large fixed block of wood, the depth of penetration of the bullet into the wood varies nearly as the square of the velocity, though as the velocity is very much increased the depth falls short of that given by this rule. Assuming this rule, show that the resistance to penetration is constant and that the time of penetration is the ratio of twice the depth to the initial velocity of the bullet. In an experiment of Dr Hutton's a ball fired with a velocity of 1500 feet per second was found to penetrate about 14 inches into a block of sound dry elm : show that the time of penetration was $\frac{1}{643}$ of a second.

THE ANEMOMETER.

126. The Anemometer called a "Robinson" consists of four hemispherical cups attached to four horizontal arms which turn round a vertical axis. The wind blows into the hollows on one side of the axis and against the convex surfaces of the cups on the other. If the anemometer start from rest, it will turn quicker and quicker until the moment of the pressures of the wind balances the moment of the resistances. Let V be the velocity of the wind and v the velocity of the centres of the cups. Let θ be the angle between the direction of motion of any one cup and that of the wind. Then the velocity of the centre of that cup relatively to the wind will be v', where

$$v'^2 = v^2 - 2Vv \cos \theta + V^2 \dots\dots\dots\dots\dots\dots\dots\dots(1).$$

The determination of the pressure of the wind on the cups is properly a problem in hydrodynamics, but no solution has yet been found. In the mean time we may assume as an approximation the law, suggested by numerous experiments, that the resistance to a body moving in a straight line in a fluid varies as the square of the relative velocity. In any one position of the anemometer the parts of any one cup have different velocities relative to the wind. We shall therefore take as our expression for the moment about the axis of the anemometer of the resultant pressure of the wind some quadratic function of V and v, such as

$$\alpha V^2 + 2\beta Vv + \gamma v^2 \dots\dots\dots\dots\dots\dots\dots\dots\dots(2),$$

where α, β, γ depend in some manner as yet unknown on the position of the cups relatively to the wind.

Thus α, β, γ are functions of θ and will change as the cups turn round the axis. What we want however is the average effect on the anemometer. The mean for space is found by multiplying this expression by $d\theta$ and integrating from $\theta = 0$ to 2π and finally dividing by 2π. If F be the mean moment about the axis of the anemometer of the wind pressure, we have

$$F = AV^2 - 2BVv - Cv^2 \dots\dots\dots\dots\dots\dots\dots\dots(3),$$

where A, B, C are constants which depend on the pattern of the anemometer. The signs of these coefficients may be determined by the following reasoning. When the anemometer starts from rest, the initial moment of the wind pressure is regarded as positive. When the cups begin to move, the pressure begins to decrease, so that $\dfrac{dF}{dv}$ must be negative when v is small; it follows that the sign of the coefficient of Vv in (3) must be negative. Finally, if the wind cease when the cups are in motion so that $V=0$, the resistance of the quiescent air must tend to stop the cups. It follows that the coefficient of v^2 in (3) must be also negative.

127. When the anemometer has attained its final state of motion, we must have F equal to the mean moment of the friction on the supports. The instrument should be so arranged that the friction due to its weight is as small as possible. We may then omit this friction, as our formula is only an approximation. The supports of the anemometer have also to sustain the lateral pressure of the wind. Probably the greater part of the friction thus produced is proportional to the pressure of the wind, and may be included in the formula (3) by an alteration of the constants. As these constants are determined by experiment, we may suppose all forces which are quadratic functions of the velocities to be included in the expression for F.

In the Observatory at Greenwich an inverted cup rotating in oil on a fixed conical point is used for the vertical bearing. No further correction is made for friction. This arrangement appears to be very successful, the instrument is very sensitive and exhibits a slow rotation with a very slight movement of the air.

When F is equated to zero, we have a quadratic to determine the ratio of V to v. Let m be the positive root thus found. Then the velocity of the centre of any cup being observed, the velocity of the wind is found by simply multiplying this observed quantity by m. We may notice that m is independent of the speed of the wind, and of the size of the machine. It depends however on the pattern of the machine.

128. A variety of experiments have been made to determine the numerical value of m. In some of these the anemometer is attached to the outer edge of a whirling-machine. The axis of the anemometer is thus made to move round with a constant velocity V. If the experiment be made on a calm day, this will represent the effects of a wind of the same velocity on a fixed anemometer. The value of v can be found by counting the number of revolutions of the anemometer in space. In a paper in 1850, published in the *Irish Transactions*, Dr Robinson gives $m=3$ as the mean value of the ratio as determined by experiments of this kind. This value of m has been generally adopted.

Other experiments made in Greenwich Park in 1860 led to the same value of m. These results were considered as confirming in a very high degree the accuracy of this ratio. See the *Greenwich Observations* for 1862. About 1872 further experiments were made with a steam merry-go-round for a whirling machine. These are described by Sir G. Stokes in the *Proceedings of the Royal Society* for May, 1881.

Another method of conducting the experiments is to have two similar anemometers rotating about *fixed* axes and to apply to one of them a known retarding force of some kind which may diminish its v. Thus we have two different machines moving with different, but known, velocities round their respective axes, from each of which we should deduce the same velocity for the wind. This leads to two equations between which we may eliminate the unknown velocity of the wind. We thus obtain an equation connecting the constants A, B, C and the known retarding force. Repeating the experiment, we may obtain a sufficient number of equations to find these constants. The value of m may then be found in the manner explained in Art. 127. The practical difficulty in this method of conducting the experiments is that of finding a known uniform retarding force which may be conveniently applied to the anemometer. The reader may consult a paper by Dr Robinson in the *Phil. Trans.* for 1880.

129. **Ex. 1.** Supposing the value of F to be represented by $AV^2 - 2BVv$, as indicated by some experiments, show that, if an anemometer start from rest, the velocity v of the cups will continually increase and tend to a certain finite limit. Show also that the time, at which the actual velocity of the cups is any given fraction of the limiting velocity, varies as the moment of inertia of the anemometer about its axis, and inversely as the velocity of the wind.

Ex. 2. When the anemometer was attached to the outer edge of a merry-go-round, as described above, it was impossible to find a perfectly calm day. If W be the velocity of the wind, which is supposed to be small, then allowance may be made for W if in the formula $F = AV^2 - 2BVv$ we write $V + \kappa \dfrac{W^2}{V}$ for V, where κ is $\frac{1}{4}$ or $\frac{3}{4}$ according as the moment of inertia of the anemometer about its axis is very small or very great. The anemometer is supposed to be without friction. This theorem is due to Sir G. Stokes: a demonstration is given in the *Proceedings of the Royal Society* for May, 1881.

Ex. 3. An anemometer without friction is acted on by a gusty wind whose velocity may be represented by the formula $V(1 + a \sin nt)$, where a is so small that its square can be neglected. Show that the velocity of any cup will be represented by an expression of the form $v\{1 + a \cos n\beta \sin n(t - \beta)\}$, so that the anemometer follows all the changes in the force of the wind after an interval β. Here $AV^2 - 2BVv - Cv^2 = 0$, and

$$\tan n\beta = \frac{In}{2a(BV + Cv)},$$

where a is the distance of the centre of a cup from the axis, and I is the moment of inertia of the machine about the axis.

The velocities of the currents of air in mines are usually determined by the aid of anemometers of a somewhat different construction. The principle of these is similar to that of Whewell's anemometer. They are formed of several light vanes placed on a horizontal axis like the sails of a windmill on a small scale but more numerous. The axis is attached to a dial or some other apparatus by which the number of revolutions made by the little windmill can be read off. If V be the velocity of the wind and v the reading of the anemometer it is found by experiment that between certain limits $V = av + b$, where a and b are two constants which depend on the pattern of the anemometer and the friction which the wind has to overcome. The reader may consult a paper by Mr Snell in the *Engineer*, June 23, 1882.

CHAPTER IV.

On the Equations of Motion.

130. THE position of a body in space of two dimensions may be determined by the co-ordinates of its centre of gravity, and the angle some straight line fixed in the body makes with some straight line fixed in space. These three have been called the co-ordinates of the body, and it is our object to determine them in terms of the time.

It will be necessary to express the effective forces of the body in terms of these co-ordinates. The resolved parts of these effective forces parallel to the axes have been already found in Art. 79, all that is now necessary is to find their moment about the centre of gravity. If (x', y') be the co-ordinates of any particle of mass m referred to rectangular axes meeting at the centre of gravity and parallel to axes fixed in space, this moment has been shown in Art. 76 to be equal to h, where

$$h = \Sigma m (x'\dot{y}' - y'\dot{x}').$$

Let θ be the " angular co-ordinate" of the body, i.e. the angle some straight line fixed in the body makes with some straight line fixed in space. Let (r', ϕ') be the polar co-ordinates of any particle m referred to the centre of gravity of the body as origin. Then r' is constant throughout the motion, and $\dot{\phi}'$ is the same for every particle of the body and equal to $\dot{\theta}$. Thus the angular momentum h, exactly as in Art. 88, is

$$h = \Sigma m (x'\dot{y}' - y'\dot{x}') = \Sigma m (r'^2 \dot{\phi}') = \Sigma m r'^2 \dot{\phi}'$$
$$= Mk^2\dot{\theta},$$

where Mk^2 is the moment of inertia of the body about its centre of gravity.

The angle θ is the angle some straight line fixed in the body makes with a straight line fixed in space. *Whatever straight lines are chosen* $\dfrac{d\theta}{dt}$ *is the same.* If this be not obvious, it may be shown thus. Let OA, $O'A'$ be any two straight lines fixed in the body inclined at an angle α to each other. Let OB, $O'B'$ be two straight lines fixed in space inclined at an angle β to each other. Let $AOB = \theta$, $A'O'B' = \theta'$, then $\theta' + \beta = \theta + \alpha$. Since α and β are independent of the time, $\dot{\theta} = \dot{\theta}'$. By this proposition we learn that the angular velocities of a body in two dimensions are the same about all points.

131. The general method of proceeding will be as follows.

Let (x, y) be the co-ordinates of the centre of gravity of any body of the system referred to rectangular axes fixed in space, M the mass of the body. Then the effective forces of the body are together equivalent to two forces measured by $M\dfrac{d^2x}{dt^2}$, $M\dfrac{d^2y}{dt^2}$ acting at the centre of gravity and parallel to the axes of co-ordinates, together with a couple measured by $Mk^2\dfrac{d^2\theta}{dt^2}$ tending to turn the body about its centre of gravity in the direction in which θ is measured. By D'Alembert's principle the effective forces of all the bodies, if reversed, will be in equilibrium with the impressed forces. The dynamical equations may then be formed according to the ordinary rules of statics. See Art. 83.

Suppose we wish to resolve the forces parallel to the axes of x and y and to take moments about the centre of gravity. Let the impressed forces acting on the body, together with the re-actions due to the other bodies if any, be equivalent to the forces X and Y acting at the centre of gravity and a couple L. The equations of motion of that body are evidently

$$M\frac{d^2x}{dt^2} = X, \qquad M\frac{d^2y}{dt^2} = Y, \qquad Mk^2\frac{d^2\theta}{dt^2} = L.$$

It is found useful in statics to be able to resolve in other directions besides the axes and to be able to take moments about any point we please. In this way we often greatly shorten and simplify the solution. Thus if we wish to avoid the introduction into our equations of some unknown reaction we take moments about the point of application or use the principle of virtual velocities. So in dynamics we are at liberty to resolve our forces and take moments at pleasure. For example, if we take moments about a point C whose co-ordinates are $(\xi\eta)$ we have an equation of the form

$$M\left\{(x - \xi)\frac{d^2y}{dt^2} - (y - \eta)\frac{d^2x}{dt^2}\right\} + Mk^2\frac{d^2\theta}{dt^2} = L',$$

where L' is the moment about C of the impressed forces. In this equation $(\xi\eta)$ may be the co-ordinates of any point whatever whether fixed or moving.

In resolving our forces we may replace the Cartesian expressions by the polar forms $M\left\{\dfrac{d^2r}{dt} - r\left(\dfrac{d\phi}{dt}\right)^2\right\}$ and $M\dfrac{1}{r}\dfrac{d}{dt}\left(r^2\dfrac{d\phi}{dt}\right)$ for the resolved parts parallel and perpendicular to the radius vector. If v be the velocity of the centre of gravity, ρ the radius of curvature of its path, we may sometimes also use with advantage the forms $M\dfrac{dv}{dt}$ and $M\dfrac{v^2}{\rho}$ for the resolved parts of the effective forces along the tangent and radius of curvature of the path of the centre of gravity.

As a guide to a proper choice of the directions in which to resolve the forces or of the points about which we should take moments we may mention two important cases.

132. First, we *should search if there be any direction fixed in space in which the resolved part of the impressed forces vanishes.* By resolving in this direction we get an equation which can be immediately integrated. Suppose the axis of x to be taken in this direction; let M, M', &c. be the masses of the several bodies, x, x', &c. the abscissæ of their centres of gravity, then by Art. 78 or 131, we have

$$M\frac{d^2x}{dt^2} + M'\frac{d^2x'}{dt^2} + \ldots = 0,$$

which by integration gives

$$M\frac{dx}{dt} + M'\frac{dx'}{dt} + \ldots\ldots = C,$$

where C is some constant to be found from the initial conditions. This equation may be again integrated if necessary.

This result might have been derived from the general principle of the conservation of the motion of translation of the centre of gravity laid down in Art. 79. For, since there is no impressed force parallel to the axis of x, the velocity of the centre of gravity of the *whole system* resolved in that direction is constant.

133. Next, we *should search if there be any point fixed in space about which the moment of the impressed forces vanishes.* By taking moments about that point we again have an equation which admits of immediate integration. Suppose the point to be taken as origin, and the letters to have their usual meaning, then by the first article of this chapter we have

$$\Sigma\left\{M\left(x\frac{d^2y}{dt^2} - y\frac{d^2x}{dt^2}\right) + Mk^2\frac{d^2\theta}{dt^2}\right\} = 0,$$

the Σ referring to summation for all the bodies of the system. Integrating we have

$$\Sigma \left\{ M \left(x \, \frac{dy}{dt} - y \, \frac{dx}{dt} \right) + Mk^2 \frac{d\theta}{dt} \right\} = C,$$

where C is some constant to be determined by the initial conditions of the question.

This equation expresses the fact that if the impressed forces have no moment about any point, the angular momentum about that point is constant throughout the motion. This result follows at once from the reasoning in Art. 78.

134. **Angular Momentum.** As we shall have so frequently to use the equation formed by taking moments, it is important to consider other forms into which it may be put. Let the point about which we are to take moments be *fixed in space*, so that it may be chosen as the origin of co-ordinates. Then the moment of the effective forces on the body M is

$$\frac{d}{dt} \left\{ M \left(x \, \frac{dy}{dt} - y \, \frac{dx}{dt} \right) + Mk^2 \frac{d\theta}{dt} \right\} = L,$$

where x and y are the co-ordinates of the centre of gravity.

The attention of the reader is directed to the meaning of the several parts of this expression. We see that, as explained in Art. 78, the moment of the effective forces is the differential coefficient of the moment of the momentum about the same point. The moment of the momentum by Art. 75 is the same as the moment about the centre of gravity together with the moment of the whole mass collected at the centre of gravity, and moving with the velocity of the centre of gravity. The moment round the centre of gravity is by the first Article either of Chap. III. or Chap. IV. equal to $Mk^2 \dfrac{d\theta}{dt}$ and the moment of the collected mass is $M \left(x \, \dfrac{dy}{dt} - y \, \dfrac{dx}{dt} \right)$. Hence in space of two dimensions we have for any body of mass M

$$\left. \begin{array}{r} \text{angular momentum round} \\ \text{the origin} \end{array} \right\} = M \left(x \, \frac{dy}{dt} - y \, \frac{dx}{dt} \right) + Mk^2 \frac{d\theta}{dt}.$$

If we prefer to use polar co-ordinates, we can put this into another form. Let (r, ϕ) be the polar co-ordinates of the centre of gravity, then

$$\left. \begin{array}{r} \text{angular momentum round} \\ \text{the origin} \end{array} \right\} = Mr^2 \frac{d\phi}{dt} + Mk^2 \frac{d\theta}{dt}.$$

If v be the velocity of the centre of gravity, and p the perpendicular from the origin on the tangent to the direction of its motion, the moment of momentum of the mass collected at the centre of gravity is Mvp, so that we have again

$$\left.\begin{array}{c}\text{angular momentum round}\\ \text{the origin}\end{array}\right\} = Mvp + Mk^2 \frac{d\theta}{dt}.$$

It is clear from Art. 75 that this is the instantaneous angular momentum of the body about the origin whether it is fixed or moving, though in the latter case its differential coefficient with regard to t is not the moment of the effective forces.

Since the instantaneous centre of rotation may be regarded as a fixed point, when we have to deal only with the co-ordinates and with their *first differential coefficients* with regard to the time, we have

$$\left.\begin{array}{c}\text{angular momentum round the}\\ \text{instantaneous centre}\end{array}\right\} = M(r^2 + k^2) \frac{d\theta}{dt}.$$

If Mk'^2 be the moment of inertia about the instantaneous centre, this last moment may be written $Mk'^2 \dfrac{d\theta}{dt}$.

In taking moments about any point, whether it be the centre of gravity or not, it should be noticed that the Mk^2 in all these formulæ is the moment of inertia with regard to the centre of gravity, and not with regard to the point about which we are taking moments. It is only when we are taking moments about the instantaneous centre or about a fixed point that we can use the moment of inertia about that point instead of the moment of inertia about the centre of gravity, and in these cases our expression for the angular momentum includes the angular momentum of the mass collected at the centre of gravity.

135. General Mode of Solution. Suppose we form the equations of motion of each body by resolving parallel to the axes of co-ordinates and by taking moments about the centre of gravity. We shall get three equations for each body of the form

$$\left.\begin{array}{l}M\ddot{x} = F\cos\phi + R\cos\psi + \ldots\\ M\ddot{y} = F\sin\phi + R\sin\psi + \ldots\\ Mk^2\ddot{\theta} = \quad Fp \quad + \quad Rq \quad + \ldots\end{array}\right\}\ldots\ldots\ldots\ldots(1),$$

where F is one of the impressed forces acting on the body, whose resolved parts are $F\cos\phi$, $F\sin\phi$, and whose moment about the centre of gravity is Fp, and R is any one of the re-actions. These we shall call the *dynamical equations* of the body.

Besides these there will be certain geometrical equations expressing the connections of the system. As every such forced connection is accompanied by a reaction, and every reaction by some forced connection, the number of geometrical equations will be the same as the number of unknown reactions in the system.

Having obtained the proper number of equations of motion we proceed to their solution. Two general methods have been proposed.

First Method of Solution. Differentiate the geometrical equations twice with respect to t, and substitute for $\ddot{x}$, $\ddot{y}$, $\ddot{\theta}$ from the dynamical equations. We shall then have a sufficient number of equations to determine the reactions. This method will be of great advantage whenever the geometrical equations are of the form

$$Ax + By + C\theta = D \quad \dots\dots\dots\dots\dots(2),$$

A, B, C, D being constants. Suppose also that the dynamical equations are such that when written in the form (1) they contain *only the reactions and constants on the right-hand side without any* x, y, *or* θ. Then, when we substitute in the equation

$$A\ddot{x} + B\ddot{y} + C\ddot{\theta} = 0,$$

obtained by differentiating (2), we have an equation containing only the reactions and constants. This being true for all the geometrical relations, it is evident that all the reactions will be constant throughout the motion and their values may be found. Hence, when these values are substituted in the dynamical equations (1), their right-hand members will all be constants and the values of x, y, and θ may be found by an easy integration.

If however the geometrical equations are not of the form (2), this method of solution will usually fail. Thus suppose a geometrical equation to take the form

$$x^2 + y^2 = c^2,$$

containing *squares* instead of *first* powers, then its second differential equation will be

$$x\ddot{x} + y\ddot{y} + \dot{x}^2 + \dot{y}^2 = 0 \ ;$$

and, though we can substitute for $\ddot{x}$, $\ddot{y}$, we cannot in general eliminate the terms $\dot{x}^2$ and $\dot{y}^2$.

136. The reactions in a dynamical problem are in many cases produced by the pressures of some smooth fixed obstacles which are touched by the moving bodies. Such obstacles can only *push*, and therefore if the equations show that such a reaction changes sign at any instant, it is clear that the body will leave the obstacle at that instant. This will occasionally introduce discontinuity into our equations. At first the system moves under certain constraints, and our equations are found on that supposition. At some instant to be determined by the vanishing

of a reaction one of the bodies leaves its constraints, and the equations of motion have to be changed by the omission of that reaction. Similar remarks apply if the reactions be produced by the pressure of one body against another.

It is important to notice that *when this first method of solution applies,* the reactions are constant throughout the motion, so that the above discontinuity can never occur. In this case, then, *if one body be in contact with another, they will either separate at the beginning of their motion or will always continue in contact.* Such reactions are also independent of the initial conditions, and are therefore the same as if the system were placed in any position at rest.

137. Suppose that in a dynamical system *we have two bodies which press on each other with a reaction* R; *let us consider how we are to form the corresponding geometrical equation.* We have clearly to express the fact that the velocities of the points of contact of the two bodies resolved along the direction of R are equal. The following proposition will be often useful. Let a body be turning about a point G with an angular velocity $\theta = \omega$ in a direction opposite to the hands of a watch, and let G be moving in the direction GA with a velocity V. It is required to find the velocity of any point P resolved in any

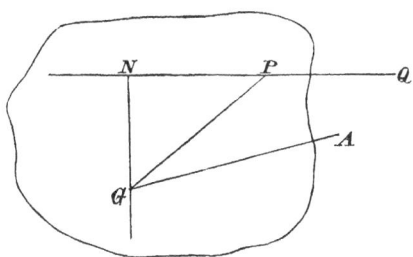

direction PQ making an angle ϕ with GA. In the time dt the whole body, and therefore also the point P, is moved through a space Vdt parallel to GA, and during the same time P is moved perpendicular to GP through a space $\omega \cdot GP \cdot dt$. Resolving parallel to PQ, the whole displacement of P

$$= (V \cos \phi - \omega \cdot GP \sin GPN) \, dt.$$

If $GN = p$ be the perpendicular from G on PQ, we see that the velocity of P parallel to PQ is $V \cos \phi - \omega p$.

It should be noticed that this expression is independent of the position of P on the straight line PQ. *It follows that the velocities of all points in any straight line PQ resolved along PQ are the same.* This result will be evident if we remember that all the points in the straight line PQ are rigidly connected together, so that if the resolved velocities of the points in it were unequal, the line PQ would alter in length.

When therefore we require the velocity of any point P in any direction PQ we may replace P by any other point in the line PQ so situated that its resolved velocity is more easily found. Usually the point N is the most convenient point to use, for without quoting a formula, its velocity resolved along PQ is seen by inspection to be $V \cos \phi - \omega p$.

If (x, y, θ), (x', y', θ') be the co-ordinates of the two bodies, q, q' the perpendiculars from the points (x, y), (x', y') on the direction of any reaction R, ψ the angle the direction of R makes with the axis of x, the required geometrical equation will be

$$\dot{x} \cos \psi + \dot{y} \sin \psi + \dot{\theta} q = \dot{x}' \cos \psi + \dot{y}' \sin \psi + \dot{\theta}' q'.$$

If the bodies be perfectly rough and roll on each other without sliding, there will be *two* resolved reactions at the point of contact, one normal and the other tangential to the common surface of the touching bodies. For each of these we shall have an equation similar to that just found. But if there be any *sliding* friction this reasoning will not apply. The latter case will be considered a little further on.

138. *Second Method of Solution.* Suppose that in a dynamical system two bodies of masses M, M' are pressing on each other with a reaction R. Let the equations of motion of M be those marked (1) in Art. 135, and let those of M' be obtained from these by accenting all the letters except R, ψ and t, and writing $-R$ for R, ψ and t being of course unaltered. Let us multiply the equations of motion of M by $2\dot{x}$, $2\dot{y}$, $2\dot{\theta}$ respectively, and those of M' by corresponding quantities. Adding all these six equations, we get

$$2M (\dot{x}\ddot{x} + \dot{y}\ddot{y} + k^2\dot{\theta}\ddot{\theta}) + \&\text{c.}$$

$$= 2F (\dot{x} \cos \phi + \dot{y} \sin \phi + p\dot{\theta}) + \&\text{c.}$$

$$+ 2R (\dot{x} \cos \psi + \dot{y} \sin \psi + q\dot{\theta}) - 2R (\dot{x}' \cos \psi + \dot{y}' \sin \psi + q'\dot{\theta}').$$

The coefficient of R will vanish by virtue of the geometrical equation obtained in the last Article. Similar reasoning will apply to all the reactions between each two of the moving bodies.

Suppose the body M to press against some external fixed obstacle, then R acts only on the body M, and the coefficient of $2R$ will be restricted to the part included in the first bracket. But the velocity of the point of contact resolved along the direction of R must vanish, and therefore the coefficient of R is again zero.

Let A be the point of application of the impressed force F, and let the velocity of A resolved along the direction of action of F be $\dot{f}$. Then we see that the coefficient of $2F$ is $\dot{f}$. It also follows from the definition of df that Fdf is what is called in statics the virtual moment of the force F.

We have thus a general method of obtaining an equation free from the unknown reactions of perfectly smooth or perfectly rough bodies. The rule is, multiply the equations having $M\ddot{x}$, $M\ddot{y}$, $Mk^2\ddot{\theta}$, &c. on their left-hand sides by $\dot{x}$, $\dot{y}$, $\dot{\theta}$, &c., and add together all the resulting equations for all the bodies. The coefficients of all the unknown reactions will be found to be zero by virtue of the geometrical equations.

The left-hand side of the equation thus obtained is clearly a perfect differential. Integrating we get

$$M\{\dot{x}^2 + \dot{y}^2 + k^2\dot{\theta}^2\} + \&\text{c.} = C + 2\int Fdf + \ldots\ldots$$

where C is the constant of integration.

In practice it is usual to omit all the intermediate steps and to write down the equation in the following manner:

$$\Sigma M\{\dot{x}^2 + \dot{y}^2 + k^2\dot{\theta}^2\} = C + 2U,$$

where U is the integral of the virtual moment of the forces.

This is called the equation of *Vis Viva*. Another proof will be given in the chapter under that heading.

139. **Vis Viva of a Body.** The left-hand side of this equation is called the *vis viva* of the whole system. Taking any one body M, we may say that

$$\text{vis viva of } M = M\left\{\left(\frac{dx}{dt}\right)^2 + \left(\frac{dy}{dt}\right)^2 + k^2\left(\frac{d\theta}{dt}\right)^2\right\}.$$

If the whole mass were collected into its centre of gravity and were to move with the velocity of the centre of gravity, k would be zero, and the vis ·viva would be reduced to the two first terms. These terms are therefore together called the *vis viva of translation*, and the last term is called the *vis viva of rotation*.

If v be the velocity of the centre of gravity, we may write this equation

$$\text{vis viva of } M = Mv^2 + Mk^2\left(\frac{d\theta}{dt}\right)^2.$$

If we wish to use polar co-ordinates, we have

$$\text{vis viva of } M = M\left\{\left(\frac{dr}{dt}\right)^2 + r^2\left(\frac{d\phi}{dt}\right)^2 + k^2\left(\frac{d\theta}{dt}\right)^2\right\},$$

where (r, ϕ) are the polar co-ordinates of the centre of gravity.

If ρ be the distance of the centre of gravity from the instantaneous centre of rotation of the body, $\rho \dfrac{d\theta}{dt}$ is clearly the velocity of the centre of gravity, and therefore

$$\text{vis viva of } M = M\left(\rho^2 + k^2\right)\left(\frac{d\theta}{dt}\right)^2.$$

140. **Force Function and Work.** The function U in the equation of vis viva is called the *force function* of the forces. It may always be obtained, when it exists, by writing down the virtual moment of the forces according to the rules of statics, integrating the result and adding a constant. This definition is sufficient for our present purpose; for a more complete explanation the reader is referred to the beginning of the chapter on Vis Viva.

When the forces are functions of several co-ordinates, it may be supposed that it will often happen that the virtual moment cannot be integrated until the relations between these co-ordinates have been found by some other means. But it will be shown in the chapter on Vis Viva that this is not so. In nearly all the cases we have to consider the virtual moment will be a perfect differential. In the remarks which follow in this and in the next three articles it will therefore be convenient to suppose that the function U exists, and is a known function of the co-ordinates of the system.

In a subsequent chapter we shall discuss more particularly the various forms which the force function may assume. For the present we shall merely show how to find its form for a system of bodies under any constraints which are falling through the action of gravity alone.

Let x, y be the horizontal and vertical co-ordinates of any particle of the system and let the latter co-ordinate be measured downwards. Let m be the mass of the particle. The virtual moment is therefore $\Sigma mgdy$. The force function may therefore be written

$$U = \int \Sigma mgdy = \Sigma mgy + C$$
$$= g\bar{y}\Sigma m + C,$$

where $\bar{y}$ is the depth of the centre of gravity of the whole system below the axis of x.

Sometimes to avoid the constant C we take the integral between limits. The force function is then called the *work of the forces* as the system passes from the position indicated by the lower limit to that indicated by the upper limit.

The result just arrived at may therefore be stated thus. *If, as a system moves from one position to another, its centre of gravity descends a vertical space* h, *the work done by gravity is* Mgh, *where* M *is the whole mass of the system.*

We notice that this result is independent of any changes in the arrangement of the bodies which constitute the system, and depends solely on the vertical space descended by the centre of gravity.

141. **Principle of Vis Viva.** Sometimes a system may move from one position to another in one of several ways. Perhaps we do not want the intermediate motion but only the motion in the later position when that in the earlier is given. In such a case we avoid the introduction of the constant C in the equation of vis viva by taking the integral in Art. 138 between limits. Thus we say that

$$\left.\begin{array}{c}\text{the change in}\\ \text{the vis viva}\end{array}\right\} = \left\{\begin{array}{c}\text{twice the work done}\\ \text{by the forces.}\end{array}\right.$$

In this equation the change in the vis viva is found by subtracting from the vis viva in the final position the vis viva in the first. In finding the work done by the forces, the upper limit of the integral (as already explained) depends on the final position of the system and the lower limit depends on the initial position.

The great importance of this equation is that we have a result free from all the reactions or constraints of the system. The manner in which the system moves from the first position to the last is a matter of indifference. So far as this equation is concerned, we may change the mode of motion in any way by introducing or removing any constraints or reactions, provided only that they are such as do not appear in the equation of virtual moments as used in statics.

We must notice that some reactions will not disappear from the equation of virtual velocities in statics, for example, friction between two surfaces which *slide over each other.* In forming the equation of vis viva in dynamics this kind of friction, when it occurs, will appear along with the other forces on the right-hand side of the equation.

As the system moves from one given position to another, it is evident that the change in the vis viva produced by each force is twice the integral of the virtual moment of that force. It follows that the whole change is the sum of the changes produced by the separate forces. Taking then any one force F, we see that, when its direction makes an acute angle with the direction of the motion of the point A of the body at which it acts, F and df have the same sign, and the integral in the equation of vis viva is positive. The effect of the force is therefore to increase the vis viva. But when the direction of the force is opposed to the direction of the motion of A, i.e. when the force makes an acute angle with the reversed direction of the motion of A, the effect of the force is to decrease the vis viva. This rule will enable us to determine the general effect of any force on the vis viva of the system.

142. Suppose, for example, a body to move or roll under the action of gravity with one point in contact with a fixed surface which is either perfectly rough or perfectly smooth, so that there can be no sliding friction. Let it be started off in any manner, so that the initial vis viva is known. The vis viva decreases or increases according as the centre of gravity rises above or falls below its original level. As the body moves the pressure on the surface will change and may possibly vanish and change sign. In this case the body will leave the surface. The centre of gravity by Art. 79 will then describe a parabola and the angular velocity of the body about its centre of gravity will be constant. Presently the body may impinge again on the surface, but until such impact occurs the equation of vis viva is in no way affected by the body leaving the surface. But the case is different when the body impinges on the surface. To make this point clearer, let F be the reaction of the surface, A the point of the body at which it acts, and Fdf its virtual moment as in Art. 138. Then as the body moves on the surface, df is zero, and when the body has left the surface, F is zero, so that during the motion before the impact occurs the virtual moment Fdf is zero for the one reason or the other. The reaction therefore does not appear in the equation of vis viva. But when the body impinges on the surface, the point A is approaching the surface and the reaction F is resisting the advance of A so that neither F nor df is zero. Here we measure F in the same manner as in the first part of the motion, regarding it as a very great force which destroys the velocity of A in a very short time (Art. 84). During the period of compression, the force F resists the advance of A, and therefore the vis viva of the body is decreased. But during the period of restitution the force assists the motion of A, and thus the vis viva is increased. We shall show further on that the vis viva is decreased by an impact except in the extreme case in which the bodies are perfectly elastic, and we shall investigate the amount lost. As a general rule we may notice that the equation of vis viva is altered by an impact.

We may find a superior limit to the altitude $\bar{y}$ to which the centre of gravity can rise above its original level. The equation of vis viva may be written

$$\begin{pmatrix} \text{vis viva in any} \\ \text{position} \end{pmatrix} - \begin{pmatrix} \text{initial vis} \\ \text{viva} \end{pmatrix} = -2Mg\bar{y},$$

where M is the mass of the body. Now the vis viva can never be negative, hence the centre of gravity cannot rise so high that

$$2Mg\overline{y} > \text{initial vis viva.}$$

In order that the centre of gravity should reach this altitude it is necessary that the vis viva of the body should vanish, i.e. both the velocity of translation of the centre of gravity and the angular velocity of the body must simultaneously vanish. This cannot in general occur if the body jump off the surface, for the angular velocity and the horizontal velocity of the centre of gravity will not usually both vanish at the moment of the jump, and both will remain constant, as explained above, during the parabolic motion. After the subsequent impact a new motion may be supposed to begin with a diminished vis viva and therefore a diminished superior limit to the altitude of the centre of gravity.

143. Sometimes there is only one way in which the system can move. In such a case all we have to find is the velocity of the motion. The geometry of the system will determine the x, y, θ of each body in terms of some one quantity which we may call ϕ. The vis viva of the body M, as given by Art. 139, will now take the form

$$\text{vis viva of } M = M \left\{ \left(\frac{dx}{d\phi}\right)^2 + \left(\frac{dy}{d\phi}\right)^2 + k^2 \left(\frac{d\theta}{d\phi}\right)^2 \right\} \left(\frac{d\phi}{dt}\right)^2 = P \left(\frac{d\phi}{dt}\right)^2,$$

where P is a known function of the co-ordinates of M. The equation of vis viva will therefore take the form

$$(\Sigma P) \left(\frac{d\phi}{dt}\right)^2 = C + 2U,$$

and thus $\dfrac{d\phi}{dt}$ can be found for any given position of the system.

It follows that, if there is only one way in which the system can move, that motion will be determined by the equation of vis viva. But, if there be more than one possible motion, we must find another integral of the equations of the second order. What should be done will depend on the special case under consideration. The discovery of the proper treatment of the equations is often a matter of great difficulty. The difficulty will be increased if, in forming the equations, care has not been taken to give them the simplest possible forms.

144. **Examples of these Principles.** The following examples have been constructed to illustrate the methods of applying the above principles to the solution of dynamical problems. In some cases more solutions than one have been given, to enable the reader to compare different methods. The mode of forming each equation has been minutely explained. Running remarks have been made

which it is hoped will clear up those difficulties which generally trouble a beginner.   The attention of the student is therefore particularly directed to the different principles used in the following solutions.

A homogeneous sphere rolls directly down a perfectly rough inclined plane under the action of gravity. Find the motion.

Let a be the inclination of the plane to the horizon, a the radius of the sphere, mk^2 its moment of inertia about a horizontal diameter.

Let. O be that point of the inclined plane which was initially touched by the sphere, and N the point of contact at the time t.   Then it is obviously convenient to choose O for origin, and ON for axis of x.

The forces which act on the sphere are, first, the reaction R perpendicularly to ON, secondly, the friction F acting at N along NO and mg acting vertically at C the centre.   The effective forces are $m\ddot{x}$, $m\ddot{y}$ acting at C parallel to the axes of x and y, and a couple $mk^2\ddot{\theta}$ tending to turn the sphere round C in the direction

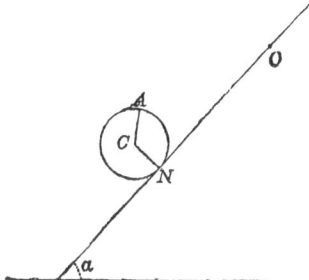

NA.   Here θ is the angle which any straight line fixed in the body makes with a straight line fixed in space.   We shall take the fixed straight line in the body to be the radius CA, and the fixed straight line in space the normal to the inclined plane.   Then θ is the angle turned through by the sphere.

Resolving along and perpendicular to the inclined plane we have

$$m\ddot{x} = mg \sin a - F \dotfill (1),$$
$$m\ddot{y} = -mg \cos a + R \dotfill (2).$$

Taking moments about N to avoid the reactions, we have

$$ma\ddot{x} + mk^2\ddot{\theta} = mga \sin a \dotfill (3).$$

Since there are two unknown reactions F and R, we shall require two geometrical relations.   Because there is no slipping at N we have

$$x = a\theta \dotfill (4).$$

Also, because there is no jumping, $y = a$ \dotfill (5).

Both these equations are of the form required in the first method.   Differentiating (4) we get $\ddot{x} = a\ddot{\theta}$.   Joining this to (3) we have

$$\ddot{x} = \frac{a^2}{a^2 + k^2} g \sin a \dotfill (6).$$

Since the sphere is homogeneous, $k^2 = \frac{2}{5} a^2$, and we have $\ddot{x} = \frac{5}{7} g \sin a$.

If the sphere had been sliding down a *smooth* plane, the equation of motion would have been $\ddot{x} = g \sin \alpha$, so that two-sevenths of gravity is used in turning the sphere, and five-sevenths in urging the sphere downwards.

Supposing the sphere to start from rest we have clearly

$$x = \frac{1}{2} \cdot \frac{5}{7} g \sin \alpha \cdot t^2,$$

and the whole motion is determined.

In the above solutions only a few of the equations of motion have been used, and if the motion only had been required it would have been unnecessary to write down any equations except (3) and (4). If the reactions also are required, we must use the remaining equations. From (1) we have

$$F = \frac{2}{7} mg \sin \alpha.$$

From (2) and (5) we have $R = mg \cos \alpha.$

It is usual to delay the substitution of the value of k^2 in the equations until the end of the investigation, for this value is often very complicated. But there is another advantage. It serves as a verification of the signs in our original equations, for if equation (6) had been

$$\ddot{x} = \frac{a^2}{a^2 - k^2} g \sin \alpha,$$

we should have expected some error to exist in the solution. For it seems clear that the acceleration could not be made infinite by any alteration of the internal structure of the sphere.

Ex. If the plane were imperfectly rough with a coefficient of friction μ less than $\frac{2}{7} \tan \alpha$, show that the angular velocity of the sphere after a time t from rest would be $\dfrac{5\mu}{2} \dfrac{g \cos \alpha}{a} t.$

145. *A homogeneous sphere rolls down another perfectly rough fixed sphere. Find the motion.*

Let a and b be the radii of the moving and fixed spheres, respectively, C and O the two centres. Let OB be the vertical radius of the fixed sphere, and $\phi = \angle BOC$. Let F and R be the friction and the normal reaction at N. Then, resolving tangentially and normally to the path of C, we have

$$m(a+b)\ddot{\phi} = mg \sin \phi - F \dotfill (1),$$
$$m(a+b)\dot{\phi}^2 = mg \cos \phi - R \dotfill (2).$$

Let A be that point of the moving sphere which originally coincided with B. Then if θ be the angle which any fixed line, as CA, in the body makes with any fixed line in space, as the vertical, we have by taking moments about C

$$mk^2\ddot{\theta} = Fa \dotfill (3).$$

It should be observed that we cannot take θ as the angle ACO because, though CA is fixed in the body, CO is not fixed in space.

The geometrical equation is clearly $a(\theta - \phi) = b\phi$ $\dotfill (4).$

No other is wanted, since in forming equations (1) and (2) the constancy of the distance CO has been already assumed.

The form of equation (4) shows that we can apply the first method. We thus obtain

$$F = \frac{k^2}{k^2 + a^2} mg \sin \phi,$$

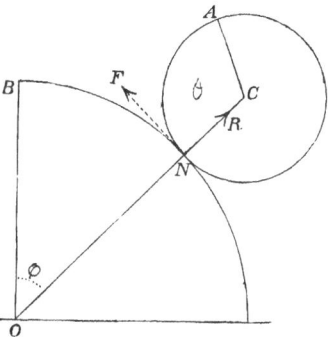

and we are finally led to the equation

$$(a + b) \ddot{\phi} = \frac{5}{7} g \sin \phi.$$

By multiplying by $2\dot{\phi}$ and integrating we get after determining the constant,

$$\dot{\phi}^2 = \frac{10}{7} \frac{g}{a + b} (1 - \cos \phi),$$

the rolling body being supposed to start from rest at a point indefinitely near B.

This result might also have been deduced from the equation of vis viva. The vis viva of the sphere is $m \{v^2 + k^2 \dot{\theta}^2\}$ and $v = (a + b) \dot{\phi}$. The force function by Art. 140 is mgy, if y be the vertical space descended by the centre. We thus have

$$(a + b)^2 \dot{\phi}^2 + k^2 \dot{\theta}^2 = 2g (a + b) (1 - \cos \phi),$$

which is easily seen to lead, by the help of (4), to the same result.

To find where the body leaves the sphere we must put $R = 0$. This gives by (2) $(a + b) \dot{\phi}^2 = g \cos \phi$; $\therefore \frac{10}{7} g (1 - \cos \phi) = g \cos \phi$; $\therefore \cos \phi = \frac{10}{17}$. It may be remarked that this result is independent of the magnitudes of the spheres.

Ex. 1. If the spheres had been smooth the upper sphere would have left the lower sphere when $\cos \phi = \frac{2}{3}$.

Ex. 2. A rod rests with one extremity on a smooth horizontal plane and the other on a smooth vertical wall at an inclination α to the horizon. If it then slips down, show that it will leave the wall when its inclination is $\sin^{-1} (\frac{2}{3} \sin \alpha)$.

Ex. 3. A beam of length a is rotating on a smooth horizontal plane about one extremity, which is fixed, under the action of no forces except the resistance of the atmosphere. Supposing the retarding effect of the resistance on a small element of length dx to be Adx (vel.)^2, then the angular velocity at the time t is given by

$$\frac{1}{\omega} - \frac{1}{\Omega} = \frac{A a^4}{4Mk^2} t. \quad \text{[Queens' Coll.]}$$

Ex. 4. An inclined plane of mass M is capable of moving freely on a smooth horizontal plane. A perfectly rough sphere of mass m is placed on its inclined face and rolls down under the action of gravity. If x' be the horizontal space advanced by the inclined plane, x the part of the plane rolled over by the sphere, prove that

$$(M+m)\, x' = mx \cos \alpha, \qquad \tfrac{7}{5}x - x' \cos \alpha = \tfrac{1}{2}gt^2 \sin \alpha,$$

where α is the inclination of the plane to the horizon.

Ex. 5. Two equal perfectly rough spheres are placed in unstable equilibrium, one on the top of the other ; the lower sphere resting on a perfectly smooth table. A slight disturbance being given to the system, show that the spheres will continue to touch each other at the same points, and that, if θ be the inclination to the vertical of the straight line joining the centres,

$$(k^2 + a^2 + a^2 \sin^2\theta)\, \theta^2 = 2ga\,(1 - \cos \theta).$$

Ex. 6. Two unequal perfectly smooth spheres are placed in unstable equilibrium one on the top of the other; the lower sphere resting on a perfectly smooth table. A very slight disturbance being given to the system, show that the spheres will separate when the straight line joining the centres makes an angle ϕ with the vertical given by the equation $\dfrac{m}{M+m} \cos^3 \phi - 3 \cos \phi + 2 = 0$, where M is the mass of the lower and m of the upper sphere.

Ex. 7. A sphere of mass M and radius a is constrained to roll on a perfectly rough curve of any form and initially the velocity of its centre of gravity is V. If the initial velocity were changed to V', show that the normal reaction would be increased by $M\dfrac{V'^2 - V^2}{\rho - a}$ and that the friction would be unaltered, ρ being the radius of curvature of the curve at the point of contact.

Ex. 8. A uniform rod is placed at an inclination α to the vertical with one extremity touching a horizontal plane. If the rod start from rest show that its angular velocity ω when it becomes horizontal is given by $a\omega^2 = \tfrac{3}{2}g \cos \alpha$ whether the plane is perfectly smooth or perfectly rough. Show also that the rod will in neither case leave the plane.

Ex. 9. A straight tunnel is constructed from London to Paris. Show that a sphere starting from rest at one terminus will arrive at the other in about forty-two minutes if the tunnel is smooth, but will take about eight minutes longer if the tunnel is perfectly rough. The sphere is supposed to move solely under the action of gravity, which inside the earth is supposed to vary as the distance of the sphere from the centre of the earth. Would the time be the same from London to Vienna ?

Ex. 10. A heavy uniform chain occupies a smooth tube of small section whose medial line is a quadrant of a circle with one bounding radius vertical. If the chain start from rest show that its velocity v on emerging from the tube is given by

$$v^2 = \frac{ga}{6\pi}\,(3\pi^2 + 16).$$

Ex. 11. A heavy chain occupies a smooth tube of small section whose form is the semi-cardioid $r = a\,(1 + \cos \theta)$ bounded by the axis. The axis is horizontal, one end of the string is at the apse and its length is $2a$, prove that the velocity of emergence is given by $10v^2 = ag\,(52 - 9\sqrt{2})$.

Ex. 12. A perfectly rough cylindrical grindstone of radius a is rotating with uniform acceleration about its axis which is horizontal. Show that, if a sphere in contact with its edge can remain with its centre at rest, the angular acceleration of the grindstone must not exceed $5g/2a$.   [Coll. Exam. 1877.]

146. *A rod OA can turn about a hinge at O, while the end A rests on a smooth wedge which can slide along a smooth horizontal plane through O. Find the motion.*

Let $a=$ the angle of the wedge, $M=$ its mass and $x=OC$. Let $l=$ the length of the beam, $m=$ its mass and $\theta=AOC$. Let $R=$ the reaction at A. Then we have *the dynamical equations,*

$$M\ddot{x} = R\sin a \dotfill (1),$$

$$mk^2\ddot{\theta} = Rl \,.\, \cos\,(a-\theta) - mg\,\frac{l}{2}\cos\theta \dotfill (2),$$

and the geometrical equation,

$$x\sin a = l \,.\, \sin\,(a-\theta) \dotfill (3).$$

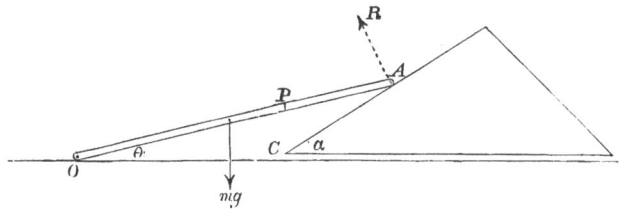

It is obvious that we must apply the second method of solution. Hence

$$2M\dot{x}\ddot{x} + 2mk^2\dot{\theta}\ddot{\theta} = -mgl\cos\theta\,\dot{\theta} + 2R\left\{\sin a\dot{x} + l\cos\,(a-\theta)\,\dot{\theta}\right\}.$$

The coefficient of R is seen to vanish by differentiating equation (3). Integrating we have

$$M\dot{x}^2 + mk^2\dot{\theta}^2 = C - mgl\sin\theta.$$

This result might have been written down at once by the principle of vis viva. For the vis viva of the wedge is clearly $M\dot{x}^2$ and that of the rod $mk^2\dot{\theta}^2$. If y be the altitude above OC of the centre of gravity of the rod OA, twice the force function is $C - 2mgy$ by Art. 140. Since $y = \frac{1}{2}l\sin\theta$, this reduces to the result already written down.

Substituting for x from (3) we have

$$\left\{M\,\frac{l^2}{\sin^2 a}\,\cos^2\,(a-\theta) + mk^2\right\}\,\dot{\theta}^2 = C - mgl\sin\theta \dotfill (4).$$

If the beam start from rest when $\theta = \beta$, then $C = mgl\sin\beta$.

This equation cannot be integrated any further. We cannot therefore find θ in terms of t. But the angular velocity of the beam, and therefore the velocity of the wedge, is given by the above equation.

147. *Two rods AB, BC are hinged together at B and can slide freely on a smooth horizontal plane. The extremity A of the rod AB is attached by another hinge to a fixed point on the table. An elastic string AC, whose unstretched length is equal to AB or BC, joins A to the extremity C of the rod BC. Initially the two rods and the string form an equilateral triangle and the system is started with an angular velocity Ω round A. Find the greatest length of the elastic string during the motion. Find also the angular velocities of the rods when they are at right angles, and the least value of Ω that this position may be possible.*

The following solution may appear at first sight rather long. The object is to illustrate the different methods of using the principles of angular momentum and vis viva. They are here minutely explained as this is the first example of the kind. It is however usual in practice to write down the equations (1) and (2) derived from these principles with but little if any explanation.

Let $2a$ be the length of either rod, mk^2 its moment of inertia about its centre of gravity, so that $k^2 = \frac{1}{3}a^2$. Let D and E be the middle points of the rods, and let (r, θ) be the polar co-ordinates of E referred to A as origin.

The only forces on the system are the reaction of the hinge at A and the tension of the elastic string AC. If we search for any direction in which the sum of the resolved parts of these vanishes, we can find none, since the direction of the reaction is at present unknown. But since the lines of action of both forces pass through A, their moments about A vanish, and therefore, by Art. 133, the angular momentum about A is constant throughout the motion and equal to its initial value. Let ω, ω' be the angular velocities of AB, BC at any instant t. The

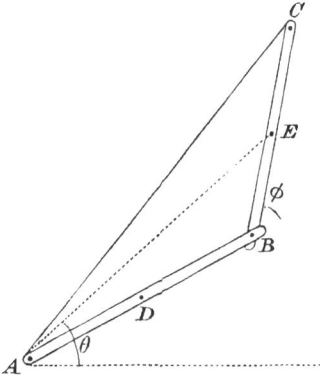

angular momentum of BC about A is by Art. 134 $m(r^2\dot\theta + k^2\omega')$. The angular momentum of AB is by the same article $m(k^2 + a^2)\omega$, since AB is turning about A as a fixed point. The initial values of these are respectively $m(3a^2 + k^2)\Omega$, and $m(k^2 + a^2)\Omega$, since ω, ω' and $\dot\theta$ are each initially equal to Ω and r is initially equal to the perpendicular from A on the opposite side of the equilateral triangle formed by the system. Hence

$$m(k^2 + a^2)\omega + mk^2\omega' + mr^2\dot\theta = m(2k^2 + 4a^2)\Omega \dots\dots\dots\dots(1).$$

We may obtain another equation by the use of the principle of vis viva. The vis viva of the rod BC is by Art. 139 $m\{\dot r^2 + r^2\dot\theta^2 + k^2\omega'^2\}$. The vis viva of AB is by the same article $m(k^2 + a^2)\omega^2$ since it is turning round A as a fixed point. The initial values of these are respectively $m(3a^2 + k^2)\Omega^2$ and $m(k^2 + a^2)\Omega^2$. If T be the tension of the string, ρ its length at time t, the force function of the tension is $\int_{2a}^{\rho}(-T)\,d\rho$. According to the rule given in statics to calculate virtual moments, the minus sign is given to the tension because it acts so as to diminish ρ; and the limits are $2a$ to ρ because the string has stretched from its initial length $2a$ to ρ. By Hooke's law $T = E\dfrac{\rho - 2a}{2a}$, so that, by integration, the force function $= -E\dfrac{(\rho - 2a)^2}{4a}$.

The reaction at A does not appear, by Art. 141. The equation of vis viva is therefore

$$m\,(k^2+a^2)\,\omega^2 + m\,\{\dot{r}^2 + r^2\dot{\theta}^2 + k^2\omega'^2\} = m\,(2k^2+4a^2)\,\Omega^2 - E\,\frac{(\rho-2a)^2}{2a} \quad\ldots\ldots(2).$$

There are only two possible independent motions of the rods. We can turn AB about A and BC about B, all motions, not compounded of these, being inconsistent with the geometrical conditions of the question. Two dynamical equations are sufficient to determine these, and we have just obtained two. All the other equations which may be wanted must be derived from geometrical considerations.

We must now express the geometrical conditions of the question. Let ϕ be the supplement of the angle ABC, then

$$r^2 = 5a^2 + 4a^2\cos\phi \quad\ldots\ldots\ldots\ldots\ldots\ldots\ldots\ldots\ldots\ldots\ldots\ldots\ldots\ldots(3).$$

Since $\dot{\phi}$ is the relative angular velocity of the rods BC, AB, $\dot{\phi}=\omega'-\omega$ $\ldots\ldots(4)$.

$$\therefore r\dot{r} = -2a^2\sin\phi\,(\omega'-\omega)\ldots\ldots\ldots\ldots\ldots\ldots\ldots\ldots\ldots\ldots(5).$$

Let ψ be the angle EAB, then $\sin\psi = \sin\phi\,\dfrac{a}{r}\ldots\ldots\ldots\ldots\ldots\ldots\ldots\ldots\ldots\ldots\ldots\ldots(6)$,

and since $\dot{\psi}=\theta-\omega$, we have

$$\cos\psi\,(\theta-\omega) = \left(\frac{a}{r}\cos\phi + \frac{2a^3}{r^3}\sin^2\phi\right)(\omega'-\omega) \quad\ldots\ldots\ldots\ldots\ldots(7).$$

Also from the triangle ABC $\qquad \rho^2 + 2a^2 = 2r^2\ldots\ldots\ldots\ldots\ldots\ldots\ldots\ldots\ldots\ldots\ldots(8).$

From these eight equations we can eliminate ω, ω', r, $\dot{r}$, ρ, ψ and θ. We shall then have a differential equation of the first order to solve, containing ϕ and $\dot{\phi}$.

It is required to find the greatest length of the elastic string during the motion. At the moment when ρ is a maximum $\dot{\rho}=0$ and the whole system is therefore moving as if it were a rigid body. We therefore have for a single moment ω, ω' and θ all equal to each other and $\dot{r}=0$. The two first equations become, when we have substituted for k^2 its value $\dfrac{a^2}{3}$,

$$\left.\begin{aligned}(5a^2+3r^2)\,\omega &= 14a^2\Omega \\ (5a^2+3r^2)\,\omega^2 &= 14a^2\Omega^2 - \frac{3E}{2am}\,(\rho-2a)^2\end{aligned}\right\}.$$

Eliminating ω and substituting for r from (8) we have the cubic

$$(3\rho^2+16a^2)\,(\rho-2a) = \frac{28m\Omega^2a^3}{E}\,(\rho+2a),$$

which has one positive root greater than $2a$.

It is also required to find the motion at the instant when the rods are at right angles. At this moment $\phi = \dfrac{\pi}{2}$, and hence by (3) $r = a\sqrt{5}$, by (5) $\dot{r} = -\dfrac{2}{\sqrt{5}}a(\omega'-\omega)$, by (7) $\theta = \tfrac{1}{5}(\omega'+4\omega)$. Substituting in equations (1) and (2) we get

$$\left.\begin{aligned}4\omega+\omega' &= \frac{7}{2}\,\Omega \\ 4\omega^2+\omega'^2 &= \frac{7}{2}\,\Omega^2 - \frac{E}{ma}\,\frac{3\,(\sqrt{2}-1)^2}{2}\end{aligned}\right\}.$$

From these two equations we may easily find ω and ω'. It is easily seen that the values of ω, ω' will not be real unless $\Omega^2 > \dfrac{10}{7}\,\dfrac{E}{ma}\,(\sqrt{2}-1)^2.$

R. D. 9

We may often save ourselves the trouble of some elimination *if we form the equations derived from the principles of angular momentum and vis viva in a slightly different manner.* The rod BC is turning round B with an angular velocity ω', while at the same time B is moving perpendicularly to AB with a velocity $2a\omega$. The velocity of E is therefore the resultant of $a\omega'$ perpendicular to BC and $2a\omega$ perpendicular to AB, both velocities, of course, being applied to the point E. When we wish our results to be expressed in terms of ω, ω' we may use these velocities to express the motion of E instead of the polar co-ordinates (r, θ).

Thus in applying the principle of angular momentum, we have to take the moment of the velocity of E about A. Since the velocity $2a\omega$ is perpendicular to AB, the length of the perpendicular from A on its direction is AB together with the projection of BE on AB, which is $2a + a \cos \phi$. Since the velocity $a\omega'$ is perpendicular to BE, the length of the perpendicular from A on its line of action is BE together with the projection of AB on BE, which is $a + 2a \cos \phi$. Hence the angular momentum of the rod BC about A is, by Art. 134,

$$mk^2\omega' + 2ma\omega (2a + a \cos \phi) + ma\omega' (a + 2a \cos \phi).$$

The principle of angular momentum for the two rods gives therefore

$$m (k^2 + 5a^2 + 2a^2 \cos \phi) \omega + m (k^2 + a^2 + 2a^2 \cos \phi) \omega' = m (2k^2 + 4a^2) \Omega.$$

The right-hand side of this equation, being the initial value of the angular momentum, is derived from the left-hand side by putting $\cos \phi = -\frac{1}{2}$ and $\omega = \omega' = \Omega$.

In applying the principle of vis viva, we require the velocity of E. Regarding it as the resultant of $2a\omega$ and $a\omega'$ we see that, if v be its value,

$$v^2 = (2a\omega)^2 + (a\omega')^2 + 2 . 2a\omega . a\omega' \cos \phi.$$

The initial value being found, as before, by putting $\cos \phi = -\frac{1}{2}$, $\omega = \omega' = \Omega$, the principle of vis viva gives, by Art. 141,

$$m (k^2 + 5a^2) \omega^2 + m (k^2 + a^2) \omega'^2 + 4ma^2 \omega\omega' \cos \phi = m (2k^2 + 4a^2) \Omega^2 - E \frac{(\rho - 2a)^2}{2a},$$

the force function being found in the same manner as before. If we join to this equation (4) given above, and substitute $\rho = 4a \cos \dfrac{\phi}{2}$, we have just three equations to find ω, ω', and ϕ. If these quantities are all that are required, as in the two cases considered above, this form of solution has the advantage of brevity. When ρ is a maximum we put $\omega = \omega'$, when the rods are at right angles we put $\cos \phi = 0$. The equations then lead to the results already given.

Ex. 1. Two rods AB, BC of equal mass are hinged together at B and the extremity A is fixed. They fall from any initial position under the action of gravity. If their lengths are respectively $2a$ and $2b$ and their inclinations to the horizon at any time θ, ϕ prove that

$$\frac{d}{dt} \{16a^2\dot{\theta} + 4b^2\dot{\phi} + 6ab \cos (\phi - \theta) (\dot{\theta} + \dot{\phi})\} = 9ag \cos \theta + 3bg \cos \phi$$

$$8a^2\dot{\theta}^2 + 2b^2\dot{\phi}^2 + 6ab \cos (\phi - \theta) \dot{\theta}\dot{\phi} = 9ag \sin \theta + 3bg \sin \phi + C.$$

The first equation is obtained by taking angular momentum about A for *both* bodies in the manner explained in Art. 78. The second is the equation of vis viva.

Ex. 2. A uniform rod of length $2a$ has a particle attached to it by a string b; the rod and string are placed in a straight line on a smooth table, and the particle is projected with a velocity V perpendicular to the string, prove that the greatest angle ϕ that the string can make with the rod is given by $\sin^2 \frac{1}{2}\phi = a (1 + n)/12b$, where n is the ratio of the mass of the rod to that of the particle. Prove also that the angular velocity then is $V/(a+b)$. [Coll. Ex.]

The common centre of gravity G moves in a straight line with uniform velocity. The vis viva and the angular momentum about G are each constant.

Ex. 3. Three equal uniform bars, formed of such material that any particle repels any other with intensity proportional to the product of their masses and directly as the distance between them, are loosely jointed at their ends so as to form an equilateral triangle. If one of the connexions at the angles be severed, prove that the angular velocity of either of the outer bars when all three are in a straight line is $\sqrt{(8\cdot4)}$ times their angular velocity when they are at right angles to the middle bar. [*Math. Tripos.*]

148. *The bob of a heavy pendulum contains a spherical cavity which is filled with water. It is required to determine the motion.*

Let O be the point of suspension, G the centre of gravity of the solid part of the pendulum, MK^2 its moment of inertia about O, and let $OG=h$. Let C be the centre of the sphere of water, a its radius and $OC=c$. Let m be the mass of the water.

If we suppose the water to be a perfect fluid, the action between it and the case must, by the definition of a fluid, be normal to the spherical boundary. There will therefore be no force tending to turn the fluid round its centre of gravity. As the pendulum oscillates to and fro the centre of the sphere will partake of its motion, but there will be no rotation of the water.

The effective forces of the water are by Art. 131 equivalent to the effective force of the whole mass collected at its centre of gravity together with a couple $mk^2\dot\omega$, where ω is the angular velocity of the water, and mk^2 its moment of inertia about a diameter. But ω has just been proved zero, hence this couple may be omitted. It follows that in all problems of this kind where the body does not turn, or turns with uniform angular velocity, we may collect the body into a single particle placed at its centre of gravity.

The pendulum and the collected fluid now form a rigid body turning about a fixed axis, hence if θ be the angle made by CO a fixed line in the body with the vertical, the equation of motion by Art. 89 is

$$(MK^2+mc^2)\,\ddot\theta+(Mh+mc)\,g\sin\theta=0,$$

where, in finding the moment of gravity, O, G and C have been supposed to lie in a straight line. The length L' of the simple equivalent pendulum is, by Art. 92,

$$L'=\frac{MK^2+mc^2}{Mh+mc}.$$

Let mk^2 be the moment of inertia of the sphere of water about a diameter. Then, if the water were to become solid and to be rigidly connected with the case, the length L of the simple equivalent pendulum would be, by similar reasoning,

$$L=\frac{MK^2+m\,(c^2+k^2)}{Mh+mc}.$$

It appears that $L'<L$, so that the time of oscillation is less than when the whole is solid.

149. **Characteristics of a body.** If we refer to the equations of motion of a body given in Art. 135, we see that the motion depends on (1) the mass of the body, (2) the position

of the centre of gravity, (3) the external forces, (4) the moments of inertia of the body about straight lines through the centre of gravity, (5) the geometrical equations. Two bodies, however different they may really be, which have these characteristics the same, will move in the same manner, *i.e.* their centres of gravity will describe the same path, and their angular motions about their centres of gravity will be the same. It is often convenient to use this proposition to change the given body into some other whose motion can be more simply found.

For example, if a sphere have an eccentric spherical cavity filled with fluid of density the same as that of the solid sphere, the motion of the sphere is independent of the position of the cavity, so that, if it be more convenient, we may put the cavity at the centre. To prove this, we may notice that since the sphere of fluid does not rotate, or rotates with uniform angular velocity, the motion is unaltered by collecting the fluid into a particle placed at its centre. This being done, the first, second, third, and fifth characteristics are clearly independent of the position of the cavity. As for the fourth characteristic, let a be the radius of the sphere, b that of the cavity, c the distance of its centre from the centre of the sphere, then the moment of inertia of the solid part of the sphere is $\frac{4}{3}\pi a^3 \cdot \frac{2}{5}a^2 - \frac{4}{3}\pi b^3 \cdot (\frac{2}{5}b^2 + c^2)$. The moment of inertia of the fluid collected into its centre is $\frac{4}{3}\pi b^3 \cdot c^2$. When we add these together c disappears, so that the whole moment of inertia is independent of the position of the cavity.

The motion of a uniform triangular area moving under the action of gravity is another example. If we replace the area by three wires forming its perimeter but without weight, the geometrical conditions of the motion will in general be unaltered, and if we also place at the middle points of these wires three particles, each one-third of the mass of the triangle, this body will have all its characteristics the same as that of the real triangle, and may replace it in any problem.

Ex. 1. A triangular area at rest is struck by a blow perpendicular to its plane at the middle point of one side, show that the instantaneous axis bisects the other two sides; but if the blow be delivered at a corner the instantaneous axis divides in the ratio of 3 : 1 each of the sides which meet at that corner.

This is not strictly a case of motion in two dimensions, but we may deduce the results from first principles, by taking moments about a straight line which passes through the point of application of the blow and one of the equivalent particles.

Ex. 2. A triangular area ABC oscillates about one side AB as a horizontal axis under the action of gravity, show that the pressures on the fixed axis are equivalent to a vertical pressure at a point O which bisects AB, and a pressure in the plane of the triangle which bisects the distance between O and the projection of C on AB.

When a string connecting two parts of a dynamical system passed over a rough pulley, it was formerly the custom to take account of the inertia of rotation by replacing the pulley by another of the same size but without mass and loaded with a particle at its circumference. If a be the radius of the pulley, k its radius of gyration about the centre, m its mass, the mass of the particle is $\dfrac{k^2}{a^2} m$, so that for a cylindrical pulley the mass of the particle is half that of the pulley. This mass must then be added on to the other particles attached to the string. For example, if two heavy masses M, M' be connected by a string passing over a cylindrical pulley of mass m, which can turn freely about its axis, the equation of motion is

$$\left(M + M' + \frac{m}{2}\right) \frac{dv}{dt} = (M - M') g$$

where v is the velocity. Here the inertia of the pulley is taken account of by simply adding $\dfrac{m}{2}$ to the mass moved. If the pulley be moveable in space as well as free to rotate, its inertia of translation is as usual taken account of by collecting the whole mass into its centre of gravity. As this representation of the inertia of rotation is not often used now, the demonstration of the above remarks, if any be needed, is left to the reader.

Ex. 3. A rod AB whose centre of gravity is at the middle point C of AB has its extremities A and B constrained to move along two straight lines Ox, Oy at right angles and is acted on by any forces. Show that the motion is the same as if the whole mass were collected into its centre of gravity and all the forces reduced in the ratio $1 + \dfrac{k^2}{a^2} : 1$, where $2a$ is the length AB and k the radius of gyration about the centre of gravity.

Ex. 4. A circular disc whose centre of gravity is in its centre rolls on a perfectly rough curve under the action of any forces, show that the motion of the centre is the same as if the curve were smooth and all the forces were reduced in the ratio $1 + \dfrac{k^2}{a^2} : 1$, where a is the radius of the disc and k its radius of gyration about the centre. But the normal pressures on the curve in the two cases are not the same. In any position of the disc they differ by $X \dfrac{k^2}{a^2 + k^2}$, where X is the force on the disc resolved along the normal to the rough curve.

On the stress at any point of a rod.

150. Suppose a rod OA to be in equilibrium under the action of any forces, it is required to determine the action across any section of the rod at P. This action may be conceived to be the resultant of the tensions positive or negative of the innumerable fibres which form the material of the rod. We know by statics

that these may be compounded into a single force R acting at any point Q which we may choose and a couple G. Since each portion of the rod is in equilibrium, these must also balance all the external forces which act on the rod on *one side* of the section at P. If the section be indefinitely small it is usual to take Q in the plane of the section, and these two, the force R and the couple G, will together measure the *stress* at the section.

If the rod be bent by the action of the forces, the fibres on one side will all be stretched and on the other compressed. The rod will begin to break as soon as these fibres have been sufficiently stretched or compressed. Let us compare the tendencies of the force R and the couple G to break the rod. Let A be the area of the section of the rod, then a force F pulling the rod will cause a resultant force $R = F$, and will produce a tension in the fibres which when referred to a unit of area is equal to $\dfrac{F}{A}$. The same force F acting on the rod at a distance p from P will cause a couple $G = Fp$, which must be balanced by the couple formed by the tensions. Let $2a$ be the mean breadth of the rod, then the mean tension produced by G referred to a unit of area is of the order $\dfrac{F}{A} \cdot \dfrac{p}{a}$. Now if the section of the rod be very small $\dfrac{p}{a}$ will be large. It appears therefore that the couple, when it exists, will generally have much more effect in breaking the rod than the force. This couple is therefore often taken to measure the whole effect of the forces to break the rod. The "*tendency to break*" at any point P of a rod OA of very small section is measured by the moment about P of all the forces which act on either of the segments OP, PA of the rod.

The resolved part of the force R perpendicular to the rod is called the *shear*. This is therefore equal to all the forces which act on either of the segments OP, PA, resolved perpendicular to the rod.

If the rod be in motion the same reasoning will, by D'Alembert's principle, be applicable; provided that we include the reversed effective forces among the forces which act on the rod.

In most cases the rod will be so little bent that in finding the moment of the impressed forces we may neglect the effects of curvature.

If the section of the rod be not very small, this measure of the "tendency to break" becomes inapplicable. It then becomes necessary to consider both the force and the couple. The case does not come within the limits of the present treatise, and the reader is referred to works on elastic solids.

In the case of a string the couple vanishes and the force acts along a tangent to the string. The stress at any point is therefore simply measured by the tension.

151. *A rod OA, of length* 2a, *and mass* m, *which can turn freely about one extremity* O, *falls in a vertical plane under the action of gravity. Find the "tendency to break" at any point* P.

Let du be any element of the rod distant u from P and on the same side of P as the end A of the rod, and let $OP = x$. Let θ be the angle the rod makes with the vertical at the time t. The effective forces on du are

$$m\frac{du}{2a}(x+u)\frac{d^2\theta}{dt^2} \quad \text{and} \quad -m\frac{du}{2a}(x+u)\left(\frac{d\theta}{dt}\right)^2$$

respectively perpendicular to and along the rod. The impressed force is $m\dfrac{du}{2a}g$ acting vertically downwards. The effective forces being reversed the tendency to break at P is equal to the moment about P of all the forces which act on the part PA of the rod. If this be called L, we have

$$L = \int m\frac{du}{2a}gu\sin\theta + \int m\frac{du}{2a}(x+u)u\frac{d^2\theta}{dt^2},$$

the limits being from $u=0$ to $u=2a-x$. Also, taking moments about O, the equation of motion is

$$m\frac{4a^2}{3}\frac{d^2\theta}{dt^2} = -mga\sin\theta.$$

Hence we easily find
$$L = -\frac{mg\sin\theta}{16a^2}x(2a-x)^2.$$

The meaning of the minus sign is that the forces tend to bend PA round P in the opposite direction to that in which θ has been measured.

To find where the rod, supposed equally strong throughout, is most likely to break, we must make L a maximum. This gives $\dfrac{dL}{dx}=0$ and therefore $x=\dfrac{2a}{3}$. The point required is at a distance from the fixed end equal to one-third of the length of the rod. Its position, it should be noticed, is independent of the initial conditions.

To find the shear at P we must resolve perpendicularly to the rod. If the result be called Y, we have

$$Y = \int m\frac{du}{2a}g\sin\theta + \int m\frac{du}{2a}(x+u)\frac{d^2\theta}{dt^2},$$

the limits being the same as before. This gives

$$Y = \frac{mg\sin\theta}{16a^2}(2a-x)(2a-3x),$$

which vanishes when the tendency to break is a maximum, and is a maximum at a distance from the fixed end equal to two-thirds of the length of the rod.

To find the tension at P we must resolve along the rod. If the result be called X, and be taken positive in the direction OA, we have

$$X = -\int m\frac{du}{2a}g\cos\theta - \int m\frac{du}{2a}(x+u)\left(\frac{d\theta}{dt}\right)^2.$$

If the rod start from rest at an inclination α to the vertical, we find, by integrating the equation of motion, $\left(\dfrac{d\theta}{dt}\right)^2 = \dfrac{3g}{2a}(\cos\theta - \cos\alpha)$. Hence

$$X = \frac{mg}{8a^2}(2a - x)\{-4a\cos\theta + 3(\cos\alpha - \cos\theta)(2a + x)\}.$$

From these equations we may deduce the following results. (1) The magnitudes of the stress couple and the shear are independent of the initial conditions. (2) The magnitude of either the couple or the shear at any given point of the rod varies as the sine of the inclination of the rod to the vertical. (3) The ratio of the magnitudes of the stress couples at any two given points of the rod is always the same, and the same proposition is also true of the shears. (4) The tension depends on the initial conditions, and, unless the rod starts from rest in the horizontal position, the ratio of the tensions at any two given points varies with the position of the rod.

152. *A rigid hoop completely cracked at one point rolls on a perfectly rough horizontal plane and is acted on by no forces but gravity. Prove that the wrench couple at the point of the hoop most remote from the crack will be a maximum whenever, the crack being lower than the centre, the inclination of the diameter through the crack to the horizon is* $\tan^{-1}2/\pi$. [Math. Tripos, 1864.]

Let ω be the angular velocity of the hoop, a its radius. The velocity of any point P of the hoop is the resultant of a velocity $a\omega$ parallel to the horizontal plane and an equal velocity $a\omega$ along a tangent to the hoop. The first is constant in direction and magnitude and therefore gives nothing to the acceleration of P. The latter is constant in magnitude but variable in direction and gives $a\omega^2$ as the acceleration, which is directed along a radius of the hoop. Let A be the cracked point, B the other end of the diameter, C the centre, θ the inclination of ACB to the horizon. Let PP' be any element on the upper half of the circle, $BCP = \phi$. Then the wrench couple, or tendency to break, at B is proportional to

$$\int_0^\pi [-a\omega^2 a\sin\phi + g\{a\cos\theta - a\cos(\phi + \theta)\}] \, ad\phi = -2a^3\omega^2 + ga^2(\pi\cos\theta + 2\sin\theta).$$

This is a maximum when $\tan\theta = 2/\pi$.

Ex. 1. A semicircular wire AB of radius a is rotating on a smooth horizontal plane about one extremity A with a constant angular velocity ω. If $a\phi$ be the arc between the fixed point A and the point where the tendency to break is greatest, prove that $\tan\phi = \pi - \phi$. If the extremity B be suddenly fixed and the extremity A let go, prove that the tendency to break is greatest at a point P where

$$\tfrac{1}{2}\tan PBA = PBA.$$

Ex. 2. Two of the angles of a heavy square lamina, a side of which is a, are connected with two points equally distant from the centre of a rod of length $2a$, so that the square can rotate about the rod. The weight of the square is equal to the weight of the rod, and the rod when supported by its extremities in a horizontal position is on the point of breaking. The rod is then held by its extremities in a vertical position, and an angular velocity ω is impressed on the square. Show that the rod will break if $a\omega^2 > 3g$. [Coll. Exam.]

Ex. 3. A wire in the form of the portion of the curve $r = a(1 + \cos\theta)$ cut off by the initial line rotates about the origin with angular velocity ω. Prove that the tendency to break at the point $\theta = \dfrac{\pi}{2}$ is measured by $m\,\dfrac{12\sqrt{2}}{5}\,\omega^2 a^3$. [St John's Coll.]

Ex. 4. A rod OA whose density varies in any manner is swung as a pendulum about a horizontal axis through O. Prove that if the rod break it will be at a point P determined by the condition that the centre of gravity of PA is the centre of oscillation of the pendulum. [Math. Tripos, 1880.]

On Friction between Imperfectly Rough Bodies.

153. **Components of a Reaction.** When one body rolls on another under pressure, the two bodies yield slightly, and are therefore in contact along a small area. At every point of this area there is a mutual action between the bodies. The elements just behind the geometrical point of contact are on the point of separation and may tend to adhere to each other, those in front may tend to resist compression. The whole of the actions across the elements are equivalent to (1) a component R, normal to the common tangent plane, and usually called the *reaction;* (2) a component F in the tangent plane usually called the *friction;* (3) a couple L about an axis lying in the tangent plane, which we shall call the *couple of rolling friction;* (4) if the bodies have any relative angular velocity about their common normal, a couple N about this normal as axis which may be called the *couple of twisting friction.*

The two couples are found by experiment to be in most cases very small and are generally neglected. But in certain cases where the friction forces are also small it may be necessary to take account of them. We shall therefore consider first the laws which relate to the friction forces, as being the most important, and afterwards those which relate to the couples.

154. **Laws of Friction.** In order to determine the laws of friction forces we must make experiments on some simple cases of equilibrium and motion. Suppose then a symmetrical body to be placed on a rough horizontal table and acted on by a force so placed that every point of the body is urged to move or does move parallel to its direction. It is found that if the force be less than a certain amount the body does not move. The first law of friction is therefore that the friction acts in such a direction and has such a magnitude as to be just sufficient to prevent sliding.

Next, let the force be gradually increased, it is found by experiment that no more than a certain amount of friction can be called into play, and that when more is required to keep the body from sliding, sliding begins. The second law of friction asserts the existence of this limit to the amount of friction which can be called into play. Its value is called the *limiting friction.*

The third law of friction found by experiment is that the magnitude of the limiting friction bears a ratio to the normal pressure which is very nearly constant for the same two bodies in contact, but is changed when either body is replaced by another of different material. This ratio is called the *coefficient of friction* of the materials of the two bodies. Its constancy is generally assumed by mathematicians.

Though all experimenters have not entirely agreed as to the absolute constancy of the coefficient of friction, yet it has been found generally that, if the relative motion of the two bodies be the same at all points of the area of contact, the *coefficient of friction is nearly independent of the extent of the area of contact and of the relative velocity.*

155. Coulomb has pointed out a distinction which exists between statical friction and dynamical friction. The friction which must be overcome to *set a body in motion* relatively to another is greater than the friction between the same bodies *when in motion* under the same pressure. He found also that if the bodies remained in contact for some time under pressure in a position of equilibrium, the friction which had to be overcome was greater than if the bodies were merely placed in contact and immediately started from rest under the same pressure. In some bodies the difference between the statical and the dynamical friction was found to be very slight, in others it was considerable*. The experiments of Morin in general confirmed its existence. According to some experiments of Fleeming Jenkin and J. A. Ewing, described in the *Phil. Trans.* for 1877, the transition from statical to dynamical friction was not abrupt. By means of an apparatus which differed essentially from any previously employed they were able to make definite measurements of the friction between surfaces whose relative velocity varied from about one hundredth of a foot per second to about one five-thousandth of a foot per second. Between the limits of these evanescent velocities the coefficient of friction was found to be decreasing gradually from its statical to its dynamical value as the velocity increased.

The experiments of Coulomb and Morin were made with bodies moving at moderate velocities, but some experiments have been lately made by Capt. Douglas Galton on the friction between cast-iron brake blocks and the steel tyres of wheels of engines moving with great velocities. These velocities varied from seven feet to eighty-eight feet per second, i.e. from five to sixty miles per hour. Two results followed from his experiments: (1) the coefficient of friction was very much less for higher than for lower velocities, (2) the coefficient of friction became smaller after the wheels had

---

* The results of Coulomb's experiments are given in his *Théorie des machines simples, Mémoires des Savants étrangers,* tome x. This paper gained the Prize of the *Académie des Sciences* in 1781 and was published separately in Paris, 1809.

been in motion for a few seconds. See the *Report of the British Association for the meeting in Dublin*, 1878. The reader will find an account of some experiments on *rolling friction* by Prof. Osborne Reynolds in the *Phil. Trans.* for 1876.

156. When bodies are said to be *perfectly rough* it is usually meant that they are so rough that the amount of friction necessary to prevent sliding under the given circumstances can certainly be called into play. The coefficient of friction is therefore practically infinite. By the first law of friction, the amount which *is* called into play is that which is just sufficient to prevent sliding.

157. **Application of the laws of Friction.** Let us now extend the theory deduced from these experiments to the case in which a body moves or is urged to move in any manner in one plane. It is a known kinematical theorem, which will be proved at the beginning of the next chapter, that such a motion may be represented by supposing the body to be turning round some instantaneous centre of rotation. Let O be the centre of rotation, then any point P of the body is moving or tends to move in a direction perpendicular to OP.

The friction at P, by the first rule just given, must also act perpendicular to OP but in the opposite direction. If P move, the amount of friction at P is limiting friction and is equal to μR, where R is the pressure at P and μ the coefficient of friction. Thus in a moving body the direction and the magnitude of the friction at every sliding point are known in terms of the co-ordinates of O and the pressure at the point.

Suppose for example that it is required to find the least couple required to move a heavy disc resting by several pins on a horizontal table, the pressures at the pins being known. By resolving in two directions and taking moments about a vertical axis we obtain three equations. From these we can find the required couple and the two co-ordinates of O.

It sometimes happens that O coincides with one of the points of support of the body. In this case the friction at *this* point of support is not limiting. It is only just sufficient in amount to prevent the point from sliding.

Ex. A heavy body rests by three pins A, B, C on a rough horizontal table, the pressures at the pins being P, Q, R. If the body be acted on by a couple so that it is just on the point of moving, show that the centre of rotation is at a point O such that the sines of the angles AOB, BOC, COA are as R, P, Q. But if the point O thus determined does not lie within the triangle ABC, the centre of rotation coincides with one of the pins.

These results follow immediately from the triangle of forces.

158. Discontinuity of Friction. The reader should particularly notice the discontinuity just mentioned. The friction at any point of support which slides is μR, where R is the normal pressure. But if the point of support does not slide, the friction is some quantity F, which is unknown, but must be less than μR. Its magnitude must be found from the equations of motion.

As this is important let us present the argument in a slightly different form by considering the case of rolling.

Suppose a body to roll on a rough surface, the friction called into play just prevents sliding, and is possibly variable in magnitude and direction. By writing down and solving the equations of motion we can find the ratio of the friction F to the normal pressure R. If this ratio be always less than the coefficient μ of friction, enough friction can always be called into play to make the body roll on the rough surface. In this case we have obtained the true motion. But if at any instant the ratio $\dfrac{F}{R}$ thus found becomes greater than the coefficient of friction, the point of contact begins to slide at that instant. In this case the equations do not represent the true motion. To correct them we must replace the unknown friction F by μR, and remove the geometrical equation which expresses the fact that there is no slipping between the bodies. The equations must now be again solved on this new supposition. It is of course possible that a second change may take place. If at any instant the velocities of the points of contact become equal to each other, all the possible friction may not be called into play. At that instant the friction ceases to be equal to μR and becomes again unknown in magnitude and direction.

159. *Discontinuity may also arise in other ways.* When, for example, one body is sliding over another, the friction is opposite to the direction of relative motion, and numerically equal to the normal reaction multiplied by the coefficient of friction. If then, during the course of the motion the direction of the normal reaction should change sign, while the direction of motion remains unaltered, or if the direction of motion should change sign while the normal reaction remains unaltered, the sign of the coefficient of friction must be changed. This may modify the dynamical equations and alter the subsequent solution. The same cause of discontinuity operates when a body moves in a resisting medium, the law of resistance being an even function of the velocity, *i.e.* any function which does not change sign when the direction of motion is changed.

160. Indeterminate Motion. In some cases the motion may be rendered indeterminate by the introduction of friction.

Thus we have seen in Art. 112 that, when a body swings on two hinges, the pressures on the hinges resolved in the direction of the straight line joining them cannot be found. The sum of these components can be found, but not either of them. But there is no indeterminateness in the motion. If however the hinges were imperfectly rough, there would be two friction couples, one at each hinge, acting on the body, their common axis being the straight line joining the hinges. The magnitude of each would be equal to the pressure multiplied by a constant depending on the roughness of the hinge. If the hinges were unequally rough, the magnitude of the resultant couple would depend on the distribution of the pressure on the two hinges. In such a case the motion of the body would be indeterminate.

161. **Examples of Friction.** *A homogeneous sphere is placed at rest on a rough inclined plane, the coefficient of friction being μ, determine whether the sphere will slide or roll.*

Let F be the friction required to make the sphere roll. The problem then becomes the same as that discussed in Art. 144. We have, therefore, $F = \frac{2}{7} R \tan \alpha$, where α is the inclination of the plane to the horizon.

If then $\frac{2}{7} \tan \alpha$ be not greater than μ, the solution given in the article referred to is the correct one. But if $\mu < \frac{2}{7} \tan \alpha$ the sphere begins to slide on the inclined plane. The subsequent motion is given by the equations

$$\left. \begin{aligned} m\ddot{x} &= mg \sin \alpha - \mu R \\ 0 &= -mg \cos \alpha + R \\ ma\ddot{x} + mk^2 \ddot{\theta} &= mga \sin \alpha \end{aligned} \right\}$$

whence we have, remembering that $k^2 = \frac{2}{5} a^2$,

$$\ddot{x} = g (\sin \alpha - \mu \cos \alpha), \qquad a\ddot{\theta} = \frac{5}{2} \mu g \cos \alpha.$$

Since the sphere starts from rest, we have by integration

$$x = \frac{1}{2} g t^2 (\sin \alpha - \mu \cos \alpha), \qquad \theta = \frac{5}{4} \mu \frac{g}{a} t^2 \cos \alpha.$$

The velocity of the point of the sphere in contact with the plane is

$$\dot{x} - a\dot{\theta} = gt (\sin \alpha - \tfrac{7}{2} \mu \cos \alpha).$$

But since, by hypothesis, μ is less than $\frac{2}{7} \tan \alpha$, this velocity can never vanish. The friction therefore will never change to rolling friction. The motion has thus been completely determined.

162. *A homogeneous sphere is rotating about a horizontal diameter, and is gently placed on a rough horizontal plane, the coefficient of friction being μ. Determine the subsequent motion.*

Since the velocity of the point of contact with the horizontal plane is not zero, the sphere evidently begins to slide, and the motion of its centre is along a straight line perpendicular to the initial axis of rotation. Let this straight line be taken as the axis of x, and let θ be the angle between the vertical and that radius of the sphere which was initially vertical. Let a be the radius of the sphere, mk^2 its moment of inertia about a diameter, and Ω the initial angular velocity. Let R be the normal reaction of the plane. Then the equations of motion are clearly

$$\left. \begin{aligned} m\ddot{x} &= \mu R \\ 0 &= mg - R \\ mk^2\ddot{\theta} &= -\mu Ra \end{aligned} \right\} \dots\dots\dots\dots\dots\dots\dots\dots\dots\dots(1),$$

whence we have $\ddot{x} = \mu g, \quad a\ddot{\theta} = -\tfrac{5}{2}\mu g$ ....(2).

Integrating, and remembering that the initial value of $\dot{\theta}$ is Ω, we have

$$x = \tfrac{1}{2}\mu g t^2, \quad \theta = \Omega t - \tfrac{5}{4}\mu \frac{g}{a} t^2 \dots\dots\dots\dots\dots\dots\dots(3).$$

But it is evident that these equations cannot represent the whole motion, for they make $\dot{x}$, the velocity of the centre of the sphere, increase continually, a result quite contrary to experience. The velocity of the point of the sphere in contact with the plane is

$$\dot{x} - a\dot{\theta} = -a\Omega + \tfrac{7}{2}\mu g t.$$

This vanishes at a time $t_1 = \tfrac{2}{7} \dfrac{a\Omega}{\mu g} \dots\dots\dots\dots\dots\dots\dots\dots\dots\dots\dots\dots(4).$

At this instant the friction suddenly changes its character. It now becomes of magnitude only sufficient to keep the point of contact of the sphere at rest. Let F be the friction required to effect this. The equations of motion will then be

$$\left. \begin{aligned} m\ddot{x} &= F \\ 0 &= mg - R \\ mk^2\ddot{\theta} &= -Fa \end{aligned} \right\} \dots\dots\dots\dots\dots\dots\dots\dots\dots\dots(5),$$

and the geometrical equation will be $x = a\theta$.

Differentiating this twice, and substituting from the dynamical equations, we get $F(a^2 + k^2) = 0$, and therefore $F = 0$. That is, no friction is required to keep the point of contact of the sphere at rest, and therefore none will be called into play. The sphere will therefore move uniformly with the velocity which it had at the time t_1. Substituting the value of t_1 in the expression for $\dot{x}$ obtained from equations (3) we find that this velocity is $\tfrac{2}{7}a\Omega$. It appears therefore that the sphere will move with a uniformly increasing velocity for a time $\tfrac{2}{7}\dfrac{a\Omega}{\mu g}$ and will then move uniformly with a velocity $\tfrac{2}{7}a\Omega$. It may be remarked that this velocity is independent of μ.

If the plane be very rough, μ is very great and the time t_1 is very small. Taking the limit when μ is infinite we find that the sphere begins immediately to move with its uniform velocity.

163. In this investigation the couple of rolling friction has been neglected. Its effect is to diminish the angular velocity. The velocity of the lowest point of the sphere tends to be no longer zero, and thus a small sliding friction is required to keep that point at rest. Suppose the moment of the friction-couple to be measured by fmg, where f is a constant. Introducing this into the equations (5) the third is changed into

$$mk^2\ddot{\theta} = -Fa - fmg,$$

the others remaining unaltered. Solving these as before we find that

$$F = -\frac{afmg}{a^2 + k^2}.$$

We see from this that F is negative and retards the sphere. The effect of the couple is to call into play a friction-force which gradually reduces the sphere to rest.

As the sphere moves we may wish to determine the effect of the resistance of the air. The chief part of this resistance may be pretty accurately represented by a force $m\beta\dfrac{v^2}{a}$ acting at the centre in the direction opposite to motion, v being the velocity of the sphere and β a constant whose magnitude depends on the density of the air. Besides this there is also a small friction between the sphere and the air whose magnitude is not known so accurately. Let us suppose it to be represented by a couple whose moment is $m\gamma v^2$ where γ is a constant of small magnitude. The equations of motion can be solved without difficulty, and we find

$$\tan^{-1} v \sqrt{\frac{\beta+\gamma}{fg}} - \tan^{-1} V \sqrt{\frac{\beta+\gamma}{fg}} = -\frac{a \sqrt{(a+\beta)\, fg}}{a^2+k^2}\, t,$$

where V is the velocity of the sphere at the epoch from which t is measured.

164. Friction couples. In order to determine by experiment the magnitude of rolling friction, let a cylinder of mass M and radius r be placed on a rough horizontal plane. Let two weights whose masses are P and $P + p$ be suspended by a fine thread passing over the cylinder and hanging down through a slit in the horizontal plane. Let F be the force of friction, L the couple at the point of contact A of the cylinder with the horizontal plane. Imagine p to be at first zero, and to be gradually increased until the cylinder just moves. When the cylinder is on the point of motion, we have by resolving horizontally $F = 0$, and by taking moments $L = pgr$. Now in the experiments of Coulomb and Morin p was found to vary as the normal pressure directly, and as r inversely. When p was great enough to set the cylinder in motion, Coulomb found that its acceleration was nearly constant, whence it followed that the rolling friction was independent of the velocity. M. Morin found that it was not independent of the length of the cylinder.

The laws which govern the couple of rolling friction are similar to those which govern the force of friction. The magnitude is just sufficient to prevent rolling. But no more than a certain amount can be called into play, and this is called the *limiting rolling couple*. The moment of this couple bears a constant ratio to the magnitude of the normal pressure. This ratio is called the *coefficient of rolling friction*. It depends on the materials in contact, it is independent of the curvatures of the bodies, and, in some cases, of the angular velocity.

No experiments seem to have been made on bodies which touch at one point only and have their curvatures in different directions unequal. But, since the magnitude of the couple is independent of the curvature, it seems reasonable to assume that the axis of the rolling couple, when there is no twisting couple, is the instantaneous axis of rotation.

165.   In order to test the laws of friction let us compare the results of the following problem with experiment.

Friction of a carriage. *A carriage on n pairs of wheels is dragged on a level horizontal plane by a horizontal force* 2P *with uniform motion.   Find the magnitude of* P.

Let the radii of the wheels be respectively r_1, r_2, &c., their weights w_1, w_2, &c., and the radii of the axles ρ_1, ρ_2, &c. Let $2W$ be the whole weight of the carriage, $2Q_1$, $2Q_2$, &c. the pressures on the several axles, so that $W = \Sigma Q$. Let the pressures between the wheels and axles be R_1, R_2, &c. and the pressures on the ground R_1', R_2', &c. Let C be the common centre of any wheel and axle, B their point of contact, and A the point of contact of the wheel with the ground. Let the angle $ACB = \theta$, supposed positive when B is behind AC. Let μ be the coefficient of the force of sliding friction at B and f the coefficient of the couple of rolling friction at A. The equations of equilibrium for any wheel, found by resolving vertically and taking moments about A, are

$$R' = Q + w \dots\dots\dots\dots\dots\dots\dots\dots\dots\dots\dots\dots\dots\dots\dots(1),$$

$$\mu R (r \cos \theta - \rho) - Rr \sin \theta = fR' \dots\dots\dots\dots\dots\dots\dots\dots(2).$$

The friction force at A does not appear because we have not resolved horizontally. The equations of equilibrium of the carriage, found by resolving vertically and horizontally, are

$$R \cos \theta + \mu R \sin \theta = Q \dots\dots\dots\dots\dots\dots\dots\dots\dots\dots\dots(3),$$

$$\Sigma (R \sin \theta - \mu R \cos \theta) + P = 0 \dots\dots\dots\dots\dots\dots\dots\dots\dots(4).$$

The effective forces have been omitted because the carriage is supposed to move uniformly, so that the $M\dot{v}$ of the carriage and the $mk^2 \dot{\omega}$ of the wheel are both zero. The first three of these equations give, by eliminating R and R',

$$\frac{\mu \left(\cos \theta - \dfrac{\rho}{r} \right) - \sin \theta}{\cos \theta + \mu \sin \theta} = \frac{f}{r} \left(1 + \frac{w}{Q} \right) \dots\dots\dots\dots\dots\dots\dots(5).$$

This gives the value of θ. In most wheels $\dfrac{\rho}{r}$ and $\dfrac{w}{Q}$ are both small as well as f. In such a case $\mu \cos \theta - \sin \theta$ is a small quantity. If therefore $\mu = \tan \epsilon$ we have $\theta = \epsilon$ very nearly. The third and fourth of the equations give, by eliminating R,

$$P = \Sigma \frac{\mu \cos \theta - \sin \theta}{\mu \sin \theta + \cos \theta} Q = \Sigma \left\{ \frac{\mu}{\mu \sin \theta + \cos \theta} \frac{\rho}{r} Q + \frac{f}{r} (Q + w) \right\},$$

the latter by equation (5). If $\dfrac{\rho}{r}$ be small, it will be sufficient to substitute for θ in the first term its approximate value ϵ. This gives

$$P = \Sigma \left\{ \sin \epsilon \frac{\rho}{r} Q + f \frac{Q + w}{r} \right\} \dots\dots\dots\dots\dots\dots\dots(6).$$

Here we have neglected terms of the order $\left(\dfrac{\rho}{r} \right)^2 Q$.

If all the wheels are equal and similar we have, since $\Sigma Q = W$,

$$P = \sin \epsilon \frac{\rho}{r} W + f \frac{W + nw}{r} \dots\dots\dots\dots\dots\dots\dots\dots(7).$$

Thus the force required to drag a carriage of given weight with any constant velocity is very nearly independent of the number of wheels.

In a gig the wheels are usually larger than in a four-wheel carriage, and therefore the force of traction is usually less.   In a four-wheel carriage the two fore wheels must be small in order to pass under the carriage when turning.   This will cause the term $\sin \epsilon \frac{\rho_1}{r_1} Q_1$ in the expression for P, depending on the radius r_1 of the fore wheel, to be large.   To diminish the effect of this term, the load should be so adjusted that its centre of gravity is nearly over the axle of the large wheels, when the pressure Q_1 in the numerator will be small.

Numerous experiments were made by a French engineer, M. Morin, at Metz in the years 1837 and 1838, and afterwards at Courbevoie in 1839 and 1841, with a view to determine with the utmost exactness the force necessary to drag carriages of different kinds over ordinary roads.   These experiments were undertaken by order of the French Minister of War, and afterwards under the direction of the Minister of Public Works.   The effect of each variation was determined separately, thus the same carriage was loaded with different weights to determine the effect of pressure, and dragged on the same road in the same state of moisture.   Then, the weight being the same, wheels of different radii but of the same breadth were used, and so on.

The general result was that for carriages on equal wheels, the resistance varied as the pressure directly, and the diameter of the wheels inversely, whilst it was independent of the number of wheels.   On wet soils the resistance increased as the breadth of the tire decreased, but on solid roads the resistance was independent of the breadth of the tire.   For velocities which varied from a foot pace to a gallop, the resistance on wet soils did not increase sensibly with the velocity, but on solid roads it did increase with the velocity if there were many inequalities on the road.   As an approximate result it was found that the resistance might be expressed by a function of the form $a + bV$, where a and b were two constants depending on the nature of the road and the stiffness of the carriage, and V was the velocity.

M. Morin's analytical determination of the value of P does not altogether agree with that given here, but it so happens that this does not materially affect the comparison between theory and observation.   See his *Notions Fondamentales de Mécanique*, Paris, 1855.   It is easy to see that M. Morin's experiments tend to confirm the laws of rolling friction stated in a previous article.

166.   **Problems on Friction.**   Ex. 1.   A homogeneous sphere is projected without rotation directly up an imperfectly rough plane, the inclination of which to the horizon is a, and the coefficient of friction μ.   Show that the whole time during which the sphere ascends the plane is the same as if the plane were smooth, and that the time during which the sphere slides is to the time during which it rolls as $2 \tan a : 7\mu$.

Ex. 2.   A uniform rod is placed at rest with one end in contact with a horizontal plane whose coefficient of friction is μ.   If the inclination of the rod to the vertical is a, show that it will begin to slide if $\mu (1 + 3 \cos^2 a) < 3 \sin a \cos a$.

Ex. 3.   A homogeneous sphere rolls down an imperfectly rough fixed sphere, starting from rest at the highest point.   If the spheres separate when the straight line joining their centres makes an angle ϕ with the vertical, prove that

$$\cos \phi + 2\mu \sin \phi = A e^{2\mu\phi},$$

where A is a function of μ only.   [Coll. Exam.]

R. D.   10

Proceeding as in Art. 145, we show that R remains positive and that the sphere rolls until $2 \sin \phi / \mu = 17 \cos \phi - 10$. The sphere then slides and R changes sign when ϕ satisfies the equation given in the question.

Ex. 4. A rough cylinder of mass $2nm$ capable of motion about its horizontal axis has a particle of mass m and coefficient of friction μ placed on it vertically above the axis. The system is then slightly disturbed. Show that the particle will slip on the cylinder after it has moved through an angle θ given by

$$(n+3) \cos \theta - 2 = n \sin \theta / \mu.$$

Ex. 5. A homogeneous sphere of mass M is placed on an imperfectly rough table, the coefficient of friction of which is μ. A particle of mass m is attached to the extremity of a horizontal diameter. Show that the sphere will begin to roll or slide according as μ is greater or less than $\dfrac{5 \left(M + m \right) m}{7M^2 + 17Mm + 5m^2}$. If μ be equal to this value, show that the sphere will begin to roll.

Ex. 6. A rod AB has two small rings at its extremities which slide on two rough horizontal rods Ox, Oy at right angles. The rod is started with an angular velocity Ω when very nearly coincident with Ox. Show that, if the coefficient of friction is less than $\sqrt{2}$, the motion of the rod is given by $\theta = \dfrac{2 - \mu^2}{3\mu} \log \left(1 + \dfrac{3\mu\Omega t}{2 - \mu^2} \right)$ until $\tan \theta = \dfrac{2}{\mu}$, and that when the rod reaches Oy its angular velocity is ω, where

$$\Omega^2 e^{-\frac{6\mu}{2 - \mu^2} \tan^{-1} \frac{2}{\mu}} = \omega^2 \frac{(2 + \mu^2)(4 + \mu^2)}{(2 - \mu^2)(4 - \mu^2)}.$$

What is the motion if $\mu^2 > 2$?

167. Rigidity of Cords.

After having used to determine the laws of friction the apparatus with a fine cord described in Art. 164, Coulomb replaced the cord by a stiffer one and repeated his experiments with a view to obtain a measure of the rigidity of cords. His general result may be stated as follows. Suppose a cord $ABCD$ to pass over a pulley of radius r, touching it at B and C, and moving in the direction $ABCD$. Then the rigidity may be represented by supposing the cord to be perfectly flexible, and the tension T of the portion AB of the cord which is about to be rolled on the pulley to be increased by a quantity R. The force R measures the rigidity and is equal to $\dfrac{a + bT}{r}$, where a and b are constants depending on the nature of the cord.

It appears therefore that, in the equation of moments about the axis of the pulley, the rigidity of the cord which is being wound on the pulley is represented by a resisting couple of magnitude $a + bT$, where T is the tension of the cord which is being bent, and a, b are two constants depending on the nature of the cord. The rigidity of the cord which is being unwound will be represented by a couple whose magnitude is a similar function of the tension of that cord. But as its magnitude is very much less than the first it is generally omitted.

Besides the experiments just alluded to, Coulomb made many others on a different system. He also constructed tables of the values of a and b for ropes of different kinds. The degrees of dryness and newness and the number of independent threads forming the cord were all considered. Rules were given for comparing the rigidities of cords of different thicknesses.

On Impulsive Forces.

168. Equations of motion. In the case in which the impressed forces are impulsive the general principle enunciated in Art. 131 of this chapter requires but slight modification.

Let (u, v), (u', v') be the velocities of the centre of gravity of any body of the system resolved parallel to any rectangular axes respectively just before and just after the action of the impulses. Let ω and ω' be the angular velocities of the body about the centre of gravity at the same instants. Let Mk^2 be the moment of inertia of the body about the centre of gravity. Then the effective forces on the body are equivalent to two impulsive forces measured by $M(u'-u)$ and $M(v'-v)$ acting at the centre of gravity parallel to the axes of co-ordinates together with an impulsive couple measured by $Mk^2(\omega'-\omega)$.

The resultant effective forces of all the bodies of the system may be found by the same rule. By D'Alembert's principle these will be in equilibrium with the impressed forces. The equations of motion may then be found by resolving in such directions and taking moments about such points as may be found most convenient.

To take an example, let a single body be acted on by a blow whose components are X, Y and whose moment round the centre of gravity is L. The equations of motion are evidently

$$M(u'-u)=X, \quad M(v'-v)=Y, \quad Mk^2(\omega'-\omega)-L.$$

In many cases it will be found that by the principle of virtual work the elimination of the unknown reactions may be effected without difficulty.

169. We notice that these expressions for the effective forces depend on the difference of the momenta just before and just after the action of the impulses. We may therefore conveniently sum up the equations obtained by resolving in any direction and taking moments about any point in the two following forms:

$$\left(\begin{array}{c}\text{Res. Lin. Mom.}\\ \text{after impulse}\end{array}\right) - \left(\begin{array}{c}\text{Res. Lin. Mom.}\\ \text{before impulse}\end{array}\right) = \left(\begin{array}{c}\text{Resolved}\\ \text{impulse}\end{array}\right),$$

$$\left(\begin{array}{c}\text{Ang. Momentum}\\ \text{after impulse}\end{array}\right) - \left(\begin{array}{c}\text{Ang. Momentum}\\ \text{before impulse}\end{array}\right) = \left(\begin{array}{c}\text{Moment of}\\ \text{impulse}\end{array}\right).$$

An elementary proof of these two results is given in Art. 87. The expression for the Linear Momentum is given in Art. 74, and various expressions used for Angular Momentum are given in Art. 134.

When a single blow or impulse acts on a system, we may conveniently take moments about some point in its line of action, and thus avoid introducing the impulse into the equations. We then deduce from the equation of moments that *the angular momentum of a system about any point in the line of action of an impulse is unaltered by that impulse.*

170. Ex. 1. *A string is wound round the circumference of a circular reel, and the free end attached to a fixed point. The reel is then lifted up and let fall so that, at the moment when the string becomes tight, it is vertical and a tangent to the reel. The whole motion being supposed to be parallel to one plane, determine the effect of the impulse.*

The reel in the first instance falls vertically without rotation. Let v be the velocity of the centre at the moment when the string becomes tight; v', ω' the velocity of the centre and the angular velocity just after the impulse. Let T be the impulsive tension, mk^2 the moment of inertia of the reel about its centre of gravity, a its radius.

In order to avoid introducing the unknown tension into the equations of motion, let us take moments about the point of contact of the string with the reel; we then have

$$m(v'-v)\,a + mk^2\omega' = 0 \dots\dots\dots\dots\dots\dots\dots\dots\dots\dots(1).$$

Just after the impact the part of the reel in contact with the string has no velocity. Hence

$$v' - a\omega' = 0 \dots\dots\dots\dots\dots\dots\dots\dots\dots\dots\dots(2).$$

Solving these we have $\omega' = \dfrac{av}{a^2 + k^2}$. If the reel be a homogeneous cylinder $k^2 = \dfrac{a^2}{2}$, and in this case we have $\omega' = \dfrac{2}{3}\dfrac{v}{a}$, $v' = \dfrac{2}{3}v$. If it be required to find the impulsive tension, we have by resolving vertically

$$m(v'-v) = -T \dots\dots\dots\dots\dots\dots\dots\dots\dots(3).$$

Hence
$$T = \frac{1}{3}mv.$$

To find the subsequent motion. The centre of the reel *begins* to descend vertically, and there is no horizontal force on it. Hence it will continue to descend in a vertical straight line, and throughout all the subsequent motion the string is vertical. The motion may therefore be easily investigated as in Art. 144. If we put $a = \frac{1}{2}\pi$, and F is the finite tension of the string, it may be shown that F is one-third of the weight, and that the reel descends with a uniform acceleration $\frac{2}{3}g$. The initial velocity of the reel has been found in this article $=v'$, so that the space descended in a time t after the impact is $v't + \dfrac{1}{2} \cdot \dfrac{2}{3}gt^2$.

Ex. 2. *A sphere with any initial conditions moves in a vertical plane which intersects a fixed inclined plane along the line of greatest slope. If the sphere be rough and elastic prove that the expression* $U = au + k^2\omega - agt\sin\alpha$ *is unaltered by any impact on the plane and is constant throughout the motion, where ω is the angular* *velocity of the sphere,* u *the velocity of its centre resolved parallel to and down the plane at any time* t, a *the radius and* a *the inclination of the plane to the horizon.*

We notice that the impulse acts at the point of contact. Taking moments about this point we have

$$au' + k^2\omega' = au + k^2\omega,$$

u', ω' being the values of u, ω after the impact. The expression U is therefore unchanged by an impact.

No geometrical equation has been used in arriving at this result. It is therefore true whether the body be elastic or not and whether it roll or slide.

If the body rebound and leave the plane, its centre of gravity will describe a parabola. We know that $u - gt \sin a$ and ω will then each be constant. The expression U therefore remains unchanged during the parabolic motion.

If the body again impinges on the plane we see as before that the expression U is unaltered by this second or any subsequent impulse.

If the body simply rolls or slides on the plane without rebounding we have as in Art. 144 $ma\ddot{x} + mk^2\dot{\omega} = mga \sin a.$

Hence by integration the expression U remains unchanged during this motion.

If after any number of rebounds the sphere pass over some part of the plane which is so rough and inelastic that the sphere rolls we have in addition the equation $u = a\omega$. Joining this equation to the condition that the expression U is equal to its initial value, we have two equations to find the values of u and ω.

171. **Impact of a single Inelastic body.** *A disc of any form is moving in its own plane in any manner. Suddenly a point* O *on it is seized and made to move in some given manner. Find the initial motion of the disc.*

Let Ox, Oy be two directions at right angles to which it is convenient to refer the motion. As explained in Art. 168, let (u, v) be the resolved velocities of the centre of gravity G in these directions and ω the angular velocity of the body just before the motion of O is changed. Thus if Ox can be chosen conveniently parallel to the direction of the motion of the centre of gravity we have the simplification $v - 0$. Let (u', v') be the resolved velocities of the centre of gravity in the same directions and ω' the angular velocity just after the change. Let (x, y) be the co-ordinates of the centre of gravity referred to the axes Ox, Oy at the instant of the change, and let $OG = r$.

Since the angular momentum of the body about the point of space through which O is passing is unchanged by the blow, we have, by Art. 134,

$$M(xv' - yu' + k^2\omega') = M(xv - yu + k^2\omega).$$

Let (U', V') be the resolved parts of the velocity of O just after the change. Then we have by Art. 137,

$$u' = U' - y\omega', \quad v' = V' + x\omega'.$$

From these three equations we easily find

$$(k^2 + r^2) \omega' = x(v - V') - y(u - U') + k^2\omega.$$

If the point O *be suddenly fixed* we have $U' = 0$, $V' = 0$, and then we find

$$(k^2 + r^2)\, \omega' = xv - yu + k^2\omega.$$

Another investigation will be given later in the chapter under the heading *relative motion*.

172. *To find the blow at* O *necessary to produce the given change.*

Let X, Y be the components of the blow parallel to the axes Ox, Oy. Then by Art. 168 we have, resolving parallel to the axes

$$M(u' - u) = X, \quad M(v' - v) = Y.$$

If we take the axis of x to pass through the centre of gravity, we have $y = 0$. We then find by substitution

$$X = -M(u - U'), \quad Y = M\frac{k^2}{k^2 + r^2}(r\omega - v + V').$$

173. **Ex. 1.** A circular area is turning about a fixed point A on its circumference, when suddenly A is loosed and another point B on the circumference is fixed. If AB is a quadrant show that the angular velocity is reduced to one-third of its value. If AB is a third of the circumference the area is reduced to rest.

Ex. 2. A disc of any form is moving in any manner. Suddenly the motion of a point O is changed, show that the increase of vis viva is equal to

$$MW'^2\left(1 - \frac{p'^2}{k^2 + r^2}\right) - MW^2\left(1 - \frac{p^2}{k^2 + r^2}\right),$$

where W, W' are the resultant velocities of O just before and just after the change; p, p' the perpendiculars from the centre of gravity on the directions of motion of O, and the rest of the notation is the same as before.

If O be reduced to rest and the loss of vis viva is to be a given quantity, then O must lie on a certain conic which becomes two coincident straight lines when the whole vis viva is lost.

174. **Examples of different kinds of Impacts.** **Ex. 1.** An inelastic sphere of radius a, sliding with a velocity V on a smooth horizontal plane, impinges on a perfectly rough fixed point or peg at a height c above the plane. Show (1) that unless the velocity V be greater than $\sqrt{2gc\dfrac{a^2 + k^2}{(a-c)^2}}$ the sphere will not jump over the peg. Supposing the velocity V to have this value show (2) that the sphere will immediately leave the peg if $\dfrac{c}{a}$ be greater than $\dfrac{a^2 + k^2}{3a^2 + k^2}$. In this latter case show (3) that the sphere will again hit the peg after a time t, given by the lesser root of the equation $\frac{1}{4}g^2t^2 - U\sin a g t + U^2 - ag\cos a = 0$, where $U^2 = 2gc\dfrac{a^2}{a^2 + k^2}$ and $\cos a = 1 - \dfrac{c}{a}$. Show also that the roots of this quadratic are real and positive.

Ex. 2. A rectangular parallelepiped of mass $3m$, having a square base $ABCD$, rests on a horizontal plane and is moveable about CD as a hinge. The height of the solid is $3a$ and the side of the base a. A particle m moving with a horizontal velocity v strikes directly the middle of that vertical face which stands on AB and lodges there without penetrating. Show that the solid will not upset unless $v^2 > \dfrac{53}{9} ga$. [King's Coll.]

Ex. 3. A vertical column in the form of a right circular cylinder rests on a perfectly rough horizontal plane. Suddenly the plane is jerked with a velocity V in a direction making an angle e with the horizon. Show that the column will not be overturned unless (1) the direction of jerk be such that a parallel to it drawn through the centre of gravity does not cut the base, and (2) the velocity of jerk be greater than U, where U is given by $U^2 = \frac{1}{6} gl \, (15 + \cos^2 \theta) \, \dfrac{1 - \cos \theta}{\cos^2 (\theta + e)}$. Here $2l$ is the length of a diagonal of the cylinder and θ is the angle any diagonal makes with the vertical.

Ex. 4. If the velocity of the jerk of the horizontal plane be exactly equal to U, find the vertical pressure of the cylinder on the plane. Show that the cylinder will not continue to touch the plane during the whole ascent of the centre of gravity unless $1 + \frac{1}{2} \sin \theta < 3 \cos \theta$. What is the general character of the motion if this condition is not satisfied?

Let the cylinder touch the ground at the point A of the rim, and let ϕ be the angle made by the diagonal through A with the vertical. Then by the principle of vis viva we have

$$(k^2 + l^2) \, \dot{\phi}^2 = C - 2gl \cos \phi,$$

where $k^2 = l^2 \left(\frac{1}{3} \cos^2 \theta + \frac{1}{4} \sin^2 \theta \right)$, by Art. 18, Ex. 8. If the angular velocity of the cylinder vanishes when the centre of gravity is at its highest we have $C = 2gl$. Let mR be the vertical reaction at A, where m is the mass of the cylinder. Then $\dfrac{d^2}{dt^2} (l \cos \phi) = R - g$. From these equations we find

$$R \, \frac{k^2 + l^2}{gl^2} = 3 \cos^2 \phi - 2 \cos \phi + \frac{1}{3} \cos^2 \theta + \frac{1}{4} \sin^2 \theta.$$

If R vanish we have $\cos \phi = \frac{1}{3} (1 \pm \frac{1}{2} \sin \theta)$. In order that R may keep one sign both these values of ϕ must be excluded by the circumstances of the case, i.e. both these values of ϕ must be greater than θ. This leads to the result given above.

175. **Earthquakes.** The last two problems are interesting from their connection with Mallet's theory of earthquakes. Let us suppose that the action of an earthquake on any building may be represented by such a motion of the base as that of the plane just described. Then the direction and the magnitude of the *equivalent jerk* are both independent of the building operated on, and depend only on the nature of the earthquake at the place.

On these principles Mr Mallet has constructed a seismometer of great simplicity. A set of six right cylinders is turned in some hard material such as boxwood. The cylinders are all of the same height but vary in diameter. They stand upright on a plank fixed to a level floor in the order of their size, with a space between each pair greater than their height, so that when one falls it does not strike its neighbour. When a shock passes some of the cylinders are overturned and some left standing. Suppose the jerk to knock over the narrow based cylinders 4, 5, 6, leaving the broader based cylinders 1, 2, 3 standing, then the jerk must have been greater than that required to overturn cylinder No. 4, but not great enough to overturn cylinder No. 3.

The formula used is that given in Ex. 3, which is ascribed by Mr Mallet to Dr Haughton. The value of e is small when the origin or focus of the earthquake is distant, so that as a first approximation we may put $e = 0$. It does not appear to have been noticed that if we are to use *this* formula for the *standing cylinders* they must be such as to satisfy the conditions given in Ex. 4.

In December, 1857, an earthquake of great violence occurred in the southern provinces of Italy. Mr Mallet visited the place early in the next year for the express purpose of determining the circumstances of the shock. The problem to be solved was to some extent a mechanical one. Given the positions of the overturned columns and buildings, to find the depth and position of the focus or origin of the earthquake, the velocity of the earthquake wave, and the magnitude of the jerk at any place. In this case the depth of the focus was about three miles below the surface of the earth, the velocity of the wave was about 800 feet per second, while the velocity of jerk, which upset several buildings, was as little as 12 feet per second. This last is about the same velocity as that acquired by a particle falling from rest under gravity through a height of between two and three feet. See *The Great Neapolitan Earthquake of* 1857, two volumes, 1862, by R. Mallet. The observations made during the earthquake of Dec. 1884 in Spain and that of August 1886 at Charleston indicated a depth of focus very much greater than that above given. *Flammarian Astronomie*, Oct. 1887.

The column seismometer described above has not been very successful in practice. The displacement of the earth is not a simple rectilinear motion, but rather a prolonged series of motions in different directions. These give rotational motions to the columns which therefore fall in different directions. A model, by means of a long copper wire, of the actual path of a point on the earth's surface during a severe earthquake in Jan. 1887 in Japan has been constructed by Prof. Sekiya and is described in *Nature* Jan. 26, 1888. Whatever degree of accuracy this may have, it tends to show the complicated nature of the displacement. For an account of modern seismometers the reader may consult Milne's *Earthquakes* 1886, *Nature* April 12, and July 26, 1888, and *Phil. Mag.*, April 1887. Some recent experiments in connection with earthquakes are described in the *Proceedings of the Royal Society* for Dec. 1881. The velocities and amplitudes of the waves of direct and transverse vibration were separately determined. The motion of a point on the earth's surface was found to be such as would result from the composition of two harmonic motions of different periods and in different directions.

176. Impact of a Compound Inelastic body. *Four equal rods each of length* 2a *and mass* m *are freely jointed so as to form a rhombus. The system falls from rest with a diagonal vertical under the action of gravity and strikes against a fixed horizontal inelastic plane. Find the subsequent motion.*

Let AB, BC, CD, DA be the rods and let AC be the vertical diagonal impinging on the horizontal plane at A. Let V be the velocity of every point of the rhombus just before impact and let α be the angle any rod makes with the vertical.

Let u, v be the horizontal and vertical velocities of the centre of gravity and ω the angular velocity of either of the upper rods just after impact. Then the effective forces on either rod are equivalent to the force $m(v - V)$ acting vertically and mu horizontally at the centre of gravity and a couple $mk^2\omega$ tending to increase the angle α. Let R be the impulse at C, the direction of which by the rule of symmetry is horizontal. To avoid introducing the reactions at B into our equations, let us take moments for the rod BC about B and we have

$$mk^2\omega + m(v - V)a\sin\alpha - mua\cos\alpha = -R \cdot 2a\cos\alpha \ldots\ldots\ldots\ldots(1).$$

Either of the lower rods begins to turn round its extremity A as a fixed point. If ω' be its angular velocity just after impact, the moment of the momentum about A just after impact is $m(k^2 + a^2)\omega'$ and just before is $mVa\sin\alpha$. The difference of these two is the moment about A of the effective forces on the rod. We may now take moments about A for the two rods AB, BC together and we have

$$m(k^2 + a^2)\omega' - mVa\sin\alpha - mk^2\omega + m(v - V)a\sin\alpha + mu \cdot 3a\cos\alpha = R \cdot 4a\cos\alpha \ldots(2).$$

The geometrical equations may be found thus. Since the two rods must make equal angles with the vertical during the whole motion we have

$$\omega' = \omega \ldots\ldots\ldots\ldots\ldots\ldots\ldots\ldots\ldots\ldots\ldots\ldots\ldots\ldots\ldots(3).$$

Again, since the two rods are connected at B, the velocities of their extremities must be the same in direction and magnitude. Resolving these horizontally and vertically, we have

$$u + a\omega\cos\alpha = 2a\omega'\cos\alpha \ldots\ldots\ldots\ldots\ldots\ldots\ldots\ldots\ldots(4),$$

$$v - a\omega\sin\alpha = 2a\omega'\sin\alpha \ldots\ldots\ldots\ldots\ldots\ldots\ldots\ldots\ldots(5).$$

These five equations are sufficient to determine the initial motion.

Eliminating R between (1) and (2), and substituting for u, v, ω' in terms of ω from the geometrical equations, we find

$$\omega = \frac{3}{2} \cdot \frac{V\sin\alpha}{a(1 + 3\sin^2\alpha)} \ldots\ldots\ldots\ldots\ldots\ldots\ldots\ldots\ldots(6).$$

In this problem we might have avoided the introduction of the unknown reaction R by the use of Virtual Velocities. Supposing we give the system such a displacement that the inclination of each rod to the vertical is increased by the same quantity $\delta\alpha$. Then the principle of Virtual Velocities gives

$$mk^2\omega\delta\alpha - m(v - V)\delta(3a\cos\alpha) + mu\delta(a\sin\alpha) + m(k^2 + a^2)\omega'\delta\alpha + mV\delta(a\cos\alpha) = 0,$$

which reduces to

$$(2k^2 + a^2)\omega - Va\sin\alpha + 3(v - V)a\sin\alpha + ua\cos\alpha = 0,$$

and the solution may be continued as before.

Ex. 1. Prove that the rhombus loses by the impact $\dfrac{1}{1 + 3\sin^2\alpha}$ of its momentum.

Ex. 2. Show that the direction of the impulsive action at the hinges B or D makes with the horizon an angle whose tangent is $\dfrac{3\sin^2\alpha - 2}{\tan\alpha}$.

To find the subsequent motion. This may be found very easily by the method of *Vis Viva.* But in order to illustrate as many modes of solution as possible, we shall proceed in a different manner. The effective forces on either of the upper rods are represented by the differential coefficients $m\dot{v}$, $m\dot{u}$, $mk^2\dot{\omega}$, and the moment for either of the lower rods is $m(k^2 + a^2)\dot{\omega}'$. Let θ be the angle any rod makes with the vertical at the time t. Taking moments in the same way as before, we have

$$mk^2\dot{\omega} + m\dot{v}a\sin\theta - m\dot{u}a\cos\theta = -R \cdot 2a\cos\theta + mga\sin\theta \ldots\ldots(1)'.$$

$$m(k^2 + a^2)\dot{\omega}' - mk^2\dot{\omega} + m\dot{u}a\sin\theta + m\dot{u} \cdot 3a\cos\theta = R \cdot 4a\cos\theta + 2mga\sin\theta \ldots(2)'.$$

The geometrical equations are the same as those given above, with θ written for a. Eliminating R and substituting for u, v, we get

$$(2k^2 + a^2)\frac{d\omega}{dt} + a^2 \left\{ 9 \sin \theta \frac{d}{dt}(\omega \sin \theta) + \cos \theta \frac{d}{dt}(\omega \cos \theta) \right\} = 4ga \sin \theta \ ;$$

then multiplying both sides by $\omega = \dot\theta$ and integrating, we get

$$\{ 2(k^2 + a^2) + 8a^2 \sin^2 \theta \} \, \omega^2 = C - 8ga \cos \theta.$$

Initially, when $\theta = a$, ω has the value given by equation (6). Hence we find that the angular velocity ω when the inclination of any rod to the vertical is θ is given by

$$(1 + 3 \sin^2 \theta) \, \omega^2 = \frac{9V^2}{4a^2} \cdot \frac{\sin^2 a}{1 + 3 \sin^2 a} + \frac{3g}{a}(\cos a - \cos \theta).$$

177. Ex. 1. A square is moving freely about a diagonal with angular velocity ω, when one of the angular points not in that diagonal becomes fixed; determine the impulsive pressure on the fixed point, and show that the instantaneous angular velocity will be $\frac{\omega}{7}$. [Christ's Coll.]

Ex. 2. Three equal rods placed in a straight line are jointed by hinges to one another; they move with a velocity v perpendicular to their lengths; if the middle point of the middle one become suddenly fixed, show that the extremities of the other two will meet in a time $\dfrac{4\pi a}{9v}$, a being the length of each rod. [Coll. Exam.]

Ex. 3. The points $ABCD$ are the angular points of a square; AB, CD are two equal similar rods connected by the string BC. The point A receiving an impulse in the direction AD, show that the initial velocity of A is seven times that of the point D. [Queens' Coll.]

Ex. 4. A series of equal beams AB, BC, CD...... is connected by hinges; the beams are placed on a smooth horizontal plane, each at right angles to the two adjacent, so as to form a figure resembling a set of steps, and an impulse is given at the end A along AB: determine the impulsive action at any hinge. [Math. Tripos.]

Result. If X_n be the impulsive action at the nth angular point, show that $X_{2n+1} - 5X_{2n+2} - 2X_{2n+3} = 0$ and that $X_{2n+2} - 5X_{2n+1} - 2X_{2n} = 0$. Thence find X_n.

Ex. 5. Two uniform rods AB, BC of equal length and mass, smoothly hinged at B, lie upon a smooth horizontal table; the end A is struck so as to begin to move with a given velocity in a direction which makes angles θ, ϕ respectively with the rods, show that, if $\sin(2\phi - \theta) = 3 \sin \theta$, AB will begin to move without rotation. [June Exam. 1880.]

Take moments for the rod BC about B and for both rods about A according to the rule in Art. 169.

178. **The kick before and behind.** *A free inelastic lamina of any form is turning in its own plane about an instantaneous centre of rotation* S, *and impinges on an obstacle at* P *situated in the straight line joining the centre of gravity* G *to* S. *To find the point* P *when the magnitude of the blow is a maximum*.*

---

* Poinsot, Sur la percussion des corps, *Liouville's Journal*, 1857; translated in the *Annals of Philosophy*, 1858.

Firstly, let the obstacle P *be a fixed point.*

Let $GP = x$, and let R be the force of the blow. Let $SG = h$, and let ω, ω' be the angular velocities about the centre of gravity before and after the impact. Then $h\omega$ is the linear velocity of G just before the impact; let v' be its linear velocity just after the impact. We have the equations

$$\omega' - \omega = \frac{-Rx}{Mk^2}, \quad v' - h\omega = -\frac{R}{M} \dots\dots\dots\dots\dots\dots(1),$$

and supposing the point of impact to be reduced to rest,

$$v' + x\omega' = 0 \dots\dots\dots\dots\dots\dots\dots\dots\dots\dots\dots(2).$$

Substituting for ω' and v' from (1) in equation (2), we get

$$R = M\omega \cdot k^2 \frac{x+h}{x^2+k^2}.$$

This is to be made a maximum. Equating to zero its differential coefficient with respect to x, we get

$$x^2 + 2hx - k^2 = 0 ; \quad \therefore \quad x = -h \pm \sqrt{h^2 + k^2} \dots\dots\dots\dots\dots(3).$$

One of these values of x is positive and the other negative. Both these correspond to points of *maximum* percussion, but in opposite directions. Thus there is a point P with which the body strikes in front and a point P' with which it strikes in rear of its own translation in space more forcibly than with any other point.

Ex. 1.   Show that the two points P, P' are equally distant from S, and if O be the centre of oscillation with regard to S as a centre of suspension, $SP^2 = SG \cdot SO$.

Ex. 2.   If P be made a point of suspension, P' is the corresponding centre of oscillation.   Also PP' is harmonically divided in G and O.

Ex. 3.   The magnitudes of the blows at P, P' are inversely proportional to their distances from G.

Secondly, let the obstacle be a free particle of mass m.

Then, besides the equations (1), we have the equation of motion of the particle m. Let V' be its velocity after impact, then $V' = \frac{R}{m}$.

The point of impact in the two bodies will have after impact the same velocity, hence instead of equation (2) we have $V' = v' + x\omega'$.

Eliminating ω', v', V', we get $R = M\omega \cdot k^2 \cdot \frac{m(x+h)}{(M+m)k^2 + mx^2}$.

This is to be made a maximum. Equating to zero its differential coefficient with respect to x, we find

$$x = -h \pm \sqrt{h^2 + k^2\left(1 + \frac{M}{m}\right)} \dots\dots\dots\dots\dots\dots(4).$$

This point does not coincide with that found when the obstacle was fixed, unless m is infinite. To find when it coincides with the centre of oscillation, we must put $k^2 = xh$. This gives $\frac{M}{m} = \frac{x+h}{h}$, or if $l = x + h$ be the length of the simple equivalent pendulum, $\frac{M}{m} = \frac{l}{h}$. Since $V' = \frac{R}{m}$, it is evident that when R is a maximum V' is a maximum. Hence the two points found by equation (4) might be called the centres of greatest communicated velocity.

There are other singular points in a moving body whose positions may be found; thus we may inquire at what points a body must impinge against a *fixed* obstacle, *firstly* that the linear velocity of the centre of gravity may be a maximum, and *secondly*, that the angular velocity may be a maximum. These points, respectively, have been called by Poinsot the centres of maximum Reflexion and Conversion. Referring to equations (1) we see that when v' is a maximum R is either a maximum or a minimum, and hence it may be shown that the first point coincides with the point of greatest impact. When ω' is a maximum, we have to make $\omega - \dfrac{Rx}{Mk^2} =$ a maximum. Substituting for R, this gives $x^2 - 2\,\dfrac{k^2}{h}\,x - k^2 = 0$. If O be the centre of oscillation, we have $h \cdot GO = k^2$. Let $GO = h'$. Then this equation becomes
$$x^2 - 2h'x - k^2 = 0 \quad\dotfill\quad (5).$$

The roots of this equation are the same functions of h' and k that those of equation (3) are of h and k, except that the signs are opposite. Now S and O are on opposite sides of G, hence the positions of the two centres of maximum Conversion bear to O and G the same relation that the positions of the two centres of maximum Reflexion do to S and G. If the point of suspension be changed from S to O, the positions of the centres of maximum Reflexion and Conversion are interchanged.

Ex. A free lamina of any form is turning in its own plane about an instantaneous centre of rotation S, and impinges on a fixed obstacle P situated in the straight line joining the centre of gravity G to S. Find the position of P, *firstly*, that the centre of gravity may be reduced to rest, *secondly*, that its velocity after impact may be the same as before but reversed in direction.

Result. In the first case, P coincides either with G, or with the centre of oscillation. In the second case if $SG = h$, $x = GP$ the points are found from the equation
$$2hx^2 = k^2\,(x - h). \quad \text{[Poinsot.]}$$

179. Elastic smooth bodies. *Two bodies impinge on each other, to explain the nature of the action which takes place between them.*

When two spheres of any hard material impinge on each other, they appear to separate almost immediately, and a finite change of velocity is generated in each by the mutual action. This sudden change of velocity is the characteristic of an impulsive force. Let the centres of gravity of the spheres be moving before impact in the same straight line with velocities u and v. Then after impact they will continue to move in the same straight line; let u', v' be their velocities. Let m, m' be the masses of the spheres, R the action between them, then we have by Article 168,
$$u' - u = -\frac{R}{m}, \qquad v' - v = \frac{R}{m'} \quad\dotfill\quad(1).$$

These equations are not sufficient to determine the three quantities u', v', R. To obtain a third equation we must consider what takes place during the impact.

Each of the balls is slightly compressed by the other, so that they are no longer perfect spheres. Each also in general tends to return to its original shape, so that there is a rebound. The period of impact may therefore be divided into two parts. Firstly, the period of compression, during which the distance between the centres of gravity of the two bodies is diminishing, and secondly the period of restitution, in which the distance between the centres of gravity is increasing. At the termination of the second period the bodies separate.

The arrangement of the particles of a body being disturbed by impact, we ought in strictness to determine the relative motions of the several parts of the body. Thus we might regard each body as a collection of free particles connected by mutual actions. These particles being set in motion might continue always in motion oscillating about some mean positions in the body.

It is however usual to assume that the changes of shape and structure are so small that the effect in altering the position of the centre of gravity and the moments of inertia of the body may be neglected; also that the whole time of impact is so short that the motion of the body in that time may be neglected. If for any bodies these assumptions are not true, the effects of their impact must be deduced from the equations of the second order. We may therefore assert that at the moment of greatest compression the centres of gravity of the two spheres are moving with equal velocities.

The ratio of the magnitude of the action between the bodies during the period of restitution to that during compression is found to be different for bodies of different materials. In some cases this ratio is so small that the force during the period of restitution may be neglected. The bodies are then said to be *in-elastic*. In this case we have just after the impact $u' = v'$. This gives $R = \dfrac{mm'}{m+m'}(u-v)$, whence $u' = \dfrac{mu+m'v}{m+m'}$.

If the force of restitution cannot be neglected, let R be the whole action between the balls, R_0 the action up to the moment of greatest compression. The magnitude of R must be found by experiment. This may be done by determining the values of u' and v', and thus determining R by means of equations (1). Such experiments were made in the first instance by Newton, and led to the result that $\dfrac{R}{R_0}$ is a *constant ratio* depending on the material of the balls. Let this constant ratio be called $1 + e$. The quantity e is never greater than unity; in the limiting case when $e = 1$ the bodies are said to be perfectly elastic.

The value of e being supposed known the velocities after impact may be easily found. The action R_0 must be first calculated as if the bodies were inelastic, when the whole value of R may be found by multiplying by $1 + e$. This gives

$$R = \frac{mm'}{m + m'} (u - v)(1 + e),$$

whence u' and v' may be found by equations (1).

180. As an example, let us consider how the motion of the reel discussed in Art. 170 would be affected if the string were so slightly elastic that we could apply this theory.

Since the point of the reel in contact with the string has no velocity at the moment of greatest compression, the impulsive tension found in the article referred to, measures the whole momentum communicated to the reel from the beginning of the impact up to the moment of greatest compression. By what has been said in the last article, the whole momentum communicated from the beginning to the termination of the period of restitution will be found by multiplying the tension found in Art. 170 by $1 + e$, if e be the measure of the elasticity of the string. This gives $T = \frac{1}{2}mv (1 + e)$. The motion of a reel acted on by this known impulsive force is easily found. Resolving vertically we find $m(v' - v) = -\frac{1}{2}mv (1 + e)$. Taking moments about the centre of gravity, $mk^2\omega' = \frac{1}{2}mva (1 + e)$, whence v' and ω' may be found.

Ex. A uniform beam is balanced about a horizontal axis through its centre of gravity, and a perfectly elastic ball is let fall from a height h on one extremity; determine the motion of the beam and ball.

Result. Let M, m be the masses of the beam and the ball, $2a$ the length of the beam, V, V' the velocities of the ball at the moments just before and after impact, ω' the angular velocity of the beam. Then $\omega' = \dfrac{6mV}{(M + 3m) a}$, $V' = V \cdot \dfrac{3m - M}{3m + M}$.

181. **Rough bodies.** Hitherto we have only considered the impulsive action normal to the common surface of the two bodies. If the bodies are rough an impulsive friction will clearly be called into play. Since an impulse is only the integral of a very great force acting for a very short time, we might suppose that impulsive friction obeys the laws of ordinary friction. But these laws are founded on experiment, and we cannot be sure that they are correct in the extreme case in which the forces are very great. This point M. Morin undertook to determine by experiment at the express request of Poisson. He found that the frictional impulse between two bodies which strike and slide bore to the normal impulse the same ratio as in ordinary friction, and that this ratio was independent of the relative velocity of the striking bodies. M. Morin's experiment is described in the following article.

182. A box AB which can be loaded with shot so as to be of any proposed weight has two vertical beams AC, BD erected on its lid; CD is joined by a cross piece and supports a weight

equal to mg attached to it by a string. The weight of the loaded box is Mg. A string AEF passes horizontally from the box over a smooth pulley E and supports a weight at F equal to $(M + m) g\mu$. The box can slide on a horizontal plane whose coefficient of friction is μ, and therefore having been once set in motion, it moves in a straight line with a uniform velocity which we will call V. Suddenly the string supporting mg is cut, and this weight falls into the box and immediately becomes fixed to the box. There clearly is an impulsive friction called into play between the box and the horizontal plane. If the velocity of the box immediately after the impulse be again equal to V, the coefficient of impulsive friction is equal to that of finite friction.

The argument may be made evident as follows. Let t be the time of the fall. When the weight strikes the box, it has a horizontal velocity equal to V and a vertical velocity equal to gt. The box itself has a horizontal velocity $V + ft$, where

$$f = \frac{\mu m g}{M + (M + m)\mu}.$$

Let F and R be the horizontal and vertical components of the impulse between the box and the horizontal plane. There will be an impulse between the falling weight and the box and an impulsive tension in the string AEF; by means of these the momenta generated by the external blows F and R are spread over the whole system. Let V' be the common velocity of the whole system just after the impulses F and R are completed. This velocity V' is found by experiment to be equal to V. Resolving horizontally and vertically as in Art. 168, we have

$$\{M + m + (M + m)\mu\} V' - \{M + (M + m)\mu\} (V + ft) - mV = -F,$$
$$mgt = R.$$

Putting $V' = V$ and substituting for f, we immediately find that $F = \mu R$.

Ex. Show that the resultant impulse between the box and the falling weight is vertical.

183. When two inelastic bodies impinge on each other at some point A, the points in contact at the beginning of the impact have a relative velocity along the common tangent plane at A and also one along the normal. Thus two reactions will be called into play, a normal force and a friction, the ratio of these two being μ, the coefficient of friction. As the impact proceeds the relative normal velocity gets destroyed, and is zero at the moment of greatest compression. Let R be the whole momentum transferred normally from one body to the other in this very short time. This force R is an unknown reaction, to determine it we have the geometrical condition that just after impact the

normal velocities of the points in contact are equal. This condition must be expressed in the manner explained in Art. 137.

The relative sliding velocity at A is also diminished. If it vanishes before the moment of greatest compression, then during the rest of the impact there is called into play only so much friction and in such a direction, as is necessary (if any be necessary) to prevent the points in contact at A from sliding, provided that this amount is less than the limiting friction. Let F be the whole momentum transferred tangentially from the one body to the other. This reaction F is to be determined by the condition that just after impact the tangential velocities of the points in contact are equal. If, however, the sliding motion does not vanish before the moment of greatest compression, the whole of the friction is called into play in the direction opposite to that of relative sliding, and we have $F = \mu R$. Generally we may distinguish these two cases in the following manner. In the first case it is necessary that the values of F and R found by solving the equations of motion should be such that $F < \mu R$. In the second case, the final relative velocity of the points in contact at A must be in the same direction after impact as before. These are however not sufficient conditions, for it is possible that, in the more complicated cases, the sliding may change, or tend to change, its direction during the impact. See Art. 187.

184. If the impinging bodies be elastic, there may be both a normal reaction and a friction during the period of restitution. Sometimes we shall have to consider this stage of the motion as a separate problem. The motion of the bodies at the moment of greatest compression having been determined, these are to be regarded as the initial conditions of a new state of motion under different impulses. The friction called into play during restitution must follow the same laws as that during compression. Just as before, two cases will present themselves; there will be sliding either during the whole period of restitution or during only a portion of it. These cases are to be treated in the manner already explained.

185. There is one very important difference between the conditions of compression and restitution. During the compression the normal reaction is unknown. The motion of the body just before compression is given, and we have a geometrical equation expressing the fact that the relative normal velocity of the points in contact is zero at the termination of the period of compression. From this geometrical equation we deduce the force of compression. The motion of the body just before restitution is thus found, but the motion just after is the thing we want to determine. For this, we have no geometrical equation, but the force

of restitution bears a given ratio to the force of compression, and is therefore known.

186. **Historical Summary.** The problem of the impact of two *smooth inelastic* bodies is considered by Poisson in his *Traité de Mécanique*, Seconde Edition, 1833. The motion of each body just before impact being supposed given, he forms six equations of motion for each body to determine the motion just after impact. These contain thirteen unknown quantities, viz., the resolved velocities of the centres of gravity of the bodies along three rectangular axes, the resolved angular velocities of the bodies about the same axes, and, lastly, the mutual reaction of the two bodies. Thus the equations are insufficient to determine the motion. A thirteenth equation is then obtained from the principle that the impact terminates at the moment of greatest compression, i.e. at the moment when the normal velocities of the points of contact of the two bodies which impinge are equal.

When the bodies are *elastic*, Poisson divides the impact into two periods. The first begins at the first contact of the bodies and terminates at the moment of greatest compression. The second begins at the moment of greatest compression and terminates when the bodies separate. The motion at the end of the first period is found exactly as if the bodies were inelastic. The motion at the end of the second period is found from the principle that the whole momentum communicated by one body to the other during the second period bears a constant ratio to that communicated during the first period of the impact. This ratio depends on the elasticity of the two bodies and can be found only by experiments made on some bodies of the same material in simple cases of impact.

When the bodies are *rough*, and *slide on each other* during the impact, Poisson remarks that there will also be a frictional impulse. This is to be found from the principle (Art. 181) that the magnitude of the friction at each instant must bear a constant ratio to the normal pressure and the direction must be opposite to that of the relative motion of the points in contact. He applies this to the case of a sphere, either inelastic or perfectly elastic, impinging on a rough plane, the sphere turning before the impact about a horizontal axis perpendicular to the direction of motion of the centre of gravity. He points out that there are several cases to be considered; (1) when the sliding is the same in direction during the whole of the impact and does not vanish, (2) when the sliding vanishes during the impact and remains zero, (3) when the sliding vanishes and changes sign. This third case, however, contains an unknown quantity and his formulæ therefore fail to determine the motion. Poisson points out that the problem becomes very complicated if the sphere have an initial rotation about an axis not perpendicular to the vertical plane in which the centre of gravity moves. This case he does not attempt to solve, but passes on to discuss at greater length the impact of *smooth* bodies.

M. Coriolis in his *Jeu de Billard* (1835) considers the impact of two *rough* spheres *sliding* on each other during the whole of the impact. He shows that if two rough spheres impinge on each other the direction of sliding is the same throughout the impact.

M. Ed. Phillips in the fourteenth volume of *Liouville's Journal*, 1849, considers the problem of the impact of two *rough inelastic* bodies of any form when the direction of the friction is not necessarily the same throughout the impact, assuming that the *sliding does not vanish* during the impact. He divides the period of impact into elementary portions and applies Poisson's rule for the magnitude and direction of the friction to each elementary period. He points out how the solution of the equations may be effected, and in particular discusses the case in which the two bodies have their principal axes at the point of contact parallel each to each and also each body has its centre of gravity on the common normal at the point of contact. He deduces for this case two results, which will be given in the chapter on *Momentum*.

M. Phillips does not examine in detail the impact of *elastic* bodies, though he remarks that the period of impact must be divided into two portions which must be considered separately. These however, he considers, do not present any further peculiarities when the same suppositions are made.

The case in which the sliding vanishes and the friction becomes discontinuous, does not appear to have been examined by him.

In this chapter we shall discuss the theory of impulses only so far as motion in one plane is concerned. In the chapter on *Momentum* the theory will be taken up again and extended to bodies of any form in space of three dimensions.

187. General Problem of impact. *Two bodies of any form impinge on each other in a given manner. It is required to find the motion just after impact. The bodies are smooth or rough, inelastic or elastic.*

Let G, G' be the centres of gravity of the two bodies, A the point of contact. Let U, V be the resolved velocities of G just

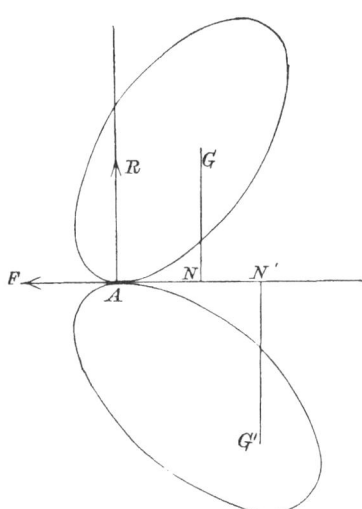

before impact, parallel respectively to the tangent and normal at A; u, v the resolved velocities at any time t after the commencement of the impact, but before its termination, so that t is indefinitely small. Let Ω be the angular velocity of the body, whose centre of gravity is G, just before impact, ω the angular velocity after the interval t. These are to be taken as positive when the rotation is like the hands of a watch. Let M be the mass of the body, k its radius of gyration about G. Let GN be a perpendicular from G on the tangent at A, and let $AN = x$, $NG = y$. Let accented letters denote corresponding quantities for the other body.

188.   Let the bodies be *perfectly rough and inelastic*, so that at the termination of the impact the relative velocity of sliding and the relative velocity of compression are both zero (see Art. 156). In this case, taking t to be equal to the whole duration of the impact, the letters u, v, ω, u', v', ω' will refer to the motion just after impact. We then have, by Art. 137,

$$u - y\omega - u' + y'\omega' = 0 \Big\}$$
$$v + x\omega - v' - x'\omega' = 0 \Big\}.$$

Resolving parallel to the tangent and normal at the point of contact we have, by Art. 169,

$$M(u - U) + M'(u' - U') = 0 \Big\}$$
$$M(v - V) + M'(v' - V') = 0 \Big\},$$

and by taking moments for each body about the point of contact

$$Mk^2(\omega - \Omega) + M(u - U)y - M(v - V)x = 0 \Big\}$$
$$M'k'^2(\omega' - \Omega') - M'(u' - U')y - M'(v' - V')x' = 0 \Big\}.$$

These six equations are sufficient to determine the motion just after impact.

189.   If the bodies are *perfectly smooth and inelastic*, the first of these six equations will no longer hold, and instead of the third we have the two equations

$$u - U = 0, \qquad u' - U' = 0,$$

obtained by resolving parallel to the tangent for each body separately.

190.   If the bodies are *smooth and elastic* we must introduce the normal reaction into the equations. We write down the equations (1) and (2) as given below in Art. 192, except that $F = 0$. Then equation (4) gives the velocity C of compression at any instant of the impact. Putting $C = 0$, we have as in equation (6) the value of R up to the moment of greatest compression, viz. $R = \dfrac{C_0}{a'}$. Multiplying this by $1 + e$ we have, by Art. 179, the complete value of R for the whole impact. Substituting this value of R in equations (1) and (2), we find the values of u, v, ω, u', v', ω'.

191.   Ex.   Two smooth perfectly elastic bodies impinge on each other. Let D, D' be the normal velocities of approach, i.e. the velocities of the point of contact of each just before impact resolved along the normal towards the other. Prove

11—2

that the vis viva lost by the body M is equal to $4\dfrac{C_0}{a'^2}\left(D'\dfrac{k^2+x^2}{Mk^2}-D\dfrac{k'^2+x'^2}{M'k'^2}\right)$, the notation being the same as in the next proposition.

Another method of finding the change in the vis viva will be given in chapter VII.

192. Next, *let the bodies be imperfectly rough and elastic.* In this case, as explained in Art. 158, the friction which can be called into play is limited in amount. The results obtained in Art. 188 will not apply to the case in which this limited amount of friction is insufficient to reduce the relative sliding to zero. To determine this, we must introduce the frictional and normal impulses into the equations.

Let R be the whole momentum communicated to the body M in the time t of the impact by the normal pressure, and let F be the momentum communicated by the frictional pressure. We shall suppose these to act on the body whose mass is M in the directions NG, NA respectively. Then they must be supposed to act in the opposite directions on the body whose mass is M'.

Since R represents the whole momentum communicated to the body M in the direction of the normal, the momentum communicated in the time dt is dR. As the bodies can only push against each other, dR must be positive, and, by Art. 136, when dR vanishes, the bodies separate. Thus the magnitude of R may be taken to measure the progress of the impact. It is zero at the beginning, gradually increases throughout, and is a maximum at the termination of the impact. It will be found more convenient to choose R rather than the time t as the independent variable.

The dynamical equations are by Art. 168

$$\left.\begin{aligned}M(u-U)&=-F\\M(v-V)&=R\\Mk^2(\omega-\Omega)&=Fy+Rx\end{aligned}\right\}\dots\dots\dots\dots(1),$$

$$\left.\begin{aligned}M'(u'-U')&=F\\M'(v'-V')&=-R\\M'k'^2(\omega'-\Omega')&=Fy'-Rx'\end{aligned}\right\}\dots\dots\dots\dots(2).$$

The relative velocity of sliding of the points in contact is by Art. 137

$$S=u-y\omega-u'-y'\omega'\dots\dots\dots\dots\dots(3),$$

and the relative velocity of compression is by the same article

$$C=v'+x'\omega'-v-x\omega\dots\dots\dots\dots\dots(4).$$

Substituting in these equations from the dynamical equations we find

$$S = S_0 - aF - bR \dots\dots\dots\dots\dots\dots(5),$$

$$C = C_0 - bF - a'R \dots\dots\dots\dots\dots\dots(6),$$

where

$$S_0 = U - y\Omega - U' - y'\Omega' \dots\dots\dots\dots\dots(7),$$

$$C_0 = V' + x'\Omega' - V - x\Omega \dots\dots\dots\dots\dots(8),$$

$$a = \frac{1}{M} + \frac{1}{M'} + \frac{y^2}{Mk^2} + \frac{y'^2}{M'k'^2} \dots\dots\dots\dots\dots(9),$$

$$a' = \frac{1}{M} + \frac{1}{M'} + \frac{x^2}{Mk^2} + \frac{x'^2}{M'k'^2} \dots\dots\dots\dots(10),$$

$$b = \frac{xy}{Mk^2} - \frac{x'y'}{M'k'^2} \dots\dots\dots\dots\dots\dots(11).$$

These may be called the constants of the impact. The first two, S_0, C_0 represent the initial velocities of sliding and compression. These we shall consider to be positive; so that the body M is sliding over the body M' at the beginning of the compression. The other three constants a, a', b are independent of the initial motion of the striking bodies. The constants a and a' are essentially positive, while b may have either sign. It will be found useful to notice that $aa' > b^2$.

193. **The Representative Point.** It often happens that $b = 0$, and in this case the discussion of the equations is very much simplified. But certainly in the general case, and even in the simple case when $b = 0$, it is found more easy to follow the changes in the forces if we adopt a graphical method.

The point which we have to consider is this. As R proceeds from zero to its final maximum value by equal continued increments dR, F proceeds also from zero by continued increments dF, which may not always be of the same sign and which are governed by a discontinuous law, viz. either $dF = \pm \mu dR$, or dF is just sufficient to prevent relative motion at the point of contact, as explained in Art. 158. We want therefore some rule to discover the value of F.

To determine the actual changes which occur in the frictional impulse as the impact proceeds, let us draw two lengths AR, AF along the normal and tangent at A in the directions NG, AN respectively, to represent the magnitudes of R and F at any moment of the impact. Then, if we consider AR and AF to be the coordinates of a point P referred to AR, AF as axes of R and F, the changes in the position of P will indicate to the eye the changes that take place in the forces during the progress of the impact. At the beginning of the impact the forces R and F are zero, the representative point P is therefore situated at the origin A.

As the impact proceeds the force R continually increases, hence the abscissa AR of P will also continually increase, i.e. the motion of the representative point resolved parallel to the axis of R will be always in the positive direction of the axis of R. The ordinate F of P is measured in the direction opposite to that in which the friction acts on the body M; it follows that the motion of the representative point resolved parallel to the axis of F will indicate to the eye the direction in which the body M is sliding. This may sometimes be in one direction during the impact and sometimes in the other.

It will be convenient to trace the two loci determined by $S = 0$, $C = 0$. By reference to (5) and (6) we see that they are both straight lines. These we shall call the straight lines of *no sliding* and of *greatest compression*. To trace them, we must find their intercepts on the axes of F and R. Take

$$AC = \frac{C_0}{a'}, \quad AS = \frac{S_0}{a}, \quad AC' = \frac{C_0}{b}, \quad AS' = \frac{S_0}{b},$$

then SS', CC' will be these straight lines. Since a and a' are necessarily positive, while b has any sign, we see that the intercepts on the axes of F and R respectively are *positive*, while their intercepts on the axes of R and F must have the *same sign*. Since $aa' > b^2$, the acute angle made by the line of no sliding with the axis of F is greater than that made by the line of greatest compression, i.e. the former line is steeper to the axis of F than the latter. It easily follows that the two straight lines cannot intersect in the quadrant contained by RA produced and FA produced.

194. In the beginning of the impact the bodies slide over each other, hence, as explained in Art. 158, the whole limiting

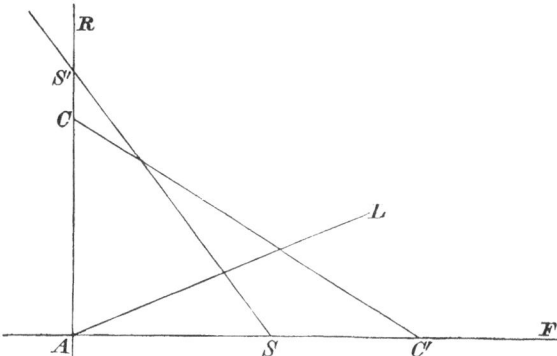

friction is called into play. The point P therefore moves along a straight line AL, defined by the equation $F = \mu R$, where μ is the coefficient of friction. The friction continues to be limiting until P reaches the straight line SS'. If R_0 be the abscissa of this point we find $R_0 = \dfrac{S_0}{a\mu + b}$. This gives the whole normal blow, from the beginning of the impact, until friction can change from sliding to rolling. If R_0 is negative, the straight lines AL and SS' do not intersect on the positive side of the axis of F. In this case the friction is limiting throughout the impact. If R_0 is positive the representative point P reaches SS'. After this only so much friction is called into play as suffices to prevent sliding, provided that this amount is less than the limiting friction. If the acute angle which SS' makes with the axis of R is less than $\tan^{-1}\mu$, the friction dF necessary to prevent sliding is less than the limiting friction μdR. Hence P must travel along SS' in such a direction that the abscissa R continues to increase positively. In this case the friction does not again become limiting during the impact.

But if the acute angle which SS' makes with the axis of R is greater than $\tan^{-1}\mu$, the ratio of dF to dR is numerically greater than μ, and more friction is necessary to prevent sliding than can be called into play. The friction therefore continues to be limiting, and P, after reaching SS', must travel along a straight line, making the same angle with the axis of R that AL does. This straight line must lie on the opposite side of SS', because the acute angle which SS' makes with AR is greater than the angle LAR. Also since the point P has crossed SS' the direction of relative sliding and therefore the direction of friction is changed. In this case it is clear that the friction continues limiting throughout the impact.

An example of each of these three cases is given in the triple diagram. The figures differ in the position of the line of no sliding. In all the three figures the representative point travels from A along a straight line AL such that the angle LAR is

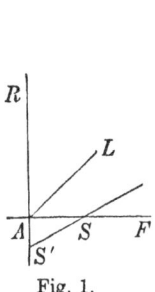

Fig. 1.

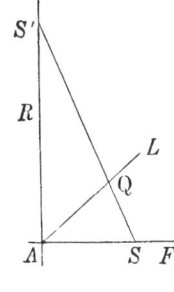

Fig. 2.

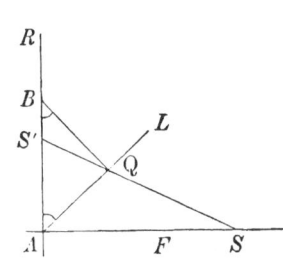

Fig. 3.

equal to $\tan^{-1} \mu$.   In fig. (1) the line of no sliding, viz. SS', makes so large an angle with AR that AL does not intersect it in the positive quadrant.   The friction therefore retains its limiting value throughout the impact.   In the other two figures AL and SS' intersect in some point Q.   In fig. (2) the angle $SS'A$ is less than the angle LAR, the representative point therefore after reaching Q travels along QS'.   In fig (3) the angle $SS'A$ is greater than the angle LAR, the representative point therefore after reaching Q travels along a straight line QB on the other side of SS' such that the angle QBA is equal to the angle QAR.

When P passes the straight line CC', compression ceases and restitution begins.   But the passage is marked by no peculiarity except this.   If R_1 be the abscissa of the point at which P crosses CC', the whole impact, for experimental reasons, is supposed to terminate when the abscissa of P is $R_2 = R_1 (1 + e)$, e being the measure of the elasticity of the two bodies.

It is obvious that a great variety of cases may occur according to the relative positions of the three straight lines AL, SS' and CC'.   But in all cases the progress of the impact may be traced by the method just explained, which may be briefly summed up in the following rule.   *The representative point* P *travels along* AL *until it meets* SS'.   *It then proceeds either along* SS', *or along a straight line making the same angle with the axis of* R *as* AL *does, but lying on the opposite side of* SS'.   *The one along which it proceeds is the steeper to the axis of* F.   *It travels along this line in such a direction as to make the abscissa* R *increase, and continues to be in this straight line to the end of the impact.   The complete value of* R *for the whole impact is found by multiplying the abscissa of the point at which* P *crosses* CC' *by* 1 + e.   *The complete value of* F *is the corresponding ordinate of* P.   *Substituting these in the dynamical equations* (1) *and* (2), *the motion just after impact may be easily found.*

If $S_0 = 0$, we have $S = - aF - bR$.   In this case the line of no sliding passes through the origin A.   If the acute angle which this straight line makes with the axis of R is less than $\tan^{-1} \mu$, i.e. if b/a is numerically less than μ, the representative point travels along this straight line in such a direction that its abscissa R continually increases.   The friction is therefore less than its limiting value throughout the impact.

If the acute angle which the line of no sliding makes with the axis of R is greater than $\tan^{-1} \mu$, i.e. if b/a is numerically greater than μ, the representative point travels along a straight line AL making with the axis of R an acute angle LAR equal to $\tan^{-1} \mu$. This straight line lies on the positive or negative side of AR according as S is positive or negative.   Since the numerical value of b is greater than $a\mu$, and $F = \pm \mu R$, the term $- bR$ governs the sign of S, hence S has the opposite sign to b.   It follows that *the straight line* AL *lies within the acute angle which the line of*

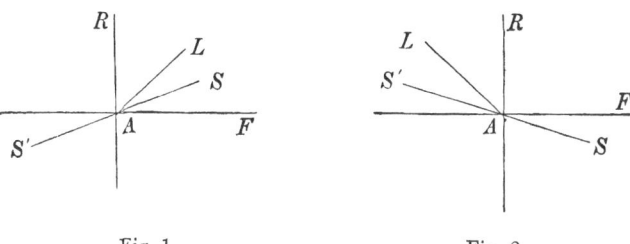

Fig. 1.   Fig. 2.

no sliding makes with AR.   Thus in fig. (1), AL is on the positive side, in fig. (2) on the negative side of AR.   As AL cannot again meet the line of no sliding the friction has its limiting value throughout the impact.

The representative point continues its journey along either SS' or AL, as the case may be, to the end of the impact.   The complete value of R for the whole impact is found by multiplying the abscissa of the point at which P crosses CC' by $1 + e$.   The complete value of F is the corresponding ordinate of P.   Substituting these in the dynamical equations the motion just after impact may be found.

195.   If the bodies are smooth, the straight line AL coincides with the axis of R.   The representative point P travels along the axis of R, and the complete value of R for the whole impact is found by multiplying the abscissa of C by $1 + e$.

If the bodies are perfectly rough (Art. 156), the straight line AL coincides with the axis of F.   The representative point P travels along the axis of F until it arrives at the point S.   It then travels along the line of no sliding SS' until it reaches the line CC' of greatest compression.   If the bodies are inelastic, the co-ordinates R_1, F_1, of this intersection are the values of R and F required.   But if the bodies are imperfectly elastic the representative point continues its journey along the line of no sliding.   The complete value of R for the whole impact is then $R_2 = R_1 (1 + e)$, and the complete value of F may be found by substituting this value for R in the equation to the line of no sliding.

196.   It is not necessary that the friction should keep the same direction during the impact.   The friction must keep one sign when P travels along AL.   But when P reaches SS', its

direction of motion changes, and the friction dF called into play in the time dt may have the same sign as before or the opposite. But it is clear that the friction can change sign only once during the impact. If $b = 0$, the straight line SS' is perpendicular to the axis of F, and in this case it is clear that the friction cannot change sign.

It is possible that the friction may continue limiting throughout the impact, so that the bodies slide on each other throughout. The necessary conditions are that either the straight line SS' must be less steep to the axis of F than AL, or the point P must not reach the straight line SS' until its abscissa has become greater than R_2. The condition for the first case is that b must be greater than μa. The abscissæ of the intersections of AL with SS' and CC' are respectively $R_0 = \dfrac{S_0}{a\mu + b}$ and $R_1 = \dfrac{C_0}{b\mu + a'}$. The necessary conditions for the second case are that R_1 must be positive, and R_0 either negative or positively greater than $R_1 (1 + e)$.

197. Ex. 1. **Rebound of a ball.** *A spherical ball moving without rotation on a smooth horizontal plane impinges with velocity* V *against a rough vertical wall whose coefficient of friction is* μ. *The line of motion of the centre of gravity before incidence making an angle* a *with the normal to the wall, determine the motion just after impact.*

This is the general problem of the motion of a spherical ball projected *without initial rotation* against any rough elastic plane. Thus it applies to a billiard ball impinging against a cushion, or to a "fives" ball projected against a wall, or to a cricket ball rebounding from the ground. When the ball has any initial rotation the problem is, in general, a problem in three dimensions and will be discussed further on.

In the figure the plane of the paper represents a horizontal plane drawn through the centre of the ball. The vertical plane against which the ball impinges intersects the plane of the paper in AS.

Let u, v be the velocities of the centre at any time t after the commencement of the impact resolved along and perpendicular to the wall. Let ω be the angular velocity at the same instant. Let R and F be the normal and frictional blows from the beginning of the impact up to that instant. Let M be the mass and r the radius of the sphere.

Then we have

$$\left. \begin{aligned} M(u - V \sin a) &= -F \\ M(v + V \cos a) &= R \\ Mk^2\omega &= Fr \end{aligned} \right\}.$$

The velocity of sliding of the point A of contact is

$$S = u - r\omega = V \sin a - \frac{r^2 + k^2}{k^2} \frac{F}{M}.$$

The velocity of compression of the point of contact is

$$C = -v = V \cos a - \frac{R}{M}.$$

Measure a length AS in the figure to represent $\frac{k^2}{r^2 + k^2} MV \sin a$, and a length

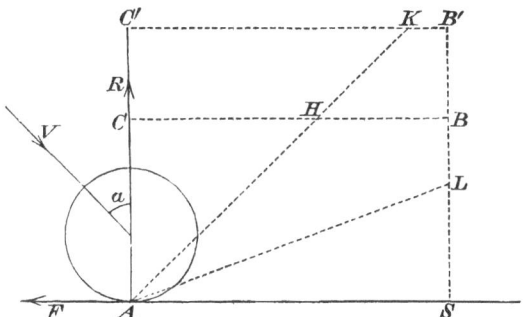

AC to represent $MV \cos a$, along the axes of F and R respectively. Then SB and CB drawn parallel to the directions of R and F will be the lines of *no sliding* and of *greatest compression*. Also we see that $\tan BAC = \frac{k^2}{r^2 + k^2} \tan a = \frac{2}{7} \tan a$. In the beginning of the impact the sphere slides on the wall, hence the *representative point* P, whose co-ordinates are R and F, begins to describe the straight line $F = \mu R$.

If $\mu > \frac{2}{7} \tan a$, this straight line cuts the line of no sliding SB in some point L before it cuts the line of greatest compression. Hence the representative point describes the broken line ALB. At the moment of greatest compression, F and R are the co-ordinates of B.

Therefore $F = \frac{2}{7} MV \sin a, \quad R = MV \cos a.$

These results are independent of μ because we see from the figure that more than enough friction could be called into play to destroy the sliding motion.

If $\mu < \frac{2}{7} \tan a$, the straight line $F = \mu R$ cuts the line of greatest compression CB in some point H before it cuts the line of no sliding. The friction is therefore insufficient to destroy the sliding. At the moment of greatest compression F and R are the co-ordinates of H,

$$F = \mu MV \cos a, \quad R = MV \cos a.$$

If the sphere be inelastic we have only to substitute these values of F and R in the equations of motion to find the values of u, v, ω just after impact.

If the sphere be imperfectly elastic with a coefficient of elasticity c, the representative point P will continue its progress until its abscissa is given by

$$R = MV \cos a (1 + e).$$

Take AC' to represent this value of R, and draw $C'B'$ parallel to CB. Then, as before, we see that $\tan B'AC' = \frac{2}{7} \frac{\tan a}{1 + e}$.

If $\mu > \frac{2}{7} \frac{\tan a}{1 + e}$, the representative point describes some broken line like ALB', and cuts SB' before it cuts $B'C'$. In this case F and R are the co-ordinates of B',

$$F = \frac{2}{7} MV \sin a, \quad R = MV \cos a (1 + e).$$

If $\mu < \dfrac{2}{7} \dfrac{\tan \alpha}{1+e}$, the representative point describes some unbroken line like AHK, and cuts $B'C'$ before it cuts SB'. In this case F and R are the co-ordinates of K,

$$F = \mu MV \cos \alpha (1+e), \quad R = MV \cos \alpha (1+e).$$

Let β be the angle the direction of motion of the centre of the ball makes with the normal to the wall after impact, then $\tan \beta = \dfrac{u}{v}$. We see therefore

$$\tan \beta = \frac{5}{7} \frac{\tan \alpha}{e}, \text{ or } = \frac{\tan \alpha - \mu (1+e)}{e},$$

according as μ is greater or less than $\dfrac{2}{7} \dfrac{\tan \alpha}{1+e}$.

Ex. 2. An imperfectly elastic cricket ball is projected so that it is rotating with an angular velocity Ω about a horizontal axis perpendicular to the plane of the parabola described by its centre. Just before it strikes the ground the velocity of the centre is V, and the direction of motion makes an angle α with the normal. Show that the angle of rebound β is given by either

$$e \tan \beta = \frac{5}{7} \tan \alpha + \frac{2}{7} \frac{r\Omega}{V \cos \alpha}, \text{ or } = \tan \alpha - \mu (1+e),$$

according as μ is greater or less than $\dfrac{2}{7} \left\{ \tan \alpha - \dfrac{r\Omega}{V \cos \alpha} \right\} \dfrac{1}{1+e}$.

Ex. 3. A sphere of radius a rolls on the ground with velocity U and impinges normally against a vertical wall whose coefficients of friction and elasticity are μ and e. If $\mu(1+e) > \frac{2}{7}$ the sliding will terminate before the end of the period of impact, and the sphere will therefore rebound with a horizontal velocity $-Ue$ and a vertical velocity $\frac{2}{7} U$ [this follows by taking moments about the point of contact]. The centre of the sphere will then describe a parabola and the sphere will afterwards impinge on the ground. If the ground be inelastic and have a coefficient of friction $\mu' < e + \frac{2}{7}$ the sliding will not terminate before the end of the impact. At the end of the impact the centre of the sphere has a velocity $- U (e - \frac{2}{7} \mu')$ and the angular velocity is $(2 - 5\mu') U/7a$. The friction continues to act as a finite force so that the sphere finally rolls on the ground with a uniform velocity equal to

$$- \tfrac{5}{7} U (e - \tfrac{4}{35}).$$

Ex. 4. A lamina whose plane is vertical falls in its own plane and impinges on an imperfectly rough elastic horizontal plane. At the moment before impact the velocity of the centre of gravity G is vertically downwards and equal to V, and the angular velocity Ω is in such a direction that the vertical velocity of the point A of impact is $V - x\Omega$, where x, y are the horizontal and vertical co-ordinates of G referred to A as origin. Let u, v be the horizontal and vertical velocities of G just after impact, v being measured upwards. Show that three cases may occur.

(1) If $V x y k^2 < \mu \{ (V - x\Omega) k^2 + V y^2 \}$,

then $(v + V)(k^2 + x^2 + y^2) = \{ (V - x\Omega) k^2 + V y^2 \} (1+e),$

$$u (k^2 + y^2) = xy (v + V) + \Omega y k^2.$$

(2) Let $A = x^2 + k^2 - \mu xy$, $B = (y^2 + k^2) \mu - xy$. If the inequality in (1) be reversed, it may be shown that A will be positive. If also B be positive and

$$A y \Omega < B (V - x\Omega)(1+e)$$

then $(v + V) A = (V - x\Omega)(1+e),$

and u is found from $v + V$ by the same formula as in (1).

(3) The inequality in (1) being still reversed, if *firstly* B be positive and the inequality in (2) be reversed, or *secondly* B be negative, then $v + V$ is found by the same formula as in (2), but $u = \mu(v + V)$.

198. Ex. 1. Show that the representative point P as it travels in the manner described in the text must cross the line of greatest compression, and that the abscissa R of the point at which it crosses this straight line must be positive.

Ex. 2. Show that the conic whose equation referred to the axes of R and F is $aF^2 + 2bFR + a'R^2 = \epsilon$, where ϵ is some constant, is an ellipse, and that the straight lines of no sliding and greatest compression are parallel to the conjugates of the axes of F and R respectively. Show also that the intersection of the straight lines of no sliding and greatest compression must lie, in that angle formed by the conjugate diameters which contains or is contained by the first quadrant.

Ex. 3. Two bodies, each turning about a fixed point, impinge on each other, find the motion just after impact.

Let G, G', in the figure of Art. 187, be taken as the fixed points. Taking moments about the fixed points, the results will be nearly the same as those given in the case considered in the text.

Ex. 4. Show that the Vis Viva lost when two bodies impinge on each other may be found from either of the formulæ

$$\text{Vis Viva lost} = 2FS_0 + 2RC_0 - aF^2 - 2bFR - a'R^2$$
$$= \frac{(aC_0{}^2 - 2bS_0C_0 + a'S_0{}^2) - (aC^2 - 2bSC + aS^2)}{aa' - b^2},$$

where F, R are the whole frictional and normal forces called into play, and C_0, S_0, C, S are the initial and final values of the velocities of compression and sliding. If the bodies are perfectly rough and inelastic C and S are both zero.

Initial Motions.

199. **Breakage of a support.** Suppose that a system of bodies is in equilibrium and that one of the supports suddenly gives way. It is required to find the initial motions of the several bodies and the initial values of the reactions which exist between them.

The problem of finding the initial motion of a dynamical system is the same as that of expanding the co-ordinates of the moving particles in powers of the time t. Let (x, y, θ) be the co-ordinates of any body of the system. For the sake of brevity let the suffix zero denote initial values. Thus $\ddot{x}_0$ denotes the initial value of $\ddot{x}$. By Taylor's theorem we have

$$x = a + \ddot{x}_0 \frac{t^2}{\underline{|2}} + \dddot{x}_0 \frac{t^3}{\underline{|3}} + \ldots\ldots\ldots\ldots\ldots\ldots(1):$$

the term $\dot{x}_0$ is omitted because we suppose the system to start from rest.

Firstly, let only the initial values of the reactions be required.
The dynamical equations contain the co-ordinates, their second
differential coefficients with regard to t, and the unknown
reactions. There are as many geometrical equations as reactions. From these we have to eliminate the second differential
coefficients and find the reactions. The process, which is really
the same as the first method of solution described in Art. 135,
is as follows.

Write down the geometrical equations, differentiate each twice
and then simplify the results by substituting for the co-ordinates
their initial values. Thus, if we use Cartesian co-ordinates, let
$\phi(x, y, \theta) = 0$ be any geometrical relation, we have since $\dot{x}_0 = 0$,
$\dot{y}_0 = 0$, $\dot{\theta}_0 = 0$,

$$\frac{d\phi}{dx}\ddot{x}_0 + \frac{d\phi}{dy}\ddot{y}_0 + \frac{d\phi}{d\theta}\ddot{\theta}_0 = 0.$$

The process of differentiating the equations may sometimes
be much simplified when the origin has been so chosen that the
initial values of some at least of the co-ordinates are zero. We
may then simplify the equations by neglecting the squares and
products of all such co-ordinates. For if we have a term x^2, its
second differential coefficient is $2(x\ddot{x} + \dot{x}^2)$, and if the initial value
of x is zero, this vanishes.

The geometrical equations must be obtained by supposing the
bodies to have their *displaced* positions, because we require to
differentiate them. But this is not the case with the dynamical
equations. These we may write down on the supposition that
each body is in its *initial* position. These equations may be
obtained according to the rules given in Art. 135. The forms
there given for the effective forces admit in this problem of some
simplifications. Thus, since $\dot{r}_0 = 0$, $\dot{\phi}_0 = 0$, the accelerations along
and perpendicular to the radius vector take the simple forms $\ddot{r}_0$
and $r\ddot{\phi}_0$. So again the acceleration $\dfrac{v^2}{\rho}$ along the normal vanishes.

If, for example, we know the initial direction of motion of the
centre of gravity of any one of the bodies, we may conveniently
resolve along the normal to the path. This will supply an equation
which contains only the impressed forces and such tensions or reactions as may act on the body. If there be only one reaction,
this equation will suffice to determine its initial value.

The rule may be shortly stated thus. *Write down the geometrical equations of the system in its general position. Differentiate
each twice and then simplify the results by substituting for the co-
ordinates their initial values. Write down the dynamical equations
of the system supposed to be in its initial position. Eliminate the
second differential coefficients and we shall have sufficient equations
to find the initial values of the reactions.*

We may also deduce from the equations the values of $\ddot{x}_0$, $\ddot{y}_0$, $\ddot{\theta}_0$, and thus by substituting in equation (1) we have found the initial motion up to terms depending on t^2.

200. *Secondly, let the initial motion be required.* As differential coefficients of a high order sometimes present themselves in this part of the problem it will be more convenient to use accents instead of dots to represent the differential coefficients with regard to the time. Thus $\ddot{x}$ will be written x''.

The number of terms of the series (1) which it may be necessary to retain depends on the nature of the problem. Suppose the radius of curvature of the path described by the centre of gravity of one of the bodies to be required. We have

$$\rho = \frac{(x'^2 + y'^2)^{\frac{3}{2}}}{x'y'' - y'x''},$$

and by differentiating equation (1)

$$x' = x_0''t + x_0''' \frac{t^2}{\lfloor 2} + x_0'''' \frac{t^3}{\lfloor 3} + \dots$$

$$x'' = x_0'' + x_0'''t + x_0'''' \frac{t^2}{\lfloor 2} + \dots$$

&c. = &c.;

$$\therefore (x'^2 + y'^2)^{\frac{3}{2}} = (x_0''^2 + y_0''^2)^{\frac{3}{2}} t^3 + \dots$$

$$x'y'' - y'x'' = (x_0''y_0''' - x_0'''y_0'') \frac{t^2}{2} + (x_0''y_0'''' - x_0'''' y_0'') \frac{t^3}{3} + \dots$$

These results may also be obtained by a direct use of Taylor's theorem.

If then the body start from rest, the radius of curvature is zero. But if $x_0''y_0''' - x_0'''y_0'' = 0$, the direction of the acceleration is stationary for a moment. We then have

$$\rho = 3 \frac{(x_0''^2 + y_0''^2)^{\frac{3}{2}}}{x_0''y_0'''' - x_0''''y_0''}.$$

To find these differential coefficients we may proceed thus. Differentiate each dynamical equation twice and then reduce it to its initial form by writing for x, y, θ, &c. their initial values, and for x', y', θ' zero. Differentiate each geometrical equation four times and then reduce each to its initial form. We shall thus have sufficient equations to determine x_0'', x_0''', x_0'''', &c., R_0, R_0', R_0'', &c., where R is any one of the unknown reactions. It is often of advantage to eliminate the unknown reactions from the equations *before* differentiation. We then have only the unknown coefficients x_0'', x_0''', &c. entering into the equations.

If we know the direction of motion of one of the centres of gravity under consideration, we can take the axis of y a tangent to its path. We then have $\rho = \dfrac{y^2}{2x}$, where x is of the second order and y of the first order of small quantities. We may therefore neglect the squares of x and the cubes of y. This will greatly simplify the equations. If the body start from rest we have $x_0' = 0$, and if $x_0'' = 0$, we may then use the formula $\rho = 3 \dfrac{y_0''^2}{x_0'''}$.

201. *Ex. A circular disc is hung up by three equal strings attached to three points at equal distances on its circumference, and fastened to a peg vertically over the centre of the disc. One of these strings being cut, determine the initial tensions of the other two.*

Let O be the peg, AB the circle seen by an eye in its plane. Let OA be the string which is cut, let C be the middle point of the chord joining the points of the

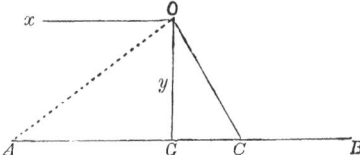

circle to which the two other strings are attached. Then the two tensions, each equal to T, are throughout the motion equivalent to a resultant tension R along CO. If 2α be the angle between the two strings, we have $R = 2T \cos \alpha$.

Let l be the length of OC, β the angle GOC, a the radius of the disc. Let (x, y) be the co-ordinates of the *displaced* position of the centre of gravity with reference to the origin O, x being measured horizontally to the left and y vertically downwards. Let θ be the angle which the displaced position of the disc makes with AB.

By drawing the disc in its displaced position it will be seen that the co-ordinates of the displaced position of C are $x - l \sin \beta \cos \theta$ and $y - l \sin \beta \sin \theta$. Hence since the length OC remains constant and equal to l, we have

$$x^2 + y^2 - 2l \sin \beta \, (x \cos \theta + y \sin \theta) = l^2 \cos^2 \beta.$$

Since the initial tensions only are required, it is sufficient to differentiate this twice. Since we may neglect the squares of small quantities, we may omit x^2, and put $\cos \theta = 1$, $\sin \theta = \theta$. The process of differentiation will not then be very long, for it is easy to see beforehand what terms will disappear when we equate the differential coefficients $(\dot{x}, \dot{y}, \theta)$ to zero, and put for (x, y, θ) their initial values $(0, l \cos \beta, 0)$. We get

$$\ddot{y}_0 \cos \beta = \sin \beta \, (\ddot{x}_0 + l \cos \beta \, \ddot{\theta}_0).$$

This equation may also be obtained by an artifice which is often useful. The motion of G is made up of the motion of C and the motion of G relatively to C. Since C begins to describe a circle from rest, its acceleration along CO is zero. Again, the acceleration of G relatively to C when resolved along CO is $GC\ddot{\theta} \cos \beta$. The resolved acceleration of G is the sum of these two, but it is also equal to $\ddot{y}_0 \cos \beta - \ddot{x}_0 \sin \beta$. Hence the equation follows at once.

In this problem we require the dynamical equations only in their initial form. These are

$$m\ddot{x}_0 = R_0 \sin\beta$$
$$m\ddot{y}_0 = mg - R_0 \cos\beta$$
$$mk^2\ddot{\theta}_0 = R_0 l \sin\beta \cos\beta$$

where m is the mass of the body. Substituting in the geometrical equation we find

$$R_0 = mg \cdot \frac{\cos\beta}{1 + \dfrac{l^2}{k^2}\sin^2\beta\cos^2\beta} \cdot$$

The tension of any string, *before* the string OA was cut, may be found by the rules of statics, and is clearly $T_1 = \dfrac{mg}{3\cos\gamma}$, where γ is the angle AOG. Hence the change of tension can be found.

202. **Ex. 1.** Two strings of equal length have each an extremity tied to a weight C and their other extremities tied to two points A, B in the same horizontal line. If one be cut the tension of the other will be instantaneously altered in the ratio $1 : 2\cos^2\dfrac{C}{2}$. [St Pet. Coll.]

Ex. 2. An elliptic lamina is supported with its plane vertical and transverse axis horizontal by two weightless pins passing through the foci. If one pin be released show that, if the eccentricity of the ellipse be $\frac{1}{5}\sqrt{10}$, the pressure on the other pin is initially unaltered. [Coll. Exam.]

Ex. 3. Three equal particles A, B, C repelling each other with any forces, are tied together by three strings of unequal length, so as to form a triangle right-angled at A. If the string joining B and C be cut, prove that the instantaneous changes of tension of the strings joining BA, CA will be $\frac{1}{2}T\cos B$ and $\frac{1}{2}T\cos C$ respectively, where B and C are the angles opposite the strings joining CA, AB respectively, and T is the repulsive force between B and C.

Ex. 4. Two uniform equal rods, each of mass m, are placed in the form of the letter X on a smooth horizontal plane, the upper and lower extremities being connected by equal strings; show that, whichever string be cut, the tension of the other is the same function of the inclination of the rods, and initially is $\frac{3}{2}mg\sin\alpha$, where α is the initial inclination of the rods. [St Pet. Coll.]

Ex. 5. A horizontal rod of mass m and length $2a$ hangs by two parallel strings of length $2a$ attached to its ends: an angular velocity ω being suddenly communicated to it about a vertical axis through its centre, show that the initial increase of tension of either string equals $\frac{1}{4}ma\omega^2$, and that the rod rises through a space $a^2\omega^2/6g$. [Coll. Exam.]

Ex. 6. A particle is suspended by three equal strings of length a from three points forming an equilateral triangle of side $2b$ in a horizontal plane. If one string be cut the tension of each of the others will be instantaneously changed in the ratio $\dfrac{3a^2 - 4b^2}{2(a^2 - b^2)}$. [Coll. Exam.]

Ex. 7. A sphere resting on a rough horizontal plane is divided into an infinite number of solid lunes and tied together again with a string; the axis through which the plane faces of the lunes pass being vertical. Show that if the string be cut the pressure on the plane will be instantaneously diminished in the ratio $45\pi^2 : 2048$. [Emm. Coll.]

R. D. 12

Ex. 8.   Three equal and similar rods moveable about one common extremity are held at right angles to each other so that the three other extremities are in a horizontal plane with the common extremity either above or below. Show that if they are dropped on a smooth inelastic horizontal plane, the velocity of their centre of gravity is diminished by one-half.

Ex. 9.   A small ring of mass p is strung on a rod, of mass m and length $2a$, capable of turning about one extremity as a fixed point. The system starts from rest with the rod horizontal and the ring at a distance c from the fixed point. Show that the polar co-ordinates of the ring referred to the fixed point are $c + r_0''''t^4/24$ and $\ddot{\theta}_0 t^2/2$. Find also $\ddot{\theta}_0$, and prove that $r_0'''' = g\ddot{\theta}_0 + 2c\ddot{\theta}_0{}^2$. Thence find the initial radius of curvature of the path of the particle. [May Exam. 1888.]

On Relative Motion or Moving Axes.

203.   In many dynamical problems the relative motion of the different bodies of the system is all that is required. In such cases it will be an advantage if we can determine this without finding the absolute motion of each body in space. Let us suppose that the motion relative to some one body (A) is required. There are then two cases to be considered, (1) when the body (A) has a motion of translation only, and (2) when it has a motion of rotation only. The case in which the body (A) has a motion both of translation and rotation may be regarded as a combination of these two cases. Let us consider them in order.

204.   **The Fundamental Theorem.**   Let it be required to find the motion of any dynamical system relative to some moving point C. We may clearly reduce C to rest by applying to every element of the system an acceleration equal and opposite to that of C. It is also necessary to suppose that an *initial* velocity equal and opposite to that of C has been applied to each element.

Let f be the acceleration of C at any time t. If every particle m of a body be acted on by the same accelerating force f parallel to any given direction, it is clear that these are together equivalent to a force $f\Sigma m$ acting at the centre of gravity. Hence to reduce any point C of a system to rest, it will be sufficient to apply to the centre of gravity of each body in a direction opposite to that of the acceleration of C a force measured by Mf, where M is the mass of the body and f the acceleration of C.

The point C may now be taken as the origin of co-ordinates. We may also take moments about it as if it were a point fixed in space.

Let us consider the equation of moments a little more minutely. Let (r, θ) be the polar co-ordinates of any element of a body whose mass is m referred to C as origin. The accelerations of the particle are $\dfrac{d^2r}{dt^2} - r\left(\dfrac{d\theta}{dt}\right)^2$ and $\dfrac{1}{r}\dfrac{d}{dt}\left(r^2\dfrac{d\theta}{dt}\right)$, along and perpendicular to the radius vector r. Taking moments about C we get

$$\Sigma m\,\frac{d}{dt}\left(r^2\frac{d\theta}{dt}\right) = \left\{ \begin{array}{l} \text{moment round } C \text{ of the impressed forces} \\ \text{plus the moment round } C \text{ of the reversed} \\ \text{effective forces of } C \text{ supposed to act at the} \\ \text{centre of gravity.} \end{array} \right.$$

If the point C be fixed in the body and move with it, $\dfrac{d\theta}{dt}$ will be the same for every element of the body, and, as in Art. 88, we have $\Sigma m\,\dfrac{d}{dt}\left(r^2\dfrac{d\theta}{dt}\right) = Mk^2\dfrac{d^2\theta}{dt^2}$.

205. From the general equation of moments about a moving point C we learn that we may use the equation

$$\frac{d\omega}{dt} = \frac{\text{moment of forces about } C}{\text{moment of inertia about } C}$$

in the following cases.

Firstly. If the point C be fixed both in the body and in space; or if the point C, being fixed in the body, move in space with uniform velocity; for the acceleration of C is zero.

Secondly. If the point C be the centre of gravity; for in that case, though the acceleration of C is not zero, yet the moment vanishes.

Thirdly. If the point C be the instantaneous centre of rotation, and the motion be a small oscillation or an initial motion which starts from rest. At the time t the body is turning about C, and the velocity of C is therefore zero. At the time $t + dt$, the body is turning about some point C' very near to C. Let $CC' = d\sigma$, then the velocity of C is $\omega d\sigma$. Hence in the time dt the velocity of C has increased from zero to $\omega d\sigma$, therefore its acceleration is $\omega\,\dfrac{d\sigma}{dt}$. To obtain the accurate equation of moments about C we must apply the effective force $\Sigma m\,.\,\omega\,\dfrac{d\sigma}{dt}$ in the reversed direction at the centre of gravity. But in small oscillations ω and $\dfrac{d\sigma}{dt}$ are both small quantities whose squares and products are to be neglected, and in an initial motion ω is zero. Hence the moment of this force must be neglected, and the equation of motion will be the same as if C had been a fixed point.

It is to be observed that we may take moments about any point very near to the instantaneous centre of rotation, but it will usually be more convenient to take moments about the centre in its disturbed position.   If there be any unknown reactions at the centre of rotation, their moments will then be zero.

206.   If the accurate equation of moments about the instantaneous centre be required, we may proceed thus.   Let L be the moment of the impressed forces about the instantaneous centre, G the centre of gravity, r the distance between the centre of gravity and the instantaneous centre C, M the mass of the body; then the moment of the impressed forces and the reversed effective forces about C is

$$L - M\omega \frac{d\sigma}{dt} . r \cos GC'C.$$

If k be the radius of gyration about the centre of gravity, the equation of motion becomes

$$M (k^2 + r^2) \frac{d\omega}{dt} = L - M\omega r \frac{dr}{dt},$$

writing for $\cos GC'C$ its value $\dfrac{dr}{d\sigma}$.

207.   **Impulsive forces.**   The argument of Art. 204 may evidently be also applied to impulsive forces.   We may thus obtain very simply a solution of the problem considered in Art. 171.

A body is moving in any manner when suddenly a point O in the body is constrained to move in some given manner, it is required to find the motion relative to O.

To reduce O to rest, we must apply at the centre of gravity G a momentum equal to Mf, where f is the resultant of the reversed velocity of O after the change and the velocity of O before the change.   If ω, ω' be the angular velocities of the body before and after the change, and $r = OG$, we have by taking moments about O,

$$(r^2 + k^2) (\omega' - \omega) = \text{moment of } f \text{ about } O.$$

Now the moment about O of a velocity at G is equal and opposite to the moment about G of the same velocity applied at O.   Hence if L, L' be the moments about G of the velocity of O just before and just after the change, and k be the radius of gyration about the centre of gravity, we have $\omega' - \omega = \dfrac{L' - L}{k^2 + r^2}$.

208.   Ex. 1.   *Two heavy particles whose masses are* m *and* m' *are connected by an inextensible string, which is laid over the vertex of a double inclined plane whose mass is* M, *and which is capable of moving freely on a smooth horizontal plane. Find the force which must act on the wedge that the system may be in a state of relative equilibrium.*

Here it will be convenient to reduce the wedge to rest by applying to every particle an acceleration f equal and opposite to that of the wedge.   Supposing this done the whole system is in equilibrium.   If F be the required force, we have by resolving horizontally $(M + m + m') f = F$.

Let α, α' be the inclinations of the sides of the wedge to the horizontal.   The particle m is acted on by mg vertically and mf horizontally.   Hence the tension of the string is $m(g \sin \alpha + f \cos \alpha)$.   By considering the particle m', we find the tension to be also $m'(g \sin \alpha' - f \cos \alpha')$.   Equating these two we have

$$f = \frac{m' \sin \alpha' - m \sin \alpha}{m' \cos \alpha' + m \cos \alpha} g.$$

Hence F is found.

209. Ex. 2. *A cylindrical cavity whose section is any oval curve and whose generating lines are horizontal is made in a cubical mass which can slide freely on a smooth horizontal plane. The surface of the cavity is perfectly rough and a sphere is placed in it at rest so that the vertical plane through the centres of gravity of the mass and the sphere is perpendicular to the generating lines of the cylinder. A momentum B is communicated to the cube by a blow in this vertical plane. Find the motion of the sphere relatively to the cube and the least value of the blow that the sphere may not leave the surface of the cavity.*

Simultaneously with the blow B there will be an impulsive friction between the cube and the sphere. Let M, m be the masses of the cube and sphere, a the radius of the sphere, k its radius of gyration about a diameter. Let V_0 be the initial velocity of the cube, v_0 that of the centre of the sphere *relatively* to the cube, ω_0 the initial angular velocity. Then by resolving horizontally for the whole system, and taking moments for the sphere alone about the point of contact, we have

$$\left. \begin{aligned} m(v_0 + V_0) + MV_0 &= B \\ a(v_0 + V_0) + k^2\omega_0 &= 0 \end{aligned} \right\} \quad \dots\dots\dots\dots\dots\dots\dots\dots\dots\dots\dots\dots(1),$$

and since there is no sliding $v_0 - a\omega_0 = 0$...(2).

To find the subsequent motion, let (x, y) be the co-ordinates of the centre of the sphere referred to rectangular axes attached to the cubical mass, x being horizontal and y vertical, then, the equation to the cylindrical cavity being given, y is a known function of x. Let ψ be the angle which the tangent to the cavity at the point of contact of the sphere makes with the horizon, then $\tan \psi = \dfrac{dy}{dx}$. Let V be the velocity of the cubical mass, then, by Art. 132, $m\left(\dfrac{dx}{dt} + V\right) + MV = B$.......................(3).

If T_0 be the initial vis viva and y_0 the initial value of y, we have by the equation of vis viva

$$m\left\{ \left(\frac{dx}{dt} + V\right)^2 + \left(\frac{dy}{dt}\right)^2 + k^2\omega^2 \right\} + MV^2 = T_0 - 2mg(y - y_0) \quad \dots\dots(4),$$

where ω is the angular velocity of the sphere at the time t. If v be the velocity of the centre of the sphere relatively to the cube, we have since there is no sliding $v = a\omega$. Eliminating V and ω from these equations, we have

$$\left(\frac{dx}{dt}\right)^2 \cdot \left\{ (1 + \tan^2 \psi)\left(1 + \frac{k^2}{a^2}\right) - \frac{m}{M + m} \right\} = Cg - 2gy \dots\dots\dots\dots(5),$$

where

$$Cg = \frac{B^2}{(M + m)\left\{ M + (M + m)\dfrac{k^2}{a^2} \right\}} + 2gy_0 \dots\dots\dots\dots\dots\dots(6).$$

This equation gives the motion of the sphere relatively to the cube.

210. To find the pressure on the cube, let us reduce the cube to rest. Let R be the normal pressure of the sphere on the cube, F the friction measured positively in the direction in which the arc is measured. The whole effective force on the cube is $X = R \sin \psi + F \cos \psi$. By Art. 204 we must apply to every particle an acceleration

$\frac{X}{M}$ opposite to this force. The sphere will therefore be acted on by a force $\frac{m}{M} X$ in a horizontal direction in addition to the reaction R, the friction F and its own weight.

Taking moments about the centre, we have $mk^2 \frac{d\omega}{dt} = Fa$ ....(7).

Resolving along a tangent to the path, $m \frac{dv}{dt} = -F - \frac{m}{M} X \cos \psi - mg \sin \psi$ ....(8).

But since there is no sliding, we have $v = a\omega$ ....(9).

Differentiating this and substituting from (7) and (8), we find

$$F = -R \frac{\gamma \sin \psi \cos \psi}{1 + \gamma \cos^2 \psi} - mg \frac{k^2}{a^2 + k^2} \frac{\sin \psi}{1 + \gamma \cos^2 \psi} \ldots\ldots\ldots\ldots(10),$$

where $\gamma = \frac{k^2}{a^2 + k^2} \frac{m}{M}$. Resolving the forces on the centre of the sphere along a normal to the path, we have

$$\frac{mv^2}{\rho} = R + \frac{m}{M} X \sin \psi - mg \cos \psi \ldots\ldots\ldots\ldots\ldots(11),$$

where ρ is the radius of curvature of the path. Substituting for v^2 its value given by (5), which may be conveniently written in the form

$$v^2 (1 - \beta \cos^2 \psi) = \frac{a^2}{a^2 + k^2} (C - 2y) g \ldots\ldots\ldots\ldots(12),$$

where $\beta = \frac{a^2}{a^2 + k^2} \frac{m}{M + m}$, we have two equations to find the reactions F and R.

Eliminating F, we get

$$C - 2y + \rho \cos \psi \frac{1 - \beta \cos^2 \psi}{1 + \gamma \cos^2 \psi} \frac{\beta + \gamma}{\beta} = \frac{R}{mg} \rho P \ldots\ldots\ldots\ldots(13),$$

where P is rather a complicated function of ψ which is not generally wanted. We have

$$P = \frac{(1 - \beta \cos^2 \psi)^2}{1 + \gamma \cos^2 \psi} \frac{\beta + \gamma}{\beta (1 - \beta)} \ldots\ldots\ldots\ldots\ldots\ldots(14).$$

We notice that, since β is necessarily less than unity, P cannot vanish and is always finite and positive.

If the sphere is to go all round the cavity, it is necessary that the value of v as given by (12) should be real for all values of y and $\cos \psi$. Hence the value of C as found by (6) must be greater than the greatest value of $2y$. It is also necessary that R should be always positive, so that the values of $\cos \psi$ given by the equation (13) when $R = 0$ must be all imaginary or numerically greater than unity. We observe that, if $C > 2y$ and ρ be always positive, R cannot vanish for any positive value of $\cos \psi$.

If the equation (13), when $R = 0$, have two equal roots which are less than unity, the pressure on the cavity vanishes but does not change sign. In this case the sphere does not leave the cavity at the point indicated by this value of $\cos \psi$. The condition for equal roots gives us

$$\frac{d}{d\psi} \left\{ \rho \cos \psi \frac{1 - \beta \cos^2 \psi}{1 + \gamma \cos^2 \psi} \right\} = \frac{2\beta}{\beta + \gamma} \rho \sin \psi \ldots\ldots\ldots\ldots(15),$$

where ρ is given as a function of ψ from the equation to the cylinder. Writing $\xi = \cos \psi$ for brevity, this reduces to

$$\frac{d \log \rho}{d\psi} \xi (1 - \beta \xi^2)(1 + \gamma \xi^2)(\beta + \gamma) = \sin \psi \{3\beta + \gamma - (3\beta^2 + \gamma^2) \xi^2 + \beta \gamma (\gamma - \beta) \xi^4\} \ldots\ldots(16).$$

If no other real value of $\cos \psi$ makes R vanish and change sign in (13), and if also $C > 2y$ the sphere is said *just to go round*. We may put this reasoning in another way. If the sphere is *just* to go round, then R must be positive throughout and must vanish at the point where it is least. In this case we have R and $\dfrac{dR}{d\psi}$ simultaneously zero. Differentiating (13) we notice that the differential coefficient of the right-hand side is zero, except at some singular points where ρ or $\dfrac{d\rho}{d\psi}$ is infinite. We notice also that the constant C which depends on the initial conditions disappears. In this way we again obtain equation (15).

It should be observed that the point where the pressure vanishes and is a minimum cannot be the highest point of the cavity unless the radius of curvature ρ is a maximum or minimum at that point. This follows at once from equation (16).

If we wish to find the blow B that the sphere may just go round we must examine the roots of the equations (13) and (16). To effect this we trace the curve whose abscissa is ξ and ordinate η, where η is the left-hand side of (13), from $\xi = 0$ to $\xi = -1$. The curve may undulate and the maxima and minima ordinates are given by (16). If the sphere goes round, the value of C must be such that every ordinate between $\xi = 0$ and $\xi = -1$ must be positive. We therefore examine the roots of (16) and select that root which makes η least. The value of C is found by equating this value of η to zero. The value of C having been found, that of B is known from (6). The result of course is subject to the limitations mentioned above.

211. Moving Axes. Next, let us consider the case in which we wish to refer the motion to two straight lines $O\xi$, $O\eta$ at right angles, turning round a fixed origin O with angular velocity ω.

Let Ox, Oy be any fixed axes at right angles and let the angle $xO\xi = \theta$. Let $\xi = OM$, $\eta = PM$ be the co-ordinates of any point P. Let u, v be the resolved velocities and X, Y the resolved accelerations of the point P in the directions $O\xi$, $O\eta$.

It is evident that the motion of P is made up of the motions of the two points M, N by simple addition. The resolved parts of the velocity of M are $\dfrac{d\xi}{dt}$ and $\xi\omega$ along and perpendicular to OM.

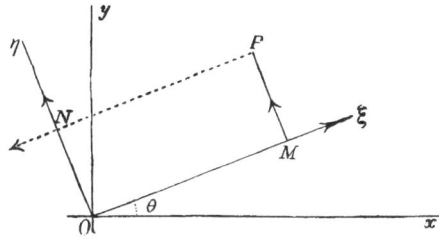

The resolved parts of the velocity of N are in the same way $\dfrac{d\eta}{dt}$

and $\eta\omega$ along and perpendicular to ON. By adding these with their proper signs we have

$$u = \frac{d\xi}{dt} - \eta\omega, \quad v = \frac{d\eta}{dt} + \xi\omega.$$

Since acceleration is the rate of increase of velocity just as velocity is the rate of increase of space, we obtain the corresponding formulæ for X, Y by writing u, v for x, y. We thus have

$$X = \frac{du}{dt} - v\omega, \quad Y = \frac{dv}{dt} + u\omega.$$

In the same way by adding the accelerations of M and N we have

$$X = \frac{d^2\xi}{dt^2} - \xi\omega^2 - \frac{1}{\eta}\frac{d}{dt}(\eta^2\omega),$$

$$Y = \frac{d^2\eta}{dt^2} - \eta\omega^2 + \frac{1}{\xi}\frac{d}{dt}(\xi^2\omega).$$

By using these formulæ instead of $\dfrac{d^2x}{dt^2}$ and $\dfrac{d^2y}{dt^2}$ we may refer the motion to the moving axes $O\xi$, $O\eta$.

212. **Ex. 1.** Let the axes $O\xi$, $O\eta$ be oblique and make an angle α with each other, prove that, if the velocity be represented by the two components u, v parallel to the axes,

$$u = \dot{\xi} - \omega\xi \cot \alpha - \omega\eta \operatorname{cosec} \alpha,$$
$$v = \dot{\eta} + \omega\eta \cot \alpha + \omega\xi \operatorname{cosec} \alpha.$$

In this case PM is parallel to $O\eta$. The velocities of M and N are the same as before. Their resultant is, by the question, the same as the resultant of u and v. By resolving in any two directions and equating the components we get two equations to find u and v. The best directions to resolve along are those perpendicular to $O\xi$ and $O\eta$, for then u is absent from one of the equations and v from the other. Thus u or v may be found separately when the other is not wanted.

Ex. 2. If the acceleration be represented by the components X and Y, prove

$$X = \dot{u} - \omega u \cot \alpha - \omega v \operatorname{cosec} \alpha,$$
$$Y = \dot{v} + \omega v \cot \alpha + \omega u \operatorname{cosec} \alpha.$$

These may be obtained in the same way by resolving velocities and accelerations perpendicular to $O\xi$ and $O\eta$.

Ex. 3. If u, v be the velocities of a point P referred to rectangular moving axes rotating with an angular velocity ω, prove that the radius of curvature of the *path of P in space* is given by

$$(u^2 + v^2)^{\frac{3}{2}}/\rho = u\dot{v} - v\dot{u} + (u^2 + v^2)\,\omega.$$

By taking fixed axes coincident for a moment with the moving axes the left side of this equation is seen to be $\dot{x}\ddot{y} - \ddot{x}\dot{y}$. Substituting $\dot{x} = u$, $\dot{y} = v$, and for $\ddot{x} = X$, $\ddot{y} = Y$ their values given above the result follows at once.

The ordinary expression for ρ in polar co-ordinates follows from this by writing $u = \dot{r}$, $v = r\dot{\theta}$, $\omega = \dot{\theta}$. If the independent variable is θ we have $\dot{\theta} = 1$.

Ex. 4.   In the case of initial motions which start from rest the formula for ρ in the last example becomes nugatory.   Show by proceeding as in Art. 200 that $\rho = 0$ unless $\dot{u}\ddot{v} - \ddot{u}\dot{v} + 2(\dot{u}^2 + \dot{v}^2)\,\omega = 0$, and that in that case

$$(\dot{u}^2 + \dot{v}^2)^{\frac{3}{2}}/\rho = \tfrac{1}{3}(\dot{u}\dddot{v} - \dot{v}\dddot{u}) + (\dot{u}\ddot{u} + \dot{v}\ddot{v})\,\omega + (\dot{u}^2 + \dot{v}^2)\,\dot{\omega}$$

where $\dot{u}$, $\ddot{u}$ &c. $\dot{v}$, $\ddot{v}$ &c. represent their initial values, the suffix zero being omitted for the sake of brevity.

213.   Ex.   *A particle under the action of any forces moves on a smooth curve which is constrained to turn with angular velocity ω about a fixed axis.   Find the motion relative to the curve.*

Let us suppose the motion to be in three dimensions.   Take the axis of Z as the fixed axis, and let the axes of ξ, η be fixed relatively to the curve.   Let the mass be the unit of mass.   Then the equations of motion are

$$\left. \begin{aligned} \frac{d^2\xi}{dt^2} - \xi\omega^2 - \frac{1}{\eta}\frac{d}{dt}(\eta^2\omega) &= X + Rl \\[2mm] \frac{d^2\eta}{dt^2} - \eta\omega^2 + \frac{1}{\xi}\frac{d}{dt}(\xi^2\omega) &= Y + Rm \\[2mm] \frac{d^2z}{dt^2} &= Z + Rn \end{aligned} \right\} \quad \dots\dots\dots\dots\dots(1),$$

where X, Y, Z are the resolved parts of the impressed accelerating forces in the directions of the axes, R is the pressure on the curve, and (l, m, n) the direction-cosines of the direction of R.   Then since R acts perpendicular to the curve

$$l\frac{d\xi}{ds} + m\frac{d\eta}{ds} + n\frac{dz}{ds} = 0.$$

Suppose the moving curve to be projected orthogonally on the plane of ξ, η, let σ be the arc of the projection, and $v' = \dfrac{d\sigma}{dt}$ the resolved part of the velocity parallel to the plane of projection.   Then the equations may be written in the form

$$\frac{d^2\xi}{dt^2} = X + \omega^2\xi + \frac{d\omega}{dt}\eta + 2\omega v'\frac{d\eta}{d\sigma} + Rl,$$

$$\frac{d^2\eta}{dt^2} = Y + \omega^2\eta - \frac{d\omega}{dt}\xi - 2\omega v'\frac{d\xi}{d\sigma} + Rm,$$

$$\frac{d^2z}{dt^2} = Z + Rn.$$

The two terms $2\omega v'\dfrac{d\eta}{d\sigma}$ and $-2\omega v'\dfrac{d\xi}{d\sigma}$ may be regarded as the resolved parts of a force $2\omega v'$ acting in a direction whose direction-cosines are

$$l' = \frac{d\eta}{d\sigma}, \quad m' = \frac{-d\xi}{d\sigma}, \quad n' = 0.$$

These satisfy the equation $l'\dfrac{d\xi}{ds} + m'\dfrac{d\eta}{ds} + n'\dfrac{dz}{ds} = 0$.

Hence the force is perpendicular to the tangent to the curve, and also perpendicular to the axis of rotation.   Ler R' be the resultant of the reaction R and of the force $2\omega v'$.   Then R' also acts perpendicularly to the tangent, let (l'', m'', n'') be the direction-cosines of its direction.

The equations of motion therefore become

$$\left.\begin{aligned}
\frac{d^2\xi}{dt^2} &= X + \omega^2\xi + \frac{d\omega}{dt}\,\eta + R'l'' \\[2mm]
\frac{d^2\eta}{dt^2} &= Y + \omega^2\eta - \frac{d\omega}{dt}\,\xi + R'm'' \\[2mm]
\frac{d^2z}{dt^2} &= Z + R'n''
\end{aligned}\right\} \quad \dots \quad \dots\dots\dots\dots\dots(2).$$

These are the equations of motion of a particle moving on a *fixed curve*, and acted on in addition to the impressed forces by two extra forces, viz. (1) a force $\omega^2 r$ tending directly from the axis, where r is the distance of the particle from the axis, and (2) a force $\dfrac{d\omega}{dt}\,r$ perpendicular to the plane containing the particle and the axis, and tending opposite to the direction of rotation of the curve.

In any particular problem we may therefore treat the curve as fixed. Thus suppose the curve to be turning round the axis with uniform angular velocity. Then resolving along the tangent we have

$$v\frac{dv}{ds} = X\frac{dx}{ds} + Y\frac{dy}{ds} + Z\frac{dz}{ds} + \omega^2 r\frac{dr}{ds}\,,$$

where r is the distance of the particle from the axis. Let V be the initial value of v, r_0 that of r. Then

$$v^2 - V^2 = 2\!\int(X\,dx + Y\,dy + Z\,dz) + \omega^2\,(r^2 - r_0{}^2).$$

Let v_0 be the velocity the particle would have had under the action of the same forces if the curve had been fixed. Then

$$v_0{}^2 - V^2 = 2\!\int(X\,dx + Y\,dy + Z\,dz).$$

Hence　　　　　　　　　　$$v^2 - v_0{}^2 = \omega^2\,(r^2 - r_0{}^2).$$

The pressure on the moving curve is not equal to the pressure on the fixed curve. The pressure R on the moving curve is clearly the resultant of the pressure R' on the fixed curve, and a pressure $2\omega v'$ acting perpendicular both to the curve and to the axis in the direction of motion of the curve.

Thus suppose the curve to be plane and revolving uniformly about an axis perpendicular to its plane, and that there are no impressed forces. We have, resolving along the normal,

$$\frac{v^2}{\rho} = -\omega^2 r \sin\phi + R',$$

where ϕ is the angle which r makes with the tangent. If p be the perpendicular drawn from the axis on the tangent, we have, therefore,

$$R = \frac{v^2}{\rho} + \omega^2 p + 2\omega v.$$

This example might also have been advantageously solved by cylindrical co-ordinates. The fixed axis might be taken as axis of z and the projection on the plane of xy referred to polar co-ordinates. This method of treating the question is left to the student as an exercise.

Ex. If ω be variable, we have in a similar manner

$$R = \frac{v^2}{\rho} + \omega^2 p + 2\omega v + \frac{d\omega}{dt}\sqrt{r^2 - p^2}.$$

EXAMPLES*.

1. A circular hoop, whose weight is nw, is free to move on a smooth horizontal plane. It carries on its circumference a small ring, weight w, the coefficient of friction between the two being μ. Initially the hoop is at rest and the ring has an angular velocity ω about the centre of the hoop. Show that the ring will be at rest on the hoop after a time $\dfrac{1+n}{\mu\omega}$.

2. A heavy circular wire has its plane vertical and its lowest point at a height h above a horizontal plane. A small ring is projected along the wire from its highest point with an angular velocity about its centre equal to $\pi n \sqrt{\dfrac{2g}{h}}$ at the instant that the wire is let go. Show that, when the wire reaches the horizontal plane, the particle will just have described n revolutions.

3. A wire in the form of a circle is capable of turning in a horizontal plane about a fixed point O in its circumference, and carries a bead P which is initially projected from the opposite end A of the diameter through O with a given velocity V. Supposing the mass of the wire to be double that of the bead, show that $(16a^4 + 4a^2r^2 - r^4)\,\dot{\phi}^2 = V^2 r^2$, where $r = OP$, $OA = 2a$, $\phi = \angle POA$.

4. Two equal uniform rods of length $2a$, loosely jointed at one extremity, are placed symmetrically upon a fixed smooth sphere of radius $\frac{1}{3}a\sqrt{2}$, and raised into a horizontal position so that the hinge is in contact with the sphere. If they be allowed to descend under the action of gravity, show that, when they are first at rest, they are inclined at an angle $\cos^{-1}\frac{1}{3}$ to the horizon, that the points of contact with the sphere are the centres of oscillation of the rods relatively to the hinge, that the pressure on the sphere at each point of contact equals one-fourth the weight of either rod, and that there is no strain on the hinge.

5. A heavy uniform circular hoop of radius a and mass $2\pi am$, which is completely broken at one point, rolls with its plane vertical with uniform angular velocity ω on a horizontal plane. Find the maximum and minimum values of the bending moment at any point Q of the hoop, and prove that if ω be so large that the bending moment never vanishes, the greatest of these values will be $2ma^2 \sin^2\theta\,(a\omega^2 + g)$, 2θ being the angular distance of Q from the point of fracture.

6. Two straight equal and uniform rods are connected at their ends by two strings of equal length a, so as to form a parallelogram. One rod is supported at its centre by a fixed axis about which it can turn freely, this axis being perpendicular to the plane of motion which is vertical. Show that the middle point of the lower rod will oscillate in the same way as a simple pendulum of length a, and that the *angular* motion of the rods is independent of this oscillation.

7. A fine string is attached to two points A, B in the same horizontal plane, and carries a weight W at its middle point. A rod whose length is AB and weight W, has a ring at either end, through which the string passes, and is let fall from the position AB. Show that the string must be at least $\frac{5}{3}AB$, in order that the weight may ever reach the rod.

Also if the system be in equilibrium, and the weight be slightly and vertically displaced, the time of its small oscillations is $2\pi\,(AB/3g\sqrt{3})^{\frac{1}{2}}$.

* These examples are taken from the Examination Papers which have been set in the University and in the Colleges.

8. A fine thread is enclosed in a smooth circular tube which rotates freely about a vertical diameter; prove that, in the position of relative equilibrium, the inclination (θ) to the vertical of the diameter through the centre of gravity of the thread will be given by the equation $\cos\theta = \dfrac{g}{a\omega^2\cos\beta}$, where ω is the angular velocity of the tube, a its radius, and $2a\beta$ the length of the thread. Explain the case in which the value of $a\omega^2\cos\beta$ lies between g and $-g$.

9. A smooth wire without inertia is bent into the form of a helix which is capable of revolving about a vertical axis coinciding with a generating line of the cylinder on which it is traced. A small heavy ring slides down the helix, starting from a point in which this vertical axis meets the helix: prove that the angular velocity of the helix will be a maximum when it has turned through an angle θ given by the equation $\cos^2\theta + \tan^2\alpha + \theta\sin 2\theta = 0$, α being the inclination of the helix to the horizon.

10. A spherical hollow of radius a is made in a cube of glass of mass M, and a particle of mass m is placed within. The cube is then set in motion on a smooth horizontal plane so that the particle just gets round the sphere, remaining in contact with it. If the velocity of projection be V, prove that $V^2 = 5ag + 4ag\,\dfrac{m}{M}$.

11. A perfectly rough ball is placed within a hollow cylindrical garden-roller at its lowest point, and the roller is then drawn along a level walk with a uniform velocity V. Show that the ball will roll quite round the interior of the roller, if V^2 be $> \frac{2}{7}g\,(b-a)$, a being the radius of the ball, and b of the roller.

12. AB, BC are two equal uniform rods loosely jointed at B, and moving with the same velocity in a direction perpendicular to their length; if the end A be suddenly fixed, show that the initial angular velocity of AB is three times that of BC. Also show that in the subsequent motion of the rods, the greatest angle between them equals $\cos^{-1}\frac{2}{3}$; and that when they are next in a straight line, the angular velocity of BC is nine times that of AB.

13. Three equal heavy uniform beams jointed together are laid in the same right line on a smooth table, and a given horizontal impulse is applied at the middle point of the centre beam in a direction perpendicular to its length; show that the instantaneous impulse on each of the other beams is one-sixth of the given impulse.

14. Three beams of like substance, joined together so as to form one beam, are laid on a smooth horizontal table. The two extreme beams are equal in length, and one of them receives a blow at its free extremity in a direction perpendicular to its length. Determine the length of the middle beam in order that the greatest possible angular velocity may be given to the other extreme beam.

Result. If m be the mass of either of the outer rods, βm that of the inner rod, P the momentum of the blow, ω the angular velocity communicated to the third rod, then $ma\omega\left(\dfrac{1}{\beta} + \dfrac{8}{3} + \dfrac{4\beta}{3}\right) = P$. Hence when ω is a maximum $\beta = \frac{1}{2}\sqrt{3}$.

15. Two rough rods A, B are placed parallel to each other and in the same horizontal plane. Another rough rod C is laid across them at right angles, its centre of gravity being half way between them. If C be raised through any angle α and let fall, determine the conditions that it may oscillate, and show that if its

length be equal to twice the distance between A and B, the angle θ through which it will rise in the nth oscillation is given by the equation $\sin \theta = \left(\dfrac{1}{7}\right)^{2n} \cdot \sin \alpha$.

16. The corners A, B of a heavy rectangular lamina $ABCD$ are moveable on two smooth fixed wires OA, OB, at right angles to each other in a vertical plane, and equally inclined to the vertical. The lamina being in a position of equilibrium with AB horizontal, find the velocity of the centre of gravity and the angular velocity produced by an impulse applied along the lowest edge CD. Having given that $AB = 2a$, $BC = 4a$, prove that AB will just rise to coincidence with a wire, if the impulse is such as would impart to a mass equal to that of the lamina the velocity whose square is $\frac{8}{3}ga\,(2 - \sqrt{2})$. Also find the impulsive stresses at A and B.

17. A ball spinning about a vertical axis moves on a smooth table and impinges directly on a perfectly rough vertical cushion; show that the vis viva of the ball is diminished in the ratio $10 + 14\tan^2\theta : \dfrac{10}{e^2} + 49\tan^2\theta$, where e is the elasticity of the ball and θ the angle of reflexion.

18. A rhombus is formed of four rigid uniform rods, each of length $2a$, freely jointed at their extremities. If the rhombus be laid on a smooth horizontal table and a blow be applied at right angles to any one of the rods, the rhombus will begin to move as a rigid body if the blow be applied at a point distant $a\,(1 - \cos a)$ from an acute angle, where a is the acute angle.

19. A rectangle is formed of four uniform rods of lengths $2a$ and $2b$ respectively, which are connected by hinges at their ends. The rectangle is revolving about its centre on a smooth horizontal plane with an angular velocity n, when a point in one of the sides of length $2a$ suddenly becomes fixed. Show that the angular velocity of the sides of length $2b$ immediately becomes $\dfrac{3a + b}{6a + 4b}\,n$. Find also the change in the angular velocity of the other sides and the impulsive action at the point which becomes fixed.

20. Three equal uniform inelastic rods loosely jointed together are laid in a straight line on a smooth horizontal table, and the two outer ones are set in motion about the ends of the middle one with equal angular velocities (1) in the same direction, and (2) in opposite directions. Prove that in the first case, when the outer rods make the greatest angle with the direction of the middle one produced on each side, the common angular velocity of the three is $\frac{2}{7}\omega$, and that in the second case after the impact of the two outer rods the triangle formed by them will move with uniform velocity $\frac{2}{3}a\omega$, $2a$ being the length of each rod.

21. An equilateral triangle formed of three equal heavy uniform rods of length a hinged at their extremities is held in a vertical plane with one side horizontal and the vertex downwards. If after falling through any height, the middle point of the upper rod be suddenly stopped, the impulsive strains on the upper and lower hinges will be in the ratio of $\sqrt{13}$ to 1. If the lower hinge would just break if the system fell through a height $\dfrac{8a}{\sqrt{3}}$, prove that if the system fell through a height $\dfrac{32a}{\sqrt{3}}$ the lower rods would just swing through two right angles.

22. A perfectly rough and rigid hoop rolling down an inclined plane comes in contact with an obstacle in the shape of a spike. Show that if the radius of the hoop $= r$, height of spike above the plane $= \frac{1}{2}r$ and velocity just before impact $= V$, then the condition that the hoop will surmount the spike is $V^2 > \frac{16}{9}gr \{1 - \sin(a + \frac{1}{6}\pi)\}$, a being the inclination of the plane to the horizon. Show that the hoop will not *remain* in contact with the spike unless $V^2 < \frac{16}{9}gr \cdot \sin(a + \frac{1}{6}\pi)$, and if it does, the hoop will leave the spike when the diameter through the point of contact makes an angle with the horizon $= \sin^{-1}\left\{\frac{9}{32}\frac{V^2}{gr} + \frac{1}{2}\sin\left(a + \frac{\pi}{6}\right)\right\}$.

23. A flat circular disc of radius a is projected on a rough horizontal table, which is such that the friction upon an element a is cV^3ma, where V is the velocity of the element, m the mass of a unit of area: find the path of the centre of the disc.

If the initial velocity of the centre of gravity and the angular velocity of the disc be u_0, ω_0, prove that the velocity u and angular velocity ω at any subsequent time satisfy the relation $\left(\dfrac{3u^2 - a^2\omega^2}{3u_0^2 - a^2\omega_0^2}\right)^2 = \dfrac{u^2\omega}{u_0^2\omega_0}$.

24. A heavy circular lamina of radius a and mass M rolls on the inside of a rough circular arc of twice its radius fixed in a vertical plane. Find the motion. If the lamina be placed at rest in contact with the lowest point, the impulse which must be applied horizontally that it may rise as high as possible (not going all round), without falling off, is $M\sqrt{3ag}$.

25. A string without weight is coiled round a rough horizontal cylinder, of which the mass is M and the radius a, and which is capable of turning round its axis. To the free extremity of the string is attached a chain of which the mass is m and the length l; if the chain be gathered close up and then let go, prove that the angle θ through which the cylinder has turned after a time t before the chain is fully stretched is given by $Ma\theta = \dfrac{m}{l}\left(\dfrac{gt^2}{2} - a\theta\right)^2$.

26. Two equal rods AC, BC are freely connected at C, and hooked to A and B, two points in the same horizontal line, each rod being inclined at an angle a to the horizon. The hook B suddenly giving way, prove that the direction of the strain at C is instantaneously shifted through an angle $\tan^{-1}\left(\dfrac{1 + 6\sin^2 a}{1 + 6\cos^2 a} \cdot \dfrac{2 - 3\cos^2 a}{3\sin a\cos a}\right)$.

27. Two particles A, B are connected by a fine string; A rests on a rough horizontal table and B hangs vertically at a distance l below the edge of the table. If A be on the point of motion and B be projected horizontally with a velocity u, show that A will begin to move with acceleration $\dfrac{\mu}{\mu+1}\dfrac{u^2}{l}$, and that the initial radius of curvature of B's path will be $(\mu + 1)l$, where μ is the coefficient of friction.

28. Two particles (m, m') are connected by a string passing through a small fixed ring and are held so that the string is horizontal; their distances from the ring being a and a'. If ρ, ρ' be the initial radii of curvature of their paths when they are let go, prove that $\dfrac{m}{\rho} = \dfrac{m'}{\rho'}$, and $\dfrac{1}{\rho} + \dfrac{1}{\rho'} = \dfrac{1}{a} + \dfrac{1}{a'}$.

29. A sphere whose centre of gravity is not in its centre is placed on a rough table; the coefficient of friction being μ, determine whether it will begin to slide or to roll.

30. A circular ring is fixed in a vertical position upon a smooth horizontal plane, and a small ring is placed on the circle, and attached to the highest point by a string, which subtends an angle α at the centre; prove that if the string be cut and the circle left free, the pressures on the ring before and after the string is cut are in the ratio $M + m \sin^2 \alpha : M \cos \alpha$, m and M being the masses of the ring and circle.

31. One extremity C of a rod is made to revolve with uniform angular velocity n in the circumference of a circle of radius a, while the rod itself is made to revolve in the opposite direction with the same angular velocity about that extremity. The rod initially coincides with a diameter, and a smooth ring capable of sliding freely along the rod is placed at the centre of the circle. If r be the distance of the ring from C at the time t, prove $r = \dfrac{2a}{5}(e^{nt} + e^{-nt}) + \dfrac{a}{5} \cos 2nt$.

32. Two equal uniform rods of length $2a$ are joined together by a hinge at one extremity, their other extremities being connected by an inextensible string of length $2l$. The system rests upon two smooth pegs in the same horizontal line, distant $2c$ from each other. If the string be cut prove that the initial angular acceleration of either rod will be $g \dfrac{8a^2 c - l^3}{\dfrac{8a^2 l^2}{3} + \dfrac{32a^4 c^2}{l^2} - 8a^2 cl}$.

33. A smooth horizontal disc revolves with angular velocity $\sqrt{\mu}$ about a vertical axis, at the point of intersection of which is placed a material particle attracted to a certain point of the disc by a force whose acceleration is $\mu \times$ distance ; prove that the path on the disc is a cycloid.

34. A hollow cylinder of radius a rests on a rough table, and contains an insect resting within it on the lowest generator ; if the insect start off and continue to walk at a uniform velocity V relative to the cylinder in a vertical plane cutting the axis of the cylinder at right angles, then the angle θ the axial plane containing the insect makes with the vertical is given by

$$a^2 \dot\theta^2 (M + 2m \sin^2 \tfrac{1}{2}\theta) = MV^2 - 2mag \sin^2 \tfrac{1}{2}\theta,$$

it being understood that the cylinder is very thin.

If the internal radius be b, prove

$$\dot\theta^2 [M(k^2 + a^2) + m(a^2 - 2ab \cos \theta + b^2)] = C - 2mgb(1 - \cos \theta),$$

where $\quad Cb^2 [M(k^2 + a^2) + m(a - b)^2] = V^2 [M(k^2 + a^2) + ma(a - b)]^2.$

CHAPTER V.

Translation and Rotation.

214. If the particles of a body be. rigidly connected, then, whatever be the nature of the motion generated by the forces, there must be some general relations between the motions of the particles of the body. These must be such that if the motion of three points not in the same straight line be known, that of every other point may be deduced. It will then in the first place be our object to consider the general character of the motion of a rigid body apart from the forces that produce it, and to reduce the determination of the motion of every particle to as few independent quantities as possible: and in the second place we shall consider how when the forces are given these independent quantities may be found.

215. *One point of a moving rigid body being fixed, it is required to deduce the general relations between the motions of the other points of the body.*

Let O be the fixed point and let it be taken as the centre of a moveable sphere which we shall suppose fixed in the body. Let the radius vector to any point Q of the body cut the sphere in P, then the motion of every point Q of the body will be represented by that of P.

If the displacements of two points A, B, on the sphere in any time be given as AA', BB', the displacement of any other point P on the sphere may clearly be found by constructing on $A'B'$ as base a triangle $A'P'B'$ similar and equal to APB. Then PP' will represent the displacement of P. It may be assumed as evident, or it may be proved as in Euclid, that on the same base and on the same side of it there cannot be two triangles on the same sphere, which have their sides terminated in one extremity of the base equal to one another, and likewise those terminated in the other extremity.

Let D and E be the middle points of the arcs AA', BB', and let DC, EC be arcs of great circles drawn perpendicular to AA', BB' respectively. Then clearly $CA = CA'$ and $CB = CB'$, and therefore since the bases AB, $A'B'$ are equal, the two triangles

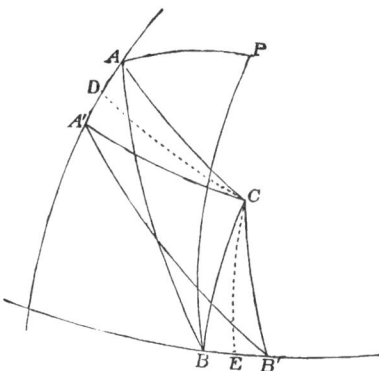

ACB, $A'CB'$ are equal and similar. Hence the displacement of C is zero. Also it is evident since the displacements of O and C are zero, that the displacement of every point in the straight line OC is also zero.

Hence a body may be brought from any position, which we may call AB, *into another* A'B' *by a rotation about* OC *as an axis through an angle* PCP' *such that any one point* P *is brought into coincidence with its new position* P'. Then every point of the body will be brought from its first to its final position.

This theorem is due to Euler. *Mémoires de l'Académie de Berlin* 1750, and the *Commentaires de Saint-Pétersbourg* 1775.

216.   If we make the radius of the sphere infinitely great, the various circles in the figure will become straight lines. We may therefore infer that if a body be moving in one plane it may be brought from any position which we may call AB into any other $A'B'$ by a rotation about some point C.

217.   Ex. 1.   A body is referred to rectangular axes x, y, z, and, the origin remaining the same, the axes are changed to x', y', z', according to the scheme in the margin. Show that this is equivalent to turning the body round an axis whose equations are any two of the following three:

	x',	y',	z'
x	$a_1,$	$a_2,$	a_3
y	$b_1,$	$b_2,$	b_3
z	$c_1,$	$c_2,$	c_3

$$(a_1 - 1)\, x + a_2 y + a_3 z = 0,$$
$$b_1 x + (b_2 - 1)\, y + b_3 z = 0,$$
$$c_1 x + c_2 y + (c_3 - 1)\, z = 0,$$

through an angle θ, where $3 - 4 \sin^2 \tfrac{1}{2}\theta = a_1 + b_2 + c_3.$

R. D.   13

The positive directions of x', y' being arbitrary, show that the condition that these three equations are consistent is satisfied, provided the positive direction of the axis of z' is properly chosen. See also a question in the *Smith's Prize Examination* for 1868.

Take two points one on each of the axes of z and z' at a distance h from the origin. Their co-ordinates are $(0, 0, h)(a_3h, b_3h, c_3h)$, therefore their distance is $h\sqrt{2(1-c_3)}$. But it is also $2h\sin\gamma\sin\frac{1}{2}\theta$; $\therefore$ $2\sin^2\frac{1}{2}\theta\sin^2\gamma = 1 - c_3$, where γ is the angle $z0z'$. Similarly $2\sin^2\frac{1}{2}\theta\sin^2 a = 1 - a_1$ and $2\sin^2\frac{1}{2}\theta\sin^2\beta = 1 - b_2$, whence the equation to find θ follows at once.

Ex. 2. Show that the equations to the axis may also be written in the form

$$\frac{x}{c_1 + a_3} = \frac{y}{c_2 + b_3} = \frac{z}{c_3 - a_1 - b_1 + 1}.$$

218. When a body is in motion we have to consider not merely its first and last positions, but also the intermediate positions. Let us then suppose AB, $A'B'$ to be two positions at any indefinitely small interval of time dt. We see that when a body moves about a fixed point O, there is, at every instant of the motion, a straight line OC, such that the displacement of every point in it during an indefinitely short time dt is zero. This straight line is called the instantaneous axis.

Let $d\theta$ be the angle through which the body must be turned round the instantaneous axis to bring any point P from its position at the time t to its position at the time $t + dt$, then the ultimate ratio of $d\theta$ to dt is called the angular velocity of the body about the instantaneous axis. The angular velocity may also be defined as the angle through which the body would turn in a unit of time if it continued to turn uniformly about the same axis throughout that unit with the angular velocity it had at the proposed instant.

219. Let us now remove the restriction that the body is moving with some one point fixed. We may establish the following proposition.

Every displacement of a rigid body may be represented by a combination of the two following motions, (1) a motion of translation, whereby every particle is moved parallel to the direction of motion of any assumed point P *rigidly connected with the body and through the same space; (2) a motion of rotation of the whole body about some axis through this assumed point* P.

This theorem and that of the central axis are given by Chasles. *Bulletin des Sciences Mathématiques par Ferussac*, vol. xiv. 1830. See also Poinsot, *Théorie Nouvelle de la Rotation des Corps* 1834.

It is evident that the change of position may be effected by moving P from its old to its new position P' by a motion of translation, and then retaining P' as a fixed point by moving any two points of the body not in one straight line with P into their final positions. This last motion has been proved to be equivalent to a rotation about some axis through P'.

Since these motions are quite independent, it is evident that their order may be reversed, *i.e.* we may first rotate the body and then translate it. We may also suppose them to take place simultaneously.

It is clear that any point P of the body may be chosen as the *base point* of the double operation. Hence the given displacement may be constructed in an infinite variety of ways.

220. **Change of Base.** *To find the relations between the axes and angles of rotation when different points* P, Q *are chosen as bases.*

Let the displacement of the body be represented by a rotation θ about an axis PR and a translation PP'. Let the same displacement be also represented by a rotation θ' about an axis QS and a translation QQ'. It is clear that any point has two displacements, (1) a translation equal and parallel to PP', and (2) a rotation through an arc in a plane perpendicular to the axis of rotation PR. This second displacement is zero only when the point is on the axis PR. Hence the only points whose displacements are the same as that of the base point lie on the axis of rotation corresponding to that base point. Through the second base point Q draw a parallel to PR. Then for all points in this parallel, the displacements due to the translation PP', and the rotation θ round PR, are the same as the corresponding displacements for the point Q. Hence this parallel must be the axis of rotation corresponding to the base point Q. We infer *that the axes of rotation corresponding to all base points are parallel.*

221. The axes of rotation at P and Q having been proved parallel, let a be the distance between them. Let the plane of the paper intersect these axes at right angles in P and Q, then

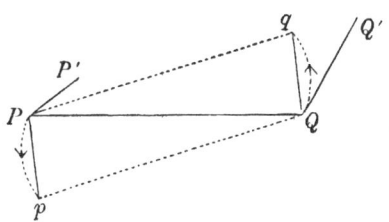

$PQ = a$. Let PP', QQ' represent the linear displacements of P and Q respectively, though these need not necessarily be in the plane of the paper.

The rotation θ about PR will cause Q to describe an arc of a circle of radius a and angle θ, the chord Qq of this arc is $2a \sin \dfrac{\theta}{2}$ and is the displacement due to rotation. The whole displacement QQ' of Q is the resultant of Qq and the displacement PP' of P. In the same way the rotation θ' about QS will cause P to describe an arc, whose chord Pp is equal to $2a \sin \dfrac{\theta'}{2}$ The whole displacement PP' of P is the resultant of Pp and the displacement QQ' of Q. But if the displacement of Q is equal to that of P together with Qq, and the displacement of P is equal to that of Q together with Pp, we must have Pp and Qq equal and opposite. This requires that the two rotations θ, θ' about PR and QS should be equal and in the same direction. We infer that *the angles of rotation corresponding to all base points are equal.*

222. Since the translation QQ' is the resultant of PP' and Qq, we may by this theorem find both the translation and rotation corresponding to any proposed base point Q when those for P are given.

Since Qq, the displacement due to rotation round PR, is perpendicular to PR, the projection of QQ' on the axis of rotation is the same as that of PP'. *Hence the projections on the axis of rotation of the displacements of all points of the body are equal.*

223. An important case is that in which the displacement is a simple rotation θ about an axis PR, without any translation. If any point Q distant a from PR be chosen as the base, the same displacement is represented by a translation of Q along a chord $Qq = 2a \sin \dfrac{\theta}{2}$ in a direction making an angle $\dfrac{\pi - \theta}{2}$ with the plane QPR, and a rotation which must be equal to θ about an axis which must be parallel to PR. Hence *a rotation about any axis may be replaced by an equal rotation about any parallel axis together with a motion of translation.*

224. When the rotation is indefinitely small, the proposition can be enunciated thus:—a motion of rotation ωdt about an axis PR is equivalent to an equal motion of rotation about any parallel axis QS, distant a from PR, together with a motion of translation $a\omega dt$ perpendicular to the plane containing the axes and in the direction in which QS moves.

225. Central axis. It is often important to choose the base point so that the direction of translation may coincide with the axis of rotation. Let us consider how this may be done.

Let the given displacement of the body be represented by a rotation θ about PR, and a translation PP'. Draw $P'N$ perpendicular to PR. If possible let this same displacement be represented by a rotation about an axis QS, and a translation QQ' *along* QS. By Arts. 220 and 221 QS must be parallel to PR and the rotation about it must be θ. This translation will move P a length equal to QQ'

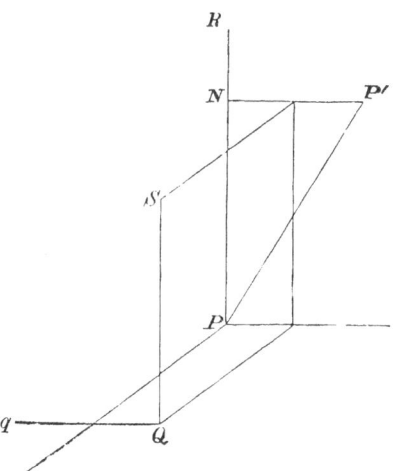

along PR, and the rotation about QS will move P along an arc perpendicular to PR. Hence QQ' must equal PN and NP' must be the chord of the arc. It follows that QS must lie on a plane bisecting NP' at right angles and at a distance a from PR where $NP' = 2a \sin \dfrac{\theta}{2}$, or, which is more convenient, at a distance y from the plane NPP' where $NP' = 2y \tan \dfrac{\theta}{2}$. The rotation θ round QS is to bring N to P' and is in the same direction as the rotation θ round PR. Hence the distance y must be measured from the middle point of NP' in the direction in which that middle point is moved by its rotation round PR.

Having found the only possible position of QS, it remains to show that the displacement of Q is really along QS. The rotation θ round PR will cause Q to describe an arc whose chord Qq is parallel to $P'N$ and equal to $2a \sin \dfrac{\theta}{2}$. The chord Qq is therefore equal to NP', and the translation NP' brings q back to its position at Q. Hence Q is moved only by the translation PN, i.e. Q is moved along QS.

226. It follows from this reasoning that any displacement of a body can be represented by a rotation about some straight line

and a translation *parallel* to that straight line. This mode of constructing the displacement is called a *screw*. The straight line is sometimes called the *central axis* and sometimes the *axis of the screw*. The ratio of the translation to the angle of rotation is called the *pitch* of the screw.

227. The same displacement of a body cannot be constructed by two different screws. For if possible let there be two central axes AB, CD. Then AB and CD by Art. 220 are parallel. The displacement of any point Q on CD is found by turning the body round AB and moving it parallel to AB, hence Q has a displacement perpendicular to the plane ABQ and therefore cannot move only along CD.

228. When the rotations are indefinitely small, the construction to find the central axis may be simply stated thus. *Let the displacement be represented by a rotation* ωdt *about an axis* PR *and a translation* Vdt *in the direction* PP'. *Measure a distance*
$$y = \frac{V \sin P'PR}{\omega}$$
from P *perpendicular to the plane* P'PR *on that side of the plane towards which* P' *is moving. A parallel to* PR *through the extremity of* y *is the central axis.*

Ex. 1. Given the displacements AA', BB', CC' of three points of a body in direction and magnitude, but not necessarily in position, find the direction of the axis of rotation corresponding to any base point P.

Through any assumed point O draw Oa, $O\beta$, $O\gamma$ parallel and equal to AA', BB', CC'. If Op be the direction of the axis of rotation, the projections of Oa, $O\beta$, $O\gamma$ on Op are all equal. Hence Op is the perpendicular drawn from O on the plane $a\beta\gamma$. This also shows that the direction of the axis of rotation is the same for all base points.

Ex. 2. If in the last example the motion be referred to the central axis, show that the translation along it is equal to Op.

Ex. 3. Given the displacements AA', BB' of two points A, B of the body and the direction of the central axis, find the position of the central axis. Draw planes through AA', BB' parallel to the central axis. Bisect AA', BB' by planes perpendicular to these planes respectively and parallel to the direction of the central axis. These two last planes intersect in the central axis.

Composition of Rotations and Screws.

229. It is often necessary to compound rotations about axes OA, OB which meet at a point O. But, as the only case which occurs in rigid dynamics is that in which these rotations are indefinitely small, we shall first consider this case with some particularity, and then indicate generally at the end of the chapter the mode of proceeding when the rotations are of finite magnitude.

230. *To explain what is meant by a body having angular velocities about more than one axis at the same time.*

A body in motion is said to have an angular velocity ω about a straight line, when, the body being turned round this straight line through an angle ωdt, every point of the body is brought from its position at the time t to its position at the time $t + dt$.

Suppose that during three successive intervals each of time dt, the body is turned successively round three different straight lines OA, OB, OC meeting at a point O through angles $\omega_1 dt$, $\omega_2 dt$, $\omega_3 dt$. Then we shall first prove that the final position is the same in whatever order these rotations are effected. Let P be any point in the body, and let its distances from OA, OB, OC, respectively be r_1, r_2, r_3. First let the body be turned round OA, then P receives a displacement $\omega_1 r_1 dt$. By this motion let r_2 be increased to $r_2 + dr_2$, then the displacement caused by the rotation about OB will be in magnitude $\omega_2 (r_2 + dr_2)\, dt$. But according to the principles of the differential calculus we may in the limit neglect the quantities of the second order, and the displacement becomes $\omega_2 r_2 dt$. So also the displacement due to the remaining rotation will be $\omega_3 r_3 dt$. And these three results will be the same in whatever order the rotations take place. In a similar manner we can prove that the *directions* of these displacements will be independent of the order. The final displacement is the diagonal of the parallelopiped described on these three lines as sides, and is therefore independent of the order of the rotations. Since then the three rotations are quite independent, they may be said to take place simultaneously.

When a body is said to have angular velocities about three different axes it is only meant that the motion may be determined as follows. Divide the whole time into a number of small intervals each equal to dt. During each of these, turn the body round the three axes successively, through angles $\omega_1 dt$, $\omega_2 dt$, $\omega_3 dt$. Then when dt diminishes without limit the motion during the whole time will be accurately represented.

231. It is clear that a rotation about an axis OA may be represented in magnitude by a length measured along the axis. This length will also represent its direction if we follow the same rule as in statics, viz. the rotation shall appear to be in some standard direction to a spectator placed along the axis so that OA is measured from his feet at O towards his head. This direction of OA is called the positive direction of the axis.

232. **Parallelogram of angular velocities.** *If two angular velocities about two axes* OA, OB *be represented in magnitude and direction by the two lengths* OA, OB; *then the diagonal* OC *of the parallelogram constructed on* OA, OB *as sides will be the*

resultant axis of rotation, and its length will represent the magnitude of the resultant angular velocity.

Let P be any point in OC, and let PM, PN be drawn perpendicular to OA, OB. Since OA represents the angular velocity about OA and PM is the perpendicular distance of P from OA, the product $OA \cdot PM$ will represent the velocity of P due to the angular velocity about OA. Similarly $OB \cdot PN$ will represent the velocity of P due to the angular velocity about OB. Since P is on the left-hand side of OA and on the right-hand side of OB, as we respectively look along these directions, it is evident that these velocities are in opposite directions.

Hence the velocity of any point P is represented by

$$OA \cdot PM - OB \cdot PN$$
$$= OP \{OA \cdot \sin COA - OB \cdot \sin COB\}$$
$$= 0.$$

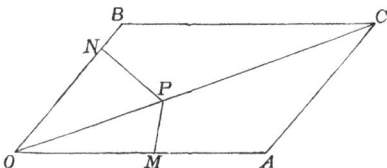

Therefore the point P is at rest and OC is the resultant axis of rotation.

Let ω be the angular velocity about OC, then the velocity of any point A in OA is perpendicular to the plane AOB and is represented by the product of ω into the perpendicular distance of A from $OC = \omega \cdot OA \sin COA$. But since the motion is also determined by the two given angular velocities about OA, OB, the motion of the point A is also represented by the product of OB into the perpendicular distance of A from $OB = OB \cdot OA \sin BOA$;

$$\therefore \omega = OB \cdot \frac{\sin BOA}{\sin COA} = OC.$$

Hence the angular velocity about OC is represented in magnitude by OC.

From this proposition we may deduce as a corollary " the parallelogram of angular accelerations." For if OA, OB represent the additional angular velocities impressed on a body at any instant, it follows that the diagonal OC will represent the resultant additional angular velocity in direction and magnitude.

233. This proposition shows that angular velocities and angular accelerations may be compounded and resolved by the same rules

and in the same way as if they were forces. Thus an angular velocity ω about any given axis may be resolved into two, $\omega \cos \alpha$ and $\omega \sin \alpha$, about axes at right angles to each other and making angles α and $\frac{1}{2}\pi - \alpha$ with the given axis.

If a body have angular velocities ω_1, ω_2, ω_3 about three axes Ox, Oy, Oz at right angles, they are together equivalent to a single angular velocity ω, where $\omega = \sqrt{\omega_1{}^2 + \omega_2{}^2 + \omega_3{}^2}$, about an axis making angles with the given axes whose cosines are respectively $\dfrac{\omega_1}{\omega}$, $\dfrac{\omega_2}{\omega}$, $\dfrac{\omega_3}{\omega}$. This may be proved, as in the corresponding proposition in statics, by compounding the three angular velocities, taking them two at a time.

It will however be needless to recapitulate the several propositions proved for forces in statics with special reference to angular velocities. We may use "the triangle of angular velocities" or the other rules for compounding several angular velocities together, without any further demonstration.

234. **The Angular Velocity couple.** *A body has angular velocities* ω, ω' *about two parallel axes* OA, O'B *distant* a *from each other, to find the resulting motion.*

Since parallel straight lines may be regarded as the limit of two straight lines which intersect at a very great distance, it follows from the parallelogram of angular velocities that the two given angular velocities are equivalent to an angular velocity about some parallel axis $O''C$ lying in the plane containing OA, $O'B$.

Let x be the distance of this axis from OA, and suppose it

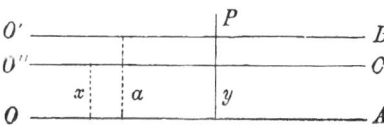

to be on the same side of OA as $O'B$. Let Ω be the angular velocity about it.

Consider any point P, distant y from OA and lying in the plane of the three axes. The velocity of P due to the rotation about OA is ωy, the velocity due to the rotation about $O'B$ is $\omega'(y - a)$. But these two together must be equivalent to the velocity due to the resultant angular velocity Ω about $O''C$, and this is $\Omega(y - x)$,

$$\therefore \ \omega y + \omega'(y - a) = \Omega(y - x).$$

This equation is true for all values of y, $\therefore \Omega = \omega + \omega'$, $x = \dfrac{a\omega'}{\Omega}$.

This is the same result we should have obtained if we had been seeking the resultant of two *forces* ω, ω' acting along OA, $O'B$.

If $\omega = -\omega'$, the resultant angular velocity vanishes, but x is infinite. The velocity of any point P is in this case $\omega y + \omega'(y-a) = a\omega$, which is independent of the position of P.

The result is that two angular velocities, each equal to ω but tending to turn the body in opposite directions about two parallel axes at a distance a from each other, are equivalent to a *linear velocity* represented by $a\omega$. This corresponds to the proposition in statics that "a couple" is properly measured by its moment.

We may deduce as a corollary, that a motion of rotation ω about an axis OA *is equivalent to an equal motion of rotation about a parallel axis* O'B *plus a motion of translation* $a\omega$ *perpendicular to the plane containing* OA, O'B, *and in the direction in which* O'B *moves.* See also Art. 223.

235. **The analogy to Statics.** *To explain a certain analogy which exists between statics and dynamics.*

All propositions in statics relating to the composition and resolution of forces and couples are founded on these theorems:

1. The parallelogram of forces and the parallelogram of couples.

2. A force F is equivalent to any equal and parallel force together with a couple Fp, where p is the distance between the forces.

Corresponding to these we have in dynamics the following theorems on the instantaneous motion of a rigid body:

1. The parallelogram of angular velocities and the parallelogram of linear velocities.

2. An angular velocity ω is equivalent to an equal angular velocity about a parallel axis together with a linear velocity equal to ωp, where p is the distance between the parallel axes.

It follows that every proposition in statics relating to forces has a corresponding proposition in dynamics relating to the motion of a rigid body, and these two may be proved in the same way.

To complete the analogy it may be stated (i) that an angular velocity like a force in statics requires, for its complete determination, five constants, and (ii) that a velocity like a couple in statics requires but three. Four constants are required to determine the line of action of the force or of the axis of rotation, and one to determine the magnitude of either. There will also be a convention in either case to determine the positive direction of the line.

Two constants and a convention are required to determine the positive direction of the axis of the couple or of the velocity and one the magnitude of either.

The discovery of this analogy is due to Poinsot.

236. In order to show the great utility of this analogy and how easily we may transform any known theorem in statics into the corresponding one in dynamics, we shall place in close juxtaposition the more common theorems which are in continual use both in statics and dynamics.

It is proved in statics that any given system of forces and couples can be reduced to three forces X, Y, Z, which act along any rectangular axes which may be convenient and which meet at any base point O we please, together with three couples which we may call L, M, N and which act round these axes. A simpler representation is then found, for it is proved that these forces and couples can be reduced to a single force which we may call R and a couple G which acts round the line of action of R. This line of action of R is called the central axis. There is but one central axis corresponding to a given system of forces. The term *wrench* has been applied to this representation of a given system of forces. Draw any straight line AB parallel to the central axis at a distance c from it. Then we may move R from the central axis to act along AB at A, provided we introduce a new couple whose moment is Rc. Combining this with the couple G, we have for the new base point A a new couple $G' = \sqrt{G^2 + R^2c^2}$, the force being the same as before. The couple G' is a minimum when $c = 0$, i.e. when AB coincides with the central axis. By taking moments round AB we see that the moment of the forces round every straight line parallel to the central axis is the same and equal to the minimum couple.

The same train of reasoning by which these results were obtained will lead to the following propositions. The instantaneous motion may be reduced to a linear velocity of any base point we please and an angular velocity round some axis through the base. These are then reduced to an angular velocity which we may call Ω about an axis called the central axis, and a linear velocity along that axis which we may call V. The term *screw* has been applied to this representation of the motion. Draw any straight line AB parallel to the central axis. Then we may move Ω from the central axis to act round AB, provided that we introduce a new linear velocity represented by Ωc. Combining this with the velocity V we have for the new base A (which is any point on AB) a new linear velocity $V' = \sqrt{V^2 + c^2\Omega^2}$, the angular velocity being the same as before. The linear velocity V' is a minimum when $c = 0$, i.e. when AB coincides with the central axis. We see that the linear velocity of any point A resolved in the direction AB, i.e. parallel to the central axis, is always the same and equal to the minimum velocity of translation.

It will be seen that most of these results have already been obtained in Arts. 219 to 228 for *finite rotations*.

237. Another useful representation depends on the following proposition. Any system of forces can be replaced by some force F which acts along a straight line which we may choose at pleasure, and some other force F' which acts along some other line and does not in general cut the first force. These are called *conjugate forces*. The shortest distance between these is proved in statics to intersect the central axis at right angles. The directions and magnitudes of the forces F, F' are such that R would be their resultant if they were moved parallel to themselves, so as to intersect the central axis. Also it is known that, if θ be the angle between the directions of the forces F, F' and a the shortest distance between them, $FF'a \sin \theta = GR$.

By help of the analogy we may obtain the corresponding propositions in the motion of a body. Any motion may be represented by two angular velocities, one ω about an axis which we may choose at pleasure and another ω' about some axis which does not in general cut the first axis. These are called *conjugate axes*. The shortest distance between these intersects the central axis at right angles. These angular velocities are such that Ω would be their resultant if their axes were placed parallel to their actual positions, so as to intersect the central axis. If θ be the angle between the axes of ω, ω' and a be the shortest distance between these axes, then $\omega\omega'a \sin \theta = V\Omega$.

238. **The velocity of any Point.** The motion of a body during the time dt may be represented, as explained in Art. 219, by a velocity of translation of a base point O, and an angular velocity about some axis through O. Let us choose any three rectangular axes Ox, Oy, Oz which may suit the particular purpose we have in view. These axes meet in O and move with O, keeping their directions fixed in space. Let u, v, w be the resolved parts along these axes of the linear velocity of O, and ω_x, ω_y, ω_z, the resolved parts of the angular velocity. These angular velocities are supposed positive when they tend the same way round the axes that positive couples tend in statics. Thus the positive directions of ω_x, ω_y, ω_z, are respectively from y to z, from z to x and from x to y.

The whole motion during the time dt of the body is known when these six quantities u, v, w, ω_x, ω_y, ω_z are given. These six quantities may be called the *components of the motion*. We now propose to find the motion of any point P whose co-ordinates are x, y, z.

Let us find the velocity of P parallel to the axis of z. Let PN be the ordinate of z and let PN be drawn perpendicular to Ox.

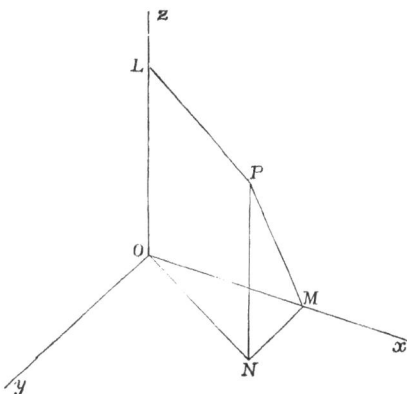

The velocity of P due to the rotation round Ox is clearly $\omega_x PM$. Resolving this along NP we get $\omega_x PM \sin NPM = \omega_x y$. Similarly that due to the rotation about Oy is $-\omega_y x$ and that due to the rotation about Oz is zero. Adding the linear velocity w of the origin, we see that the whole velocity of P parallel to Oz is

$$w' = w + \omega_x y - \omega_y x.$$

Similarly the velocities parallel to the other axes are

$$u' = u + \omega_y z - \omega_z y,$$
$$v' = v + \omega_z x - \omega_x z.$$

239. It is sometimes necessary to change our representation of a given motion from one base point to another. These formulæ will enable us to do so. Thus suppose we wish our new base point to be at a point O', the axes at O' being parallel to those at O. Let (ξ, η, ζ) be the co-ordinates of O' and let u', v', w', ω_x', ω_y', ω_z' be the linear and angular components of motion for the base O'. We have now two representations of the same motion, both these must give the same result for the linear velocities of any point P. Hence

$$u + \omega_y z - \omega_z y = u' + \omega_y' (z - \zeta) - \omega_z' (y - \eta),$$
$$v + \omega_z x - \omega_x z = v' + \omega_z' (x - \xi) - \omega_x' (z - \zeta),$$
$$w + \omega_x y - \omega_y x = w' + \omega_x' (y - \eta) - \omega_y' (x - \xi),$$

must be true for all values of x, y, z.

These equations give $\omega_x' = \omega_x$, $\omega_y' = \omega_y$, $\omega_z' = \omega_z$; so that whatever base is chosen the angular velocity is always the same in direction and magnitude. See Art. 221. We also see that u', v' w'

are given by formulæ analogous to those in Art. 238, as indeed might have been expected.

The reader should compare these with the corresponding formulæ in statics. If all the forces of any system be equivalent to three forces X, Y, Z acting at a base point along three rectangular axes together with three couples round those axes, then we know that the corresponding forces and couples for any other base point ξ, η, ζ are

$$X' = X, \qquad L' = L + Y\zeta - Z\eta,$$
$$Y' = Y, \qquad M' = M + Z\xi - X\zeta,$$
$$Z' = Z, \qquad N' = N + X\eta - Y\xi.$$

240. **To find the equivalent Screw.** *The motion being given by the linear velocities* (u, v, w) *of some base* O, *and the angular velocities* (ω_x, ω_y, ω_z), *find the central axis, the linear velocity along it and the angular velocity round it, i.e. find the equivalent screw.*

Let P be any point on the central axis, then if P were chosen as base, the components of the angular velocity would be the same as at the base O. If then Ω be the resultant of the angular velocities ω_x, ω_y, ω_z we see that

(1) The direction-cosines of the central axis are

$$\cos\alpha = \frac{\omega_x}{\Omega}, \quad \cos\beta = \frac{\omega_y}{\Omega}, \quad \cos\gamma = \frac{\omega_z}{\Omega}.$$

(2) The angular velocity about the central axis is Ω.

(3) The velocity of every point resolved in a direction parallel to the central axis is the same and equal to that along the central axis. See Art. 222 or Art. 236. If then V be the linear velocity along the central axis we have

$$V = u\cos\alpha + v\cos\beta + w\cos\gamma;$$
$$\therefore \ V\Omega = u\omega_x + v\omega_y + w\omega_z.$$

(4) Let (x, y, z) be the co-ordinates of P, i.e. of any point on the central axis. Then the linear velocity of P is along the axis of rotation. Hence

$$\frac{u + \omega_y z - \omega_z y}{\omega_x} = \frac{v + \omega_z x - \omega_x z}{\omega_y} = \frac{w + \omega_x y - \omega_y x}{\omega_z}.$$

These are therefore the equations to the central axis.

If we multiply the numerator and denominator of each of these fractions by ω_x, ω_y, ω_z respectively and add them together, we see that each fraction is

$$= \frac{u\omega_x + v\omega_y + w\omega_z}{\omega_x{}^2 + \omega_y{}^2 + \omega_z{}^2} = \frac{V}{\Omega}.$$

This ratio is called the *pitch of the screw*.

241. **The Invariant.** It follows from the third result just proved that whatever base be chosen and whatever be the direction of the axes, the quantity $u\omega_x + v\omega_y + w\omega_z$ is invariable and equal to $V\Omega$. This quantity may therefore be called the *invariant* of the components. The resultant angular velocity Ω is also invariable and may be called the *invariant* of the rotation.

If the motion be such that the first of these invariants is zero, it follows that either $V = 0$, or $\Omega = 0$. This therefore is *the condition that the motion is equivalent to either a simple translation or a simple rotation*. If we wish the motion to be equivalent to a simple rotation, we must also have ω_x, ω_y, ω_z not all zero.

The corresponding invariant in statics is $LX + MY + NZ = GR$. When this vanishes, the forces are equivalent to either a single resultant or a single couple.

242. When the motion is equivalent to a simple rotation, it may be required to find the axis of rotation. But this is obviously only the central axis under another name, and has been found above.

243. A screw motion may thus be given in two ways. We may have given the six components of motion, which we have called $(u, v, w, \omega_x, \omega_y, \omega_z)$, which also depend on the point chosen as base. Or it may be given by the equations to the central axis the velocity V along it, and the angular velocity Ω round it.

In this last case a convention is necessary to prevent confusion as to the directions implied by the velocities V and Ω. One direction of the axis is called the positive direction, and the opposite the negative direction. Then V is taken positive when it implies a velocity in the positive direction. So also Ω is positive when the rotation appears to be in the direction of the hands of a watch, when viewed by a person placed with his back along the axis, so that the positive direction is from his feet to his head. This of course is only the ordinary definition of a positive couple as given in statics. See Art. 231.

The method of determining the positive direction of the axis is easy to understand, though it takes long to explain. Describe a sphere of unit radius with its centre at the origin, and let the *positive* directions of the axes cut this sphere in x, y, z. Let a parallel to the central axis drawn through the origin cut the sphere in L and L'. Let the direction-cosines of the axis be given say, l, m, n. Then (lmn) are the cosines of certain arcs drawn on the sphere which begin at xyz, and terminate say at L, while $(-l, -m, -n)$ are the cosines of supplementary arcs which begin at the same points xyz, and terminate at L'. Then OL is the positive direction of the axis and OL' the negative direction.

244. *The position of the central axis being given, together with the linear velocity along it and the angular velocity round it, it is required to find the components of the motion when the origin is taken as the base.*

This is of course the converse proposition to that just discussed. Let the equation to the central axis be $\dfrac{x-f}{l} = \dfrac{y-g}{m} = \dfrac{z-h}{n}$, where (lmn) are the actual direction cosines of the axis. Let V be the linear and Ω the angular velocity.

If (fgh) were taken as the base, the components of the linear velocities would be lV, mV, nV, and the components of the angular velocities would be $l\Omega$, $m\Omega$, $n\Omega$. Hence by Art. 238, writing $-f$, $-g$, $-h$ for x, y, z, the components of the motion when the origin is the base point are

$$u = lV - \Omega (mh - ng), \qquad \omega_x = l\Omega,$$
$$v = mV - \Omega (nf - lh), \qquad \omega_y = m\Omega,$$
$$w = nV - \Omega (lg - mf), \qquad \omega_z = n\Omega.$$

245. **Composition and Resolution of Screws.** *Given two screw motions to compound them into a single screw and conversely given any screw motion to resolve it into two screws.*

Two screws being given, let us choose some convenient base and axes. By Art. 244 we may find the six components of motion of each screw for this base. Adding these two and two, we have the six components of the resultant screw. Then by Art. 240 the central axis together with the linear and angular velocities of the screw may be found.

Conversely, we may resolve any given screw motion into two screws in an infinite number of ways. Since a screw motion is represented by six components at any base we have in the two screws twelve quantities at our disposal. Six of these are required to make the two screws equivalent to the given screw. We may therefore in general satisfy six other conditions at pleasure.

Thus we may choose the axis of one screw to be any given straight line we please with any linear velocity along it and any angular velocity round it. The other screw may then be found by reversing this assumed screw and joining it thus changed to the given motion. The screw equivalent to this compound motion is the second screw, and it may be found in the manner just explained.

Or again, we may represent the motion by two screws whose pitches are both chosen to be zero, the axis of one being arbitrary. These are the conjugate axes spoken of in Art. 237.

246. **Examples.** **Ex. 1.** The locus of points in a body moving about a fixed point which at any instant have the same resultant velocity is a circular cylinder.

Ex. 2. A body has an angular velocity Ω about an axis whose equation is $\frac{x-f}{l} = \frac{y-g}{m} = \frac{z-h}{n}$, find the resolved velocities parallel to the axes of any point whose co-ordinates are (ξ, η, ζ).

Ex. 3. A body has equal angular velocities about two axes which neither meet nor are parallel. Prove that the central axis is equally inclined to each. Find also the linear velocity along the central axis and the angular velocity round it.

Ex. 4. If radii vectores be drawn from a fixed point O to represent in direction and magnitude the velocities of all points of a rigid body in motion, prove that the extremities of these radii vectores at any one instant lie in a plane. [Coll. Exam.]

This plane is evidently perpendicular to the central axis, and its distance from O measures the velocity along the axis.

Ex. 5. The locus of the tangents to the trajectories of different points of the same straight line in the instantaneous motion of a body is a hyperbolic paraboloid.

Let AB be the given straight line, CD its conjugate. The points on AB are turning round CD, and therefore the tangents all pass through two straight lines, viz. AB and its consecutive position $A'B'$, and are also all parallel to a plane which is perpendicular to CD.

Ex. 6. Two screws (V, Ω), (V', Ω') have their axes inclined at an angle θ. If the axes intersect, the invariant of the components of the motion is

$$V\Omega + V'\Omega' + (V\Omega' + V'\Omega) \cos \theta.$$

If the axes do not intersect, let D be their shortest distance, then we add to the above expression $\Omega\Omega'D \sin \theta$.

Ex. 7. A motion is represented by angular velocities Ω_1, Ω_2, &c. about any axes. If D is the shortest distance between any two axes, say with angular velocities Ω, Ω', and θ the angle between these axes, then the invariant of the motion is $\Sigma\Omega\Omega'D \sin \theta$, where Σ implies summation for every combination of axes taken two and two.

Ex. 8. Let the restraints on a body be such that it admits of two motions A and B, each of which may be represented by a screw motion, and let m, m' be the pitches of these screws. Then the body must admit of a screw motion compounded of any indefinitely small rotations ωdt, $\omega'dt$ about the axes of these screws accompanied of course by the translations $m\omega dt$, $m'\omega'dt$. Prove that (1) the locus of the axes of all these screws is the surface $z(x^2 + y^2) = 2axy$. (2) If the body be screwed along any generator of this surface the pitch is $c + a \cos 2\theta$, where c is a constant which is the same for all generators and θ is the angle the generator makes with the axis of x. (3) The size and position of the surface being chosen so that the two given screws A and B lie on the surface with their appropriate pitch, show that only one surface can be drawn to contain two given screws. (4) If any three screws of the surface be taken and a body be displaced by being screwed along each of these through a small angle proportional to the sine of the angle between the other two, the body after the last displacement will occupy the same position that it did before the first.

This surface has been called the *cylindroid* by Sir R. Ball, to whom these four theorems are due.

R. D.   14

Ex. 9. If the instantaneous motion of a body be represented by two conjugate angular velocities ω, ω', the axis of the resultant screw intersects at right angles the shortest distance between the conjugate axes. Let γ, γ' be the angles the conjugate axes make with the axis of their resultant, a the angle they make with each other, c, c' the shortest distances between the conjugate axes and the axis of the screw, V and Ω the linear and angular velocities of the screw, then prove that

$$\frac{\omega}{\sin \gamma'} = \frac{\omega'}{\sin \gamma} = \frac{\Omega}{\sin a}, \quad \frac{c\omega}{\cos \gamma'} = \frac{c'\omega'}{\cos \gamma} = \frac{V}{\sin a}; \quad c \tan \gamma' = c' \tan \gamma = \frac{V}{\Omega}.$$

The first set follows from Art. 237. The second expresses the fact that the direction of the linear motion of the point where the axis cuts the shortest distance is along the axis of the screw.

Ex. 10. An instantaneous motion is given by the linear velocities (u, v, w) along, and the angular velocities $(\omega_x, \omega_y, \omega_z)$ round the co-ordinate axes. It is required to represent this by two conjugate angular velocities, one being about the straight line $\dfrac{x-f}{l} = \dfrac{y-g}{m} = \dfrac{z-h}{n}$.

If Ω be the angular velocity about the given axis, then

$$\frac{u\omega_x + v\omega_y + w\omega_z}{\Omega} = lu + mv + nw - \begin{vmatrix} f, & g, & h \\ \omega_x, & \omega_y, & \omega_z \\ l, & m, & n \end{vmatrix},$$

where (l, m, n) are the actual direction-cosines.

The equations to the conjugate axis are

$$\begin{vmatrix} x, & y, & z \\ \omega_x, & \omega_y, & \omega_z \\ l, & m, & n \end{vmatrix} = lu + mv + nw, \qquad \begin{vmatrix} x, & y, & z \\ \omega_x, & \omega_y, & \omega_z \\ f, & g, & h \end{vmatrix} = (f-x)u + (g-y)v + (h-z)w.$$

These general equations will be simplified if the circumstances of any problem permit the co-ordinate axes to be so chosen that some of the constants may be zero. Thus, if the central axis of the instantaneous motion is taken as the axis of z and the shortest distance between that axis and the given straight line as the axis of x, we have $u=0$, $v=0$, $\omega_x = 0$, $\omega_y = 0$; $g=0$, $h=0$, and $l=0$. After these substitutions the results become equivalent to those given in the last example.

The first of these equations may be obtained as indicated in Art. 245. Reverse Ω and join it to the given motion, then the invariant of this compound motion vanishes. If the angular velocity Ω be thus supposed known, the conjugate axis is the central axis of the compound motion and may be found as in Art. 245. But if the conjugate axis be required independently of Ω, we may use the second and third equations.

The second equation follows from the fact that the direction of motion of any point on the conjugate is perpendicular to the given axis.

The third follows from the fact that the direction of motion is also perpendicular to the straight line joining the point to (f, g, h).

There is an apparent exception to these results when the given motion and the given axis are such that Ω, as found from the first equation, is infinite. This is a limiting case rather than an exception. It is easy to see that both the second and third equations are, in this case, satisfied by substituting $x=f+lt$, $y=g+mt$, $z=h+nt$; i.e. the conjugate axis coincides with the given axis. If Ω' be the angular velocity about the conjugate axis, Ω and Ω' are together equivalent to the resultant angular velocity of the given motion; it follows that Ω' is also infinite. In this limiting case, therefore, the motion is represented by two infinite angular velocities about two coincident lines.

Another limiting case is when the given axis is parallel to the central axis of the given motion and the invariant of the motion is not zero. In this case l, m, n are proportional to ω_x, ω_y, ω_z, and the second equation represents a plane at infinity. The conjugate axis is therefore at infinity and the angular velocity about it is zero.

There is a third limiting case when the invariant of the given motion is zero. If the given motion is a simple rotation about some axis, say Oz, and the given axis is not parallel to Oz and does not intersect it, $\Omega = 0$ and the conjugate axis coincides with Oz. If the given axis is parallel to Oz or intersects it, Ω may have any value and the conjugate axis is the resultant axis of the given rotation and the reversed Ω.

If the given motion is a simple translation parallel to some axis Oz and the given axis is not perpendicular to Oz, $\Omega = 0$ and the conjugate is at infinity. If the given axis is perpendicular to Oz, Ω may have any value, and the conjugate axis is found as before; see Art. 234.

In discussing these limiting cases analytically, it will be convenient to choose the simplified form of axes described above.

Ex. 11. If one conjugate of an instantaneous motion is at right angles to the central axis the other meets it, and conversely. If one conjugate is parallel to the central axis the other is at an infinite distance, and conversely.

Ex. 12. A body is moved from any position in space to any other, and every point of the body in the first position is joined to the same point in the second position. If all the straight lines thus found be taken which pass through a given point, they will form a cone of the second order. Also if the middle points of all these lines be taken, they will together form a body capable of an infinitesimal motion, each point of it along the line on which the same is situate. Cayley's *Report to the British Assoc.*, 1862.

247. **Characteristic and focus.** If the instantaneous motion of a body be represented by two conjugate rotations about two axes *at right angles*, a plane can be drawn through either axis perpendicular to the other. The axis in the plane has been called the *characteristic* of that plane, and the axis perpendicular to the plane is said to cut the plane in its *focus*. These names were given by M. Chasles in the *Comptes Rendus* for 1843. Some of the following examples were also given by him, though without demonstrations.

Ex. 1. Show that every plane has a characteristic and a focus.

Let the central axis cut the plane in O. Resolve the linear and angular velocities in two directions Ox, Oz, the first in the plane and the second perpendicular to it. The translations along Ox, Oz may be removed if we move the axes of rotation Ox, Oz parallel to themselves, by Art. 234. Thus the motion is represented by a rotation about an axis in the plane and a rotation about an axis perpendicular to it. It also follows that the characteristic of a plane is parallel to the projection of the central axis.

Ex. 2. If a plane be fixed in the body and move with the body, it intersects its consecutive position in its characteristic. The velocity of any point P in the plane when resolved perpendicular to the plane is proportional to its distance from the characteristic, and when resolved in the plane is proportional to its distance from the focus and is perpendicular to that distance.

Ex. 3. If two conjugate axes cut a plane in F and G, then FG passes through the focus.

If two conjugate axes be projected on a plane, they meet in the characteristic of that plane.

Ex. 4. If two axes CM, CN meet in a point C, their conjugates lie in a plane whose focus is C and intersect in the focus of the plane CMN.

This follows from the fact that if a straight line cut an axis the direction of motion of every point on it is perpendicular to the straight line only when it also cuts the conjugate.

Ex. 5. Any two axes being given and their conjugates, the four straight lines lie on the same hyperboloid.

Ex. 6. If the instantaneous motion of a body be given by the linear and angular velocities (u, v, w), $(\omega_1, \omega_2, \omega_3)$, prove that the characteristic of the plane

$$Ax + By + Cz + D = 0$$

is its intersection with $A(u + \omega_2 z - \omega_3 y) + B(v + \omega_3 x - \omega_1 z) + C(w + \omega_1 y - \omega_2 x) = 0$,

and its focus may be found from $\dfrac{u + \omega_2 z - \omega_3 y}{A} = \dfrac{v + \omega_3 x - \omega_1 z}{B} = \dfrac{w + \omega_1 y - \omega_2 x}{C}$.

For the characteristic is the locus of the points whose directions of motion are perpendicular to the normal to the plane, and the focus is the point whose direction of motion is perpendicular to the plane.

What do these equations become when the central axis is the axis of z?

Ex. 7. The locus of the characteristics of planes which pass through a given straight line is a hyperboloid of one sheet; the shortest distance between the given straight line and the central axis being the direction of one principal diameter, and the other two being the internal and external bisectors of the angle between the given straight line and the central axis. Prove also that the locus of the foci of the planes is the conjugate of the given straight line.

Ex. 8. Let any surface A be fixed in a body and move with it, the normal planes to the trajectories of all its points envelope a second surface B. Prove that if the surface B be fixed in the body and move with it, the normal planes to the trajectories of its points will envelope the surface A: so that the surfaces A and B have conjugate properties, each surface being the locus of the foci of the tangent planes to the other. Prove that if one surface is a quadric the other is also a quadric.

Euler's Equations.

248. In Euler's equations the motion of the body is referred to a system of axes which move in space. It might therefore be thought that the consideration of these equations should be postponed until we treat generally of such axes. Yet as his axes are fixed in the body and move only with it, they really form a special case. On account of their importance and the fact that they are in more general use than any other form of moving axes, we shall treat them separately, even though this may necessitate a little repetition of the arguments concerning such axes.

The following proof starts from first principles and is intended to be very elementary. Several other proofs will be found further on.

To determine the general equations of motion of a body about a fixed point.

Let the fixed point O be taken as origin, and let x, y, z be the co-ordinates at time t of any particle m referred to any rectangular axes fixed in space. Let Xm, Ym, Zm be the impressed forces acting on this element parallel to the axes of co-ordinates, and let L, M, N be the moments of all these forces about the axes.

Then by D'Alembert's Principle, if the effective forces $m\ddot{x}$, $m\ddot{y}$, $m\ddot{z}$ be applied to every particle m in a reversed direction, there will be equilibrium between these forces and the impressed forces. Taking moments therefore about the axes, we have

$$\Sigma m\,(x\ddot{y} - y\ddot{x}) = N \dots\dots\dots\dots\dots\dots(1),$$

and two similar equations.

To simplify these equations, let ω_x, ω_y, ω_z be the angular velocities about the axes. Then $\dot{x} = \omega_y z - \omega_z y$, $\dot{y} = \omega_z x - \omega_x z$, $\dot{z} = \omega_x y - \omega_y x$;

$$\therefore\ \ddot{x} = z\dot{\omega}_y - y\dot{\omega}_z + \omega_y\,(\omega_x y - \omega_y x) - \omega_z\,(\omega_z x - \omega_x z),$$

$$\ddot{y} = x\dot{\omega}_z - z\dot{\omega}_x + \omega_z\,(\omega_y z - \omega_z y) - \omega_x\,(\omega_x y - \omega_y x).$$

Substituting in equation (1) we get

$$\left.\begin{array}{l}\Sigma m\,(x^2 + y^2)\,\dot{\omega}_z - \Sigma myz\,.\,\dot{\omega}_y - \Sigma mxz\,.\,\dot{\omega}_x + \Sigma mxz\,.\,\omega_y\omega_z \\ -\,\Sigma mxy\,.\,(\omega_x{}^2 - \omega_y{}^2) + \Sigma m\,(x^2 - y^2)\,\omega_x\omega_y - \Sigma myz\,.\,\omega_x\omega_z\end{array}\right\} = N.$$

The other two equations may be treated in the same manner.

The coefficients in this equation are the moments and products of inertia of the body with regard to axes fixed in space and are therefore variable as the body moves about. Let us then take a second set of rectangular axes OA, OB, OC fixed in the body, and let ω_1, ω_2, ω_3 be the angular velocities about these axes. Since the axes Ox, Oy, Oz are perfectly arbitrary, let them be so chosen that the axes OA, OB, OC are passing through them at the moment under consideration.

249. The axes of reference OA, OB, OC move in space. We suppose the motion determined by the three angular velocities ω_1, ω_2, ω_3 in the same manner as if the axes were fixed for an instant in space. The position of the body at the time $t + dt$ may be constructed from that at the time t by turning the body through the angles $\omega_1 dt$, $\omega_2 dt$, $\omega_3 dt$ successively round the instantaneous positions of the axes.

Let Ω be the angular velocity of the body and let OR be the instantaneous axis of rotation. Then since the two sets of axes coincide at this instant, we have the resolved angular velocities also equal, i.e. $\omega_x = \omega_1$, $\omega_y = \omega_2$, $\omega_z = \omega_3$. But at the time $t + dt$ the two sets of axes have separated, so that we can no longer assert that any two components such as $\omega_z + d\omega_z$ and $\omega_3 + d\omega_3$ are necessarily equal.

We shall now show that *if the moving axes are fixed in the body, then $d\omega_3 = d\omega_z$* as far as the first order of small quantities. Let OR, OR' be the resultant axes of rotation of the body at the times t and $t + dt$, i.e. let a rotation Ωdt about OR bring the body into the position in which OC coincides with Oz at the time t; and let a further rotation $\Omega' dt$ about OR' bring the body into some adjacent position at the time $t + dt$ while in the same interval dt, OC moves into the position OC'. Then according to the definition of a differential coefficient

$$\frac{d\omega_3}{dt} = \text{limit } \frac{\Omega' \cos R'C' - \Omega \cos RC}{dt},$$

$$\frac{d\omega_z}{dt} = \text{limit } \frac{\Omega' \cos R'z - \Omega \cos Rz}{dt}.$$

The angles RC and Rz are equal by hypothesis. Since OC is fixed in the body, it makes a constant angle with OR' as the body turns round OR', hence the angles $R'C'$ and $R'z$ are also equal. Hence these differential coefficients are also equal.

250. The following demonstration of this equality has been given by the late Professor Slesser, and is instructive as founded on a different principle. Let A, B, C be the points in which the principal axes cut a sphere whose centre is at the fixed point. Let OL be any other axis, and let Ω be the angular velocity about it. Let the angles LOA, LOB, LOC be called respectively α, β, γ. Then by Art. 233

$$\Omega = \omega_1 \cos \alpha + \omega_2 \cos \beta + \omega_3 \cos \gamma;$$

$$\therefore \frac{d\Omega}{dt} = \dot{\omega}_1 \cos \alpha + \dot{\omega}_2 \cos \beta + \dot{\omega}_3 \cos \gamma - \omega_1 \sin \alpha \dot{\alpha} - \omega_2 \sin \beta \dot{\beta} - \omega_3 \sin \gamma \dot{\gamma}.$$

Now let the line OL be fixed in space and coincide with OC at the moment under consideration. Then $\alpha = \frac{\pi}{2}$, $\beta = \frac{\pi}{2}$, $\gamma = 0$; therefore $\dot{\Omega} = \dot{\omega}_3 - \omega_1 \dot{\alpha} - \omega_2 \dot{\beta}$.

Also $\dot{\alpha}$ is the angular rate at which A separates from a *fixed* point at C, this is clearly ω_2. Similarly $\dot{\beta} = -\omega_1$. Hence $\dot{\Omega} = \dot{\omega}_3$. Thus $\dot{\omega}_x = \dot{\omega}_1$, $\dot{\omega}_y = \dot{\omega}_2$, $\dot{\omega}_z = \omega_3$.

251. **Euler's dynamical equations.** We have now proved that we may substitute in the equations of motion for ω_x, &c. the angular velocities ω_1, &c. about a set of axes OA, OB, OC fixed in the body and moving with it. *The advantage of this transformation is that all the moments and products of inertia which occur in the equation are now constants.*

We can make a further simplification by properly choosing these axes in the body. Let us choose as the axes fixed in the body the principal axes at the fixed point O. In this case the products of inertia are all zero. If A, B, C be the principal moments the equations take the simple form

$$C \frac{d\omega_3}{dt} - (A - B) \omega_1\omega_2 = N.$$

Similarly
$$A \frac{d\omega_1}{dt} - (B - C) \omega_2\omega_3 = L,$$

$$B \frac{d\omega_2}{dt} - (C - A) \omega_3\omega_1 = M.$$

These are called Euler's equations.

252. We know by D'Alembert's principle that the moment of the effective forces about any straight line is equal to that of the impressed forces. The equations of Euler therefore indicate that the moments of the effective forces about the principal axes at the fixed point are expressed by the left-hand sides of the above equations. If there is no point of the body which is fixed in space, the motion of the body about its centre of gravity is the same as if that point were fixed. In this case, if A, B, C be the principal moments at the centre of gravity, the left-hand sides of Euler's equations give the moments of the effective forces about the principal axes at the centre of gravity. If we want the moment about any other straight line passing through the fixed point, we may find it by simply resolving these moments by the rules of statics.

253. Ex. 1. If $2T = A\omega_1^2 + B\omega_2^2 + C\omega_3^2$ and G be the moment of the impressed forces about the instantaneous axis, Ω the resultant angular velocity, prove that $\frac{dT}{dt} = G\Omega$.

Ex. 2. A body turning about a fixed point is acted on by forces which tend to produce rotation about an axis at right angles to the instantaneous axis, show that the angular velocity cannot be uniform unless two of the principal moments at the fixed point are equal. The axis about which the forces tend to produce rotation is that axis about which it would begin to turn if the body were placed at rest.

254. *To determine the pressure on the fixed point.*

Let x, y, z be the co-ordinates of the centre of gravity referred to rectangular axes fixed in space meeting at the fixed point, and let P, Q, R be the resolved parts of the pressures on the body in these directions. Let μ be the mass of the body. Then we have

$$\mu\ddot{x} = P + \Sigma mX$$

and two similar equations. Substituting for $\ddot{x}$ its value in terms ω_x, ω_y, ω_z we have

$$\mu \left\{ z\dot{\omega}_y - y\dot{\omega}_z + \omega_y \left(\omega_x y - \omega_y x \right) - \omega_z \left(\omega_z x - \omega_x z \right) \right\} = P + \Sigma m X$$

and two similar equations.

If we now take the axes fixed in space to coincide with the principal axes at the fixed point at the moment under consideration we may substitute for $\dot{\omega}_y$ and $\dot{\omega}_z$ from Euler's equations. We then have

$$\mu \left\{ \omega_1 (B+C-A) \left(\frac{y\omega_2}{C} + \frac{z\omega_3}{B} \right) - (\omega_2{}^2 + \omega_3{}^2) x \right\} = P + \Sigma m X - \mu \left(\frac{M}{B} z - \frac{N}{C} y \right),$$

with similar expressions for Q and R.

255. **Ex.** If G be the centre of gravity of the body, show that the terms on the left-hand sides of the equations which give the pressures on the fixed point are the components of two forces, one $\Omega^2 \cdot GH$ parallel to GH which is a perpendicular on the instantaneous axis OI, Ω being the resultant angular velocity, and the other $\Omega'^2 \cdot GK$ perpendicular to the plane OGK, where GK is a perpendicular on a straight line OJ whose direction-cosines are proportional to $\dfrac{B-C}{A} \omega_2 \omega_3$, $\dfrac{C-A}{B} \omega_3 \omega_1$, $\dfrac{A-B}{C} \omega_1 \omega_2$, and Ω'^4 is the sum of the squares of these quantities.

256. **Euler's geometrical equations.** *To determine the geometrical equations connecting the motion of the body in space with the angular velocities of the body about the three moving axes,* OA, OB, OC.

Let the fixed point O be taken as the centre of a sphere of radius unity; let X, Y, Z and A, B, C be the points in which the sphere is cut by the fixed and moving axes respectively. Let ZC, BA produced if necessary, meet in E. Let the angle $XZC = \psi$,

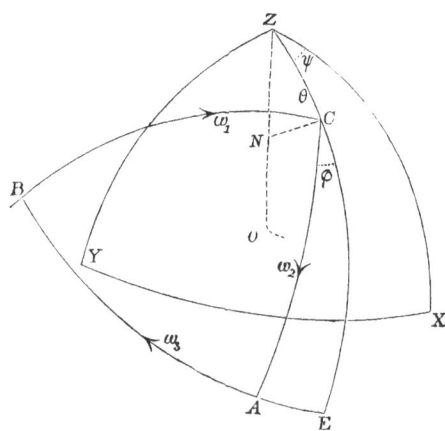

$ZC = \theta$, $ECA = \phi$. It is required to determine the geometrical relations between θ, ϕ, ψ, and ω_1, ω_2, ω_3.

Draw CN perpendicular to OZ. Then since ψ is the angle the plane COZ makes with a plane XOZ fixed in space, the velocity of C perpendicular to the plane ZOC is $CN \dfrac{d\psi}{dt}$, which is the same as $\sin \theta \dfrac{d\psi}{dt}$, the radius OC of the sphere being unity. Also the velocity of C along ZC is $\dfrac{d\theta}{dt}$. Thus the motion of C is represented by $\dfrac{d\theta}{dt}$ and $\sin \theta \dfrac{d\psi}{dt}$ respectively along and perpendicular to ZC. But the motion of C is also expressed by the angular velocities ω_1 and ω_2 respectively along BC and CA. These two representations of the same motion must therefore be equivalent. Hence resolving along and perpendicular to ZC we have

$$\left. \begin{aligned} \frac{d\theta}{dt} &= \omega_1 \sin \phi + \omega_2 \cos \phi \\ \sin \theta \frac{d\psi}{dt} &= -\omega_1 \cos \phi + \omega_2 \sin \phi \end{aligned} \right\}$$

Similarly by resolving along CB and CA we have

$$\left. \begin{aligned} \omega_1 &= \frac{d\theta}{dt} \sin \phi - \frac{d\psi}{dt} \sin \theta \cos \phi \\ \omega_2 &= \frac{d\theta}{dt} \cos \phi + \frac{d\psi}{dt} \sin \theta \sin \phi \end{aligned} \right\}$$

These two sets of equations are precisely equivalent to each other and one may be deduced from the other by an algebraic transformation.

In the same way by drawing a perpendicular from E on OZ we may show that the velocity of E perpendicular to ZE is $\dfrac{d\psi}{dt} \sin ZE$, and this is the same as $\dfrac{d\psi}{dt} \cos \theta$. Also the velocity of A relative to E along EA is in the same way $\dfrac{d\phi}{dt} \sin CA$, and this is the same as $\dfrac{d\phi}{dt}$. Hence the whole velocity of A in space along AB is represented by $\dfrac{d\psi}{dt} \cos \theta + \dfrac{d\phi}{dt}$. But this motion is also expressed by ω_3. As before these two representations of the same motion must be equivalent. Hence we have

$$\omega_3 = \frac{d\psi}{dt} \cos \theta + \frac{d\phi}{dt}.$$

If in a similar manner we had expressed the motion of any other point of the body as B, both in terms of ω_1, ω_2, ω_3 and θ, ϕ, ψ, we should have obtained other equations. But as we cannot have more than three independent relations, we should only arrive at equations which are algebraic transformations of those already obtained.

257. It is sometimes necessary to express the angular velocities of the body about the *fixed* axes OX, OY, OZ in terms of θ, ϕ, ψ. This may be effected in the following manner. Let ω_x, ω_y, ω_z be the angular velocities about the fixed axes, Ω the resultant angular velocity. If we impress on space and also on the body in addition to its existing motion, an angular velocity equal to $-\Omega$ about the resultant axis of rotation, the axes OA, OB, OC will become fixed, and the axes OX, OY, OZ will move with angular velocities $-\omega_x$, $-\omega_y$, $-\omega_z$. Hence, in the formulæ of the text, if we change ϕ into $-\psi$, θ into $-\theta$, ψ into $-\phi$, ω_1, ω_2, ω_3 will become $-\omega_x$, $-\omega_y$, $-\omega_z$, and we have

$$\omega_x = -\frac{d\theta}{dt}\sin\psi + \frac{d\phi}{dt}\sin\theta\cos\psi,$$

$$\omega_y = \frac{d\theta}{dt}\cos\psi + \frac{d\phi}{dt}\sin\theta\sin\psi,$$

$$\omega_z = \frac{d\phi}{dt}\cos\theta + \frac{d\psi}{dt}.$$

Sometimes it will be more convenient to measure the angular co-ordinates θ, ϕ, ψ in a different manner. Suppose, for example, we wish to refer the axes fixed in space to the axes fixed in the body as co-ordinate axes. To obtain the standard figure corresponding to this case, we must in the figure of Art. 256 interchange the letters X, Y, Z with A, B, C, each with each. The angles θ, ϕ, ψ being measured as indicated in the figure after this change, the relations connecting them with the angular velocities about the axes fixed in space, are obtained from those in Art. 256 by simply changing ω_1, ω_2, ω_3 into $-\omega_x$, $-\omega_y$, $-\omega_z$. If we choose to measure θ in the opposite direction to that indicated in the figure, the expressions for ω_x, ω_y, become identical with those for ω_1, ω_2, in Art. 256.

258. Ex. 1. If p, q, r be the direction cosines of OZ with regard to the axes OA, OB, OC, show that two of Euler's geometrical equations may be put into the symmetrical form

$$\frac{dp}{dt} - q\omega_3 + r\omega_2 = 0, \qquad \frac{dq}{dt} - r\omega_1 + p\omega_3 = 0, \qquad \frac{dr}{dt} - p\omega_2 + q\omega_1 = 0.$$

Any one of these may be obtained by differentiating one of the expressions $p = -\sin\theta\cos\phi$, $q = \sin\theta\sin\phi$, $r = \cos\theta$. The others may be inferred by the rule of symmetry.

Ex. 2. Prove that the direction cosines of either set of Euler's axes with regard to the other are given by the formulæ

$$\begin{aligned}\cos XA &= -\sin\psi\sin\phi + \cos\psi\cos\phi\cos\theta \\ \cos YA &= \cos\psi\sin\phi + \sin\psi\cos\phi\cos\theta \\ \cos ZA &= -\sin\theta\cos\phi\end{aligned}\Bigg\},$$

$$\begin{aligned}\cos XB &= -\sin\psi\cos\phi - \cos\psi\sin\phi\cos\theta \\ \cos YB &= \cos\psi\cos\phi - \sin\psi\sin\phi\cos\theta \\ \cos ZB &= \sin\theta\sin\phi\end{aligned}\Bigg\},$$

$$\begin{aligned}\cos XC &= \sin\theta\cos\psi \\ \cos YC &= \sin\theta\sin\psi \\ \cos ZC &= \cos\theta\end{aligned}\Bigg\}.$$

To prove the first three, produce XY to cut AB in M, then the angle $XMA = \theta$, $MY = \psi$, $MX = 90 + \psi$, $MA = 90 - \phi$. To deduce the second set from the first, write $\phi + \frac{1}{2}\pi$ for ϕ.

These results are given here for reference as they are useful in the higher problems of dynamics.

Ex. 3. If OC describe a right cone in space with uniform angular velocity about its axis Oz and if the angular velocity of the body about the axis OC be also uniform, find the expressions for ω_1 and ω_2.

259. It is clear that instead of referring the motion of the body to the principal axes at the fixed point, as Euler has done, we may use any axes fixed in the body. But these are in general so complicated as to be nearly useless. When, however, a body is making small oscillations about a fixed point, so that some three rectangular axes fixed in the body never deviate far from three axes fixed in space, it is often convenient to refer the motion to these even though they are not principal axes. In this case ω_1, ω_2, ω_3 are all small quantities, and we may neglect their products and squares. The general equation of Art. 248 reduces in this case to

$$C\dot{\omega}_3 - D\dot{\omega}_2 - E\dot{\omega}_1 = N,$$

where the coefficients have the usual meanings given to them in Chap. I. We have thus three linear equations which may be written thus:

$$A\dot{\omega}_1 - F\dot{\omega}_2 - E\dot{\omega}_3 = L,$$
$$- F\dot{\omega}_1 + B\dot{\omega}_2 - D\dot{\omega}_3 = M,$$
$$- E\dot{\omega}_1 - D\dot{\omega}_2 + C\dot{\omega}_3 = N.$$

260. **The centrifugal forces.** It appears from Euler's Equations that the whole changes of ω_1, ω_2, ω_3 are not due merely to the direct action of the forces, but are in part due to the centrifugal forces of the particles tending to carry them away from the axis about which they are revolving. For consider the equation

$$\frac{d\omega_3}{dt} = \frac{N}{C} + \frac{A - B}{C} \omega_1\omega_2.$$

Of the increase $d\omega_3$ in the time dt, the part $\dfrac{N}{C} dt$ is due to the direct action of

the forces whose moment is N, and the part $\dfrac{A - B}{C} \omega_1\omega_2 dt$ is due to the centrifugal

forces. This may be proved as follows.

If a body be rotating about an axis OI with an angular velocity ω, then the moment of the centrifugal forces of the whole body about the axis Oz is $(A - B)\,\omega_1\omega_2$.

Let P be the position of any particle m, and let x, y, z be its co-ordinates. Then $x = OR$, $y = RQ$, $z = QP$. Let PS be a perpendicular on OI, let $OS = u$, and $PS = r$. Then the centrifugal force of the particle m is $\omega^2 rm$ tending from OI.

The force $\omega^2 rm$ is evidently equivalent to the *four* forces $\omega^2 xm$, $\omega^2 ym$, $\omega^2 zm$, and $-\omega^2 um$ acting at P parallel to x, y, z, and u respectively.

The moment of $\omega^2 xm$ round $Oz = -\omega^2 xym$
.... $\omega^2 ym$ .... $= \omega^2 xym$;
.... $\omega^2 zm$ .... $= 0$

these three therefore produce no effect.

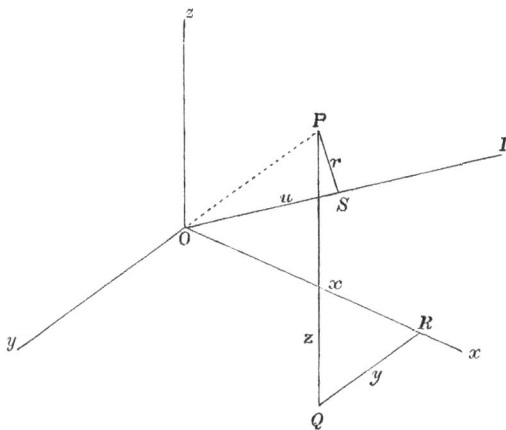

The force $-\omega^2 um$ parallel to OI is equivalent to the three, $-\omega\omega_1 um$, $-\omega\omega_2 um$, $-\omega\omega_3 um$, acting at P parallel to the axes, and their moment round Oz is evidently $\omega u m (\omega_1 y - \omega_2 x)$. Now the direction cosines of OI being $\dfrac{\omega_1}{\omega}$, $\dfrac{\omega_2}{\omega}$, $\dfrac{\omega_3}{\omega}$, we get by

projecting the broken line x, y, z on OI, $u = \dfrac{\omega_1}{\omega} x + \dfrac{\omega_2}{\omega} y + \dfrac{\omega_3}{\omega} z$; therefore substituting for u, the moment of centrifugal forces about Oz is

$$= (\omega_1 y - \omega_2 x)(\omega_1 x + \omega_2 y + \omega_3 z)\, m,$$
$$= (\omega_1^2 xy + \omega_1\omega_2 y^2 + \omega_1\omega_3 yz - \omega_1\omega_2 x^2 - \omega_2^2 xy - \omega_2\omega_3 xz)\, m.$$

Writing Σ before each term, and supposing the axes of x, y, z to be principal axes, then the moment of the centrifugal forces about the principal axis Oz

$$= \omega_1\omega_2 \Sigma m (y^2 - x^2) = \omega_1\omega_2 (A - B).$$

Let the moments of the centrifugal forces about the principal axes of the body be represented by L', M', N', so that

$$L' = (B - C)\, \omega_2\omega_3, \quad M' = (C - A)\, \omega_3\omega_1, \quad N' = (A - B)\, \omega_1\omega_2,$$

and let G be their resultant couple. The couple G is usually called the *centrifugal couple*.

Since $L'\omega_1 + M'\omega_2 + N'\omega_3 = 0$, it follows that the axis of the centrifugal couple is at right angles to the instantaneous axis.

Describe the momental ellipsoid at the fixed point O and let the instantaneous axis cut its surface in I. Let OH be a perpendicular from O on the tangent plane at I. The direction cosines of OH are proportional to $A\omega_1$, $B\omega_2$, $C\omega_3$. Since $A\omega_1 L' + B\omega_2 M' + C\omega_3 N' = 0$, it follows that the axis of the centrifugal couple is at right angles to the perpendicular OH.

The plane of the centrifugal couple is therefore the plane IOH.

If μk^2 be the moment of inertia of the body about the instantaneous axis of rotation, we have $k^2 = \dfrac{t^4}{OI^2}$, and $T = \mu k^2 \omega^2$ is the Vis Viva of the body. We may then easily show that the magnitude G of the centrifugal couple is $G = T \tan \phi$, where ϕ is the angle IOH.

This couple will generate an angular velocity of known magnitude about the diametral line of its plane. By compounding this with the existing angular velocity, the change in the position of the instantaneous axis may be found.

Expressions for Angular Momentum.

261.   We may now investigate convenient formulæ for the angular momentum of a body about any axis. The importance of these has been already pointed out in Art. 75. In fact, the general equations of a motion of a rigid body as given in Art. 78, cannot be completely expressed until these formulæ have been found.

When the body is moving in space of two dimensions about either a fixed point, or its centre of gravity regarded as a fixed point, the angular momentum about that point has been proved in Art. 88 to be $Mk^2\omega$ where Mk^2 is the moment of inertia, and ω the angular velocity about that point. Our object is to find corresponding formulæ when the body is moving in space of three dimensions. We shall show first how to find the angular momentum about a straight line which is such that one axis of reference (say, the axis of z) can be chosen parallel to it. We shall then find an expression for the angular momentum when the straight line is inclined to all three axes of reference. The former result has of course the advantage of simplicity and is therefore more generally useful.

It is particularly important to find a simple form for the angular momentum of the moving body about a *straight line fixed in space*, for then we may use the general principle proved in Art. 78, viz.

$$\frac{d}{dt} \begin{pmatrix} \text{Angular momentum about} \\ \text{a fixed straight line} \end{pmatrix} = \begin{pmatrix} \text{Moment of im-} \\ \text{pressed forces} \end{pmatrix}.$$

262.   **Angular Momentum about the axis of z.**   *The instantaneous motion of a body about a fixed point is given by the angular velocities* ω_x, ω_y, ω_z *about three axes which meet at the point, find the angular momentum about the axis of z.*

Let x, y, z be the co-ordinates of any particle m of the body, and u', v', w' the resolved velocities of that particle parallel to the axes. Then by Art. 77 the moment of the momentum about the axis of z is

$$h_3 = \Sigma m (xv' - yu').$$

Substituting $u' = \omega_y z - \omega_z y$, $v' = \omega_z x - \omega_x z$ from Art. 238, we have

$$h_3 = \Sigma m \, (x^2 + y^2) \, \omega_z - (\Sigma mxz) \, \omega_x - (\Sigma myz) \, \omega_y.$$

Similarly the angular momenta about the axes of x and y are

$$h_1 = \Sigma m \, (y^2 + z^2) \, \omega_x - (\Sigma mxy) \, \omega_y - (\Sigma mxz) \, \omega_z,$$
$$h_2 = \Sigma m \, (z^2 + x^2) \, \omega_y - (\Sigma myz) \, \omega_z - (\Sigma myx) \, \omega_x.$$

Here the coefficients of ω_x, ω_y, ω_z are the moments and products of inertia about the axes which meet at the fixed point.

263. *If there be no fixed point in the body* we must use all the six components of motion. The form of the result depends on the point which is chosen as the base. The form is much simplified by choosing the centre of gravity as the base point, and for the reasons given in Arts. 74, 75 this is generally the most convenient point.

Let Oz be the axis about which the angular momentum is required, and let Ox, Oy be two other axes, thus forming a set of rectangular axes. Let $\bar{x}$, $\bar{y}$, $\bar{z}$ be the co-ordinates of the centre of gravity. Let the instantaneous motion of the body be constructed (as in Art. 238) by the linear velocities u, v, w of the centre of gravity parallel to the axes of reference and the angular velocities ω_x, ω_y, ω_z round three parallel axes meeting at the centre of gravity.

By Art. 75 the angular momentum about Oz is equal to that about a parallel axis through the centre of gravity regarded as a fixed point together with the angular momentum of the whole mass collected at the centre of gravity. The former of these has been found in the last Article and the latter is obviously $M (\bar{x} v - \bar{y} u)$. The required angular momentum is therefore

$$M (\bar{x} v - \bar{y} u) + \Sigma m \, (x^2 + y^2) \, \omega_z - (\Sigma mxz) \, \omega_x - (\Sigma myz) \, \omega_y.$$

Here M is the whole mass of the body, and the coefficients of ω_x, ω_y, ω_z are the moments and products of inertia about axes which meet at the centre of gravity.

264. **Moving axes.** When the axes of reference are moving in space, the motion of the body during any time dt is constructed by using the components of motion as if the axes were fixed for the moment in space. See Art. 249. In the expressions just given for the angular momentum the axes, regarded as fixed in space, may be any whatever. Let them be chosen so that any set of moving axes coincides with them at the time t. Then these formulæ will express the angular momentum about the moving axis of z at that particular moment, whether the axis of z continues to occupy the same position in space or not. The formulæ are therefore quite general and give the instantaneous angular momentum whether the axis be fixed or not.

If the axes chosen be fixed in space the coefficients of ω_x, ω_y, ω_z in the expression for h_3 will generally be variable and their changes may be governed by complicated laws. In such a case it is more convenient to choose axes fixed in the body, and this is the choice made by Euler in his equations of motion, Art. 251.

Suppose a body to be moving about a fixed point O, and let its instantaneous motion be given by the angular velocities ω_1, ω_2, ω_3 about axes Ox', Oy', Oz' fixed in the body. Then the angular momentum about the axis of z' is

$$h_3' = C\omega_3 - E\omega_1 - D\omega_2,$$

where C, E and D are absolute constants, viz.

$$C = \Sigma m \, (x'^2 + y'^2), \quad E = \Sigma mx'z', \quad D = \Sigma my'z'.$$

If the axes fixed in the body be principal axes, then the products of inertia vanish. These expressions for the moments of the momentum will then take the simple form

$$h_1' = A\omega_1, \quad h_2' = B\omega_2, \quad h_3' = C\omega_3,$$

where A, B, C are the principal moments of the body at the origin supposed to be fixed in space.

265. From these results we may deduce *a rule to find the angular momentum about an axis fixed in space in a form which is often more convenient than that given in Art.* 262.

Supposing a body to be turning about a fixed point O, we look for a set of axes Ox', Oy', Oz' such that we may easily find the angular momenta of the body about them. These will generally be some axes fixed in the body. Let the axes fixed in space about which the angular momenta are required be Ox, Oy, Oz. Let the direction cosines of either with regard to the other be given by the diagram; where for example b_3 is the cosine of the angle between the axes of z and y' (see Art. 217). Let the momenta of all the particles of the body be equivalent to the three "couples" h_1', h_2', h_3' about the axes Ox', Oy', Oz'. Then the moment of the momentum about the axis Oz may be written in the form

	x	y	z
x'	a_1	a_2	a_3
y'	b_1	b_2	b_3
z'	c_1	c_2	c_3

$$h_3 = h_1'a_3 + h_2'b_3 + h_3'c_3 \dots\dots\dots\dots\dots(1).$$

In the same way we have

$$h_3' = h_1c_1 + h_2c_2 + h_3c_3 \dots\dots\dots\dots\dots(2).$$

The simplicity of this process depends on the proper choice of the subsidiary set of axes Ox', Oy', Oz'. Generally the most convenient axes to choose are the principal axes of the body at the point O. In this case the equation (1) takes the form

$$h_3 = A\omega_1a_3 + B\omega_2b_3 + C\omega_3c_3.$$

We may now substitute for ω_1, ω_2, ω_3 their values given by Euler's geometrical equations (Art. 256). The values of the direction cosines a_3, b_3, c_3 are written at length in Ex. 2, Art. 258.

When the body is uniaxal, so that the two principal moments of inertia A and B are equal, these results take a very simple form. The substitutions are rather long though not difficult, but we may greatly shorten them by taking the axes Ox', Oz' to coincide with OE, OC in the figure of Art. 256. Oy' will then be perpendicular to both OE and OC. These also are principal axes since the body is uniaxal, thus h_1', h_2', h_3' have the simple forms $A\omega_1$, $A\omega_2$, $C\omega_3$ while the direction cosines are formed by very simple trigonometrical formulæ. In fact the direction cosines are those found by putting $\phi = 0$ in the general forms given in Ex. 2, Art. 258. In this way we find

$$h_1 = A \left\{ -\sin\psi \, \frac{d\theta}{dt} - \sin\theta \cos\theta \cos\psi \, \frac{d\psi}{dt} \right\} + C\omega_3 \sin\theta \cos\psi,$$

$$h_2 = A \left\{ \cos\psi \, \frac{d\theta}{dt} - \sin\theta \cos\theta \sin\psi \, \frac{d\psi}{dt} \right\} + C\omega_3 \sin\theta \sin\psi,$$

$$h_3 = A \, \sin^2\theta \, \frac{d\psi}{dt} + C\omega_3 \cos\theta.$$

We might substitute for ω_3 its value given by Euler's third geometrical equation, but this would introduce $d\phi/dt$ into the equation, and it will generally be found more convenient to retain ω_3.

In this way the *angular momenta of a uniaxal body about any straight lines are expressed in terms of the direction angles of the axis of the body and the angular velocity about it.*

We may find a geometrical meaning for these results which will at once supply us with an easy proof of them and enable us to write them down when wanted.

The angular momenta of the body about the principal axes at the fixed point O are $A\omega_1$, $A\omega_2$, $C\omega_3$. Suppose we attach to the axis OC one or more imaginary particles so that their united moment of inertia about any axis through O perpendicular to OC is equal to A. Let these particles move about with the axis. The motion of the axis is given by the angular velocities ω_1, ω_2 and therefore the angular momenta of these particles about the axes OA, OB are clearly $A\omega_1$, $A\omega_2$. These are the same as those of the body. The angular momentum of the particles about OC is zero. Hence the angular momenta of the body about OA, OB, OC are the same as those of the particles together with an angular momentum $C\omega_3$ about OC. It follows by the "parallelogram law" that the same equality holds for all axes.

Hence the *angular momentum of a uniaxal body about any axis through O is the same as that of one or more particles arranged along its axis (so that their united moment of inertia about O is equal to A) together with the angular momentum $C\omega_3$ about the axis.*

Let a single particle be placed on the axis of the body at a distance unity from the origin. Its mass is therefore represented by A. Let $(\xi\eta\zeta)$ be the co-ordinates of this particle referred to the axes x, y, z, then $(\xi\eta\zeta)$ are also the direction cosines of the axis. The angular momenta about the axes are therefore

$$h_1 = A \left(\eta \frac{d\zeta}{dt} - \zeta \frac{d\eta}{dt} \right) + C\omega_3\, \xi,$$

$$h_2 = A \left(\zeta \frac{d\xi}{dt} - \xi \frac{d\zeta}{dt} \right) + C\omega_3\, \eta,$$

$$h_3 = A \left(\xi \frac{d\eta}{dt} - \eta \frac{d\xi}{dt} \right) + C\omega_3\, \zeta.$$

We have now to write for ξ, η, ζ their values $\xi = \sin\theta\cos\psi$, $\eta = \sin\theta\sin\psi$, $\zeta = \cos\theta$. The substitution in the last equation is easily effected if we remember the rule in the differential calculus $\xi d\eta - \eta d\xi = r^2 d\psi$. See Art. 77.

In this way we arrive at the same results for the angular momenta h_1, h_2, h_3 as before.

If the uniaxal body is making small oscillations and the axis OC is always so nearly coincident with the axis Oz that we can reject the squares of θ, we have

$$\xi = \theta\cos\psi \qquad \eta = \theta\sin\psi \qquad \zeta = 1,$$

$$\left. \begin{aligned} h_1 &= -A\,\frac{d\eta}{dt} + C\omega_3\,\xi \\ h_2 &= A\,\frac{d\xi}{dt} + C\omega_3\,\eta \\ h_3 &= C\omega_3 \end{aligned} \right\}.$$

These are very simple formulæ for the angular momenta about the fixed axes.

If the body is moving freely in space we use the centre of gravity instead of the fixed point. In this case it is convenient to attach to the axis *two equal particles* at equal distances on each side of the centre of gravity, so that the centre of gravity of the imaginary system is the same as that of the body. The angular momentum of the free body about any straight line is then the same as that of the system of particles together with the couple $C\omega_3$ about the axis.

Ex. 1. A body not necessarily uniaxal is turning about a fixed point O. Three particles are attached to the principal axes at such distances a, b, c from O that

$$Ma^2 = \tfrac{1}{2}(B + C - A), \quad Mb^2 = \tfrac{1}{2}(C + A - B), \quad Mc^2 = \tfrac{1}{2}(A + B - C).$$

Prove that the angular momentum of the body about any straight line through O is equal to that of these particles.

This follows at once from Art. 76.

Ex. 2. A rod is constrained to remain on the surface of a smooth cone of revolution having its vertex at the point of suspension of the rod. Show that the angular motion of the rod round the axis of the cone is the same as that of a simple pendulum of length $\tfrac{2}{3} a \sin\alpha / \sin\beta$ where a is the length of the rod, α the semivertical angle of the cone and β the angle the axis of the cone makes with the vertical. [St. John's Coll.]

To find the moments of the effective forces, collect the mass at the centre of gyration. To find the moments of the impressed forces collect the mass at the centre of gravity. Equating the moments about the axis of the cone the result follows at once.

R. D.   15

Ex. 3. A body is turning about a fixed point O and has all its principal moments of inertia at O equal. If θ, ϕ, ψ be the Eulerian coordinates of the axes OA, OB, OC, fixed in the body, show that the angular momenta about the axes fixed in space are respectively

$$h_1 = A \left(-\sin\psi \frac{d\theta}{dt} + \sin\theta \cos\psi \frac{d\phi}{dt} \right),$$

$$h_2 = A \left(\cos\psi \frac{d\theta}{dt} + \sin\theta \sin\psi \frac{d\phi}{dt} \right),$$

$$h_3 = A \left(\frac{d\psi}{dt} + \cos\theta \frac{d\phi}{dt} \right).$$

266. **Angular momentum about any axis.** *The motion of a body is given by the linear velocities* (u, v, w) *of the centre of gravity and the angular velocities* (ω_x, ω_y, ω_z), *prove that the angular momentum about the straight line* $\dfrac{x-f}{l} = \dfrac{y-g}{m} = \dfrac{z-h}{n}$ *is equal to*

$$lh_1 + mh_2 + nh_3 + M \begin{vmatrix} l & m & n \\ u & v & w \\ f & g & h \end{vmatrix}$$

where M *is the mass of the body,* h_1, h_2, h_3 *have the values given in Art.* 262, *and* (l, m, n) *are the actual direction cosines of the given straight line.*

This may be done by the use of the principle proved in Art. 75. The angular momentum about a parallel to the given axis is clearly $lh_1 + mh_2 + nh_3$. We must now find the angular momentum of the whole mass collected at the centre of gravity round the given straight line and add these two results together.

Referring to the figure in Art. 238, let P be the point (fgh). Let us find the angular momentum about a set of axes parallel to the given co-ordinate axes with P for origin. It is clear that NP produced will be the new axis of z. The moment of the velocity of the origin O about NP is seen to be $u . MN - v . OM$, which is the same as $ug - vf$; this tends in the positive direction round NP. Similarly the moments of the velocities of O about the parallels to x and y will be $vh - wg$ and $wf - uh$. If we multiply these three by (n, l, m) respectively, we have the moment of the velocity of the centre of gravity about the straight line. Multiplying this by M we have the angular momentum of the mass at the centre of gravity. The required result follows at once.

267. *To find the angular momentum of a body about the instantaneous axis and also about any perpendicular axis which intersects the instantaneous axis.*

Taking the instantaneous axis for the axis of z, we may use the expressions for h_1, h_2, h_3 given in Art. 262.

In our case $\omega_x = 0$, $\omega_y = 0$, and $\omega_z = \Omega$, where Ω is the resultant angular velocity of the body. The angular momenta about the axes of x, y, z are therefore respectively

$$h_1 = -\left(\Sigma mxz\right)\Omega, \quad h_2 = -\left(\Sigma myz\right)\Omega, \quad h_3 = \Sigma m\left(x^2 + y^2\right)\Omega.$$

It appears therefore that the angular momentum about any straight line Ox perpendicular to the instantaneous axis Oz is not zero unless the product of inertia about those two axes is zero.

To understand this properly we must remember that the angular velocities ω_x, ω_y, ω_z are used merely to construct the motion of the body during the time dt. Referring to the figure of Art. 238, let Oz be the instantaneous axis, then the particle of the body at P is moving perpendicular to the plane PLO, and therefore the direction of its velocity is not parallel to Ox and does not intersect Ox. The velocity of this particle has therefore a moment about Ox, although Ox is perpendicular to the instantaneous axis. Let θ be the angle PMN, $r = PM$, then

$$r^2 \frac{d\theta}{dt} = y\frac{dz}{dt} - z\frac{dy}{dt} = r^2\omega_x - xz\omega_z - xy\omega_y,$$

so that the angular velocity $\dfrac{d\theta}{dt}$ of the particle P about Ox vanishes when $\omega_x = 0$ and $\omega_y = 0$ only when the particle lies in either of the planes xy or yz.

Examples. A triangular area ACB whose mass is M is turning round the side CA with an angular velocity ω. Show that the angular momentum about the side CB is $\frac{1}{12}Mab \sin^2 C\omega$, where a and b are the sides containing the angle C.

Ex. 2. Two rods OA, AB, are hinged together at A and suspended from a fixed point O. The system turns with angular velocity ω about a vertical straight line through O so that the two rods are in a vertical plane. If θ, ϕ be the inclinations of the rods to the vertical, a, b their lengths, M, M' their masses, show that the angular momentum about the vertical axis is

$$\omega\left[\left(\tfrac{1}{3}M + M'\right)a^2\sin^2\theta + M'ab\sin\theta\sin\phi + \tfrac{1}{3}M'b^2\sin^2\phi\right].$$

Ex. 3. A right cone, whose vertex O is fixed, has an angular velocity ω communicated to it about its axis OC, while at the same time its axis is set moving in space. The semi-angle of the cone is $\tfrac{1}{4}\pi$ and its altitude is h. If θ be the inclination of the axis to a fixed straight line Oz and ψ the angle the plane zOC makes with a fixed plane through Oz, prove that the angular momentum about Oz is $\tfrac{3}{4}Mh^2\left(\sin^2\theta\dfrac{d\psi}{dt} + \tfrac{2}{3}\omega\cos\theta\right)$, where M is the mass of the cone.

Ex. 4. A rod AB is suspended by a string from a fixed point O and is moving in any manner. If (l, m, n) (p, q, r) be the direction cosines of the string and rod referred to any rectangular axes Ox, Oy, Oz, show that the angular momentum about the axis of z is

$$Mb^2\left(l\frac{dm}{dt} - m\frac{dl}{dt}\right) + M\frac{a^2}{3}\left(p\frac{dq}{dt} - q\frac{dp}{dt}\right) + M\frac{ab}{2}\left(p\frac{dm}{dt} - m\frac{dp}{dt} + l\frac{dq}{dt} - q\frac{dl}{dt}\right),$$

where M is the mass of the rod, and a, b are the lengths of the rod and string.

268. As examples of the use of the expressions for the angular momentum of a body we shall apply them to the solution of two problems on the motion of a body in three dimensions. In these the axes of reference are fixed in space, the use of moving axes being reserved for the present. For further information we must refer the reader to the second volume where a whole chapter is devoted to examples and illustrations of the different methods of finding the motion of a body in three dimensions.

PROBLEM 1. *A uniaxal top spins on a perfectly rough table with its axis nearly vertical, find the small oscillations of the top*.

Let O be the apex, OC the axis of the top. Let C and A be the moments of inertia about the axis OC and any perpendicular to OC through O. Since the

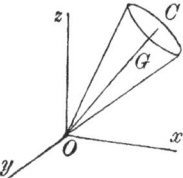

centre of gravity G of the top is in its axis, the impressed forces have no moment about OC. Also $A = B$, hence by Euler's third dynamical equation

$$C\dot{\omega}_3 = 0.$$

Thus the angular velocity of the top about its axis is always the same. Let $\omega_3 = n$ be this constant angular velocity.

Let ξ, η, ζ be the direction cosines of OC referred to fixed axes in space, viz. Ox, Oy, Oz where Oz is vertical. Since the axis of the top is to be always very nearly vertical we have $\zeta = 1$ while ξ, η are small quantities whose squares will be neglected. Let $l = OG$, and let the mass be represented by unity.

The moments of gravity acting at G round the axes are found by the usual formulæ

$$L = yZ - zY = -lg\eta$$
$$M = zX - xZ = lg\xi$$

where $X = 0$, $Y = 0$, $Z = -g$ are the components of gravity. The angular momenta of the body about these axes are by Art. 265,

$$\begin{aligned} h_1 &= -A\dot{\eta} + Cn\xi \\ h_2 &= A\dot{\xi} + Cn\eta \end{aligned}.$$

Hence by Art. 261 we have

$$\begin{aligned} -A\ddot{\eta} + Cn\dot{\xi} &= -gl\eta \\ A\ddot{\xi} + Cn\dot{\eta} &= gl\xi \end{aligned}.$$

---

* The general motion of a top under the action of gravity will be considered in the second volume. The small oscillations of unsymmetrical and inclined tops will be found in that volume. A slight historical account will also be given.

The equation obtained by using the angular momentum about the axis of z merely shows over again that ω_3 is constant, a result already deduced from Euler's equations.

To solve these we put

$$\xi = P \cos (\mu t + f) \qquad \eta = Q \sin (\mu t + f)$$

substituting we find

$$(A\mu^2 + gl) Q - Cn\mu P = 0 \rbrace$$
$$Cn\mu Q - (A\mu^2 + gl) P = 0 \rbrace \quad.$$

These give

$$A\mu^2 + gl = \pm Cn\mu.$$

It is unnecessary to take both the signs on the right hand side. If we choose one sign the effect of the other sign is merely to change the sign of μ and this merely alters the as yet undetermined constants Q and f. Without loss of generality we may choose the *upper sign*. This makes both the resulting values of μ positive. It also gives $P = Q$. The values of μ are

$$\mu = \frac{Cn}{2A} \pm \frac{(C^2 n^2 - 4gAl)^{\frac{1}{2}}}{2A}.$$

Representing these two by $\mu = \mu_1$ and μ_2 we have

$$\xi = P_1 \cos (\mu_1 t + f_1) + P_2 \cos (\mu_2 t + f_2)$$
$$\eta = P_1 \sin (\mu_1 t + f_1) + P_2 \sin (\mu_2 t + f_2)$$

where P_1, P_2, f_1, f_2 are four constants to be determined by the initial values of $\xi, \eta, \dot{\xi}, \dot{\eta}$. Let us represent the initial values of the co-ordinates by the suffix zero. Then

$$\xi_0 = P_1 \cos f_1 + P_2 \cos f_2$$
$$\eta_0 = P_1 \sin f_1 + P_2 \sin f_2$$
$$-\dot{\xi}_0 = P_1 \mu_1 \sin f_1 + P_2 \mu_2 \sin f_2$$
$$\dot{\eta}_0 = P_1 \mu_1 \cos f_1 + P_2 \mu_2 \cos f_2.$$

These give

$$P_1^{\,2} (\mu_1 - \mu_2)^2 = (\dot{\eta}_0 - \mu_2 \xi_0)^2 + (\dot{\xi}_0 + \mu_2 \eta_0)^2 \rbrace$$
$$P_2^{\,2} (\mu_1 - \mu_2)^2 = (\dot{\eta}_0 - \mu_1 \xi_0) + (\dot{\xi}_0 + \mu_1 \eta_0)^2 \rbrace \quad.$$

If θ, ψ be the angular co-ordinates of the axis we have

$$\theta^2 = \xi^2 + \eta^2 = P_1^2 + P_2^2 + 2P_1 P_2 \cos \{(\mu_1 - \mu_2) t + f_1 - f_2)\}$$
$$\theta^2 \dot{\psi} = \xi \dot{\eta} - \dot{\xi} \eta = P_1^2 \mu_1 + P_2^2 \mu_2 + P_1 P_2 (\mu_1 + \mu_2) \cos \{(\mu_1 - \mu_2) t + f_1 - f_2\}.$$

Supposing P_1 and P_2 not to be equal we see that θ can never vanish, i.e. the axis of the top can never become strictly vertical. Also $\dot{\psi}$ will never vanish unless $P_1 P_2 (\mu_1 + \mu_2)$ is greater than $P_1^2 \mu_1 + P_2^2 \mu_2$ i.e. the plane ZOC will revolve round OZ always in the same direction or with temporary reversals of direction according as P_1/P_2 does not or does lie between μ_2/μ_1 and unity.

In order that $P_1 = P_2$ it is necessary that initially

$$2 (\mu_1 - \mu_2) (\xi \dot{\eta} - \dot{\xi} \eta) = (\mu_1^2 - \mu_2^2) (\xi^2 + \eta^2).$$

This requires that $\dot{\psi}$ should initially differ from $\frac{1}{2} (\mu_1 + \mu_2)$ by small quantities of the order P. In this case $\dot{\psi}$ will keep one sign throughout the motion and the axis will become vertical at a constant interval equal to $2\pi/(\mu_1 - \mu_2)$.

We have assumed that the values of μ are both real and unequal. If the value of n be so small that the values of μ are imaginary, the values of ξ and η will contain real exponentials. In this case the values of ξ and η do not in general remain small. This indicates that the top has not sufficient rotation about its axis to keep the axis vertical. It will fall away from its initial position.

If $C^2 n^2 = 4gAl$ the two values of μ are real and equal. In this case it will be seen that the equations are satisfied by

$$\xi = P_1 \cos(\mu t + f_1) + P_2 t \cos(\mu t + f_2)$$
$$\eta = P_1 \sin(\mu t + f_1) + P_2 t \sin(\mu t + f_2),$$

so that the motion is in general unstable. The axis of the top cannot remain nearly vertical unless the initial conditions are such that $P_2 = 0$.

Ex. A uniaxal body rotates about its axis with an angular velocity n. Two inextensible strings are attached to two points on the axis at distances, each equal to b, from the centre of gravity G of the body. The other extremities of the strings are attached to two points fixed in space. The length of each string is a and its tension is T. The mass of the body is unity. Prove that the period $2\pi/p$ of the linear oscillations of G is given by $ap^2 = 2T$, while the periods $2\pi/q$ of the angular oscillations of the axis are given by $Aq^2 - Cnq = 2T(a+b)b/a$.

269. PROBLEM II. *To find the motion of a sphere on a perfectly rough plane.*

Let the plane be taken as the plane of xy and let F, F' be the frictions at the point of contact resolved parallel to these axes. Let X, Y be the resolved impressed forces which we shall suppose to act through the centre. Let a be the radius of the sphere, k its radius of gyration about a diameter and let its mass be taken as unity.

Consider the diameters parallel to the axes of x and y. The angular momenta about them are $k^2 \omega_1$ and $k^2 \omega_2$. These directions are fixed in space, hence we have by Art. 78 or 261,

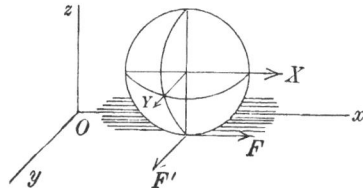

$$k^2 \frac{d\omega_1}{dt} = F'a, \qquad k^2 \frac{d\omega_2}{dt} = -Fa.$$

If u and v be the velocities of the centre of gravity parallel to the axes

$$\frac{du}{dt} = X + F, \qquad \frac{dv}{dt} = Y + F'.$$

Also since the point of contact with the plane does not slide

$$u - a\omega_2 = 0, \qquad v + a\omega_1 = 0.$$

Eliminating F, F' ω_1 and ω_2 we find

$$\frac{du}{dt} = \frac{a^2}{a^2 + k^2} X, \qquad \frac{dv}{dt} = \frac{a^2}{a^2 + k^2} Y.$$

These are the equations of motion of a sphere moving as a particle without rotation on a *smooth* plane under the action of the same forces but reduced in the ratio $a^2/(a^2 + k^2)$. Since $k^2 = \tfrac{2}{5} a^2$ we may enunciate this result as follows.

If a homogeneous sphere* roll on a perfectly rough fixed plane under the action of any forces whatever, whose resultant passes through the centre of the sphere, the motion of the centre is the same as if the plane were smooth, and all the forces were reduced to five-sevenths of their former value.

Ex. 1. If the plane is not perfectly rough yet if the coefficient of friction is greater than $\frac{2}{7} R/Z$ where R is the resultant force parallel to the plane and Z the normal force, prove that the friction will always be sufficient to prevent the sphere from sliding.

Ex. 2. A sphere is placed on an inclined plane sufficiently rough to prevent sliding, and a velocity in any direction is communicated to it. Show that the path of the centre will be a parabola. If V be the initial horizontal velocity of the centre, a the inclination of the plane to the horizon, the latus rectum will be $\frac{14}{5} V^2/g \sin a$.

Ex. 3. A homogeneous sphere rolls on a perfectly rough plane under the action of a force varying inversely as the square of the distance from a point in the plane of motion of the centre, prove that its centre describes a conic section; and if, when the distance of its centre from the centre of force is one-quarter of the major axis of its orbit, the sphere come to a smooth part of the plane, the major axis of the orbit will be suddenly reduced in the ratio $7:13$. [Trin. Coll.]

Ex. 4. A homogeneous sphere moves, without rotation, on a smooth horizontal plane, under the action of a central force such that the centre of the sphere describes an ellipse with the centre of force in the focus. If the sphere arrive at a part of the plane which is perfectly rough when the distance of its centre from the centre of force is $1/n$th of the major axis of its orbit, show that the major axis is diminished in the ratio $7 : 5 + 2n$. If the sphere come again to the smooth part of the plane when the distance of its centre from the focus is the same fraction as before of the major axis, the major axis is again diminished in the same ratio.

Ex. 5. Two spheres equal in volume but of different masses attract each other according to the law of nature and roll on a rough plane. Show that they each describe ellipses relatively to their common centre of gravity with that point for a focus.

270. The principal axes are generally chosen as the axes of reference because the moments of the effective forces for these are extremely simple. Thus the somewhat long equations of Art. 248 reduce to the simple Eulerian forms when referred to principal axes. But sometimes it is important to choose other axes which suit better the geometrical conditions of the problem. The discussion of such axes is reserved for the second volume of this treatise. But when the motion is steady, so that the angular velocities are constant, the equations of Art. 248 will sometimes take so simple a form that an easy solution can be found.

Ex. *A heavy body is attached by two hinges to a horizontal axis about which it is capable of moving freely. The axis is made to rotate with a uniform angular velocity ω about a vertical axis intersecting it in a point O. It is required to find the conditions that the body may be inclined at a constant angle to the vertical.*

* This theorem was given by the author as a problem in the Mathematical Tripos 1860. See the solutions for that year. Another demonstration is given in the second volume by which a corresponding theorem is obtained for the case in which the sphere rolls on another sphere.

Let the horizontal axis which is fixed in the body be taken as the axis of z. Then the vertical lies in the plane of xy, let it make angles θ and $\frac{1}{2}\pi - \theta$ with the axes of x and y. The whole system turns round the vertical with an angular velocity ω. Hence by resolution $\omega_x = \omega \cos \theta$, $\omega_y = \omega \sin \theta$, $\omega_z = 0$. Remembering that these angular velocities are constant, the general equation of moments of Art. 248 becomes

$$- \Sigma mxy \, (\omega_x{}^2 - \omega_y{}^2) + \Sigma m \, (x^2 - y^2) \, \omega_x \omega_y = N.$$

To find N, we resolve the weight Mg parallel to the axes, then $X = -Mg \cos \theta$, $Y = -Mg \sin \theta$, $Z = 0$. If $(x \, y \, z)$ be the coordinates of the centre of gravity we have $N = xY - yX$. The required relation between ω and θ is therefore

$$\omega^2 \left\{ \cos 2\theta \Sigma mxy - \tfrac{1}{2} \sin 2\theta \Sigma m \, (x^2 - y^2) \right\} = Mg \, (x \sin \theta - y \cos \theta).$$

The integrals Σmxy and $\Sigma m \, (x^2 - y^2)$ can be expressed in terms of the moments and products of inertia of the body in the usual manner.

Problems on steady motion may often be easily solved by a direct application of D'Alembert's principle. Thus, in the problem just discussed, each element of the body describes with uniform angular velocity a horizontal circle whose centre is in the vertical axis. If r be the radius of this circle the effective force on the element is $m\omega^2 r$ and its direction is along the radius. The body may therefore be regarded as being in equilibrium under the action of its weight and a system of forces acting directly from the vertical axis and varying as the distance from that axis. The equation found above may be obtained by taking moments about Oz.

Ex. 1. If the body be pushed along the axis of z and made to rotate about the vertical with the same angular velocity as before, show that no effect is produced on the inclination of the body to the vertical.

Ex. 2. If the body be a heavy disc capable of turning about a horizontal axis Oz in its own plane, show that the plane of the disc will be vertical unless $k^2\omega^2 > gh$, where h is the distance of the centre of gravity of the disc from Oz and k the radius of gyration about Oz.

Ex. 3. If the body be a circular disc capable of turning about a horizontal axis perpendicular to its plane and intersecting the disc in its circumference, show that if the tangent to the disc at the hinge make an angle θ with the vertical, the angular velocity ω must be $\sqrt{\dfrac{g}{a \sin \theta}}$.

Ex. 4. Two equal balls A and B are attached to the extremities of two equal thin rods Aa, Bb. The ends a and b are attached by hinges to a fixed point O and the whole is set in rotation about a vertical through O as in the governor of the steam-engine. *If the mass of the rods be neglected show that the time of rotation is equal to the time of oscillation of a pendulum whose length is the vertical distance of either sphere below the hinges at O.*

Ex. 5. If in the last example m be the mass of either thin rod and M that of either sphere, l the length of a rod, r the radius of a sphere, h the depth of either centre below the hinge, then the length of the pendulum is $\dfrac{h}{l+r} \cdot \dfrac{M(l+r)^2 + \frac{1}{3}ml^2}{M(l+r) + \frac{1}{2}ml}$.

ON FINITE ROTATIONS.

271. When the rotations to be compounded are finite in magnitude, the rule to find the resultant is somewhat complicated. As already mentioned in Art. 229 such rotations are not very important in rigid dynamics. We shall therefore only briefly mention a few propositions which may throw light on those already discussed when the motion is infinitely small. We begin with the proposition corresponding to the parallelogram of angular velocities.

Rodrigues' Theorem. *A body has two rotations, (1) a rotation about an axis OA through an angle θ; (2) a subsequent rotation about an axis OB through an angle θ′, and both these axes are fixed in space. It is required to compound the rotations.*

Let lengths measured along OA, OB represent these rotations in the manner explained in Art. 231.

Let the directions of the axes OA, OB cut a sphere whose centre is at O in A and B. On this sphere measure the angle BAC equal to $\frac{1}{2}\theta$ in a direction opposite

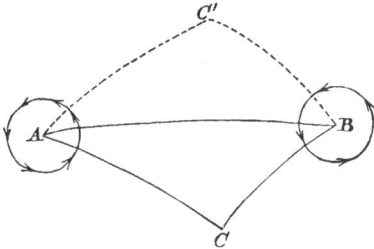

to the rotation round OA and also the angle ABC equal to $\frac{1}{2}\theta′$ in the same direction as the rotation round OB, and let the arcs intersect in C. Lastly, measure the angles $BAC′$, $ABC′$ respectively equal to BAC, ABC, but on the other side of AB.

The rotation $θ$ round OA will then carry any point P in OC into the straight line $OC′$, and the subsequent rotation $θ′$ about OB will carry the point P back into OC. Thus the points in OC are unmoved by the double rotation and OC is there-fore the axis of the single rotation by which the given displacement of the body may be constructed. The straight line OC is called the resultant axis of rotation. If the order of the rotations were reversed, so that the body was rotated first about OB and then about OA, the resultant axis would be $OC′$.

If the axes OA, OB were fixed in the body, the rotation $θ$ about OA would bring OB into a position $OB′$. Then the body may be brought from its first into its last position by rotations $θ$, $θ′$ about the axes OA, $OB′$ *fixed in space*. Hence the same construction will again give the position of the resultant axis and the rotation about it.

To find the magnitude $θ″$ of the rotation about the resultant axis OC we notice that if a point P be taken in OA, it is unmoved by the rotation $θ$ about OA, and the subsequent rotation $θ′$ about OB will bring it into the position $P′$, where $PP′$ is bisected at right angles by the plane OBC. But the rotation $θ″$ about OC must give P the same displacement, hence in the standard case $θ″$ is twice the external angle between the planes OCA, OCB. If the order of the rotations be reversed, the rotation about the resultant axis $OC′$ would be twice the external angle at $C′$, which is the same as that at C. So that though the position of the resultant axis of rotation depends on the order of rotation the resultant angle of rotation is independent of that order.

272. A rotation represented by twice any internal angle of the spherical triangle ABC is equal and opposite to that represented by twice the corresponding external angle. For since the sum of the internal and external angles is π, these two rotations only differ by 2π; and it is evident that a rotation through an angle 2π cannot alter the position of any point of the body. This is merely another way of saying that when a body turns about a fixed axis it may be brought from one given position to another by turning the body either way round the axis.

273. The rule for compounding finite rotations may be stated thus:

If ABC *be a spherical triangle, a rotation round* OA *from* C *to* B *through twice the internal angle at* A, *followed by a rotation round* OB *from* A *to* C *through twice the internal angle at* B, *is equal and opposite to a rotation round* OC *from* B *to* A *through twice the internal angle at* C.

It will be noticed that the order in which the axes are to be taken as we travel round the triangle is opposite to that of the rotations.

As the demonstrations in Art. 271 are only modifications of those of Rodrigues, we may call this theorem after his name. Rodrigues' paper may be found in the fifth volume of Liouville's Journal.

Ex. 1. If two rotations θ, θ' about two axes OA, OB at right angles be compounded into a single rotation ϕ about an axis OC, then

$$\tan COA = \tan \frac{\theta'}{2} \operatorname{cosec} \frac{\theta}{2}, \quad \tan COB = \tan \frac{\theta}{2} \operatorname{cosec} \frac{\theta'}{2}, \quad \text{and} \quad \cos \frac{\phi}{2} = \cos \frac{\theta}{2} \cos \frac{\theta'}{2}.$$

274. **Sylvester's Theorem.** From Rodrigues' theorem we may deduce Sylvester's theorem by drawing the polar triangle $A'B'C'$. Since a side $B'C'$ is the supplement of the angle A, a rotation represented in direction and magnitude by $2B'C'$ differs from that represented by $2A$ in the opposite direction by a rotation through an angle 2π. But a rotation through 2π cannot alter the position of the body, hence the two rotations $2B'C'$ and $2A$ are equivalent in magnitude but opposite in direction. *If therefore* A'B'C' *be any spherical triangle, a rotation represented by twice* B'C' *followed by a rotation twice* C'A' *produces the same displacement of the body as a rotation twice* B'A'. By a rotation $B'C'$ is meant a rotation about an axis perpendicular to the plane of $B'C'$ which will bring the point B' to C'.

275. The following proof of the preceding theorem was given by Prof. Donkin in the *Phil. Mag.* for 1851. Let ABC be any triangle on a sphere fixed in space,

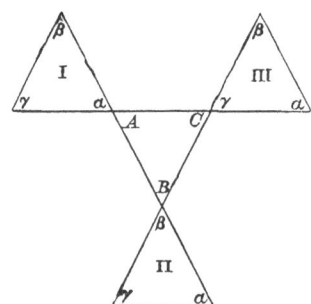

$\alpha\beta\gamma$ a triangle on an equal and concentric sphere moveable about its centre. The sides and angles of $\alpha\beta\gamma$ are equal to those of ABC, but differently arranged, one triangle being the inverse or reflection of the other. If the triangle $\alpha\beta\gamma$ be placed in the position I, so that the sides containing the angle α may be in the same great circles with those containing A, it is obvious that it may slide along AB into the position II, and then along BC into the position III; into which last position it might also be brought by sliding along AC. To slide $\alpha\beta\gamma$ along AB is equivalent to moving β and α each through an arc twice the arc AB about an axis perpendicular to the plane of AB. A similar remark applies when the triangle slides along BC or AC. Hence, twice the rotation AB followed by twice the rotation BC produces the same displacement as twice the rotation AC.

276. **Rotation Couples.** If it be required to compound the rotations about two parallel axes, the construction of Rodrigues requires only a slight modification. Instead of arcs drawn on a sphere, let planes be drawn through the axes making with the plane containing the axes the same angles as before; their intersection will be the resultant axis. One case deserves notice. If $\theta = -\theta'$, the resultant axis is at infinity. A rotation about an axis at infinity is evidently equivalent to a translation. Hence a rotation θ about any axis OA followed by an equal and opposite rotation about a parallel axis $O'B$ distant a from OA is equivalent to a translation $2a \sin \frac{1}{2}\theta$ perpendicular to a plane through OA making an angle $\frac{1}{2}\theta$ with the plane containing the axes and in the direction of the chord of the arc described by any point in OA. These results also follow easily from Art. 223.

277. **Conjugate Rotations.** *Any given displacement of a body may be represented by two finite rotations, one about any given straight line and the other about some other straight line which does not necessarily intersect the first.* When a displacement is thus represented, the axes are called *conjugate axes* and the rotations are called *conjugate rotations.*

Let OA be the given straight line, and let the given displacement be represented by a rotation ϕ about a straight line OR and a translation OT. We wish to resolve this rotation about OR into two rotations, one about OA to be followed by a rotation about OB, where OB is some straight line perpendicular to OT. To do this we follow the rule in Art. 271, we describe a sphere whose centre is O and radius unity and let it intersect OA, OR, OT in A, R and T. Make the angle ARB equal to the supplement of $\frac{\phi}{2}$, and produce RB to B so that $TB = \frac{\pi}{2}$, and join AB. By the triangle of rotations the rotation ϕ is now represented by a rotation about OA which we may call θ, followed by a rotation about OB which we may call θ'.

By Art. 276 the rotation θ' is equivalent to an equal rotation θ' about a parallel axis CD, together with a translation, which may be made to destroy the translation OT. This will be the case if the angle OT makes with the plane of OB, CD be $\frac{1}{2}(\pi - \theta')$ on the one side or the other of OT according to the direction of the rotation, and if the distance r between OB, CD be such that $2r \sin \frac{1}{2}\theta' = OT$.

The whole displacement has thus been reduced to a rotation θ about OA followed by a rotation θ' about CD.

278. **Composition of Screws.** *Any two successive displacements of a body may be represented by two successive screw motions. It is required to compound these.*

Let the body be screwed first along the axis OA with linear displacement a and angle of rotation θ, and secondly along the axis CD with displacement a' and angle

θ'. Let OC be the shortest distance between OA and CD, and for the sake of the perspective let it be called the axis of y. Let O be the origin and let the axis of x

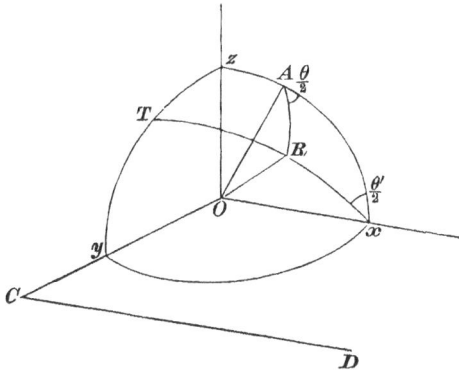

be parallel to CD, so that OA lies in the plane of xz. Let $OC=r$, and the angle $AOx=a$. Draw a plane xOT making with the plane of xz an angle $\frac{1}{2}\theta'$, and let it cut yz in OT. Draw another plane AOR making with xz an angle $\frac{1}{2}\theta$, and cutting the plane xOT in OR.

Produce AO to a point P, not marked in the figure, so that $PO=a$, and let us choose P as a base point to which the whole displacement of the body may be referred. The rotation θ' is equivalent to a rotation θ' about Ox together with a translation along $OT=2r\sin\frac{1}{2}\theta'$ by Art. 223. By Art. 271 the rotation θ about OA followed by θ' about Ox is equivalent to a rotation Ω about OR where Ω is twice the angle ART, so that $\sin\dfrac{\Omega}{2}=\sin\dfrac{\theta}{2}\cdot\dfrac{\sin Ax}{\sin Rx}$. The whole displacement is now represented by (1) a translation of the base point P to O, (2) the rotation Ω, (3) a further linear translation which is the resultant of the translations $2r\sin\frac{1}{2}\theta'$ along OT and a' along Ox. By Art. 219 these displacements may be made in any order, being all connected with the same base point. They may therefore be compounded into a single screw by the rule given in Art. 225. This is called the *resultant screw*. A screw equal and opposite to the resultant screw will bring the body back to its original position.

The angle of rotation of the resultant screw is Ω and its axis is parallel to OR by Art. 220. It follows by Art. 272 that the sine of half the angle of rotation of each screw is proportional to the sine of the angle between the axes of the other two screws.

To find the linear displacement along the axis of the resultant screw, we must by Art. 222 add together the projections on OR of the three displacements OT, a, a'. The projection of $OT=2r\sin\frac{1}{2}\theta'\cos TR=2r\cos Ty$. $\cos TR$, which is twice the projection of the shortest distance r on the axis of rotation. If T be the linear displacement, we have $T=2r\cos Ry+a\cos RA+a'\cos Rx$.

279. If the component screws be simple rotations, we have $a=0$, $a'=0$, and it may be shown without difficulty that $T\sin\dfrac{\Omega}{2}=2r\sin\dfrac{\theta}{2}\sin\dfrac{\theta'}{2}\sin a$. It has been shown in Art. 277 that any displacement may be represented by two conjugate rotations in an infinite number of ways, but it now follows that in all these

$r \sin \dfrac{\theta}{2} \sin \dfrac{\theta'}{2} \sin \alpha$ is the same. When the rotations are indefinitely small, and equal to ωdt, $\omega' dt$ respectively, this becomes $\frac{1}{4} r \omega \omega' (dt)^2 \sin \alpha$; that is, the product of an angular velocity into the moment of its conjugate angular velocity about its axis is the same for all conjugates representing the same motion.

Ex. 1. If the component screws be simple finite rotations, show that the equations to the axis of the resultant screw are

$$- x \tan \phi' + y \sin \frac{\theta'}{2} + z \cos \frac{\theta'}{2} = r \sin \frac{\theta'}{2}, \quad y \cos \frac{\theta'}{2} - z \sin \frac{\theta'}{2} = r \sin \frac{\theta'}{2} \cos \phi' \cot \frac{\Omega}{2},$$

where ϕ' is the angle xOR and Ω is the resultant rotation. The first equation expresses the fact that the central axis lies in a plane which bisects at right angles a straight line drawn from O perpendicular to OR in the plane xOR to represent the linear translation in that direction. The second expresses the fact that the central axis lies in a plane parallel to TOR at a distance from it determined by Art. 225.

These equations may also be deduced from those of Rodrigues given in Art. 281. To effect this we must write for (a, b, c) the resolved parts of the translation along OT. Since however the positive direction of the rotation in Rodrigues' formulæ has been taken opposite to that chosen in the preceding article, we must write for (l, m, n) the direction cosines of OR with their signs changed.

The equations to the central axis of any two screws may be found by either of these methods.

Ex. 2. Let the motion be constructed by two finite rotations θ, θ' taken in order round axes OA, CD at right angles to each other, and let CO be the shortest distance between the axes. Let the two straight lines OP, CP be drawn in the plane DCO such that the angle $POC = \dfrac{\theta}{2}$ and $\tan PCO = \sin^2 \dfrac{\theta'}{2} \cot \dfrac{\theta}{2}$. Then if P be moved backwards by the rotation θ or forwards by the rotation θ', in either case its new position is a point on the central axis.

Ex. 3. If OA, OB be the axes of two screws at right angles, with linear displacements a and b, the point P is the intersection of two parallels to the straight lines described in the last example; these parallels being drawn respectively at distances $\dfrac{a}{2} \tan \phi$ and $\dfrac{b}{2} \left(1 + \cot^2 \phi' \sin^2 \dfrac{\theta''}{2}\right)^{-\frac{1}{2}}$, where ϕ, ϕ' are the angles the resultant axis of rotations makes with OA and CD. Then if P be screwed backwards by the first screw or forwards by the second, in either case its new position is a point on the central axis.

280. **The Velocity of any Point.** The formulæ corresponding to those given in Art. 238 for infinitely small motions are rather more complicated.

A displacement of a body is given by a rotation through a finite angle θ about an axis passing through the origin whose direction cosines are (l, m, n). *It is required to find the changes produced in the co-ordinates* (x, y, z) *of any point P.*

Let PP' be the chord of the arc described by P and let Q be the middle point of PP'. Let $x + \delta x$, $y + \delta y$, $z + \delta z$ be the co-ordinates of P' and ξ, η, ζ those of Q. Since the abscissa of Q is the arithmetic mean of those of P and P', we have $\xi = x + \frac{1}{2} \delta x$; similarly $\eta = y + \frac{1}{2} \delta y$, $\zeta = z + \frac{1}{2} \delta z$. Let QM be a perpendicular from Q on the axis, then $PP' = 2QM \tan \frac{1}{2} \theta$.

Let $(\lambda,\ \mu,\ \nu)$ be the direction cosines of PP', then since PP' is perpendicular to the axis, we have $l\lambda + m\mu + n\nu = 0$, and since it is also perpendicular to OQ we have $\xi\lambda + \eta\mu + \zeta\nu = 0$, hence

$$\frac{\lambda}{m\zeta - n\eta} = \frac{\mu}{n\xi - l\zeta} = \frac{\nu}{l\eta - m\xi}.$$

The sum of the squares of the denominators is

$$(\xi^2 + \eta^2 + \zeta^2)(l^2 + m^2 + n^2) - (l\xi + m\eta + n\zeta)^2,$$

which is $OQ^2 - OM^2 = QM^2$. Hence each of these ratios is $= \dfrac{1}{QM}$.

Now δx is the projection of PP' on the axis of x,

$$\therefore\ \delta x = 2QM \cdot \tan\frac{\theta}{2}\lambda = 2\tan\frac{\theta}{2}(m\zeta - n\eta)\,;$$

similarly $\delta y = 2\tan\dfrac{\theta}{2}(n\xi - l\zeta)$, $\delta z = 2\tan\dfrac{\theta}{2}(l\eta - m\xi)$, which are the required formulæ.

If the origin have a linear displacement whose resolved parts parallel to the axes are $(a,\ b,\ c)$, we must add those displacements to the values of δx, δy, δz found by solving these equations. Let the co-ordinates of the middle point of the *whole* displacement of P be represented by ξ', η', ζ'. Then we have, as before, $\xi' = x + \dfrac{\delta x}{2}$ &c.,

but since δx, δy, δz, are increased by a, b, c we must write $\xi' - \dfrac{a}{2}$, $\eta' - \dfrac{b}{2}$, $\zeta' - \dfrac{c}{2}$ for ξ, η, ζ. We thus obtain

$$\delta x = a + 2\tan\frac{\theta}{2}\left\{ m\left(\zeta' - \frac{c}{2}\right) - n\left(\eta' - \frac{b}{2}\right) \right\},$$

with similar expressions for δy and δz.

281. The equations to the central axis follow from these expressions without difficulty. The whole displacement of any point in the central axis is along the axis, so that $(\xi',\ \eta',\ \zeta')$ the co-ordinates of the middle point of the displacement are co-ordinates of a point in the axis, and δx, δy, δz are proportional to $(l,\ m,\ n)$ the direction-cosines of the axis. Hence

$$\frac{a + 2\tan\dfrac{\theta}{2}\left\{ m\left(\zeta' - \dfrac{c}{2}\right) - n\left(\eta' - \dfrac{b}{2}\right) \right\}}{l} = \frac{b + 2\tan\dfrac{\theta}{2}\left\{ n\left(\xi' - \dfrac{a}{2}\right) - l\left(\zeta' - \dfrac{c}{2}\right) \right\}}{m}$$

$$= \frac{c + 2\tan\dfrac{\theta}{2}\left\{ l\left(\eta' - \dfrac{b}{2}\right) - m\left(\xi' - \dfrac{a}{2}\right) \right\}}{n}.$$

Each of these is evidently equal to $la + mb + nc$, which is the linear displacement along the central axis. The results of this and the preceding Article are due to Rodrigues.

CHAPTER VI

282. THE term Momentum has been given as the heading of this Chapter, though it only expresses a portion of its contents. The object of the Chapter may be enunciated in the following problem. The circumstances of the motion of a system at any time t_0 are given. At the time t_1 the system is moving under other circumstances. It is required to determine the relations which may exist between these two motions. The manner in which these changes are effected by the forces is not the subject of enquiry. We only wish to determine what changes have been effected in the time $t_1 - t_0$. If the time $t_1 - t_0$ be very small, and the forces very great, this becomes the general problem of impulses. This also will be considered in the Chapter.

Let us refer the system to any fixed axes Ox, Oy, Oz. Then the six general equations of motion may, by Art. 72, be written in the form

$$\Sigma m \, \frac{d^2 z}{dt^2} = \Sigma m Z$$
$$\Sigma m \left(x \frac{d^2 y}{dt^2} - y \frac{d^2 x}{dt^2} \right) = \Sigma m \, (xY - yX) \Bigg\}$$

Integrating these from $t = t_0$ to $t = t_1$, we have

$$\left[\Sigma m \, \frac{dz}{dt} \right]_{t_0}^{t_1} = \Sigma m \int_{t_0}^{t_1} Z dt,$$
$$\left[\Sigma m \left(x \frac{dy}{dt} - y \frac{dx}{dt} \right) \right]_{t_0}^{t_1} = \Sigma m \int_{t_0}^{t_1} (xY - yX) \, dt.$$

Let an accelerating force F act on a moving particle m during any time $t_1 - t_0$, and let this time be divided into intervals each equal to dt. At the middle of each of these intervals let a line be drawn from the position of m at that instant, to represent, at the same instant, the value of $mFdt$ both in direction and magnitude. Then the resultant of these forces, found by the rules of statics, may be called the *whole force* expended in the time $t_1 - t_0$. Thus $\int_{t_0}^{t_1} mZdt$ is the whole force resolved parallel to the axis of Z. These equations then show that

(1) *The change produced by any forces in the resolved part of the momentum of any system is equal in any time to the whole resolved force in that direction.*

(2) *The change produced by any forces in the moment of the momentum of the system about any straight line is, in any time, equal to the whole moment of these forces about that straight line.*

When the interval $t_1 - t_0$ is very small, the "whole force" expended is the usual measure of an impulsive force, and the preceding equations are identical with those given in Art. 86.

It is not necessary to deduce these two results from the equations of motion. The following general theorem, which is really equivalent to the two theorems enunciated above, may be easily obtained by an application of D'Alembert's principle.

283. **Fundamental Theorem.** *If the momentum of any particle of a system in motion be compounded and resolved, as if it were a force acting at the instantaneous position of the particle, according to the rules of statics, then the momenta of all the particles at any time t_1 are together equivalent to the momenta at any previous time t_0 together with the whole forces which have acted during the interval.*

The argument from D'Alembert's principle may be made clearer by being put at greater length. If we multiply the mass m of any particle P by its velocity v, the product is the momentum mv of the particle. Let us represent this in direction and magnitude by a straight line PP' drawn from the particle in the direction of its motion. For the purposes of composition and resolution this representative straight line (in accordance with the rules of statics) may be moved to any position in the line of motion. It may therefore move with the particle. If the particle be acted on at any instant by an external force mF, a new momentum equal to $mF dt$ is generated in the time dt. This also can be represented by a straight line and compounded with the mv of the particle. If two particles act and re-act on each other with a force R for a time dt, two equal and opposite momenta (viz. $R dt$) are communicated to the particles. Taking all the particles, we see that the change in their momenta is equal to the resultant of every $mF dt$ which has acted on the system. This being true for each element of time is true for any finite interval $t_1 - t_0$. Since the resultant of every $mF dt$ has been defined to be the whole force, the theorem follows at once.

In the case in which no forces act on the system, except the mutual actions of the particles, we see that the momenta of all the particles of a system at any two times are equivalent.

The two principles of the Conservation of Linear Momentum and Conservation of Areas may be enunciated as follows.

If the forces which act on a system be such that they have no component along a certain fixed straight line, then the motion is such that the linear momentum resolved along this line is constant.

If the forces be such that they have no moment about a certain fixed straight line, then the moment of the momentum or the area conserved about this straight line is constant.

It is evident that these principles are only particular cases of the results proved in Art. 79.

284. Example of a central force. Suppose that a single particle m describes an orbit about a centre of force O. Let v, v' be its velocities at any two points P, P' of its course. Then mv' supposed to act along the tangent at P' if reversed would be in equilibrium with mv acting along the tangent at P together with the whole central force from P to P'. If p, p' be the lengths of the perpendiculars from O on the tangents at P, P', we have, by taking moments about O, $vp = v'p'$, and hence vp is constant throughout the motion. Also if the tangents meet in T, the whole central force expended must act along the line TO, and may be found in terms of v, v' by the rules for compounding velocities.

Ex. Two particles of masses m, m' move about the same centre of force. If h, h' be the double areas described by each per unit of time, prove that $mh + m'h'$ is unaltered by an impact between the particles.

285. Example of three particles. Suppose three particles to start from rest attracting each other, but under the action of no external forces. Then the momenta of the three particles at any instant are together equivalent to the three initial momenta and are therefore in equilibrium. Hence at any instant the tangents to their paths must meet in some point O, and if parallels to their directions of motion be drawn so as to form a triangle, the momenta of the several particles are proportional to the sides of that triangle.

If there are n particles it may be shown in the same way that the n forces represented by mv, $m'v'$, &c. are in equilibrium, and if parallels be drawn to the directions of motion and proportional to the momenta of the particles beginning at any point, they will form a closed polygon.

If F, F', F'' be the resultant attraction on the three particles, the lines of action of F, F', F'' also meet in a point. For let X, Y, Z be the actions between the particles $m'm''$, $m''m$, mm', taken in order. Then F is the resultant of $-Y$ and Z; F' of $-Z$ and X; F'' of $-X$ and Y. Hence the three forces F, F', F'' are in equilibrium*, and therefore their lines of action must meet in a point O'. Also the magnitude of each is proportional to the sine of the angle between the directions of the other two. This point is not generally fixed, and does not coincide with O.

---

* This proof is merely an amplification of the following. The three forces F, F', F'', being the internal reactions of a system of three bodies, are in equilibrium by D'Alembert's Principle.

If the attraction be directly proportional to the distance, the two points O, O' coincide with the centre of gravity G, and are fixed in space throughout the motion. For it is a known proposition in statics that, with this law of attraction, the whole attraction of a system of particles on one of the particles is the same as if the whole system were collected at its centre of gravity. Hence O' coincides with G. Also, since each particle starts from rest, the initial velocity of the centre of gravity is zero, and therefore, by Art. 79, G is a fixed point. Again, since each particle starts from rest and is urged towards a fixed point G, it will move in the straight line joining its initial position with G. Hence O coincides with G. When the attraction is directly proportional to the distance, it is proved in dynamics of a particle, that the time of reaching the centre of force from a position of rest is independent of the distance of that position of rest. Hence all the particles of the system will reach G at the same time, and meet there. If Σm be the sum of the masses, measured by their attractions in the usual manner, this time is known to be $\dfrac{1}{4}\dfrac{2\pi}{\sqrt{\Sigma m}}$.

286. Example of Laplace's Three Particles. *Three particles whose masses are* m, m', m'', *mutually attracting each other, are so projected that the triangle formed by joining their positions at any instant remains always similar to its original form. It is required to determine the conditions of projection.*

The centre of gravity will be either at rest or will move uniformly in a straight line. We may therefore consider the centre of gravity at rest and may afterwards generalise the conditions of projection by impressing on each particle an additional velocity parallel to the direction in which we wish the centre of gravity to move. Let O be the centre of gravity, P, P', P'' the positions of the particles at any time t. Then, by the conditions of the question, the lengths OP, OP', OP'' are always to be proportional, and their angular velocities about O are to be equal. Since the moment of the momenta of the system about O is always the same, we have

$$mr^2n + m'r'^2n + m''r''^2n = \text{constant},$$

where r, r', r'' are the distances OP, OP', OP'', and n is their common angular velocity. Since the ratios $r : r' : r''$ are constant, it follows from this equation that mr^2n is constant, i.e. OP traces out equal areas in equal times. Hence by Newton, Section II, the resultant force on P tends towards O.

Let ρ, ρ', ρ'' be the sides $P'P''$, $P''P$, PP' of the triangle formed by the particles, and let the law of attraction be $\dfrac{\text{mass}}{(\text{dist.})^k}$. Then, since the resultant attraction of m', m'' on m passes through O,

$$\frac{m'}{\rho''^k}\sin P'PO = \frac{m''}{\rho'^k}\sin P''PO,$$

but, since O is the centre of gravity,

$$m'\rho''\sin P'PO = m''\rho'\sin P''PO.$$

Hence either the three particles are in one straight line or $\rho''^{k+1} = \rho'^{k+1}$. If $k = -1$ the law of attraction is "as the distance." If k be not $= -1$, we have $\rho' = \rho''$, and the triangle must be equilateral.

Suppose the particles to be projected in directions making equal angles with their distances from the centre of gravity with velocities proportional to those distances, and suppose also the resultant attractions towards the centre of gravity to be proportional to those distances, then in all the three cases the same conditions will hold at the end of a time dt, and so on continually. The three particles will therefore describe similar orbits about the centre of gravity in a similar manner.

Firstly, let us suppose that the three particles are to be in one straight line. To fix our ideas, let m' lie between m and m'', and O between m and m'. Then since the attraction on any particle must be proportional to the distance of that particle from O, the three attractions

$$\frac{m'}{(PP')^k} + \frac{m''}{(PP'')^k}, \qquad \frac{m''}{(P''P')^k} - \frac{m}{(PP')^k}, \qquad -\frac{m}{(PP'')^k} - \frac{m'}{(P'P'')},$$

must be proportional to OP, OP', OP''. Since $\Sigma mOP = 0$, these two equations amount to but one on the whole. Let $z = \dfrac{P'P''}{PP'}$, so that $\dfrac{OP}{PP'} = \dfrac{m' + m''(1+z)}{m + m' + m''}$, $\dfrac{OP'}{PP'} = \dfrac{-m + m''z}{m + m' + m''}$.

Then we have

$$\left(m' + \frac{m''}{(1+z)^k} \right)(-m + m''z) = \left(\frac{m''}{z^k} - m \right) \{ m' + m''(1+z) \},$$

which agrees with the result given by Laplace, by whom this problem was first considered.

In the case in which the attraction follows the law of nature $k = 2$ and the equation becomes

$$mz^2 \{ (1+z)^3 - 1 \} - m'(1+z)^2(1 - z^3) - m'' \{ (1+z)^3 - z^3 \} = 0.$$

This is an equation of the fifth degree, and it has therefore always one real root. The left side of the equation has opposite signs when $z = 0$ and $z = \infty$, and hence this real root is positive. It is therefore always possible to project the three masses so that they shall remain in a straight line. Laplace remarks that if m be the sun, m' the earth, m'' the moon, we have very nearly $z = \sqrt[3]{\dfrac{m' + m''}{3m}} = \dfrac{1}{100}$. If then, originally, the earth and moon had been placed in the same straight line with the sun, at distances from the sun proportional to 1 and $1 + \dfrac{1}{100}$, and if their velocities had been initially parallel and proportional to those distances, the moon would have always been in opposition to the sun. The moon would have been too distant to have been in a state of continual eclipse, and thus would have been full every night. It has however been shown by Liouville, in the *Additions à la Connaissance des Temps*, 1845, that such a motion would be unstable.

The paths of the particles will be similar ellipses having the centre of gravity for a common focus.

Secondly. Let us suppose that the law of attraction is "as the distance." In this case the attraction on each particle is the same as if all the three particles were collected at the centre of gravity. Each particle will describe an ellipse having this point for centre in the same time. The necessary conditions of projection are that the velocities of projection should be proportional to the initial distances from the centre of gravity, and that the directions of projection should make equal angles with those distances.

Thirdly. Let us suppose the particles to be at the angular points of an equilateral triangle. The resultant force on the particle m is

$$\frac{m'}{\rho''^k} \cos P'PO + \frac{m''}{\rho'^k} \cos P''PO.$$

The condition that the forces on the particles should be proportional to their distances from O shows that the ratio of this force to the distance OP is the same for all the particles. Since

$$m'\rho'' \cos P'PO + m''\rho' \cos P''PO = (m + m' + m'') \; OP,$$

it is clear that the condition is initially satisfied when $\rho = \rho' = \rho''$. Hence, by the same reasoning as before, if the particles be projected with equal velocities in directions making equal angles with OP, OP', OP'' respectively, they will always remain at the angular points of an equilateral triangle.

A discussion of the stability of this motion will be given in a later part of this work.

Ex. 1. Show that if the three particles attract each other according to the law of nature, the paths of the particles, when at the corners of an equilateral triangle, are equal ellipses having O for a common focus. Find the periodic time.

Ex. 2. If four particles be placed at the corners of a quadrilateral whose sides taken in order are a, b, c, d and diagonals ρ, ρ', then the particles cannot move under their mutual attractions so as to remain always at the corners of a similar quadrilateral unless

$$(\rho^n \rho'^n - b^n d^n)(c^n + a^n) + (a^n c^n - \rho^n \rho'^n)(b^n + d^n) + (b^n d^n - a^n c^n)(\rho^n + \rho'^n) = 0,$$

where the law of attraction is the inverse $(n-1)^{\text{th}}$ power of the distance.

Show also that the mass at the intersection of b, c divided by the mass at the intersection of c, d is equal to the product of the area formed by a, ρ', d divided by the area formed by a, b, ρ and the difference $\dfrac{1}{\rho'^n} - \dfrac{1}{d^n}$ divided by $\dfrac{1}{\rho^n} - \dfrac{1}{b^n}$.

These results may be conveniently arrived at by reducing one angular point, as A, of the quadrilateral to rest. The resolved part of all the forces which act on each particle perpendicular to the straight line joining it to A will then be zero. The case of three particles may be treated in the same manner. The process is a little shorter than that given in the text, but does not illustrate so well the subject of the chapter.

287. When the system under consideration consists of rigid bodies we must use the results of Art. 74 to find the resolved part of the momentum in any direction. The moment of the momentum about any straight line may also be found by Art. 75 in Chap. II., combined with Art. 134 in Chap. IV., if the motion be in two dimensions, or with Art. 262 in Chap. V., if the motion be in three dimensions.

288. **Sudden Fixtures.** A rigid body is moving freely in space in a known manner. Suddenly a straight line in the body becomes fixed, or has its motion changed in some given manner. It is required to find the changes which occur in the motion of the rest of the body.

Such problems as these are all solved by one mechanical principle. The change in the motion is produced by impulsive forces acting at points situated in this straight line. Hence, by Art. 283, the *angular momentum of the body about the axis is the same after as before the change takes place.* This dynamical principle will supply one equation which is sufficient to determine the subsequent motion of the body round the straight line.

We may also use this principle in a more general case. Suppose we have any system of moving bodies which suddenly become rigidly connected together and are constrained to turn round some axis. Then the subsequent angular velocity about this axis may be found by equating the angular momentum of the system about this axis after the change to that before the change.

In applying this principle to various bodies it is convenient to use different methods of finding the angular momentum. The following list will assist the reader in choosing the method best adapted to each particular case.

289. Case 1. Suppose the body to be a disc moving in any manner in its own plane, and let the axis whose motion is changed be perpendicular to its plane. This case has been already solved in Art. 171.

290. Case 2. Suppose the body to be a disc turning about an instantaneous axis Ox in its own plane with an angular axis ω. Let an axis Ox' also in its own plane be suddenly fixed.

In this case the calculation of the angular momentum is so simple that we may with advantage recur to first principles.

Let $d\sigma$ be any element of the area of the disc; y, y' its distances from Ox, Ox'. Then $y\omega$, $y'\omega'$ are the velocities of $d\sigma$ just

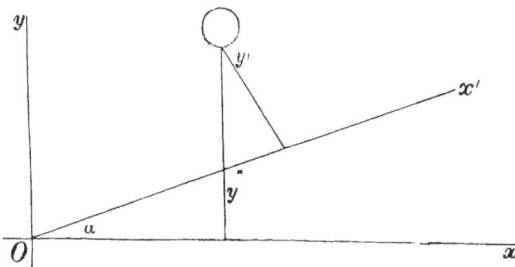

before and just after the impact. The moments of the momentum about Ox' just before and just after are therefore $yy'\omega d\sigma$ and $y'^2\omega'd\sigma$. Summing these for the whole area of the disc, we have

$$\omega'\Sigma y'^2 d\sigma = \omega\Sigma yy'd\sigma \quad\ldots\ldots\ldots\ldots\ldots\ldots(1).$$

Firstly, let Ox, Ox' be parallel, so that the point O is at infinity. Let h be the distance between the axes, then $y' = y - h$. Hence we have

$$\omega' \Sigma y'^2 d\sigma = \omega \left\{ \Sigma y^2 d\sigma - h \Sigma y d\sigma \right\}.$$

Let A, A' be the moments of inertia of the disc about Ox, Ox' respectively, $\bar{y}$ the distance of the centre of gravity from Ox, M the mass of the disc. Then we have

$$A' \omega' = \omega \left(A - Mh\bar{y} \right).$$

Secondly, let Ox, Ox' not be parallel. Let O be the origin and let the angle $xOx' = \alpha$, then $y' = y \cos \alpha - x \sin \alpha$. Let F be the product of inertia of the disc about Ox, Oy where Oy is perpendicular to Ox. Then by substitution in (1) we have

$$A' \omega' = \omega \left(A \cos \alpha - F \sin \alpha \right).$$

Ex. 1. An elliptic area of eccentricity e is turning about one latus rectum. Suddenly this latus rectum is loosed and the other fixed. Show that the angular velocity is $\dfrac{1 - 4e^2}{1 + 4e^2}$ of its former value.

Ex. 2. A right-angled triangular area ACB is turning about the side AC. Suddenly AC is loosed and BC fixed. If C be the right angle, the angular velocity is $\dfrac{BC}{2 \cdot AC}$ of its former value.

Ex. 3. A rectangle $ABCD$ has its plane vertical and its lower edge AB horizontal and fixed in space. A slight disturbance being given the rectangle turns round AB, but when its plane becomes horizontal the side AD is fixed and AB released. It then begins to turn round AD and when the plane is again vertical AB is fixed and AD released. Show that the final angular velocity about AB is given by

$$\omega^2 = 27g \left(16a + 9b \right) / 512b^2,$$

where $AB = 2a$ and $AD = 2b$.

291. Case 3. Let the body be turning round an instantaneous axis OI with a known angular velocity ω, and let some axis OI' which intersects the former in a point O be suddenly fixed.

Let l, m, n be the direction-cosines of OI referred to the principal axes at O, and l', m', n' the direction-cosines of OI'. Then by Art. 264, the angular momenta about these principal axes just before the change are $A\omega l$, $B\omega m$, $C\omega n$. The angular momentum about OI' just before the change is therefore (by Art. 265) $(All' + Bmm' + Cnn') \omega$. If ω' be the angular velocity of the body about OI' just after OI' becomes fixed in space the angular momentum is $(Al'^2 + Bm'^2 + Cn'^2) \omega'$. Equating these we have ω'.

Ex. 1. A solid right cone of semi-vertical angle a is rotating about a generating line. Suddenly another generating line is fixed, the axial planes through the generating lines being inclined at an angle ϕ. Show that the ratio of the angular velocities is equal to $(2 + (4 + n) \cos \phi) : (6 + n)$, where $n = \tan^2 a$.

Ex. 2. When a body turns about a fixed point the product of the moment of inertia about the instantaneous axis into the square of the angular velocity is called the vis viva. Let $2T$ be the vis viva of the body when it is turning freely about the axis OI, and $2T'$ its vis viva when the axis OI' is suddenly fixed. Construct the momental ellipsoid at the point O, and let θ be the angle between the eccentric lines of the two axes OI, OI'. Prove that $T' = T\cos^2\theta$. It follows that the vis viva is always lessened by fixing a new axis.

292. **Case 4.** Let the motion of the body be given by its components of motion u, v, w, ω_x, ω_y, ω_z, the centre of gravity being the base point. Let the equation to the straight line whose motion is suddenly changed be $\dfrac{x-f}{l} = \dfrac{y-g}{m} = \dfrac{z-h}{n}$, where l, m, n are the actual direction-cosines.

Suppose this straight line to be suddenly fixed in space. The angular momentum before the "fixing" is given in Art. 266. If ω' be the angular velocity about this straight line after the "fixing," the angular momentum is $I\omega'$, where I is given in Art. 18, Ex. 9. Equating these we have ω'.

293. Suppose the sudden motion forced on the straight line to be represented by the velocities U, V, W of some point P on the straight line, and the angular velocities θ, ϕ, ψ. Then the motion of the body may be represented by the linear velocities U, V, W of the same base P and the angular velocities $\theta + \Omega l$, $\phi + \Omega m$, $\psi + \Omega n$, where Ω is the only unknown quantity.

The angular velocities θ, ϕ, ψ may be chosen in an infinite variety of ways to represent the given motion of the straight line, because an angular velocity about the straight line does not move the line itself. If θ, ϕ, ψ have been chosen to make the component $l\theta + m\phi + n\psi$ about the line equal to zero, and if (l, m, n) be the actual direction-cosines of the straight line, then Ω will be the angular velocity of the body about the axis just after the change.

This quantity Ω, whatever meaning it may have, is to be found by equating the angular momenta about the axis before and after the change. These momenta may be written down as explained in Art. 266.

294. Suppose the sudden motion forced on the straight line to be represented by giving the velocities of two points P, P' on the line. And let the required motion of the body after the change be represented by the components of motion u', v', w', ω_x', ω_y', ω_z' at the centre of gravity taken as the base. The angular momentum both before and after the change may be written down by Art. 266. Equating these we have the dynamical equation. The resolved velocities of P and P' may be found by Art. 238 and equated to their given forced values. Thus we have on the whole six *independent* equations to find the six components of motion after the change.

Ex. 1. An elliptic disc is at rest. Suddenly one extremity of the major axis and one extremity of the minor are made to move perpendicularly to the plane of the disc with velocities U and V. Show that the centre of gravity will begin to move with a velocity equal to $\frac{1}{6}(U+V)$.

Ex. 2. An elliptic disc is at rest. Suddenly one extremity of the latus rectum is made to move parallel to the major axis with a velocity U, while the other extremity is made to move perpendicularly to the plane of the disc with a velocity W. Show that the velocities of the centre resolved parallel to the axes of the disc are

$$\frac{U}{2}, \quad \frac{Ue}{2(1-e^2)}, \quad \frac{W}{2(1+4e^2)}.$$

Ex. 3. A circular disc turning freely in its own plane which is vertical falls on another equal circular disc whose plane is horizontal and which is turning about a fixed vertical axis through its centre. At the moment of impact the two discs become rigidly connected. If the point of impact bisect a radius of the horizontal circle, show that the angular velocity about the fixed vertical axis is reduced one half.

Ex. 4. Let the motion of a free body be given by the components u, v, w, ω_x, ω_y, ω_z referred to any base. Let the sudden motion given to a straight line be represented by the components U, V, W, θ, ϕ, ψ referred to the same base. Then the relative motion is given by the components $u - U$, $v - V$, &c. Taking these as the given quantities, find the components of motion after the change on the supposition that the straight line is suddenly *fixed*. Let these results be u', v', &c. Then prove that the required motion is represented by the components $U + u'$, $V + v'$, &c.

This process of solution may be called *reducing the straight line to rest*.

295. Case 5. In some cases, instead of a straight line, a single point P in the body is seized and made to move in some given manner. In this case the angular momentum about every straight line through the fixed point is unchanged. Choosing some three convenient axes through the point and equating the angular momentum about each before the change to that after the change we have three dynamical equations. Besides these we have the geometrical equations, supplied by Art. 238, expressing the fact that the resolved velocities of P are equal to the given forced velocities. In this way we may form six equations to find the six components of motion.

296. Let us consider an example of this process. Suppose the motion of the body to be given by the components u, v, w, ω_x, ω_y, ω_z, the centre of gravity being the base; and let the point P whose co-ordinates are f, g, h be suddenly *fixed*. Let A, B, C, D, E, F be the moments and products of inertia of the body about the axes at the centre of gravity, and let accented letters represent the corresponding quantities for parallel axes at P. Let Ω_x, Ω_y, Ω_z be the required angular velocities of the body about the axes meeting at P parallel to those at the centre of gravity. Then the equations of momenta give

$$A\omega_x - F\omega_y - E\omega_z + M(vh - wg) = \quad A'\Omega_x - F'\Omega_y - E'\Omega_z,$$
$$-F\omega_x + B\omega_y - D\omega_z + M(wf - uh) = -F'\Omega_x + B'\Omega_y - D'\Omega_z,$$
$$-E\omega_x - D\omega_y + C\omega_z + M(ug - vf) = -E'\Omega_x - D'\Omega_y + C'\Omega_z.$$

It is obvious that these equations may be greatly simplified by choosing the axes so that one set may be principal axes.

297. If the body be turning about an axis GI through the centre of gravity G just before the point P is fixed, the terms which contain the velocities of the centre of gravity disappear from the equations. They now admit of an easy geometrical interpretation. The equation to the momental ellipsoid at the centre of gravity is

$$AX^2 + BY^2 + CZ^2 - 2DYZ - 2EZX - 2FXY = M\epsilon^4.$$

It is therefore clear that the left-hand sides of these equations are proportional to the direction-cosines of the diametral plane of a straight line whose direction-cosines are proportional to $(\omega_x, \omega_y, \omega_z)$. In the same way if we construct the momental ellipsoid at P, the right-hand sides are proportional to the direction-cosines of the diametral plane of the axis $(\Omega_x, \Omega_y, \Omega_z)$. *Thus the instantaneous axes of rotation, before and after* P *is fixed, are so related that their diametral planes with regard to the momental ellipsoids at* G *and* P *respectively are parallel.*

We may also deduce this result, without difficulty, from Art. 118. The motion of the body about the axis GI may be produced by an impulsive couple in the plane diametral to GI with regard to the momental ellipsoid at G. Let us then suppose the body at rest and P fixed, and let it be acted on by this couple. It follows from the same article, that the body will begin to turn about an axis PI' which is such that its diametral plane with regard to the momental ellipsoid at P is parallel to the plane of the couple.

The direction of the blow at P may also be easily found. The centre of gravity being at rest suddenly begins to move perpendicularly to the plane containing it and the axis PI'. This is obviously the direction of the blow.

298. Ex. 1. *A sphere, in co-latitude* θ, *hung up by a point* O *in its surface, is in equilibrium under the action of gravity. Suddenly the rotation of the earth is stopped, it is required to determine the motion of the sphere.*   [Math. Tripos, 1857.]

Let G be the centre of its sphere, O its point of suspension, and a its radius. Let C be the centre of the earth. Let us suppose the figure so drawn that the sphere is moving away from the observer.

Let ω = angular velocity of the earth, then if $CG = \mu a$, the sphere is turning about an axis Gp parallel to CP, the axis of the earth, with angular velocity ω, while the centre of gravity is moving with velocity $\mu a \sin \theta . \omega$.

Let OC, Op, and the perpendicular to the plane of OC, Op be taken as the axes of x, y, z respectively, and let Ω_x, Ω_y, Ω_z be the angular velocities about them just after the rotation of the earth is stopped.

By Art. 295, the angular momenta about Ox, just before and just after the rotation is stopped, are equal to each other;

$$\therefore\ Mk^2\omega \cos \theta = Mk^2\Omega_x,$$

where Mk^2 is the moment of inertia of the sphere about a diameter.

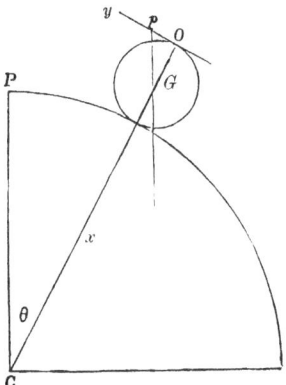

Again, the angular momenta about Oy are equal to each other;

$$\therefore\ -Mk^2\omega \sin \theta + M\mu a^2\omega \sin \theta = M\left(k^2 + a^2\right)\Omega_y.$$

Lastly, the angular momenta about Oz are equal; $\therefore\ 0 = Mk^2\Omega_z$.

Solving these equations, we get $\Omega_y = \omega \sin \theta \,\dfrac{-k^2 + \mu a^2}{k^2 + a^2} = \omega \sin \theta \,\dfrac{-2 + 5\mu}{7}$

But $\Omega_x = \omega \cos \theta$. Adding together the squares of Ω_x, Ω_y, Ω_z we have

$$\Omega^2 = \omega^2 \left\{ \cos^2 \theta + \left(\frac{-2 + 5\mu}{7} \right)^2 \sin^2 \theta \right\},$$

where Ω is the angular velocity of the sphere about its instantaneous axis.

Ex. 2. A particle of mass M, without velocity, is suddenly attached to the surface of the earth at the extremity of a radius vector making an angle θ with the axis of the earth. If E be the mass of the earth before the addition of M, A and C its principal moments of inertia at the centre, ω the angular velocity about its axis, prove that

$$\frac{\omega}{\Omega} = 1 + \frac{EMAr^2 \sin^2 \theta}{(E + M)\,AC + EMCr^2 \cos^2 \theta},$$

$$\cot \phi = \cot \theta + \frac{E + M}{E} \cdot \frac{A}{Mr^2 \sin \theta \cos \theta},$$

where Ω is the initial angular velocity about an axis parallel to the axis of the earth, and ϕ the angle that the initial axis of rotation makes with the axis of the earth.

Ex. 3. A regular homogeneous prism whose normal section is a regular polygon of n sides rolls on a perfectly rough plane. Prove that, when the axis of rotation changes from one edge to another, the angular velocity is reduced in the ratio of

$$2 + 7 \cos \frac{2\pi}{n} \ :\ 8 + \cos \frac{2\pi}{n}$$

299. **Gradual Changes.** In these examples the changes produced in the motion were sudden, but the method of proceeding is the same if the changes are gradual.

Ex. 1. A bead of mass m slides on a circular wire of mass M and radius a, and the wire can turn freely about a vertical diameter. Prove that, if ω, Ω be the angular velocities of the wire when the bead is respectively at the extremities of a horizontal and a vertical diameter, $\dfrac{\Omega}{\omega} = 1 + 2\dfrac{m}{M}$.

Ex. 2. If the earth gradually contracted by radiation of heat, so as to be always similar to itself as regards its physical constitution and form, prove that when every radius vector has contracted an nth part of its length, where n is small, the angular velocity has increased a $2n^{\text{th}}$ part of its value.

Ex. 3. If two railway trains each of mass M were to travel in opposite directions from the pole along a meridian and to arrive at the equator at the same time, prove that the angular velocity of the earth would be decreased by $\dfrac{2Ma^2}{Ek^2}$ of itself, where a is the equatorial radius of the earth and Ek^2 its moment of inertia about its axis of figure.

What would be the effect if one train only were to travel from the pole to the equator?

Ex. 4. A fly alights perpendicularly on a sheet of paper lying on a smooth horizontal plane and proceeds to describe the curve $r = f(\theta)$ traced on the sheet of paper, the equation to the curve being referred to the centre of gravity of the paper as origin. Supposing the fly to be able to prevent himself from slipping on the paper, show that his angular velocity in space about the common centre of gravity of the paper and fly is equal to $\dfrac{(M+m)\,k^2}{(M+m)\,k^2 + mr^2}\dfrac{d\theta}{dt}$, where M and m are the masses of the paper and the fly, and k is the radius of gyration of the paper about its centre of gravity. Hence find the path of the fly in space.

Ex. 5. Suppose the ice to melt from the polar regions twenty degrees round each pole to the extent of something more than a foot thick, enough to give $1\frac{1}{10}$ feet over those areas or ·066 of a foot of water spread over the whole globe, which would in reality raise the sea-level by only some such undiscoverable difference as $\frac{3}{4}$th of an inch or an inch, then this would slacken the earth's rate as a time-keeper by one-tenth of a second per year. This and the next example are taken from the *Phil. Mag.* They are both due to Sir W. Thomson.

If E be the mass of the earth, a its radius, k its radius of gyration about the polar axis, ω its angular velocity before the melting, we have by the principle of angular momentum $\dfrac{\delta\omega}{\omega} = -\dfrac{Ma^2}{3Ek^2}\cos\theta\,(1+\cos\theta)$, where M is the mass of the ice melted and θ is twenty degrees. Substituting for the letters their known numerical values, the value of $\delta\omega$ is easily found.

Ex. 6. A layer of dust is formed on the earth h feet thick, where h is small, by the fall of meteors reaching the earth from all directions. Show that the change in the length of the day is nearly $\dfrac{5h}{a}\dfrac{\rho}{D}$ of a day, where a is the radius of the earth in feet, ρ and D the densities of the dust and earth respectively. If the density of the dust be twice that of water and $h = \frac{1}{10}$, express this result numerically.

Oppolzer in a paper in the *Astronomische Nachrichten* (No. 2573) and more recently H. A. Newton in the *American Journal of Science*, vol. xxx, 1884, have considered the effects on the earth both of the impact of meteors and the gravitation attraction of those which pass near the earth without hitting it.

Ex. 7. A spiral tube of small uniform section can turn freely about a vertical axis and has its two extremities on the axis. A variable quantity Q of fluid per second enters at its upper extremity and flows out at the lower. If M be the mass of the tube, m that of the fluid contained, show that $(M+m) k^2 \omega + Q \int r \sin \phi ds$ is constant, where ϕ is the angle the tangent to the tube at any point P makes with the plane containing P and the axis. One form of this experiment was used by Maxwell to determine if electricity had momentum. See *Electricity*, Vol. ii, Art. 574.

300. The principle of linear momentum may also be used, like that of angular momentum, to determine the gradual changes produced by alterations of mass. The general theory is as follows.

Let a body of mass M, whose resolved velocity parallel to x is v, be acted on by a finite force X. Let this body lose a small portion $m = -dM$ of its mass in each element of time dt. It is required to find its equation of motion. In this time the force increases the linear momentum by Xdt, while the momentum lost by diminution of mass is mv. But the gain of momentum is $d(Mv)$. The equation of motion is therefore

$$d(Mv) = Xdt + vdM.$$

$$\therefore M \frac{dv}{dt} = X.$$

This equation may also be obtained by taking M to represent the mass of the body just after the loss of the element m. Then equating the two expressions for the gain of momentum in the next element of time, we have $Mdv = Xdt$.

Next, let us suppose that the body gains a mass $m = dM$ in the time dt, and let the resolved velocity of this increment *just before* it is attached to M be v'. The total gain of momentum is now, Xdt due to the force, and mv' due to the impact produced by the sudden junction of the masses M and m with different velocities. The equation of motion is therefore

$$d(Mv) = Xdt + v'dM.$$

If $v' = v$ this reduces to the former result.

According to the rule given in Art. 85 the finite force X should be neglected in determining the effect of an impulse. But since m is infinitely small, being equal to dM, the change of momentum produced by the impulse is of the same order of small quantities as Xdt. We must therefore include the force X in the equation.

These principles may be illustrated by the solution of some problems on the rectilinear motion of strings. The curvilinear motion of strings will be discussed further on.

Ex. 1. One portion of a heavy uniform string is coiled up on a table in a small heap A, the other portion, viz. ACB, passes over a small pulley C (which is situated vertically over A) and hangs freely down on the other side of the pulley to a depth $CB = b$. If $CA = a$ and b is greater than a, find the motion when the system starts from rest.   [Tait and Steele's *Dynamics*, 1856.]

Let $x = CB$. As the end B descends, successive links (with velocity $v' = 0$) are taken from the heap and added to the moving chain. We therefore have

$$d\left[(x + a)\,v\right] = (x - a)\,g\,dt.$$

Multiply by $(x + a)\,v$ and integrate, we find

$$v^2 = \tfrac{2}{3}g\,(x - b)\cdot\frac{x^2 + bx + b^2 - 3a^2}{(x + a)^2}.$$

Ex. 2. A flexible chain $ABCDE$ hangs in equilibrium over a smooth vertical circle with one end A fixed to the extremity of a horizontal diameter. One portion ABC hangs vertically on one side and another portion DE hangs vertically on the other side of the circle. If the fixed end A be set free, show that the equation for determining the distance (viz. y) of the lowest point of the chain from the horizontal diameter during the first part of the motion is

$$(l - y + \tfrac{1}{2}gt^2)\,\ddot{y} - (\dot{y} + gt)^2 = g\,(y + \tfrac{1}{2}c),$$

where l is the whole length of the string and $2c$ the circumference of the circle.
                                    [Math. Tripos, 1870.]

Before A is set free the lengths AB, BC and DE are all equal and ABC forms a catenary whose parameter is zero. When A is set free, AB begins first to descend. Each element of AB falls freely under gravity; if therefore $x = AB$ we have $y - x = \tfrac{1}{2}gt^2$. The successive elements of AB are transferred to BC each with a velocity $v' = gt$, the length of each element being $- dx$. Thus as BC descends and DE ascends the equation of motion of $BCDE$ is

$$d\left[(l - x)\,v\right] = (- dx)\,v' + g\,(2y + x + c - l)\,dt.$$

Here v is the velocity of the chain $BCDE$ and this is equal to the velocity of E upwards (not that of B downwards). Since $DE = l - c - x - y$ we have $v = \dot{x} + \dot{y}$. Substituting for v and v' the result follows without difficulty.

Ex. 3. An inelastic string of length l is attached by one end to the lower surface of the edge of a smooth horizontal table with a fine edge on which the rest of the string lies, being held taut at right angles to the edge by a force at the other end. If this end be set free, show that the velocity with which it will leave the table will be $\sqrt{\{2gl\,(\log 4 - 1)\}}$.   [June Exam.]

Ex. 4. A fine uniform chain is collected in a heap on a horizontal table, and to one end is attached a fine string which passes over a smooth pulley vertically above the chain and carries a weight equal to the weight of a length a of the chain. Prove that the length of the chain raised before the weight comes to rest is $a\sqrt{(3)}$, and find the length suspended when the weight next comes to rest.   [May Exam.]

Ex. 5. A chain of length a is coiled up on a ledge at the top of a rough inclined plane and one end is allowed to slide down. Show that if the inclination of the plane is double the angle of friction (viz. λ), the chain will be moving freely at the end of a time t given by $gt^2 = 6a \cot \lambda$.   [Coll. Exam.]

Ex. 6. A balloon is at a certain moment at a height h, descending with velocity V, and moving horizontally with a velocity V' equal to the velocity of the wind at that height. If the velocity of the wind be proportional to the height, and if with a view of descending at a particular spot the escape of the gas be regulated so as to keep the velocity of descent constant, prove that a miscalculation dh in the initial height will produce in the point reached an error

$$= \frac{V'dh}{c^2V}\{1 + \tfrac{1}{2}c^2 - e^{-c}(1+c)\},$$

where $V^2c = gh.$ [Math. Tripos.]

Ex. 7. A spherical raindrop, descending by the action of gravity, receives continually by the precipitation of vapour an accession of mass proportional to its surface; c being its radius when it begins to descend, and r its radius after the interval t, show that its velocity V is given by the equation

$$V = \frac{gt}{4}\left(1 + \frac{c}{r} + \frac{c^2}{r^2} + \frac{c^3}{r^3}\right).$$

the resistance of the air being left out of account. [Smith's Prize Ex., 1853.]

The Invariable Plane.

301. Let us represent the momentum mv of a particle P by a straight line PP' drawn from the particle in the direction of its motion; see Art. 283. By the rules of statics, this momentum is equivalent to an equal and parallel linear momentum applied at any arbitrary point O, together with a couple whose moment is mvp, where p is the perpendicular from O on PP'. Let us represent this transferred linear momentum by the straight line OM, which of course is equal and parallel to PP'. The plane of the couple is the plane containing OM and P, and it may be represented in direction and magnitude by an axis ON perpendicular to its plane.

Taking all the particles of the system we may compound the linear and couple momenta of the several particles into a single resultant linear momentum applied at the arbitrary point O, together with a single couple momentum. Let OV and OH be two straight lines drawn from O to represent in direction and magnitude these two resultants. Then these two straight lines will represent graphically the instantaneous momenta of the particles considered as one system.

Let us refer the system to Cartesian co-ordinates. Since $m\dot{x}$, $m\dot{y}$, $m\dot{z}$ are the resolved parts of the momentum of the particle m, the vector OV is the resultant of $\Sigma m\dot{x}$, $\Sigma m\dot{y}$, $\Sigma m\dot{z}$. Again, as in Art. 75, $m(y\dot{z} - z\dot{y})$ is the moment of the momentum of the same particle about the axis of x. Hence OH is the resultant of the three couple-momenta

$$h_1 = \Sigma m \, (y\dot{z} - z\dot{y}),$$
$$h_2 = \Sigma m \, (z\dot{x} - x\dot{z}),$$
$$h_3 = \Sigma m \, (x\dot{y} - y\dot{x}).$$

Let us now suppose that no external forces act on the system, so that it moves subject only to the mutual actions and reactions of its several parts. In this case, since no additional momentum is given to the system, both the straight lines OV and OH are fixed in magnitude and direction throughout the motion; see Art. 283.

Their resolved parts also must be constant. It follows therefore that each of the quantities h_1, h_2, h_3 is constant. If we represent' by h the angular momentum about OH, we have $h^2 = h_1{}^2 + h_2{}^2 + h_3{}^2$. The ratios $\dfrac{h_1}{h}$, $\dfrac{h_2}{h}$, $\dfrac{h_3}{h}$ are therefore the direction-cosines of a straight line (viz. OH) which is fixed throughout the motion.

That the resolved angular momenta h_1, h_2, h_3 are constant follows also at once from Art. 78. Referring to the second equation given in that article, we see that, when the moment of the external forces about any straight line fixed in space is zero, the angular momentum about that line is constant.

The straight line OH is called the *invariable line* at O. A plane perpendicular to OH is called the *invariable plane* at O. The straight line OH is sometimes called the resultant axis of angular or couple momentum at O.

If any straight line OL be drawn through O making an angle θ with the invariable line OH at O, the angular momentum about OL is $h \cos \theta$. For the axis of the resultant momentum-couple is OH, and the resolved part about OL is therefore $OH \cos \theta$. Hence the invariable line at O may also be defined as that axis through O about which the moment of the momentum is greatest.

At different points of the system the positions of the invariable line are different. But the rules by which they are connected are the same as those which connect the axes of the resultant couple of a system of forces when the origin of reference is varied. These have been already stated in Art. 235 of Chap. v., and it is unnecessary here to do more than generally to refer to them.

If the system is acted on by any external forces, the straight lines OV and OH may not both be fixed in space. Consider first any one particle, let OM', ON' repre- sent its linear and couple momenta after an interval of time dt. Then MM', NN' will represent in direction and magnitude the linear momentum and the couple or angular momentum added on in the time dt. Hence the effective force on any particle m is equivalent to a single linear effective force acting at O, represented by $MM \, / dt$, and a single effective couple represented by NN' / dt.

Taking next the whole system of particles, let OV', OH' represent its linear and couple momenta after an interval dt. Thus OV is the resultant of the group OM corresponding to all the particles of the system, OV' the resultant of the group OM'. Hence VV'/dt represents the whole linear effective force of the system at the time t. By similar reasoning HH' dt represents the resultant effective couple of the system.

It appears therefore that the points V and H trace out two curves in space whose properties are analogous to those of the hodograph in dynamics of a particle.

From this reasoning it follows also, that if V_x be the resolved part of the momentum of a system in the direction of any straight line Ox and H_x the moment of the momentum about that straight line, then $\dot{V}_x$ and $\dot{H}_x$ are respectively the resolved part along and the moment about that straight line of the effective forces of the system.

Now D'Alembert's principle asserts that the effective forces of a system are equivalent to the impressed forces. Hence, whatever coordinates are used, if X and L be the resolved parts and the moment of the impressed forces about any straight line which we may call the axis of x, then $\dot{V}_x = X$ and $\dot{H}_x = L$. These equations correspond respectively to those marked (A) and (B) in Art. 72.

We may notice the following cases:

(1) If all the impressed forces pass through a fixed point, let this point be chosen as origin, then, though OV may be variable, OH is fixed in position and magnitude.

(2) If all the impressed forces be equivalent to a system of couples, then, though OH may be variable, OV is fixed in position and magnitude.

In a memoir on the *differential coefficients and determinants of lines*, Mr Cohen has discussed some properties of these resultant lines. *Phil. Trans.* 1862.

302. The position of the invariable plane at the centre of gravity of the solar system may be found in the following manner. Let the system be referred to any rectangular axes meeting in the centre of gravity. Let ω be the angular velocity of any body about its axis of rotation. Let Mk^2 be its moment of inertia about that axis and (α, β, γ) the direction-angles of that axis. The axis of revolution and two perpendicular axes form a system of principal axes at the centre of gravity. The angular momentum about the axis of revolution is $Mk^2\omega$, hence the angular momentum about an axis parallel to the axis of z is $Mk^2\omega \cos\gamma$. The moment of the momentum about the axis of z of the whole mass collected at the centre of gravity is $M\left(x\dfrac{dy}{dt} - y\dfrac{dx}{dt}\right)$, hence we have

$$h_3 = \Sigma Mk^2\omega \cos\gamma + \Sigma M\left(x\frac{dy}{dt} - y\frac{dx}{dt}\right).$$

The values of h_1, h_2 may be found in a similar manner. The position of the invariable plane is then known.

303. The Invariable Plane may be used in Astronomy as a standard of reference. We may observe the positions of the heavenly bodies with the greatest care, determining the co-ordi-

nates of each with regard to any axes we please. It is, however, clear that, unless these axes are fixed in space, or if in motion unless their motion is known, we have no means of transmitting our knowledge to posterity. The planes of the ecliptic and the equator have been generally made the chief planes of reference. Both these are in motion, and their motions are known to a near degree of approximation, and will hereafter probably be known more accurately. It might, therefore, be possible to calculate at some future time what their positions in space were when any set of valuable observations were made. But in a very long time some error may accumulate from year to year and finally become considerable The present positions of these planes in space may also be transmitted to posterity by making observations on the fixed stars. These bodies, however, are not absolutely fixed, and, as time goes on, the positions of the planes of reference can be determined from these observations with less and less accuracy. A third method, which has been suggested by Laplace, is to make use of the Invariable Plane. If we suppose the bodies forming our system, viz. the sun, planets, satellites, comets, &c., to be subject only to their mutual attractions, it follows from the preceding articles that the direction in space of the Invariable Plane at the centre of gravity is absolutely fixed. It also follows from Art. 79 that the centre of gravity either is at rest or moves uniformly in a straight line. We have here neglected the attractions of the stars. These, however, are too small to be taken account of in the present state of our astronomical knowledge. We may, therefore, determine to some extent the positions of our co-ordinate planes in space, by referring them to the Invariable Plane, as being a plane which is more nearly fixed than any other known plane in the solar system. The position of this plane may be calculated at the present time from the present state of the solar system, and at any future time a similar calculation may be made founded on the then state of the system. Thus a knowledge of its position cannot be lost. A knowledge of the co-ordinates of the Invariable Plane is not, however, sufficient to determine conversely the position of our planes of reference. We must also know the co-ordinates of some straight line in the Invariable Plane whose direction is fixed in space. Such a line, as Poisson has suggested, is supplied by projecting on the Invariable Plane the direction of motion of the centre of gravity of the system. If the centre of gravity of the solar system is at rest or moves perpendicularly to the Invariable Plane, this method fails. In any case our knowledge of the motion of the centre of gravity is not at present sufficient to enable us to make much use of this fixed direction in space.

304. If the planets and bodies forming the solar system can be regarded as spheres whose strata of ‧ equal density are concentric spheres, their mutual attractions act along the straight

lines joining their centres. In this case the motion of their centres is the same as if each mass were collected into its centre of gravity, while the motion of each about its centre of gravity would continue unchanged for ever. Thus we may obtain another fixed plane by omitting these latter motions altogether. This is what Laplace has done, and in his formulæ the terms depending on the rotations of the bodies in the preceding values of h_1, h_2, h_3 are omitted. This plane may be called the Astronomical Invariable Plane to distinguish it from the true Dynamical Invariable Plane. The former is perpendicular to the axis of the momentum couple due to the motions of translation of the several bodies, the latter is perpendicular to the axis of the momentum couple due to the motions of translation and rotation.

The Astronomical Invariable Plane is not strictly fixed in space, because the mutual attractions of the bodies do not strictly act along the straight lines joining their centres of gravity, so that the terms omitted in the expressions for h_1, h_2, h_3 are not absolutely constant. The effect of precession is to make the axis of rotation of each body describe a cone in space, so that, even though the angular velocity is unaltered, the position in space of the Astronomical Invariable Plane must be slightly altered. A collision between two bodies of the system, if such a thing were possible, or an explosion of a planet similar to that by which Olbers supposed the planets Pallas, Ceres, Juno and Vesta, &c., to have been produced, might make a considerable change in the sum of the terms omitted. In this case there would be a change in the position of the Astronomical Invariable Plane, but the Dynamical Invariable Plane would be altogether unaffected. It might be supposed that it would be preferable to use in Astronomy the true Invariable Plane. But this is not necessarily the case, for the angular velocities and moments of inertia of the bodies forming our system are not all known, so that the position of the Dynamical Invariable Plane cannot be calculated to any near degree of approximation, while we do know that the terms into which these unknown quantities enter are all very small or nearly constant. All the terms rejected being small compared with those retained, the Astronomical Invariable Plane must make only a small angle with the Dynamical Invariable Plane. Although the plane is very nearly fixed in space, yet its intersection with the Dynamical Invariable Plane, owing to the smallness of the inclination, may undergo considerable changes of position.

In the *Mécanique Céleste*, Laplace calculated the position of the Astronomical Invariable Plane at the two epochs, 1750 and 1950, assuming the correctness for this period of his formulæ for the variations of the eccentricities, inclinations and nodes of the planetary orbits. At the first epoch the inclination of this plane to the ecliptic was $1°\cdot7689$, and the longitude of the ascending node 114°·3979; at the second epoch the inclination will be the same as before, and the longitude of the node 114°·3934.

305. **Ex. 1.** Show that the invariable plane at any point of space in the straight line described by the centre of gravity of the solar system is parallel to that at the centre of gravity.

Ex. 2. If the invariable planes at all points in a certain straight line are parallel, then that straight line is parallel to the straight line described by the centre of gravity.

Impulsive Forces in Three Dimensions.

306. **Constrained single body.** *To determine the general equations of motion of a body about a fixed point under the action of given impulses.*

Let the fixed point be taken as the origin, and let the axes of co-ordinates be rectangular. Let $(\Omega_x, \Omega_y, \Omega_z)$, $(\omega_x, \omega_y, \omega_z)$ be the angular velocities of the body just before and just after the impulse, and let the differences $\omega_x - \Omega_x$, $\omega_y - \Omega_y$, $\omega_z - \Omega_z$ be called $\omega_x', \omega_y', \omega_z'$. Then $\omega_x', \omega_y', \omega_z'$ are the angular velocities *generated by the impulse*. By D'Alembert's Principle, see Art. 87, the difference between the moments of the momenta of the particles of the system just before and just after the action of the impulses is equal to the moment of the impulses. Hence by Art. 262

$$\begin{aligned} A\omega_x' - (\Sigma mxy)\,\omega_y' - (\Sigma mxz)\,\omega_z' &= L \\ B\omega_y' - (\Sigma myz)\,\omega_z' - (\Sigma myx)\,\omega_x' &= M \\ C\omega_z' - (\Sigma mzx)\,\omega_x' - (\Sigma mzy)\,\omega_y' &= N \end{aligned} \right\} \dots\dots\dots(1),$$

where L, M, N are the moments of the impulsive forces about the axes.

These three equations will suffice to determine the values of $\omega_x', \omega_y', \omega_z'$. By adding these to the angular velocities before the impulse, the initial motion of the body after the impulse is found.

307. **Ex. 1.** Show that these equations are independent of each other, and that none of the angular velocities $\omega_x, \omega_y, \omega_z$ is infinite.

This follows from Art. 20, where it is shown that the eliminant of the equations cannot vanish.

Ex. 2. Show that, if the body be acted on by a finite number of given impulses following each other at infinitely short intervals, the final motion is independent of their order.

308. It is to be observed that these equations leave the axes of reference undetermined. They should be so chosen that the values of A, Σmxy, &c., may be most easily found. If the

· 17—2

positions of the principal axes at the fixed point are known, these will in general be found the most suitable.

In that case the equations reduce to the simple forms

$$A\omega_x' = L \atop B\omega_y' = M \atop C\omega_z' = N \right\} \quad \dots\dots\dots\dots\dots\dots\dots(2).$$

The values of ω_x', ω_y', ω_z' being known, we can find the pressures on the fixed point. For by D'Alembert's Principle the change in the linear momentum of the body in any direction is equal to the resolved part of the impulsive forces. Hence if F, G, H be the pressures of the fixed point on the body

$$\Sigma X + F = M \cdot \frac{d\bar{x}}{dt} \text{ by Art. 86}$$
$$= M\,(\omega_y'\bar{z} - \omega_z'\bar{y}) \text{ by Art. 238} \right\} \dots\dots\dots(3).$$
$$\Sigma Y + G = M\,(\omega_z'\bar{x} - \omega_x'\bar{z})$$
$$\Sigma Z + H = M\,(\omega_x'\bar{y} - \omega_y'\bar{x})$$

309. **Ex.** *A uniform disc bounded by an arc* OP *of a parabola, the axis* ON, *and the ordinate* PN, *has its vertex* O *fixed. A blow* B *is given to it perpendicular to its plane at the extremity* P *of the curved boundary. Supposing the disc to be at rest before the application of the blow, find the initial motion.*

Let the equation to the parabola be $y^2 = 4ax$, and let the axis of z be perpendicular to its plane. Then $\Sigma mxz = 0$, $\Sigma myz = 0$. Let μ be the mass of a unit of area and let $ON = c$. Also $\Sigma mxy = \mu \iint xy\,dx\,dy = \mu \int_0^c x\frac{y^2}{2}\,dx = 2\mu \int_0^c ax^2\,dx = \frac{2}{3}\mu ac^3$,

$A = \frac{1}{3}\mu \int_0^c y^3\,dx = \frac{16}{15}\mu a^{\frac{3}{2}}c^{\frac{5}{2}}$, $\quad B = \mu \int_0^c x^2y\,dx = \frac{4}{7}\mu a^{\frac{1}{2}}c^{\frac{7}{2}}$, and $C = A + B$, by Art. 7.

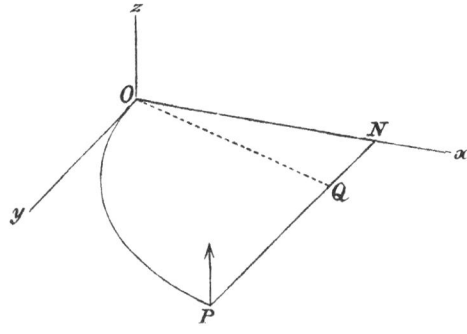

The moments of the blow B about the axes are $L = B\sqrt{4ac}$, $M = -Bc$, $N = 0$. The equations of Art. 306 will become after substitution of these values

$$\frac{16}{15}\mu a^{\frac{3}{2}}c^{\frac{5}{2}}\omega_x - \frac{2}{3}\mu ac^3 \omega_y = 2Ba^{\frac{1}{2}}c^{\frac{1}{2}} \left.\vphantom{\begin{matrix}1\\1\\1\end{matrix}}\right\}$$
$$\frac{4}{7}\mu a^{\frac{1}{2}}c^{\frac{7}{2}}\omega_y - \frac{2}{3}\mu ac^3 \omega_x = -Bc$$
$$\omega_z = 0$$

From these ω_x, ω_y may be found. By eliminating B we have $\dfrac{\omega_y}{\omega_x} = \dfrac{7}{25}\dfrac{2\sqrt{ac}}{c}$.

Hence, if NQ be taken equal to $\dfrac{7}{25}NP$, the disc will begin to rotate about OQ. The

resultant angular velocity will be $\dfrac{75}{26}\dfrac{B}{\mu ac^3}OQ$.

310. **New statement of the Problem.** When a body free to turn about a fixed point is acted on by any number of impulses, each impulse is equivalent to an equal and parallel impulse acting at the fixed point together with an impulsive couple. The impulse at the fixed point can have no effect on the motion of the body, and may therefore be left out of consideration if only the motion is wanted. Compounding all the couples, we see that the general problem may be stated thus:—A body moving about a fixed point is acted on by a given impulsive *couple*, find the change produced in the motion. The analytical solution is comprised in the equations which have been written down in Art. 306. The following examples express the result in a geometrical form.

Ex. 1. Show from these equations that the resultant axis of the angular velocity generated by the couple is the diametral line of the plane of the couple with regard to the momental ellipsoid. See also Art. 118.

Ex. 2. Let G be the magnitude of the couple, p the perpendicular from the fixed point on the tangent plane to the momental ellipsoid parallel to the plane of the couple G. Let Ω be the angular velocity generated, r the radius vector of the ellipsoid which is the axis of Ω. Let $M\epsilon^4$ be the parameter of the ellipsoid. Prove that $\dfrac{G}{\Omega} = \dfrac{M\epsilon^4}{pr}$.

Ex. 3. If Ω_x, Ω_y, Ω_z be angular velocities about three conjugate diameters of the momental ellipsoid at the fixed point, such that their resultant is the angular velocity generated by an impulsive couple G, A', B', C' the moments of inertia about these conjugate diameters, prove that

$$A'\Omega_x = G\cos\alpha, \qquad B'\Omega_y = G\cos\beta, \qquad C'\Omega_z = G\cos\gamma,$$

where α, β, γ are the angles the axis of G makes with the conjugate diameters.

Ex. 4. If a body free to turn about a fixed point O be acted on by an impulsive couple G, whose axis is the radius vector r of the ellipsoid of gyration at O, and if p be the perpendicular from O on the tangent plane at the extremity of r, then the axis of the angular velocity generated by the blow will be the perpendicular p, and the magnitude Ω is given by $G = Mpr\Omega$.

Ex. 5. Show that, if a body at rest be acted on by any impulses, we may take moments about the initial axis of rotation, according to the rule given in Art. 89, as if it were a fixed axis.

Ex. 6. When a body turns about a fixed point, the product of the moment of inertia about the instantaneous axis and the square of the angular velocity is called the Vis Viva. Let the vis viva generated from rest by any impulse be $2T$, and let the vis viva generated by the same impulse when the body is constrained to turn about a fixed axis passing through the fixed point be $2T'$. Then prove that $T' = T \cos^2 \theta$, where θ is the angle between the eccentric lines of the two axes of rotation with regard to the momental ellipsoid at the fixed point.

Ex. 7. Hence deduce Euler's theorem, that the vis viva generated from rest by an impulse is greater when the body is free to turn about the fixed point than when constrained to turn about any axis through the fixed point. This theorem was afterwards generalized by Lagrange and Bertrand in the second part of the first volume of the *Mécanique Analytique*.

311. **Free single body.** *To determine the motion of a free body acted on by any given impulse.*

Since the body is free, the motion round the centre of gravity is the same as if that point were fixed. Hence, the axes being any three straight lines at right angles meeting at the centre of gravity, the angular velocities of the body may still be found by equations (1) and (2) of Art. 306.

To find the motion of the centre of gravity, let (U, V, W), (u, v, w) be the resolved velocities of the centre of gravity just before and just after the impulse. Let X, Y, Z be the components of the blow, and let M be the whole mass. Then by resolving parallel to the axes we have

$$M(u - U) = X, \quad M(v - V) = Y, \quad M(w - W) = Z.$$

If we follow the same notation as in Art. 306, the differences $u - U, v - V, w - W$ may be called u', v', w'.

312. Ex. 1. A body at rest is acted on by an impulse whose components parallel to the principal axes at the centre of gravity are (X, Y, Z) and the co-ordinates of whose point of application referred to these axes are (p, q, r). Prove that if the resulting motion be one of rotation only about some axis,

$$A(B - C)pYZ + B(C - A)qZX + C(A - B)rXY = 0.$$

Is this condition sufficient as well as necessary? See Art. 241.

Ex. 2. A homogeneous cricket-ball is set rotating about a horizontal axis in the vertical plane of projection with an angular velocity Ω. When it strikes the ground, supposed perfectly rough and inelastic, the centre is moving with velocity V in a direction making an angle α with the horizon, prove that the direction of the motion of the ball after impact will make with the plane of projection an angle $\tan^{-1}\dfrac{2}{5}\dfrac{a\Omega}{V\cos\alpha}$, where a is the radius of the ball.

313. **Motion of any point of the body.** *The components of the change of velocity of any point of the body are linear functions of the components of the blow.* The equations of Art. 311 completely determine the motion of a free body acted on by a given

impulse, and from these by Art. 238 we may determine the initial motion of any point of the body. Let (p, q, r) be the co-ordinates of the point of application of the blow, then the moments of the blow round the axes are respectively $qZ - rY$, $rX - pZ$, $pY - qX$. These must be written on the right-hand sides of the equations of Art. 306. Let (p', q', r') be the co-ordinates of the point whose initial velocities parallel to the axes are required. Let (u_1, v_1, w_1), (u_2, v_2, w_2) be its velocities just before and just after the impulse. Let the rest of the notation be the same as that used in Art. 306. Then

$$u_2 - u_1 = u' + \omega_y' r' - \omega_z' q',$$

with similar equations for $v_2 - v_1$, $w_2 - w_1$. Substituting in these equations the value of u', v', w', ω_x', ω_y', ω_z' given by Art. 311 we see that $u_2 - u_1$, $v_2 - v_1$, $w_2 - w_1$ are all linear functions of X, Y, Z of the first degree of the form

$$u_2 - u_1 = FX + GY + HZ,$$

where F, G, H depend on the structure of the body and the co-ordinates of the two points.

314. When the point whose initial motion is required is the point of application of the blow, and the axes of reference are the principal axes at the centre of gravity, these expressions take the simple forms

$$u_2 - u_1 = \left(\frac{1}{M} + \frac{r^2}{B} + \frac{q^2}{C}\right) X - \frac{pq}{C} Y - \frac{pr}{B} Z,$$

$$v_2 - v_1 = -\frac{pq}{C} X + \left(\frac{1}{M} + \frac{p^2}{C} + \frac{r^2}{A}\right) Y - \frac{qr}{A} Z,$$

$$w_2 - w_1 = -\frac{pr}{B} X - \frac{qr}{A} Y + \left(\frac{1}{M} + \frac{q^2}{A} + \frac{p^2}{B}\right) Z.$$

The right-hand sides of these equations are the differential coefficients of a quadratic function of X, Y, Z, which we may call E. It follows that *for all blows at the same point* P *of the same body the resultant change in the velocity of the point* P *of application is perpendicular to the diametral plane of the direction of the blow with regard to a certain ellipsoid, whose centre is at* P, *and whose equation is* E = *constant.*

The expression for E may be written in either of the equivalent forms:

$$2E = \frac{X^2 + Y^2 + Z^2}{M} + \frac{1}{ABC}\{(Ap^2 + Bq^2 + Cr^2)(AX^2 + BY^2 + CZ^2) - (ApX + BqY + CrZ)^2\}$$

$$= \frac{X^2 + Y^2 + Z^2}{M} + \frac{1}{A}(qZ - rY)^2 + \frac{1}{B}(rX - pZ)^2 + \frac{1}{C}(pY - qX)^2.$$

In this latter form we see that it is

$$= M(u'^2 + v'^2 + w'^2) + A\omega_x'^2 + B\omega_y'^2 + C\omega_z'^2,$$

which is the vis viva of the motion generated by the impulse.

Impact of any two bodies.

315.   *Two bodies moving in any manner impinge on each other.*
To find the motion after impact.

Inelastic Bodies.   If the bodies be *inelastic and either*
perfectly smooth or perfectly rough, it is unnecessary to introduce
the reactions into the equations.   In such a case we take the
point of contact as the origin.   Let the axes of x and y be in
the tangent plane, and that of z be normal.   Let U, V, W be the
resolved velocities of the centre of gravity of one body just before
the impact, and u, v, w the resolved velocities just after the impact.
Let Ω_x, Ω_y, Ω_z, ω_x, ω_y, ω_z be the angular velocities just before and
just after.   Let A, B, C, D, E, F be the moments and products of
inertia at the centre of gravity.   Let M be the mass of the body,
and x, y, z the co-ordinates of its centre of gravity.   Let accented
letters denote the same quantities for the other body.

Then taking moments about the axes for one body we have,
by Arts. 306 and 76,

$$A\,(\omega_x-\Omega_x) - F\,(\omega_y-\Omega_y) - E\,(\omega_z-\Omega_z) - (v - V)\,z + (w - W)y = 0,$$
$$-F\,(\omega_x-\Omega_x) + B\,(\omega_y-\Omega_y) - D(\omega_z-\Omega_z) - (w - W)\,x + (u - U)\,z = 0,$$
$$-E\,(\omega_x-\Omega_x) - D\,(\omega_y-\Omega_y) + C\,(\omega_z-\Omega_z) - (u - U)\,y + (v - V)x = 0.$$

Three similar equations apply for the other body, differing from
these only in having all the letters accented.

Resolving along the axis of z for both bodies, we have

$$M\,(w - W) + M'\,(w' - W') = 0.$$

The relative velocity of compression is zero at the moment of
greatest compression, we have therefore

$$w - \omega_x y + \omega_y x = w' - \omega_x' y' + \omega_y' x'.$$

We thus have eight equations between the twelve unknown resolved velocities and angular velocities.

316.   If the *bodies be smooth* we obtain four more equations by
resolving for each body parallel to the axes of x and y.   For the
one body we have

$$u - U = 0, \quad v - V = 0,$$

with similar equations for the other body.

317.   If the *bodies be perfectly rough* we obtain two of the
four equations by resolving the linear momenta parallel to the
axes of x and y, viz.

$$\left.\begin{array}{l} M\,(u - U) + M'\,(u' - U') = 0 \\ M\,(v - V) + M'\,(v' - V') = 0 \end{array}\right\}.$$

We have also two geometrical equations obtained by equating to zero the resolved relative velocity of sliding, viz.

$$u - \omega_y z + \omega_z y = u' - \omega_y' z' + \omega_z' y' \Big\}$$
$$v - \omega_z x + \omega_x z = v' - \omega_z' x' + \omega_x' z' \Big\}.$$

318. **Smooth Elastic Bodies.** If the bodies be *smooth and imperfectly elastic*, we must introduce the normal reaction into the equations. In this case we proceed exactly as in the general case when the bodies are rough and elastic, which we shall consider in the following articles. The process is of course simplified by putting the fractions P and Q both equal to zero in the twelve equations of motion (1), (2), (3) and (4). We also have the velocity C of compression equal to zero at the moment of greatest compression. Thus we have one more equation from which the normal reaction R may be found. Multiplying this value of R by $1 + e$, where e has the meaning given to it in Art. 179, we have the complete value of R for the whole impact. Substituting this last value of R in the twelve equations of motion (1) and (2), (3) and (4), the motion of both bodies just after impact is found.

319. **Rough Elastic Bodies.** The problem of determining the motion of any two rough bodies after a collision involves some rather long analysis and yet in some points it differs essentially from the corresponding problem in two dimensions. We shall, therefore, first consider a special problem which admits of being treated briefly, and will then apply the same principles to the general problem in three dimensions.

320. *Two rough ellipsoids moving in any manner impinge on each other so that the extremity of a principal diameter of one strikes the extremity of a principal diameter of the other, at an instant when the three principal diameters of one are parallel to those of the other. Find the motion just after impact.*

Let us refer the motion to co-ordinate axes parallel to the principal diameters of either ellipsoid at the beginning of the impact. Then since the duration of the impact is indefinitely small and the velocities are finite, the bodies will not have time to change their position, and therefore the principal diameters will be parallel to the co-ordinate axes throughout the impact.

Let U, V, W be the resolved velocities of the centre of gravity of one body just before impact; u, v, w the resolved velocities at any time t, after the beginning of the impact, but before its termination. Let Ω_x, Ω_y, Ω_z be the angular velocities of the body just before impact about its principal diameters at the centre of gravity; ω_x, ω_y, ω_z the angular velocities at the time t. Let a, b, c be the semiaxes of the ellipsoid, and A, B, C the moments of inertia at the centre of gravity about these axes respectively. Let M be the mass of the body. Let accented letters denote the same quantities for the other body. Let the bodies impinge at the extremities of the axes c, c'.

Let P, Q, R be the resolved parts parallel to the axes of the momentum generated in the body M by the blow during the time t. Then $-P$, $-Q$, $-R$ are the resolved parts of the momentum generated in the other body in the same time.

The equations of motion of the body M are

$$\left.\begin{aligned} A\,(\omega_x - \Omega_x) &= Qc \\ B\,(\omega_y - \Omega_y) &= -Pc \\ C\,(\omega_z - \Omega_z) &= 0 \end{aligned}\right\}\dots\dots\dots\dots\dots(1)$$

$$\left.\begin{aligned} M\,(u - U) &= P \\ M\,(v - V) &= Q \\ M\,(w - W) &= R \end{aligned}\right\}\dots\dots\dots\dots\dots(2).$$

There are six corresponding equations for the other body which may be derived from these by accenting all the letters on the left-hand side and writing $-P$, $-Q$, $-R$, $-c'$ for P, Q, R and c on the right-hand side. Let us call these new equations respectively (3) and (4).

Let S be the velocity with which one ellipsoid slides along the other, and θ the angle which the direction of sliding makes with the axis of x, then

$$S\cos\theta = u' + c'\omega_y' - u + c\omega_y \dots\dots\dots\dots\dots(5),$$

$$S\sin\theta = v' - c'\omega_x' - v - c\omega_x \dots\dots\dots\dots\dots(6).$$

Let C be the relative velocity of compression, then

$$C = w' - w \dots\dots\dots\dots\dots\dots\dots(7).$$

Substituting in these equations from the dynamical equations we have

$$S\cos\theta = S_0\cos\theta_0 - pP\dots\dots\dots\dots\dots(8),$$

$$S\sin\theta = S_0\sin\theta_0 - qQ\dots\dots\dots\dots\dots(9),$$

$$C = C_0 - rR\dots\dots\dots\dots\dots\dots\dots(10),$$

where

$$\left.\begin{aligned} S_0\cos\theta_0 &= U' + c'\Omega_y' - U + c\Omega_y \\ S_0\sin\theta_0 &= V' - c'\Omega_x' - V - c\Omega_x \\ C_0 &= W' - W \end{aligned}\right\}\dots\dots\dots(11),$$

$$\left.\begin{aligned} p &= \frac{1}{M} + \frac{1}{M'} + \frac{c^2}{B} + \frac{c'^2}{B'} \\[1mm] q &= \frac{1}{M} + \frac{1}{M'} + \frac{c^2}{A} + \frac{c'^2}{A'} \\[1mm] r &= \frac{1}{M} + \frac{1}{M'} \end{aligned}\right\}\dots\dots\dots(12).$$

These are the constants of the impact. S_0, C_0 are the initial velocities of sliding, and θ_0 the angle which the direction of initial sliding makes with the axis of x. Let us take as the standard case that in which the body M' is sliding along and compressing the body M, so that S_0 and C_0 are both positive. The other three constants p, q, r are independent of the initial motion and are essentially positive quantities.

321.   Exactly as in two dimensions, we shall adopt a graphical method of tracing the changes which occur in the frictions. Let us measure along the axes of x, y, z three lengths OP, OQ, OR to represent the three reactions P, Q, R. Then, if these be regarded as the co-ordinates of a point T, the motion of T will represent the changes in the forces. It will be convenient to trace the loci given by $S = 0$, $C = 0$. The locus given by $S = 0$ is a straight line parallel to the axis of R, which we may call the *line of no sliding*. The locus given by $C = 0$ is a plane parallel to the plane POQ, which we may call the *plane of greatest compression*. At the beginning of the impact one ellipsoid is sliding along the other, so that according to Art. 154 the friction called into play is limiting. Since P, Q, R are the whole resolved momenta generated in the time t, dP, dQ, dR are the resolved momenta generated in the time dt, the two former being due to the frictional, and the latter to the normal blow. Then the direction of the resultant of dP, dQ must be opposite to the direction in which one point of contact slides over the other, and the magnitude of the resultant must be equal to μdR, where μ is the coefficient of friction. We have therefore

$$\frac{dP}{dQ} = \cot\theta = \frac{S_0 \cos\theta_0 - pP}{S_0 \sin\theta_0 - qQ} \quad\ldots\ldots\ldots\ldots\ldots(13),$$

$$(dP)^2 + (dQ)^2 = \mu^2 (dR)^2 \ldots\ldots\ldots\ldots\ldots(14).$$

The solution of these equations will indicate the manner in which the representative point T approaches the line of no sliding.

The equation (13) can be solved by separating the variables. We get

$$(S_0 \cos\theta_0 - pP)^{\frac{1}{p}} = \alpha (S_0 \sin\theta_0 - qQ)^{\frac{1}{q}},$$

where α is an arbitrary constant. At the beginning of the motion P and Q are zero, hence we have

$$\left(\frac{S_0 \cos\theta_0 - pP}{S_0 \cos\theta_0}\right)^{\frac{1}{p}} = \left(\frac{S_0 \sin\theta_0 - qQ}{S_0 \sin\theta_0}\right)^{\frac{1}{q}} \ldots\ldots\ldots(15),$$

which may also be written

$$\left(\frac{S \cos\theta}{S_0 \cos\theta_0}\right)^{\frac{1}{p}} = \left(\frac{S \sin\theta}{S_0 \sin\theta_0}\right)^{\frac{1}{q}} \ldots\ldots\ldots\ldots\ldots(16),$$

or $$S = S_0 \left(\frac{\sin \theta}{\sin \theta_0}\right)^{\frac{p}{q-p}} \cdot \left(\frac{\cos \theta_0}{\cos \theta}\right)^{\frac{q}{q-p}} \ldots\ldots\ldots\ldots(17).$$

This equation gives the relation between the direction and the velocity of sliding.

322. If the direction of sliding does not change during the impact, θ must be constant and equal to θ_0. We see from (16) that, if $p = q$, then $\theta = \theta_0$; and that conversely if $\theta = \theta_0$, S is constant unless $p = q$. Also, if $\sin \theta_0$ or $\cos \theta_0$ be zero, S must be zero or infinite unless $\theta = \theta_0$. *The necessary and sufficient condition that the direction of friction should not change during the impact is therefore* $p = q$ *or* $\sin 2\theta_0 = 0$. The former of these two conditions, by (12), leads to

$$c^2 \left(\frac{1}{A} - \frac{1}{B}\right) + c'^2 \left(\frac{1}{A'} - \frac{1}{B'}\right) = 0 \ldots\ldots\ldots\ldots(18).$$

If this condition holds, we have by (13) $P = Q \cot \theta_0$ and therefore by (14)

$$\left.\begin{array}{l} P = \mu R \cos \theta_0 \\ Q = \mu R \sin \theta_0 \end{array}\right\} \ldots\ldots\ldots\ldots\ldots\ldots\ldots(19).$$

It follows from these equations that, when the friction is limiting, the representative point T moves along a straight line making an angle $\tan^{-1}\mu$ with the axis of R, in such a direction as to meet the straight line of no sliding.

323. If the condition $p = q$ does not hold, we may, by differentiating (8) and (9) and eliminating P, Q, and S, reduce the determination of R in terms of θ to an integral.

By substituting for S from (17) in (8) and (9), we then have P, Q, R expressed as functions of θ. Thus we have the equations to the curve along which the representative point T travels. The curve along which T travels may more conveniently be defined by the property that its tangent, by (14), makes a constant angle $\tan^{-1}\mu$ with the axis of R and its projection on the plane of PQ is given by (15). And it follows that this curve must meet the straight line of no sliding, for the equation (15) is satisfied by $pP = S_0 \cos \theta_0$, $qQ = S_0 \sin \theta_0$.

324. The whole progress of the impact may now be traced exactly as in the corresponding problem in two dimensions. *The representative point* T *travels along a certain known curve, until it reaches the line of no sliding. It then proceeds along the line of no sliding, in such a direction that the abscissa* R *increases. The complete value* R_2 *of* R *for the whole impact is found by multiplying the abscissa* R_1 *of the point at which* T *crosses the plane of greatest compression by* $1 + e$, *so that* $R_2 = R_1 (1 + e)$, *if* e *be*

the measure of the elasticity of the two bodies. The complete values of the frictions called into play are the ordinates of the positions of T *corresponding to the abscissa* R = R$_2$. *Substituting these in the dynamical equations* (1), (2), (3), (4), *the motion of the two bodies just after impact may be found.*

325. Since the line of no sliding is perpendicular to the plane of PQ, P and Q are constant when T travels along this line. So that, when once the sliding friction has ceased, no more friction is called into play. If therefore sliding ceases at any instant before the termination of the impact, as when the bodies are either very rough or perfectly rough, the whole frictional impulses are given by

$$P = \frac{S_0 \cos \theta_0}{p}, \quad Q = \frac{S_0 \sin \theta_0}{q}.$$

If σ be the arc of the curve whose equation is (15) from the origin to the point where it meets the line of no sliding, then the representative point T cuts the line of no sliding at a point whose abscissa is $R = \dfrac{\sigma}{\mu}$. If the bodies be so rough that $\dfrac{\sigma}{\mu} < \dfrac{C_0}{r}$, the point T will not cross the plane of greatest compression until after it has reached the line of no sliding. The whole normal impulse is therefore given by $R = \dfrac{C_0}{r}(1 + e)$. Substituting these values of P, Q, R in the dynamical equations, the motion just after impact may be found.

326. **Ex. 1.** If θ be the angle which the direction of sliding of one ellipsoid over the other makes with the axis of x, prove that θ continually increases or continually decreases throughout the impact. And if the initial value of θ lie between 0 and $\dfrac{\pi}{2}$, then θ approaches $\dfrac{\pi}{2}$ or zero according as $p >$ or $< q$. Show also that the representative point reaches the line of no sliding when θ has either of these values.

Ex. 2. If the bodies be such that the direction of sliding continues unchanged during the impact and the sliding ceases before the termination of the impact, the roughness must be such that $\mu > \dfrac{S_0 r}{C_0 p (1 + e)}$

Ex. 3. If two rough spheres impinge on each other, prove that the direction of sliding is the same throughout the impact. This proposition was first given by Coriolis. *Jeu de billard*, 1835. See Art. 322.

Ex. 4. If two inelastic solids of revolution impinge on each other, the vertex of each being the point of contact, prove that the direction of sliding is the same throughout the impact. This and the next proposition have been given by M. Phillips in the fourteenth volume of *Liouville's Journal*.

Ex. 5. If two bodies having the principal axes at their centres of gravity parallel impinge, so that these centres of gravity are in the common normal at the point of contact, and if the initial direction of sliding be parallel to a principal axis at either centre of gravity, then the direction of sliding will be the same throughout the impact.

Ex. 6. If two ellipsoids of equal mass impinge on each other at the extremities of their axes c, c', and if $aa' = bb'$ and $ca' = bc'$, prove that the direction of friction is constant throughout the impact.

Ex. 7. A billiard ball rolls without sliding on the table and impinges against a cushion, find the subsequent motion.

Let the planes of the cushion and table be called the planes of xy and xz respectively. Let the initial velocity of the centre of gravity resolved parallel to x and z be $-u$ and $-w$ and let the angular velocity about the vertical be n. After rebounding the ball will describe a series of very small parabolic jumps which are hardly perceptible. Finally the ball may be regarded as rolling on the table. This final motion is given by

$$U' = -u + \tfrac{5}{7}\gamma(u + an)$$
$$W' = -w + \tfrac{5}{7}(1 + \gamma + e)w$$

where γ is the smaller of the two quantities $\tfrac{2}{7}$ and $\mu(1+e)\,w/\{w^2 + (u+an)^2\}^{\frac{1}{2}}$.

327. *Two rough bodies moving in any manner impinge on each other. Find the motion just after impact.*

Let us refer the motion to co-ordinate axes, the axes of x, y being in the tangent plane at the point of impact and the axis of z along the normal. Let U, V, W be the resolved velocities of the centre of gravity of one body just before impact, u, v, w the resolved velocities at any time t after the beginning, but before the termination of the impact. Let Ω_x, Ω_y, Ω_z be the angular velocities of the same body just before impact about axes parallel to the co-ordinate axes, meeting at the centre of gravity; ω_x, ω_y, ω_z the angular velocities at the time t. Let A, B, C, D, E, F be the moments and products of inertia about axes parallel to the co-ordinate axes meeting at the centre of gravity. Let M be the mass of the body. Let accented letters denote the same quantities for the other body.

Let P, Q, R be the resolved parts parallel to the axes of the momentum generated in the body M from the beginning of the impact, up to the time t. Then $-P$, $-Q$, $-R$ are the resolved parts of the momentum generated in the other body in the same time.

Let $(x, y, z)\,(x', y', z')$ be the co-ordinates of the centres of gravity of the two bodies referred to the point of contact as origin. The equations of motion are therefore

$$
\left.
\begin{aligned}
A(\omega_x - \Omega_x) - F(\omega_y - \Omega_y) - E(\omega_z - \Omega_z) &= -yR + zQ \\
-F(\omega_x - \Omega_x) + B(\omega_y - \Omega_y) - D(\omega_z - \Omega_z) &= -zP + xR \\
-E(\omega_x - \Omega_x) - D(\omega_y - \Omega_y) + C(\omega_z - \Omega_z) &= -xQ + yP
\end{aligned}
\right\} \dots(1),
$$

$$M\left(u-U\right)=P$$
$$M\left(v-V\right)=Q \left.\vphantom{\begin{matrix}1\\1\\1\end{matrix}}\right\}\ \ldots\ldots\ldots\ldots\ldots\ldots\ldots(2).$$
$$M\left(w-W\right)=R$$

We have six similar equations for the other body, which differ from these in having all the letters, except P, Q, R, accented, and in having the signs of P, Q, R changed. These we shall call equations (3) and (4). Let S be the velocity with which one body slides along the other and θ the angle which the direction of sliding makes with the axis of x. Also let C be the relative velocity of compression, then

$$S\cos\theta = u' - \omega_y'z' + \omega_z'y' - u + \omega_y z - \omega_z y$$
$$S\sin\theta = v' - \omega_z'x' + \omega_x'z' - v + \omega_z x - \omega_x z \left.\vphantom{\begin{matrix}1\\1\\1\end{matrix}}\right\}\ \ldots\ldots(5).$$
$$C = w' - \omega_x'y' + \omega_y'x' - w + \omega_x y - \omega_y x$$

If we substitute from (1) (2) (3) (4) in (5) we find

$$S_0\cos\theta - S\cos\theta = aP + fQ + eR$$
$$S_0\sin\theta - S\sin\theta = fP + bQ + dR \left.\vphantom{\begin{matrix}1\\1\\1\end{matrix}}\right\}\ \ldots\ldots\ldots\ldots(6),$$
$$C_0 - C = eP + dQ + cR$$

where S_0, θ_0, C_0 are the initial values of S, θ, C and are found from (5) by writing for the letters their initial values. The expressions for a, b, c, d, e, f are rather complicated, but it is unnecessary to calculate these.

328.   We may now trace the whole progress of the impact by the use of a graphical method. Let us measure from the point of contact O, along the axes of co-ordinates, three lengths OP, OQ, OR to represent the three reactions, P, Q, R. Then if, as before, these be regarded as the co-ordinates of a point T, the motion of T will represent the changes in the forces. The equations to the line of no sliding are found by putting $S = 0$ in the first two of equations (6). We see that it is a straight line.

The equation to the plane of greatest compression is found by putting $C = 0$ in the third of equations (6).

At the beginning of the impact one body is sliding along the other, so that the friction called into play is limiting. The path of the representative point as it travels from O is given, as in Art. 321, by

$$\frac{dP}{\cos\theta} = \frac{dQ}{\sin\theta} = \mu dR \ \ldots\ldots\ldots\ldots\ldots\ldots(7).$$

When the representative point T reaches the line of no sliding, the sliding of one body along the other ceases for the instant. After this, only so much friction is called into play as will suffice to prevent sliding, provided that this amount is less than the limiting friction. If therefore the angle which the line of no sliding makes with the axis of R be less than $\tan^{-1}\mu$, the point T travels along it. But if the angle be greater than $\tan^{-1}\mu$, more friction is necessary to prevent sliding than can be called into play. Accordingly the friction continues to be limiting, but its direction is changed if S changes sign. The point T then travels along a curve given by equations (7) with θ increased by π.

The complete value R_2 of R for the whole impact is found by multiplying the abscissa R of the point at which T crosses the plane of greatest compression by $1 + e$, where e is the measure of elasticity, so that $R_2 = R_1(1 + e)$. The complete values of P and Q are represented by the ordinates corresponding to the abscissa R_2. Substituting in the dynamical equations, the motion just after impact may be found.

329. The path of the representative point before it reaches the line of no sliding must be found by integrating (7). By differentiating (6) we have

$$\frac{d(S\cos\theta)}{d(S\sin\theta)} = \frac{adP + fdQ + edR}{fdP + bdQ + ddR} = \frac{a\mu\cos\theta + f\mu\sin\theta + e}{f\mu\cos\theta + b\mu\sin\theta + d} \quad ...(8),$$

which reduces to

$$\frac{1}{S}\frac{dS}{d\theta} = \frac{\dfrac{a+b}{2} + \dfrac{a-b}{2}\cos 2\theta + f\sin 2\theta + \dfrac{e}{\mu}\cos\theta + \dfrac{d}{\mu}\sin\theta}{-\dfrac{a-b}{2}\sin 2\theta + f\cos 2\theta + \dfrac{d}{\mu}\cos\theta - \dfrac{e}{\mu}\sin\theta} \quad ...(9).$$

From this equation we may find S as a function of θ in the form $S = Af(\theta)$, the constant A being determined from the condition that $S = S_0$ when $\theta = \theta_0$. Differentiating the first of equations (6) and substituting from (7) we get

$$-Ad\{\cos\theta f(\theta)\} = (\mu a\cos\theta + \mu f\sin\theta + e)dR......(10),$$

whence we find $R = AF(\theta) + B$, the constant B being determined from the condition that R vanishes when $\theta = \theta_0$. By substituting these values of S and R in the first two equations of (6) we find P and Q in terms of θ. The three equations giving P, Q, R as functions of θ are the equations to the path of the representative point. It should be noticed that the tangent to the path at any point makes with the axis of R an angle equal to $\tan^{-1}\mu$.

330. If the direction of friction does not change during the impact, θ is constant and equal to θ_0, so that θ cannot be chosen as the independent variable. In this case $P = \mu R\cos\theta_0$, $Q = \mu R\sin\theta_0$ and the representative point moves along a straight line making with the axis of R an angle $\tan^{-1}\mu$. Substituting these values of

P and Q in the first two of equations (6) we have

$$- \frac{a-b}{2} \sin 2\theta_0 + f \cos 2\theta_0 + \frac{d}{\mu} \cos \theta_0 - \frac{e}{\mu} \sin \theta_0 = 0 \ldots (11)$$

as a necessary condition that the direction of friction should not change. Conversely, if this condition is satisfied, the equations (6) and (7) may all be satisfied by making θ constant. In this case it is also easy to see that the path of the representative point intersects the line of no sliding.

If S_0 be zero, the representative point is situated on the line of no sliding. If the angle made by this straight line with the axis of R be less than $\tan^{-1}\mu$, the representative point travels along it. But if the angle be greater than $\tan^{-1}\mu$, more friction is necessary to prevent sliding than can be called into play. Since S_0 is zero, the initial value of θ is unknown. In this case, differentiating the first two equations of (6) and putting $S = 0$, we see by division that the initial value of θ must satisfy equation (11). The condition that the direction of friction does not change is therefore satisfied. This value of θ makes the subject of integration in (9) infinite, so that the reasoning there given must be modified. But, by what has just been said, we see that the path of the representative point is a straight line, which makes with the axis of R an angle equal to $\tan^{-1}\mu$, and has the proper initial value of θ.

331. Ex. 1. Let $G = \begin{vmatrix} A & -F & -E & yR-zQ \\ -F & B & -D & zP-xR \\ -E & -D & C & xQ-yP \\ yR-zQ & zP-xR & xQ-yP & 0 \end{vmatrix}$

and let Δ be the determinant obtained by leaving out the last row and the last column. Let G', Δ' be corresponding expressions for the other body. Then a, b, c, d, e, f are the coefficients of $P^2, Q^2, R^2, 2QR, 2RP, 2PQ$ in the quadric

$$\left(\frac{1}{M} + \frac{1}{M'}\right)(P^2 + Q^2 + R^2) + \frac{1}{\Delta}G + \frac{1}{\Delta'}G' = 2E,$$

where $2E$ is a constant, which may be shown to be the sum of the vires vivæ of the motions generated in the two bodies, as explained in Art. 314.

This quadric may be shown to be an ellipsoid by comparing its equation with that given in Art. 28, Ex. 3.

Show also that a, b, c are necessarily positive, and that $ab > f^2$, $bc > d^2$, $ca > e^2$.

Show that, by turning the axes of reference round the axis of R through the proper angle, we can make f zero.

Ex. 2. Prove that the line of no sliding is parallel to the conjugate diameter of the plane containing the frictions P, Q. Prove also that the plane of greatest compression is the diametral plane of the reaction R.

Ex. 3. The line of no sliding is the intersection of the polar planes of two points situated on the axes of P and Q, at distances from the origin respectively $\dfrac{2E}{S_0 \cos \theta_0}$ and $\dfrac{2E}{S_0 \sin \theta_0}$. The plane of greatest compression is the polar plane of a point on the axis of R, distant $\dfrac{2E}{C_0}$ from the origin.

Ex. 4.   The plane of PQ cuts the ellipsoid of Ex. 1 in an ellipse, whose axes divide the plane into four quadrants; the line of no sliding cuts the plane of PQ in that quadrant in which the initial sliding S_0 occurs.

Ex. 5.   A parallel to the line of no sliding through the origin cuts the plane of greatest compression in a point whose abscissa R has the same sign as C_0.   Hence show, from geometrical considerations, that the representative point T must cross the plane of greatest compression.

EXAMPLES*.

1.   A cone revolves round its axis with a known angular velocity.   The altitude begins to diminish and the angle to increase, the volume being constant.   Show that the angular velocity is proportional to the altitude.

2.   A circular disc is revolving in its own plane about its centre; if a point in the circumference becomes fixed, find the new angular velocity.

3.   A uniform rod of length $2a$ lying on a smooth horizontal plane passes through a ring which permits the rod to rotate freely in the horizontal plane.   The middle point of the rod being indefinitely near the ring, any angular velocity is impressed on it, show that when it leaves the ring the radius vector of the middle point has swept out an area equal to $\dfrac{a^2}{6}$.

4.   An elliptic lamina is rotating about its centre on a smooth horizontal table. If ω_1, ω_2, ω_3 be its angular velocities when the extremity of its major axis, its focus, and the extremity of its minor axis respectively become fixed, prove that

$$\frac{7}{\omega_1} = \frac{6}{\omega_2} + \frac{5}{\omega_3}.$$

5.   A rigid body moveable about a fixed point O at which the principal moments are A, B, C is struck by a blow of given magnitude at a given point.   If the angular velocity thus impressed on the body be the greatest possible, prove that, (a, b, c) being the co-ordinates of the given point referred to the principal axes at O, and (l, m, n) the direction cosines of the blow,

$$al + bm + cn = 0,$$

$$\frac{a}{l}\left(\frac{1}{B^2} - \frac{1}{C^2}\right) + \frac{b}{m}\left(\frac{1}{C^2} - \frac{1}{A^2}\right) + \frac{c}{n}\left(\frac{1}{A^2} - \frac{1}{B^2}\right) = 0.$$

6.   Any triangular lamina ABC has the angular point C fixed and is capable of free motion about it.   A blow is struck at B perpendicularly to the plane of the triangle.   Show that the initial axis of rotation is that trisector of the side AB which is furthest from B.

7.   A cone of mass m and vertical angle $2a$ can move freely about its axis, and has a fine smooth groove cut along its surface so as to make a constant angle β with the generating lines of the cone.   A heavy particle of mass P moves along the groove under the action of gravity, the system being initially at rest with the

* These examples are taken from Examination Papers which have been set in the University or in the Colleges.

particle at a distance c from the vertex. Show that, if θ be the angle through which the cone has turned when the particle is at any distance r from the vertex,

$$\frac{mk^2 + Pr^2 \sin^2 a}{mk^2 + Pc^2 \sin^2 a} = \epsilon^{2\theta \sin a \cdot \cot \beta},$$

k being the radius of gyration of the cone about its axis.

8. A body is turning about an axis through its centre of gravity when a point in it becomes suddenly fixed. If the new instantaneous axis be a principal axis with respect to the point, show that the locus of the point is a rectangular hyperbola.

9. A cube is rotating with angular velocity ω about a diagonal when one of its edges which does not meet the diagonal suddenly becomes fixed. Show that the angular velocity about this edge as axis $= \dfrac{\omega}{4\sqrt{3}}$.

10. Two masses m, m' are connected by a fine smooth string which passes round a right circular cylinder of radius a. The two particles are in motion in one plane under no impressed forces, show that, if A be the sum of the absolute areas swept out in a time t by the two unwrapped portions of the string,

$$\frac{d^2A}{dt^2} = \frac{1}{2} a \left(\frac{1}{m} + \frac{1}{m'} \right) T,$$

T being the tension of the string at any time.

11. A piece of wire in the form of a circle lies at rest with its plane in contact with a smooth horizontal table, when an insect on it suddenly starts walking along the arc with uniform relative velocity. Show that the wire revolves round its centre with uniform angular velocity, while that centre describes a circle in space with uniform angular velocity.

12. A uniform circular wire of radius a, movable about a fixed point in its circumference, lies on a smooth horizontal plane. An insect of mass equal to that of the wire crawls along it, starting from the extremity of the diameter opposite to the fixed point, its velocity relative to the wire being uniform and equal to V. Prove that after a time t the wire will have turned through an angle

$$\frac{Vt}{2a} - \frac{1}{\sqrt{3}} \tan^{-1} \left(\frac{1}{\sqrt{3}} \tan \frac{Vt}{2a} \right).$$

13. A small insect moves along a uniform bar, of mass equal to itself, and of length $2a$, the extremities of which are constrained to remain on the circumference of a fixed circle, whose radius is $\dfrac{2a}{\sqrt{3}}$. Supposing the insect to start from the middle point of the bar, and its velocity relatively to the bar to be uniform and equal to V; prove that the bar in time t will turn through an angle $\dfrac{1}{\sqrt{3}} \tan^{-1} \dfrac{Vt}{a}$.

14. A rough circular disc can revolve freely in a horizontal plane about a vertical axis through its centre. An equiangular spiral is traced on the disc, having the centre for pole. An insect whose mass is an nth that of the disc crawls along the curve, starting from the point at which it cuts the edge. Show that, when the insect reaches the centre, the disc has revolved through an angle $\dfrac{\tan a}{2} \log \left(1 + \dfrac{2}{n} \right)$, where a is the angle between the tangent and the radius vector at any point of the spiral.

15.   A uniform circular disc moveable about its centre in its own plane (which is horizontal) has a fine groove in it cut along a radius, and is set rotating with an angular velocity ω.   A small rocket whose weight is an nth of the weight of the disc is placed at the inner extremity of the groove and discharged; when it has left the groove the same is done with another equal rocket, and so on.   Find the angular velocity after n of these operations, and, if n be indefinitely increased, show that the limiting value of the same is ωe^{-2}.

16.   A rigid body is rotating about an axis through its centre of gravity, when a certain point of the body becomes suddenly fixed, the axis being simultaneously set free; find the equations of the new instantaneous axis; and prove that, if it be parallel to the originally fixed axis, the point must lie in the line represented by the equations $a^2 lx + b^2 my + c^2 nz = 0$, $(b^2 - c^2)\dfrac{x}{l} + (c^2 - a^2)\dfrac{y}{m} + (a^2 - b^2)\dfrac{z}{n} = 0$; where the principal axes through the centre of gravity are taken as axes of co-ordinates, a, b, c are the radii of gyration about these lines, and l, m, n the direction-cosines of the originally fixed axis referred to them.

17.   A solid body rotating with uniform velocity ω about a fixed axis contains a closed tubular channel of small uniform section, filled with an incompressible fluid in relative equilibrium; if the rotation of the solid body were suddenly destroyed the fluid would move in the tube with a velocity $\dfrac{2A\omega}{l}$, where A is the area of the projection of the axis of the tube on a plane perpendicular to the axis of rotation, and l is the length of the tube.

18.   A gate without a latch, in the form of a rectangular lamina, is fitted with a universal joint at the upper corner, and at the lower corner there is a short bar, normal to the plane of the gate and projecting equally on both sides of it.   As the gate swings to either side from its stable position of rest, one or other end of the bar becomes a fixed point.   If h be the height of the gate, $h \tan \alpha$ its length, and 2β the angle which the bar subtends at the upper corner, show that the angular velocity of the gate as it passes through the position of rest is impulsively diminished in the ratio $\dfrac{\sin^2\alpha - \tan^2\beta}{\sin^2\alpha + \tan^2\beta}$, and that the time between successive impacts when the oscillations become small decreases in the same ratio, the weights of the bar and joint being neglected.

CHAPTER VII.

The Force-function and Work.

332. **Time and space integrals.** If a particle of mass m is projected along the axis of x with an initial velocity V and is acted on by a force F in the same direction, the motion is given by the equation $m\dfrac{d^2x}{dt^2} = F$.

Integrating this with regard to t, if v be the velocity after a time t, we have

$$m(v - V) = \int_0^t F dt.$$

If we multiply both sides of the differential equation of the second order by $\dfrac{dx}{dt}$ and integrate, we get*

$$\frac{1}{2}m(v^2 - V^2) = \int_0^x F dx.$$

---

* It is seldom that Mathematicians can be found engaged in a controversy such as that which raged for forty years in the last century. The object of the dispute was to determine how the force of a body in motion was to be measured. Up to the year 1686, the measure taken was the product of the mass of the body and its velocity. Leibnitz, however, thought he perceived an error in the common opinion, and undertook to show that the proper measure should be the product of the mass and the square of the velocity. Shortly all Europe was divided between the rival theories. Germany took part with Leibnitz and Bernoulli; while England, true to the old measure, combated their arguments with great success. France was divided, an illustrious lady, the Marquise du Chatelet, being first a warm supporter and then an opponent of Leibnitzian opinions. Holland and Italy were in general favourable to the German philosopher. But what was most strange in this great dispute was, that the same problem, solved by geometers of opposite opinions, had the same solution. However the force was measured, whether by the first or by the second power of the velocity, the result was the same. The arguments and replies advanced on both sides are briefly given in Montucla's *History*, and are most interesting. For these however we have no space. The controversy was at last closed by D'Alembert, who showed in his treatise on Dynamics that the whole dispute was a mere question of words. When we speak, he says, of the force of a moving body, we either attach no clear meaning to the word or we understand

The first of these integrals shows that the change of the momentum is equal to the time-integral of the force. By applying similar reasoning to the motion of a dynamical system we have been led in the last chapter to the general principle enunciated in Art. 283, and afterwards to its application in determining the changes produced by very great forces acting for a very short time. The second integral shows that half the change of the vis viva is equal to the space-integral of the force. It is our object in this chapter to extend this result also, and to apply it to the general motion of a system of bodies.

333. **Vis viva.** For purposes of description it is convenient to give names to the two sides of this equation. Twice the left-hand side is usually called the vis viva of the particle, a term introduced by Leibnitz about the year 1695. Half the vis viva is also called the kinetic energy of the particle. Many names have been given to the right-hand side at various times. It is now commonly called the work of the force F. When the force does not act in the direction of the motion of its point of application the term "work" requires a more extended definition. This we shall discuss in the next article.

334. **Work.** Let a force F act at a point A of a body in the direction AB, and let us suppose the point A to move into any other position A' very near A. If ϕ be the angle made by the direction AB of the force with the direction AA' of the displacement of the point of application, then the product $F \cdot AA' \cdot \cos\phi$ is called the work done by the force. If for ϕ we write the angle made by the direction AB of the force with the direction $A'A$, opposite to the displacement, the product is called the work done against the force. If we drop a perpendicular $A'M$ on AB, the work done by the force is also equal to the product $F \cdot AM$, where AM is to be estimated as positive when in the direction of the force. If F' be the resolved part of F in the direction of the displacement, the work is also equal to $F' \cdot AA'$. If several forces act, we can in the same way find the work done by each. The sum of all these is the work done by the whole system of forces.

only the property that certain resistances can be overcome by the moving body. It is not then by any simple considerations of merely the mass and the velocity of the body that we must estimate this force, but by the nature of the obstacles overcome. The greater the resistance overcome, the greater we may say is the force; provided we do not understand by this word a pretended existence inherent in the body, but simply use it as an abridged mode of expressing a fact. D'Alembert then points out that there are different kinds of obstacles and examines how their different kinds of resistances may be used as measures. It will perhaps be sufficient to observe, that the resistance may in some cases be more conveniently measured by a space-integral and in others by a time-integral. See Montucla's *History*, Vol. III. and Whewell's *History*, Vol. II.

Thus defined, the work done by a force, corresponding to any indefinitely small displacement, is the same as the virtual moment of the force. In statics we are only concerned with the small hypothetical displacements given to the system in applying the principle of virtual work, and this definition is therefore sufficient. But in dynamics the bodies are in motion, and we must extend our definition of work to include the case of a displacement of any magnitude. When the points of application of the forces receive finite displacements we must divide the path of each into elements. The work done in each element may be found by the definition given above. The sum of all these is the whole work.

It should be noticed that the work done by given forces, as the body moves from one given position to another, is independent of the time of transit. As stated in Art. 332, the work is a space-integral and not a time-integral.

335. *If two systems of forces be equivalent, the work done by one in any small displacement is equal to that done by the other.* This follows at once from the principle of virtual work in statics. For if every force in one system be reversed in direction without altering its point of application or its magnitude, the two systems will be in equilibrium, and the sum of their virtual moments will therefore be zero. Restoring the system of forces to its original state, we see that the virtual moments of the two systems are equal. If the displacements are finite the same remark applies to each successive element of the displacement, and therefore to the whole displacement.

336. We may now find an analytical expression for the work done by a system of forces. Let (x, y, z) be the rectangular co-ordinates of a particle of the system and let the mass of this particle be m. Let (X, Y, Z) be the accelerating forces acting on the particle resolved parallel to the axes of co-ordinates. Then mX, mY, mZ are the dynamical measures of the acting forces. Let us suppose the particle to move into the position $x + dx$, $y + dy$, $z + dz$; then according to the definition the work done by the forces will be

$$\Sigma \left(mX dx + mY dy + mZ dz \right) \dots \dots \dots \dots (1),$$

the summation extending to all the forces of the system. If the bodies receive any finite displacements, the whole work will be

$$\Sigma m \int (X dx + Y dy + Z dz) \dots \dots \dots \dots (2),$$

the limits of the integral being determined by the extreme positions of the system.

337. Force-function. When the forces are such as generally occur in nature, it will be proved that the summation (1) of
the last Article is a complete differential, *i.e.* it can be integrated
independently of any relation between the co-ordinates x, y, z. The
summation (2) can therefore be expressed as a function of the co-
ordinates of the system. *When this is the case the indefinite integral
of the summation* (2) *is called the force-function.* This name was
given to the function by Sir W. R. Hamilton and Jacobi independently of each other.

If the force-function be called U, the work done by the forces
when the bodies move from one given position to another is the
definite integral $U_2 - U_1$, where U_1 and U_2 are the values of U
corresponding to the two given positions of the bodies. It follows
that the work is independent of the mode in which the system
moves from the first given position to the second. *In other words,
the work depends on the co-ordinates of the two given extreme
positions, and not on the co-ordinates of any intermediate position.*
When the forces are such as to possess this property, *i.e.* when
they possess a force-function, they have been called a *conservative
system of forces.* This name was given to the system by Sir W.
Thomson.

338. *There will be a force-function, firstly, when the external
forces tend to fixed centres at finite distances and are functions
of the distances from those centres; and, secondly, when the forces
due to the mutual attractions or repulsions of the particles of the
system are functions of the distances between the attracting or
repelling particles.*

Let $m\phi(r)$ be the action of any fixed centre of force on a
particle m distant r, estimated positive in the direction in which r
is measured, *i.e. from* the centre of force. Then the summation
(1) in Art. 336 is clearly $\Sigma m\phi(r)\,dr$. This is a complete differential. Thus the force-function exists and is equal to $\Sigma m \int \phi(r)dr$.

Let $mm'\phi(r)$ be the action between two particles m, m' whose
distance apart is r, and as before let this force be considered
positive when repulsive. Then the summation (1) becomes
$\Sigma mm'\phi(r)dr$. The force-function therefore exists, and is equal
to $\Sigma mm'\int \phi(r)dr$.

If the law of attraction be the inverse square of the distance,
$\phi(r) = -\dfrac{1}{r^2}$ and the integral is $\dfrac{1}{r}$. Thus the force-function differs
from the Potential by a constant quantity.

339. It is clear that there is nothing in the definition of the
force-function to compel us to use Cartesian co-ordinates. If

P, Q, &c. be forces acting on a particle, Pdp, Qdq, &c. their virtual moments, m the mass of the particle, then the force-function is

$$U = \Sigma m \int (Pdp + Qdq + \&\text{c.}),$$

the summation extending to all the forces of the system.

Ex. 1. If (ρ, ϕ, z) be the cylindrical or semi-polar co-ordinates of the particle m; P, Q, Z the resolved parts of the forces respectively along and perpendicular to ρ and along z, prove that $dU = \Sigma m (Pd\rho + Q\rho d\phi + Zdz)$.

Ex. 2. If (r, θ, ϕ) be the polar co-ordinates of the particle m; P, Q, R the resolved parts of the forces respectively along the radius vector, perpendicular to it in the plane of θ, and perpendicular to that plane, prove that

$$dU = \Sigma m (Pdr + Qrd\theta + Rr \sin \theta d\phi).$$

Ex. 3. If (x, y, z) be the oblique Cartesian co-ordinates of m; X, Y, Z the components along the axes, prove that

$$dU = \Sigma m \{ X (dx + \nu dy + \mu dz) + Y (\nu dx + dy + \lambda dz) + Z (\mu dx + \lambda dy + dz) \},$$

where (λ, μ, ν) are the cosines of the angles between the axes yz, zx, xy respectively. This result is due to Poinsot.

340. *If a system receive any small displacement* ds *parallel to a given straight line and an angular displacement* $d\theta$ *round the line, then the partial differential coefficients* $\dfrac{dU}{ds}$ *and* $\dfrac{dU}{d\theta}$ *represent respectively the resolved part of all the forces along the line and the moment of the forces about it.*

Since dU is the sum of the virtual moments of all the forces due to any displacement, it is independent of any particular co-ordinate axes. Let the straight line along which ds is measured be taken as the axis of z. Taking the same notation as before,

$$dU = \Sigma m (Xdx + Ydy + Zdz).$$

But $dx = 0$, $dy = 0$, and $dz = ds$, hence we have

$$dU = ds \cdot \Sigma mZ; \ \therefore \ \frac{dU}{ds} = \Sigma mZ.$$

Here dU means the change produced in U by the single displacement of the system, taken as one body, parallel to the given straight line, through a space ds.

Again, the moment of all the forces about the axis of z is $\Sigma m (xY - yX)$, but $dx = -yd\theta$, $dy = xd\theta$, and $dz = 0$. Hence the above moment

$$= \Sigma m \frac{Ydy + Xdx + Zdz}{d\theta} = \frac{dU}{d\theta} .$$

Here dU is the change produced in U by the single rotation of the system, taken as one body, round the given axis, through an angle $d\theta$.

341. As considerable use will be made of the force-function, the student will find it advantageous to acquire a facility in writing down its form. The following examples have therefore been chosen as likely to be most useful.

342. Work done by gravity. *A system of bodies falls under the action of gravity. If* M *be the whole mass,* h *the space descended by the centre of gravity of the whole system, the work done by gravity is* Mgh. See Art. 140.

Let the axis of z be vertical and let the positive direction be downwards. Then in the summation (1) of Art. 336, $X=0$, $Y=0$ and $Z=g$. Hence $dU=\Sigma mgdz$. If $\bar{z}$ be the depth of the centre of gravity below the plane of xy, and C be any constant, we find $U=Mg\bar{z}+C$. Taking this between limits we easily obtain the result given.

Units of work. The theoretical unit of work is the work done by a dynamical unit of force acting through a unit of space. We may use the result of this example to supply a practical unit. The work required to raise the centre of gravity of a given mass a given height at a given place may be taken as the unit of work. English engineers use a pound for the mass and a foot for the height, and the unit is then called a *foot-pound*. The term *Horse-power* is used to express the work done per unit of time. The unit of horse-power is usually taken to be 33000 foot-pounds per minute. The *duty* of a steam-engine is the actual work done by the consumption of a unit quantity, usually a bushel, of coal.

Ex. 1. A force communicates to a particle whose mass is equal to that of a cubic foot of water a velocity of one foot per minute. Find the work done in foot-pounds.

Ex. 2. Determine the resistance of a steamer in tons when 8000 effective horse-power is required to drive it at $17\frac{1}{4}$ knots (of 6080 feet) per hour. [Univ. of London, 1886.]

Ex. 3. Supposing a tricycle and rider weighing together 200 lbs. to run uniformly at 8 miles an hour down an incline of one in 100 against the resistances of the air and of the road, without working the pedals; prove that to go up an incline of one in 200 at the same speed the rider must be working at the rate of ·064 of a horse-power; and that the mean pressure on each pedal will then be about 12·672 lbs., supposing the cranks to be 5 inches long and to make 100 revolutions a minute. [Univ. of London, 1886.]

Ex. 4. Prove that the amount of work required to raise to the surface of the earth the homogeneous contents of a very small conical cavity, whose vertex is at the centre of the earth, is equal to that which would be expended in raising the whole mass of the contents through a space from the surface equal to one-fifth of the earth's radius, supposing the force of gravity to remain constant. [Coll. Exam.]

343. Work of an elastic string. Ex. If the length of an elastic string or rod *which is uniformly stretched* be altered, the *work done by the tension is the product of the compression of the length and the arithmetic mean of the initial and final tensions.*

Let the length be altered from r to r'. Let ρ be any length between these two, let l be the unstretched length, and let E be the constant of elasticity. The tension is $T = E\,\dfrac{\rho - l}{l}$ and acts opposite to the direction in which ρ is measured. The work done while ρ becomes $\rho + d\rho$ is therefore equal to $-T d\rho$. If we integrate this from $\rho = r$ to $\rho = r'$ we find that the work required is $-\dfrac{E}{2l}\{(r' - l)^2 - (r - l)^2\}$. This leads at once to the result given.

If a string becomes slack, the tension is supposed to vanish, and no work is done until the string again becomes tight. In applying the rule, the compression is the difference between the two terminal lengths if the string be tight in both, whether it has been slack or not during the various changes of length which may have occurred during the process. If the string be slack in either terminal state we must in calculating the compression suppose the string to have its unstretched length in that terminal state.

In the case of a rod the tension becomes negative when the rod is compressed, and the rule applies so long as the rod remains straight, and we can suppose Hooke's law to be true.

If the string is not straight but is uniformly stretched over a surface or in a fine tube, the same rule to find the work is still true. To prove this, we divide the string into elements, each of which may be considered as straight. When the whole string is now uniformly stretched the work done is the mean of the tensions into the sum of the contractions of all the elements. This last is clearly the contraction of the whole string.

If the surface be fixed the string cannot contract without one, at least, of the extremities moving, and in this case the work is done at that extremity.

If the surface move, and the extremities of the string be fixed in space, the work is transferred to the surface by means of the reactions. If the string have no effective forces, these reactions are in equilibrium with the tensions at the points A, B where the string leaves the surface. Now let the surface receive any small displacement. By the principle of virtual work the work done by the reactions on the surface is equal to that done by the two equal tensions at the points A, B. But this work is the instantaneous tension into the contraction of the string, i.e. it is $-T d\rho$. If the surface receive a finite displacement, the work done is the integral of this expression, and the rule is of course the same as before.

Whether the string have mass or not, we may consider each separate element of it as one of the moving bodies whose motion enters into the equation of vis viva. The work done by the contraction of all the elements is to be regarded as distributed over all the bodies. The work done by the equal and opposite reactions between the string and surface will then be zero.

344. Work of collecting a body. Ex. 1. If m, m' be the masses of two particles attracting each other with a force $\dfrac{mm'}{r^2}$ where r is the distance between them, show that *the work done by the mutual force when they have moved from an infinite distance apart to a distance* r *is* $\dfrac{mm'}{r}$. This follows from Art. 338. If the particles repel each other we regard either m or m' as negative.

Ex. 2. Let two finite masses M, M' attract each other and occupy given positions. Prove that the work of bringing the particles of one from infinite distances apart into their given positions under the attraction of the second, supposed fixed in its given position, is the same as that of bringing the particles of the second from infinity into their positions under the attraction of the first. Prove also that *this work may be found by taking both bodies in their final positions and multiplying the mass of each element of one body by the potential of the other at that element, then integrating throughout the volume of the former body.*

This integral is sometimes called the mutual work or the mutual potential of the two bodies.

Let there be two sets of attracting particles which we may represent by m_1, m_2, &c., m_1', m_2', &c., and let the particles of each set attract the particles of the other set, but not the particles of its own set. Suppose the particles m_1, m_2, &c. to occupy any given positions, and let one particle m' of the second set be brought from an infinite distance to any given position, say to a position at distances r_1, r_2, &c. from the particles m_1, m_2, &c. The work done is $m'\left(\dfrac{m_1}{r_1} + \dfrac{m_2}{r_2} + \&c.\right) = m'V$, where V is the potential of the attracting masses at the given position of m'.

Let us now bring in succession all the particles m_1', m_2', &c. from infinite distances to their given final positions under the attraction solely of the masses m_1, m_2, &c. The whole work is $\Sigma m'V$, which may also be written in the symmetrical form $\Sigma\dfrac{mm'}{r}$, where r is the distance between the particles m, m', and the Σ implies summation for every combination of each particle of one set with each particle of the other. This symmetrical form proves the first part of the proposition.

The particles may be elementary, and in that case we see that the work of collecting any mass M' into a given position under the attraction of a mass M placed in a given position is equal to $\int V dm'$, where V is the potential of the mass M at the final position of dm' and the integration extends over the whole mass of M'.

Ex. 3. If the particles composing any mass were separated from each other, work might be obtained from their mutual attractions by allowing the particles to approach each other. The work thus obtained is greatest when the particles are collected together from infinite distances. If dv be an element of volume of a solid mass attracting according to the law of nature, ρ the density of the element, V the potential of the solid mass at the element dv, prove that the *work performed in collecting the particles composing the mass from infinite distances is* $\dfrac{1}{2}\int V\rho dv$.

The problem of determining how much work can be obtained from the bodies forming the solar system by allowing them to

consolidate into a solid mass has been considered by several philo-
sophers. Sir W. Thomson has calculated that the potential energy
or the work which can be obtained from the existing solar system
is 38×10^{37} foot-pounds. *Edin. Trans.* 1854.

Let m_1, m_2, m_3, &c. be the masses of any particles, r_{12}, r_{23}, &c. the distances
between the masses m_1, m_2; m_2, m_3; &c. in any arrangement. Then, as before,
the work done in collecting them from infinite distances is $U = \dfrac{m_1 m_2}{r_{12}} + \dfrac{m_2 m_3}{r_{23}} + $ &c.,
which may be written $U = \Sigma \dfrac{mm'}{r}$.   Now if V_1 be the potential at the particle m_1 of
all the particles except m_1 in the given arrangement, $V_1 = \dfrac{m_2}{r_{12}} + \dfrac{m_3}{r_{13}} + \dots$ If V_2, V_3, &c.
have similar meanings we may write the work in the form

$$U = \frac{1}{2}(V_1 m_1 + V_2 m_2 + \dots) = \frac{1}{2}\Sigma V m.$$

In finding the potential of any solid mass at any point P we may omit the
matter within any indefinitely small element enclosing P if its density be finite.
For, since potential is "mass divided by distance," and the mass varies as the cube
of the linear dimensions, it follows that the potential of similar figures at points
similarly situated must vary as the square of the linear dimensions and must vanish
when the mass becomes elementary and the distance indefinitely small. In
applying, therefore, the form $U = \frac{1}{2}\Sigma V m$ to a solid body we may write ρdv for m, and
take V to be the potential of the whole mass at the element dv.

Ex. 4.   The particles composing a homogeneous sphere of mass M and radius r
were originally at infinite distances from each other.   Prove that the work done by
their mutual attractions is $\dfrac{3}{5}\dfrac{M^2}{r}$.

Ex. 5.   The particles of a homogeneous ellipsoid, whose mass is M and semiaxes
are a, b, c, are collected from infinite distances, show that the work done is

$$\frac{3}{10}M^2 \int_0^\infty \frac{d\lambda}{\sqrt{(a^2+\lambda)(b^2+\lambda)(c^2+\lambda)}}.$$

Ex. 6.   The work of collecting the particles of two masses
which are wholly external to each other from infinite distances is
the sum of the works of collecting each separately, plus their
mutual potential.

If one mass be wholly internal to the other, prove that the
work of collecting the difference is the sum of the works of collecting each separately, minus their mutual potential.

If the first proposition be not evident, let M, M' be the masses already collected,
and let us bring an additional particle from an infinite distance to the mass M.
The work on this particle is evidently that due to the attraction of M together with
that due to the attraction of M'. The first is an addition to the work of collecting
M, and the second is an addition to the mutual potential of M and M'.

From the first proposition we deduce by transposition that the work of collecting
M is equal to the work of collecting $(M + M')$ minus the work of collecting M' minus

the mutual potential of M and M'. Now the mutual potential of M and M' is equal to the mutual potential of $(M + M')$ and M' minus twice the work of collecting M'. The second proposition follows at once.

Ex. 7. A quantity of homogeneous matter is bounded by two spheres which do not intersect, one sphere being wholly within the other. The radii of the spheres are a and b, and the distance between the centres is c. Show that the work of collecting this matter from infinite distances is $\dfrac{(4\pi\rho)^2}{3} \left\{ \dfrac{a^5}{5} - \dfrac{a^2b^3}{2} + \dfrac{3b^5}{10} + \dfrac{b^3c^2}{6} \right\}$.

345. Work of a gaseous pressure. Ex. 1. An envelope of any shape, whose volume is v, contains gas at a uniform pressure p. Assuming that the pressure of the gas per unit of area is some function of the volume occupied by it, prove that the *work done by the pressures when the volume increases from* v = a *to* v = b *is* $\displaystyle\int_a^b p\,dv$.

Divide the surface into elementary areas each equal to $d\sigma$, then $p\,d\sigma$ is the pressure on $d\sigma$. When the volume has increased to $v + dv$, let any element $d\sigma$ take the position $d\sigma'$, and let dn be the length of the perpendicular drawn from the central point of $d\sigma'$ on the plane of $d\sigma$, then $p\,d\sigma\,dn$ is the work done by the pressure on $d\sigma$ and $p\int d\sigma\,dn$ is the work done over the whole area. But $d\sigma\,dn$ is the volume of the oblique cylinder whose base is $d\sigma$ and opposite face $d\sigma'$; so that $\int d\sigma\,dn$ is the whole increment of volume. The whole work done when the volume increases by dv is therefore $p\,dv$.

Ex. 2. A spherical envelope of radius a contains gas at pressure P. Assuming that the pressure of the gas per unit of area is inversely proportional to the volume occupied by it, prove that the work required to compress the envelope into a sphere of radius b is $4\pi a^3 P \log a/b$.

Ex. 3. An envelope of any shape contains gas and the shape is altered without altering the volume. Show that the work done over the whole surface is zero.

Ex. 4. A hollow cylinder contains equal masses of two different elastic fluids at the same pressure P separated by a piston without weight. Show that the work done in moving the piston till the densities of the two fluids are interchanged is $PA\,(a - b)\log a/b$, where A is the area of the piston, and a, b are the lengths of the portions of the cylinder occupied by the fluid. [Pembroke College, 1868.]

Ex. 5. A mass of air of uniform density $\rho\,(1 + s)$ is enclosed in an envelope and surrounded by air of atmospheric density ρ. If the mass expand until its density is equal to that of the atmosphere, prove that the work done is $k\left(\log(1 + s) - \dfrac{s}{1+s}\right)$ where k is the product of the pressure and the volume. If s be small the work is very nearly $\frac{1}{2}ks^2$.

346. Work of an Impulse. Ex. 1. An impulsive force acts on a body in a fixed direction in space. Show that, if F be the whole momentum communicated by the force, u_0, u_1 the velocities of the point of application, resolved in the direction of the force, just before and just after the impulse, *then the work done by* *the impulse is* $\dfrac{u_0 + u_1}{2}$ F. This result is given in Thomson and Tait's *Natural Philosophy.*

When a force is measured in the usual way by the momentum generated per unit of time, the work is measured by the product of the force into the resolved displacement. But impulses are not so measured, we cannot therefore directly apply this rule to find the work of an impulse.

Let us regard the impulse as the limit of a finite force acting in the fixed direction for a very short time T. Let the direction of the axis of x be taken parallel to the fixed direction and let X be the whole momentum communicated during a time t measured from the commencement of the impulse. Here t is any time less than T, and X varies from zero to F as t varies from 0 to T. Also, since X is the whole momentum up to the time t, $\dot{X}$ is the moving force on the body at the time t. Let u *be the resolved velocity of the point of application at the time* t, then u_0 and u_1 are the values of u when $t=0$ and $t=T$. Since udt is the space described in the time dt by the point of application of the force X, the work done in the time T is $\displaystyle\int_0^F udX$. To integrate this we must know what function u is of X.

If the body be a particle of mass m, we know that, when the time of action is very small, $m(u-u_0)=X$, hence, substituting for u, we find after integrating $u_0 F + \frac{1}{2}F^2/m$. When $X=F$ we have by definition $u=u_2$, $\therefore m(u_2-u_0)=F$. Eliminating F, we find the work is $\frac{1}{2}F(u_0+u_1)$.

If the body be moving in two dimensions, let $\bar{u}$ be the velocity of the centre of gravity at the time t resolved parallel to the direction of the impulse, and ω the angular velocity; we then have by Arts. 168 and 137

$$m(\bar{u}-u_0)=X, \quad mk^2\omega=Xp, \quad u=\bar{u}+\omega p.$$

Hence $u=u_0+LX$ where L is a quantity independent of X and therefore constant during the integration. Substituting for u, the integral takes the form $F(u_0+\frac{1}{2}LF)$. But as before $u_2=u_0+LF$. Eliminating L the result follows at once.

If the body be moving in three dimensions, the velocity u is known by Art. 313 to be a linear function of X; so that we may write $u=u_0+LX$, where L is a constant depending on the nature of the body. Substituting this value of u, we have the work equal to $\displaystyle\int_0^F (u_0+LX)\, dX = u_0 F + L\frac{F^2}{2}$. But $u_1=u_0+LF$. Eliminating L we find that the work $=\dfrac{1}{2}(u_0+u_1) F$.

Ex. 2. If one blow F_1 be followed immediately by a second blow F_2 at the same point in the same straight line, and if u_0, u_1, u_2 be the resolved velocities of the point of application before and after the blows, verify that the work $\frac{1}{2}(u_0+u_2)(F_1+F_2)$ of the whole blow is the sum of the works of the separate blows, viz. $\frac{1}{2}(u_0+u_1) F_1$ and $\frac{1}{2}(u_1+u_2) F_2$.

This follows at once, since $u_1=u_0+LF_1$ and $u_2=u_1+LF_2$. The results of Ex. 3 may be deduced from Ex. 1 in this manner.

Ex. 3. Find the work done by an impulse whose direction is not necessarily the same during the indefinitely short duration of the force.

Let X, Y, Z be the components of the whole momentum given to the body in any time t measured from the commencement of the impulse. Let u, v, w be the

resolved velocities of the point of application at the time t. Then, by the same
reasoning as before, the work done $= \int_0^T (\dot{X}u + \dot{Y}v + \dot{Z}w)\, dt$. But by Art. 314 when T
is indefinitely small $u = u_0 + \dfrac{dE}{dX}$, $\quad v = v_0 + \dfrac{dE}{dY}$, $\quad w = w_0 + \dfrac{dE}{dZ}$, where E is a known
quadratic function of (X, Y, Z) depending on the nature of the body. Substituting
we have

$$\text{the work} = u_0 X_1 + v_0 Y_1 + w_0 Z_1 + \int \left(\frac{dE}{dX} dX + \frac{dE}{dY}\, dY + \frac{dE}{dZ}\, dZ \right)$$
$$= u_0 X_1 + v_0 Y_1 + w_0 Z_1 + E_1,$$

where X_1, Y_1, Z_1, E_1 are the values of X, Y, Z, E when $t = T$.

We may eliminate the form of the body and express the work in terms of
the resolved velocities of the point of application just after the termination of the
impulse. Since E_1 is a homogeneous quadratic function of X_1, Y_1, Z_1, we have

$$2E_1 = \frac{dE_1}{dX_1} X_1 + \frac{dE_1}{dY_1} Y_1 + \frac{dE_1}{dZ_1} Z_1 = (u_1 - u_0) X_1 + (v_1 - v_0) Y_1 + (w_1 - w_0) Z.$$

Substituting we find the work $= \dfrac{u_0 + u_1}{2} X_1 + \dfrac{v_0 + v_1}{2} Y_1 + \dfrac{w_0 + w_1}{2} Z_1.$

347. Work of a membrane equally stretched in all directions. Consider
a rectangle whose sides are a and b, which may be considered as an element. Let
T be the *tension across any line referred as usual to a unit of length*. The tension
across the side a is Ta, and when the side b has increased to b' the work done by
these will be $Ta\,(b' - b)$. Supposing the tension across the side b' to be still T,
(which is true when the rectangle is an element) the tension across the whole
length will be Tb', and, when the side a becomes a', the work will be $Tb'\,(a' - a)$.

The whole work is therefore $T(a'b' - ab)$, i.e. *the work is the product of the
tension and the change of area.*

If the membrane is spherical, the area is $4\pi r^2$. The increase of area is therefore
$8\pi r\,dr$. Hence the work done by the tensions when the radius is increased from
$r = a$ to $r = b$ is $8\pi \displaystyle\int_a^b Tr\,dr.$

If the membrane be such that we may apply Hooke's law to the tension T,
we have $T = E\,\dfrac{r - a}{a}$, where a is the natural radius of the membrane and E is the
coefficient of elasticity. Substituting this value of T we find that the work done
by the tensions, when the radius increases from a to b, is $\dfrac{4}{3}\,\dfrac{E}{a}\,(b - a)^2 (2b + a)$.

If we assume that for a soap-bubble T is constant, we find that the work done
when the radius increases from a to b is $4\pi T\,(b^2 - a^2)$.

If we suppose the spherical membrane to be slowly stretched by filling it with
gas at a pressure p, we have by a theorem in hydrostatics $pr = 2T$. In this case the
work required has been shown to be $\int p\,dv$, and, since $v = \dfrac{4}{3}\pi r^3$, this leads to the same
result as before.

348. Work of a couple. Ex. A given couple is moved
in its own plane from one position to another; show that the
work is the product of its moment by the angle turned through.

Any displacement of a couple is equivalent to a rotation round one extremity of its arm and a transference of the whole couple parallel to itself. The work done by the two forces during the transference is clearly zero. We need therefore only consider the work done during the rotation.

Let F be the force, a the length of the arm, and let the couple be turned round one extremity A of its arm through an angle $d\theta$. The force at A does no work, and the work done by the other force is $F \cdot a d\theta$. Integrating this we have the work done by the couple when it turns through any finite angle.

349. Work of Bending a rod. Ex. 1. A rod originally straight is bent in one plane. If L be the stress couple at any point, ρ the radius of curvature, it is known both by experiment and by theory that $L = \dfrac{E}{\rho}$, where E is a constant depending on the nature of the material, and the form of a section of the rod. Assuming this, prove (1) when the *rod is bent into a given form*, so that ρ is a known function of s (whether the forces are known or not) *the work is* $\frac{1}{2}\displaystyle\int \dfrac{E}{\rho^2} ds$, (2) when the *rod is bent by known forces* so that L is a known function of s, (whether the form of the rod is known or not) *the work is* $\frac{1}{2}\displaystyle\int \dfrac{L^2}{E} ds$. The limits of integration are from one end of the rod to the other.

Let PQ be any element of the rod and let its length be ds. As PQ is being bent, let ψ be the indefinitely small angle between the tangents at its extremities, then the stress couple is $E\dfrac{\psi}{ds}$. As ψ increases from 0 to $\dfrac{ds}{\rho}$ the work done is $\dfrac{E}{ds}\displaystyle\int \psi d\psi$, which is the same as $\dfrac{E ds}{2\rho^2}$. The work done on the whole rod is therefore $\dfrac{1}{2}\displaystyle\int \dfrac{E}{\rho^2} ds$.

Ex. 2. A uniform heavy rod of length l and weight w is supported at its two extremities so as to be horizontal. Show the work done by gravity in bending it is $\dfrac{w^2 l^3}{240 E}$.

Ex. 3. A uniform light rod is supported at its extremities A and B, and supports a weight w at any point C. If $AC = a$, $BC = b$ and $l = a + b$, the work done by gravity in bending the rod is $\dfrac{w^2 a^2 b^2}{6El}$.

Conservation of Vis Viva and Energy.

350. DEF. The *Vis Viva* of a particle is the product of its mass and the square of its velocity.

The principle of vis viva. *If a system be in motion under the action of finite forces, and if the geometrical relations of the parts of the system be expressed by equations which do not contain the time explicitly, the change in the vis viva of the system in* *passing from any one position to any other is equal to twice the corresponding work done by the forces.*

In determining the force-function all forces may be omitted which do not appear in the equation of virtual work.

Let x, y, z be the co-ordinates of any particle m, and let X, Y, Z be the resolved parts in the directions of the axes of the impressed accelerating forces acting on the particle.

The effective forces acting on the particle m at any time t are

$$m\frac{d^2x}{dt^2}, \quad m\frac{d^2y}{dt^2}, \quad m\frac{d^2z}{dt^2}.$$

If the effective forces on all the particles be reversed, they will be in equilibrium with the whole group of impressed forces, by Art. 67. Hence, by the principle of virtual work,

$$\Sigma m\left\{\left(X - \frac{d^2x}{dt^2}\right)\delta x + \left(Y - \frac{d^2y}{dt^2}\right)\delta y + \left(Z - \frac{d^2z}{dt^2}\right)\delta z\right\} = 0,$$

where δx, δy, δz are any small arbitrary displacements of the particle m consistent with the geometrical relations at the time t.

Now if the geometrical relations are expressed by equations which do not contain the time explicitly, the geometrical relations which hold at the time t will hold throughout the time δt; and, therefore, we can take the *arbitrary* displacements δx, δy, δz to be respectively equal to the *actual* displacements $\frac{dx}{dt}\delta t$, $\frac{dy}{dt}\delta t$, $\frac{dz}{dt}\delta t$, of the particle in the time δt.

Making this substitution, the equation becomes

$$\Sigma m\left(\frac{d^2x}{dt^2}\frac{dx}{dt} + \frac{d^2y}{dt^2}\frac{dy}{dt} + \frac{d^2z}{dt^2}\frac{dz}{dt}\right) = \Sigma m\left(X\frac{dx}{dt} + Y\frac{dy}{dt} + Z\frac{dz}{dt}\right).$$

Integrating, we get

$$\Sigma m\left\{\left(\frac{dx}{dt}\right)^2 + \left(\frac{dy}{dt}\right)^2 + \left(\frac{dz}{dt}\right)^2\right\} = C + 2\Sigma m\int(Xdx + Ydy + Zdz),$$

where C is a constant to be determined by the initial conditions of motion.

Let v and v' be the velocities of the particle m at the times t and t'. Also let U_1, U_2 be the values of the force-function for the system in the two positions which it has at the times t and t'. Then

$$\Sigma mv'^2 - \Sigma mv^2 = 2(U_2 - U_1).$$

351. The following illustration, taken from Poisson, may show more clearly why it is necessary that the geometrical relations

should not contain the time explicitly.   Let, for example,

$$\phi\,(x,\,y,\,z,\,t) = 0\ldots\ldots\ldots\ldots\ldots\ldots\ldots\ldots(1)$$

be any geometrical relation connecting the co-ordinates of the particle m. This may be regarded as the equation to a moving surface on which the particle is constrained to rest. The quantities $\delta x,\,\delta y,\,\delta z$ are the projections on the axes of any arbitrary displacement of the particle m consistent with the geometrical relations which hold at the time t. They must therefore satisfy the equation

$$\frac{d\phi}{dx}\,\delta x + \frac{d\phi}{dy}\,\delta y + \frac{d\phi}{dz}\,\delta z = 0.$$

The quantities $\dfrac{dx}{dt}\,\delta t,\ \dfrac{dy}{dt}\,\delta t,\ \dfrac{dz}{dt}\,\delta t$ are the projections on the axes of the displacement of the particle due to its motion in the time δt. They must therefore satisfy the equation

$$\frac{d\phi}{dx}\,\frac{dx}{dt}\,\delta t + \frac{d\phi}{dy}\,\frac{dy}{dt}\,\delta t + \frac{d\phi}{dz}\,\frac{dz}{dt}\,\delta t + \frac{d\phi}{dt}\,\delta t = 0.$$

Hence, unless $\dfrac{d\phi}{dt}$ is zero throughout the whole motion, we cannot take $\delta x,\,\delta y,\,\delta z$ to be respectively equal to $\dfrac{dx}{dt}\,\delta t,\ \dfrac{dy}{dt}\,\delta t,\ \dfrac{dz}{dt}\,\delta t$. The equation $\dfrac{d\phi}{dt} = 0$ expresses the condition that the geometrical equation (1) should not contain the time explicitly.

352.   The great advantage of this principle is that it gives at once a relation between the velocities of the bodies considered and the variables or co-ordinates which determine their positions in space, so that when, from the nature of the problem, the positions of all the bodies may be made to depend on one variable, the equation of vis viva is sufficient to determine the motion. In general the principle of vis viva will give a first integral of the equations of motion of the second order. If, at the same time, some of the other principles enunciated in Art. 282 can be applied to the bodies under consideration, so that the whole number of equations thus obtained is equal to the number of independent co-ordinates of the system, it becomes unnecessary to write down any equations of motion of the second order. See Art. 143.

The principle of vis viva was first used by Huyghens in his determination of the centre of oscillation of a body, but in a form different from that now used. *See the note to page* 73. The principle was extended by John Bernoulli and applied by his son, Daniel Bernoulli, to the solution of a great variety of problems, such as the motion of fluids in vases, and the motion of rigid bodies under certain given conditions. See Montucla, *Histoire des Mathématiques*, Tome III.

353. Initial motion. Suppose the system to begin to move from rest under the action of the forces X, Y, Z &c. After a time dt the vis viva is given by

$$\Sigma mv'^2 = 2\Sigma m (Xdx + Ydy + Zdz).$$

The left-hand side of this equation is necessarily positive. We therefore infer that if a system start from rest, the initial motion must be such that the virtual work of the forces for that motion must be positive.

There may be several different ways (geometrically considered) in which the system could begin to move from its initial state of rest. Let the system be compelled to take any one of these ways of motion by obliging a sufficient number of its points to describe certain smooth curves, or by introducing any forces which have no virtual work for that particular mode of displacement. The system can now move only in one way, or as we often express it, the system has only one path open. There are two directions in which it can travel along this path. The question arises—in which direction will it begin to move? Since the virtual work of the forces is in general positive for one of these directions and negative for the other, the system must begin to move along the former.

354. Examples of the principle. If a system be under the action of no external forces, we have $X = 0$, $Y = 0$, $Z = 0$, and hence the vis viva of the system is constant.

If, however, the mutual reactions between the particles of the system are such as do appear in the equation of virtual work, then the vis viva of the system will not be constant. Thus, even if the solar system were not acted on by any external forces, its vis viva would not be constant. For the mutual attractions between the several planets are reactions between particles whose distances do not remain the same, and hence the sum of the virtual works is not zero.

Again, if the earth be regarded as a body rotating about an axis and in course of time slowly contracting from loss of heat, the vis viva will not be constant, for the same reason as before. The increase of angular velocity produced by this contraction can be easily found by the principle of angular momentum. See Art. 299.

355. Let gravity be the only force acting on the system. Let the axis of z be vertical, then we have $X = 0$, $Y = 0$, $Z = -g$. Hence the equation of vis viva becomes

$$\Sigma mv'^2 - \Sigma mv^2 = -2Mg (z' - z).$$

Thus the vis viva of the system depends only on the altitude of the centre of gravity. If any horizontal plane be drawn, the vis viva of the system is the same whenever the centre of gravity passes through the plane. See Art. 142.

356. Ex. If a system in motion pass through a position of equilibrium, *i.e.* a position in which, if placed at rest, it would remain in equilibrium under the action of the forces, prove that the vis viva of the system is either a maximum or a minimum. De Courtivron's Theorem, *Mém. de l'Acad.* 1748 and 1749.

357. The equation of virtual work in statics is known to contain in one formula all the conditions of equilibrium. In the same way the general equation

$$\Sigma m \left(\frac{d^2x}{dt^2} \delta x + \frac{d^2y}{dt^2} \delta y + \frac{d^2z}{dt^2} \delta z\right) = \Sigma m (X\delta x + Y\delta y + Z\delta z),$$

may be made to give all the equations of motions by properly choosing the arbitrary displacements δx, δy, δz. In Article 350 we made one choice of these displacements and thus obtained an equation in an integrable form.

If we give the whole system a displacement parallel to the axis of z we have $\delta x = 0$, $\delta y = 0$, and δz is arbitrary. The equation then becomes $\Sigma m \dfrac{d^2 z}{dt^2} = \Sigma mZ$, which represents any one of the three first general equations of motion in Art. 72.

If we give the whole system a displacement round the axis of z through an angle $\delta\theta$, we have $\delta x = -y\delta\theta$, $\delta y = x\delta\theta$, $\delta z = 0$. The equation then becomes $\Sigma m \left(x \dfrac{d^2 y}{dt^2} - y \dfrac{d^2 x}{dt^2} \right) = \Sigma m\, (xY - yX)$, which represents any one of the last three general equations of motion in Art. 72.

358. **Potential and kinetic energy***. Suppose a weight mg to be placed at any height h above the surface of the earth. As it falls through a height z, the force of gravity does work which is measured by mgz. The weight acquires a velocity v, half of its vis viva is $\frac{1}{2}mv^2$, which is known to be equal to mgz. If the weight fall through the remainder of the height h, gravity may be made to do more work, measured by $mg\,(h - z)$. When the weight has reached the ground, it has fallen as far as the circumstances of the case permit, and no more work can be done by gravity until the weight has been lifted up again. Throughout the motion we see that, when the weight has descended any space z, half its vis viva, together with the work that can be done during the rest of the descent, is independent of z and equal to the work done by gravity during the whole descent h.

If we complicate the motion by making the weight work some machine during its descent, the same theorem is still true. By the principle of vis viva, proved in Art. 350, half the vis viva of the particle, when it has descended any space z, is equal to the work mgz which has been done by gravity during this descent, diminished by the work done on the machine. Hence, as before, half the vis viva together with the difference between the work done by gravity and that done on the machine during the remainder of the descent is constant and equal to the excess of the work done by gravity over that done on the machine during the whole descent.

---

* Coriolis, Helmholtz and others have suggested that it would be more con-venient if the *vis viva* were defined to be half the sum of the products of the masses into the squares of the velocities. See *Phil. Trans.* 1854, p. 89. But this change in the meaning of a term so widely established in Europe would be very likely to cause some confusion. It seems better for the present to use another name, such as *kinetic energy*.

Let us now extend this principle to the general case of a system of bodies acted on by any conservative system of forces.

359. Let us select some position of a moving system of bodies as a position of reference. This may be an actual final position passed through by the system in its motion, or any position which it may be convenient to choose, into which the system could be moved. Suppose the system to start from some position which we may call A, and at the time t, to occupy some position P. Then at the time t, half the vis viva generated is equal to the work done from A to P. Hence half the vis viva at P together with the work which can be done from P to the position of reference is constant for all positions of P.

To express this, the word *energy* has been used. Half the vis viva is called the *kinetic energy* of the system. The work which the forces can do as the system is moved from its existing position to the position of reference is called the *potential energy* of the system. The sum of the kinetic and potential energies is called the energy of the system. The principle of the conservation of energy may be thus enunciated:—

When a system moves under any conservative forces, the sum of the kinetic and potential energies is constant throughout the motion.

360. The distinction between work done and potential energy may be analytically stated thus. The force-function has been defined in Art. 337 to be the indefinite integral of the virtual work of the forces. As the system moves the work done is the definite integral taken with its lower limit determined by some standard position of reference, which we may call C, and its upper limit determined by the instantaneous position of the system. The potential energy is the definite integral taken with its upper limit determined by some fixed position of reference which we may call D, and its lower limit determined by the instantaneous position of the system. If the two fixed positions of reference which we have distinguished by the letters C and D are identical, the work integral is the same as the potential integral with its sign changed. But this is not generally the case; the positions of reference are chosen each to suit the particular integral in connection with which it is used.

361. **Examples of Potential Energy.** Ex. 1. *A particle describes an ellipse freely about a centre of force in its centre. Find the whole energy of its motion.*

Let m be the mass of the particle, r its distance at any time from the centre, μr the accelerating force on the particle. If coincidence of the particle with the centre of force be taken as the position of reference, the potential energy by Art. 360 is $\int (-m\mu r)\, dr = \frac{1}{2}m\mu r^2$ when taken between the limits $r=r$ to $r=0$. If r' be the semi-conjugate of r, the velocity of the particle is $r' \sqrt{\mu}$ and the kinetic energy is therefore $\frac{1}{2} m \mu r'^2$. As the particle describes its ellipse round the centre of force, the sum of the potential and kinetic energies is equal to $\frac{1}{2} m \mu (a^2 + b^2)$ where a and b are the semi-axes of the ellipse.

Ex. 2. A particle describes an ellipse freely about a centre of force in the centre. Show that the mean kinetic energy during a complete revolution is equal to the mean potential energy; the means being taken with regard to time.

Ex. 3. If in the last example the means be taken with regard to the angle described round the centre, the difference of the means is $\frac{1}{2} m \mu (a - b)^2$.

Ex. 4. A mass M of fluid is running round a circular channel of radius a with velocity u, another equal mass of fluid is running round a channel of radius b with velocity v, the radius of one channel is made to increase and the other to decrease until each has the original value of the other, show that the work required to produce the change is $\frac{1}{2} \left(\dfrac{v^2}{a^2} - \dfrac{u^2}{b^2} \right) (b^2 - a^2) M$. [Math. Tripos, 1866.]

362. **List of Forces to be omitted.** In applying the principle of vis viva to any actual cases, it is important to know beforehand what forces and internal reactions may be disregarded in forming the equation. The general rule is that all forces may be neglected which do not appear in the equation of virtual work. These forces may be enumerated as follows :

A. Those reactions whose virtual displacements are zero.

1. Any force whose line of action passes through an instantaneous axis; as *rolling friction*, but not sliding friction or the resistance of any medium.

2. Any force whose line of action is perpendicular to the direction of motion of the point of application ; as the reaction of a *smooth fixed surface*, but not that of a moving surface.

B. Those reactions whose virtual displacements are not zero and which therefore would enter into the equation, but disappear when joined to other reactions.

1. The reaction between two particles whose distance apart remains the same; as the tension of an *inextensible string*, but not that of an elastic string.

2. The reaction between two rigid bodies, parts of the same system, which roll on each other. It is necessary however to include both these bodies in the *same* equation of vis viva.

C. All tensions which act along inextensible strings, even though the strings are bent by passing through smooth fixed rings.

For let a string whose tension is T connect the particles m, m', and pass through a ring distant respectively r, r' from the particles. The virtual work is clearly $- T\delta r - T\delta r'$, because the tension acts along the string. But, since the string is inextensible, $\delta r + \delta r' = 0$; therefore the virtual work is zero.

363. **Expressions for the vis viva of a rigid body in motion.** *If a body move in any manner its vis viva at any instant is equal to the vis viva of the whole mass collected at its centre of gravity, together with the vis viva due to motion round the centre of gravity considered as a fixed point : or*

the vis viva of a body = vis viva due to translation

+ vis viva due to rotation.

Let x, y, z be the co-ordinates of a particle whose mass is m and velocity v, and let $\bar{x}$, $\bar{y}$, $\bar{z}$ be the co-ordinates of the centre of gravity G of the body. Let $x = \bar{x} + \xi$, $y = \bar{y} + \eta$, $z = \bar{z} + \zeta$. Then, by a property of the centre of gravity, $\Sigma m\xi = 0$, $\Sigma m\eta = 0$, $\Sigma m\zeta = 0$. Hence $\Sigma m\dfrac{d\xi}{dt} = 0$, $\Sigma m\dfrac{d\eta}{dt} = 0$, $\Sigma m\dfrac{d\zeta}{dt} = 0$. Now the vis viva of a body is

$$\Sigma m v^2 = \Sigma m \left\{ \left(\frac{dx}{dt}\right)^2 + \left(\frac{dy}{dt}\right)^2 + \left(\frac{dz}{dt}\right)^2 \right\}.$$

Substituting for x, y, z, this becomes

$$\Sigma m \left\{ \left(\frac{d\bar{x}}{dt}\right)^2 + \left(\frac{d\bar{y}}{dt}\right)^2 + \left(\frac{d\bar{z}}{dt}\right)^2 \right\} + \Sigma m \left\{ \left(\frac{d\xi}{dt}\right)^2 + \left(\frac{d\eta}{dt}\right)^2 + \left(\frac{d\zeta}{dt}\right)^2 \right\}$$

$$+ 2 \frac{d\bar{x}}{dt} \Sigma m \frac{d\xi}{dt} + 2 \frac{d\bar{y}}{dt} \Sigma m \frac{d\eta}{dt} + 2 \frac{d\bar{z}}{dt} \Sigma m \frac{d\zeta}{dt}.$$

All the terms in the last line vanish, as they should, by Art. 14. The first term in the first line is the vis viva of the whole mass Σm, collected at the centre of gravity. The second term is the vis viva due to rotation round the centre of gravity.

This expression for the vis viva may be put into a more convenient shape.

364.   Firstly.   *Let the motion be in two dimensions.* See Art. 139. Let $\bar{v}$ be the velocity of the centre of gravity, $\bar{r}$, θ its polar co-ordinates referred to any origin in the plane of motion. Let r_1 be the distance from the centre of gravity of any particle whose mass is m, and let v_1 be its velocity relatively to the centre of gravity. Let ω be the angular velocity of the whole body about the centre of gravity, and Mk^2 its moment of inertia about the same point.

The vis viva of the whole mass collected at G is $M\bar{v}^2$, which may be put into either of the forms

$$M\bar{v}^2 = M \left\{ \left(\frac{d\bar{x}}{dt}\right)^2 + \left(\frac{d\bar{y}}{dt}\right)^2 \right\} = M \left\{ \left(\frac{d\bar{r}}{dt}\right)^2 + \bar{r}^2 \left(\frac{d\bar{\theta}}{dt}\right)^2 \right\}.$$

The vis viva about G is $\Sigma m v_1^2$. But since the body is turning about G, we have $v_1 = r_1\omega$. Hence $\Sigma m v_1^2 = \omega^2 . \Sigma m r_1^2 = \omega^2 . Mk^2$.

The whole vis viva of the body is therefore

$$\Sigma m v^2 = M\bar{v}^2 + Mk^2\omega^2.$$

If the body be turning about an instantaneous axis, whose distance from the centre of gravity is r, we have $\bar{v} = r\omega$. Hence

$$\Sigma mv^2 = M\omega^2 (r^2 + k^2) = Mk'^2\omega^2,$$

where Mk'^2 is the moment of inertia about the instantaneous axis.

Secondly. *Let the body be in motion in space of three dimensions.*

Let $\bar{v}$ be the velocity of G; $\bar{r}$, $\bar{\theta}$, $\bar{\phi}$ its polar co-ordinates referred to any origin. Let ω_x, ω_y, ω_z be the angular velocities of the body about any three axes at right angles meeting in G, and let A, B, C be the moments of inertia of the body about the axes. Let ξ, η, ζ be the co-ordinates of a particle m referred to these axes.

The vis viva of the whole mass collected at G is $M\bar{v}^2$, which may be put equal to

$$M\left\{\left(\frac{d\bar{x}}{dt}\right)^2 + \left(\frac{d\bar{y}}{dt}\right)^2 + \left(\frac{d\bar{z}}{dt}\right)^2\right\} \text{ or } M\left\{\left(\frac{d\bar{r}}{dt}\right)^2 + \bar{r}^2\sin^2\theta\left(\frac{d\bar{\phi}}{dt}\right)^2 + \bar{r}^2\left(\frac{d\bar{\theta}}{dt}\right)^2\right\},$$

according as we wish to use Cartesian or polar co-ordinates.

The vis viva due to the motion about G is

$$\Sigma mv_1^2 = \Sigma m \left\{\left(\frac{d\xi}{dt}\right)^2 + \left(\frac{d\eta}{dt}\right)^2 + \left(\frac{d\zeta}{dt}\right)^2\right\}.$$

But $\dfrac{d\xi}{dt} = \omega_y\zeta - \omega_z\eta, \quad \dfrac{d\eta}{dt} = \omega_z\xi - \omega_x\zeta, \quad \dfrac{d\zeta}{dt} = \omega_x\eta - \omega_y\xi.$

Substituting these values, we get, since $A = \Sigma m (\eta^2 + \zeta^2)$, $B = \Sigma m (\zeta^2 + \xi^2)$, $C = \Sigma m (\xi^2 + \eta^2)$,

$$\Sigma mv_1^2 = A\omega_x^2 + B\omega_y^2 + C\omega_z^2$$
$$- 2(\Sigma m\xi\eta)\,\omega_x\omega_y - 2(\Sigma m\eta\zeta)\,\omega_y\omega_z - 2(\Sigma m\zeta\xi)\,\omega_z\omega_x.$$

We may find the vis viva of the motion about G in another manner. Let Ω be the angular velocity about the instantaneous axis, I the moment of inertia about it. The vis viva is then clearly $I\Omega^2$. Now I is found in Art. 15, and in our case $\omega_1 = \Omega\alpha$, $\omega_2 = \Omega\beta$, $\omega_3 = \Omega\gamma$, following the notation of that article. Eliminating α, β, γ we get the same result as before.

If the axes of co-ordinates be the principal axes at G, this reduces to

$$\Sigma mv_1^2 = A\omega_x^2 + B\omega_y^2 + C\omega_z^2.$$

If the body be turning about a point O, whose position is fixed for the moment, the vis viva may be proved in the same way to be

$$\Sigma mv^2 = A'\omega_x^2 + B'\omega_y^2 + C'\omega_z^2,$$

where A', B', C' are the principal moments of inertia at the point

O, and ω_x, ω_y, ω_z are the angular velocities of the body about the principal axes at O.

365. **Examples of vis viva.** Ex. 1. A rigid body of mass M is moving in space in any manner, and its position is determined by the co-ordinates of its centre of gravity and the angles θ, ϕ, ψ which the principal axes at the centre of gravity make with some fixed axes, in the manner explained in Art. 256. Show that its vis viva is given by

$$2T = M (\dot{x}^2 + \dot{y}^2 + \dot{z}^2) + C (\dot{\phi} + \dot{\psi} \cos \theta)^2 + (A \sin^2 \phi + B \cos^2 \phi) \, \dot{\theta}^2$$
$$+ \sin^2 \theta \, (A \cos^2 \phi + B \sin^2 \phi) \, \dot{\psi}^2 + 2 \, (B - A) \sin \theta \sin \phi \cos \phi \, \dot{\theta} \dot{\psi}.$$

Show also that, when two of the principal moments A and B are equal, this expression takes the simpler form

$$2T = M (\dot{x}^2 + \dot{y}^2 + \dot{z}^2) + C (\dot{\phi} + \dot{\psi} \cos \theta)^2 + A \, (\dot{\theta}^2 + \sin^2 \theta \dot{\psi}^2).$$

This result will be often found useful.

Ex. 2. A body moving freely about a fixed point is expanding under the influence of heat, so that in structure and form it remains always similar to itself. If the law of expansion be that the distance between any two particles at the temperature θ is equal to their distance at temperature zero multiplied by $f(\theta)$, show that the vis viva of the body $= A\omega_x^2 + B\omega_y^2 + C\omega_z^2 + \frac{1}{2} (A + B + C) \left(\dfrac{d \log f(\theta)}{dt} \right)^2$, where A, B, C are the principal moments at the fixed point.

Ex. 3. A body is moving about a fixed point and its vis viva is given by the equation

$$2T = A\omega_x^2 + B\omega_y^2 + C\omega_z^2 - 2D\omega_y\omega_z - 2E\omega_z\omega_x - 2F\omega_x\omega_y.$$

Show that the angular momenta about the axes are $\dfrac{dT}{d\omega_x}$, $\dfrac{dT}{d\omega_y}$, $\dfrac{dT}{d\omega_z}$.

Let the body be moving freely and let $2T_0$ be the vis viva of translation. Prove that, if x, y, z be the co-ordinates of the centre of gravity referred to any rectangular axes fixed or moving about a fixed point, and if accents denote differential coefficients with regard to the time, the linear momenta parallel to the axes will be

$$\frac{dT_0}{dx'}, \quad \frac{dT_0}{dy'}, \quad \frac{dT_0}{dz'}.$$

Thus the vis viva, like the force-function, is a scalar function whose differential coefficients are the components of vectors. See Arts. 262 and 340. In the case of the semi-vis viva, these are the resultant linear momentum and the resultant angular momentum round the centre of gravity.

366. **Problems on the Principle of vis viva.** Ex. 1. *A circular wire can turn freely about a vertical diameter as a fixed axis, and a bead can slide freely along it under the action of gravity. The whole system being set in rotation about the vertical axis, find the subsequent motion.*

Let M and m be the masses of the wire and bead, ω their common angular velocity about the vertical. Let a be the radius of the wire, Mk^2 its moment of inertia about the diameter. Let the centre of the wire be the origin, and let the axis of y be measured vertically downwards. Let θ be the angle which the axis of y makes with the radius drawn from the centre of the wire to the bead.

It is evident, since gravity acts vertically and since all the reactions at the fixed axis must pass through the axis, that the moment of all the forces about the vertical diameter is zero.   Hence, taking moments about the vertical, we have

$$Mk^2\omega + ma^2\omega \sin^2 \theta = h.$$

And by the principle of vis viva,

$$Mk^2\omega^2 + m\{a^2\dot\theta^2 + a^2 \sin^2 \theta\omega^2\} = C + 2mga \cos \theta.$$

These two equations will suffice for the determination of $\dot\theta$ and ω.   Solving them, we get $$\frac{h^2}{Mk^2 + ma^2 \sin^2 \theta} + ma^2 \left(\frac{d\theta}{dt}\right)^2 = C + 2mga \cos \theta.$$

This equation cannot be integrated, and hence θ cannot be found in terms of t. To determine the constants h and C we must recur to the initial conditions of motion.   Supposing that initially $\theta = \pi$, and $\dot\theta = 0$ and $\omega = a$, then $h = Mk^2a$ and $C = 2mga + Mk^2a^2$.   See Art. 352.

Ex. 2.   A lamina of any form rolls on a perfectly rough straight line under the action of no forces; prove that the velocity v of the centre of gravity G is given by $v^2 = c^2 \dfrac{r^2}{r^2 + k^2}$, where r is the distance of G from the point of contact, k the radius of gyration of the lamina about an axis through G perpendicular to its plane, and c some constant.

Ex. 3.   Two equal beams connected by a hinge at their centres of gravity so as to form an X are placed symmetrically on two smooth pegs in the same horizontal line, the distance between which is b.   Show that, if the beams be perpendicular to each other at the commencement of the motion, the velocity of their centre of gravity, when in the line joining the pegs, is equal to $\sqrt{\dfrac{b^3 g}{b^2 + 4k^2}}$, where k is the radius of gyration of either beam about a line perpendicular to it through its centre of gravity.

Ex. 4.   A uniform rod is moving on a horizontal table about one extremity, and driving before it a particle of mass equal to its own, which starts from rest indefinitely near to the fixed extremity; show that, when the particle has described a distance r along the rod, its direction of motion makes with the rod an angle $\tan^{-1} \dfrac{k}{\sqrt{r^2 + k^2}}$.   [Christ's Coll.]

Ex. 5.   A thin uniform smooth tube is balancing horizontally about its middle point, which is fixed; a uniform rod such as just to fit the base of the tube is placed end to end in a line with the tube, and then shot into it with such a horizontal velocity that its middle point shall only just reach that of the tube; supposing the velocity of projection to be known, find the angular velocity of the tube and rod at the moment of the coincidence of their middle points.   [Math. Tripos.]

Result.   If m be the mass of the rod, m' that of the tube, and $2a$, $2a'$ their respective lengths, v the velocity of the rod's projection, ω the required angular velocity, then $\omega^2 = \dfrac{3mv^2}{ma^2 + m'a'^2}$.

Ex. 6.   If an elastic string, whose natural length is that of a uniform rod, be attached to the rod at both ends and suspended by the middle point, prove by means of vis viva that the rod will sink until the strings are inclined to the horizon at an angle θ, which satisfies the equation $\cot^3 \dfrac{\theta}{2} - \cot \dfrac{\theta}{2} - 2n = 0$, where the tension of the string, when stretched to double its length, is n times the weight. [Math. Tripos.]

Ex. 7. The centre C of a circular wheel is fixed and the rim is constrained to roll in a uniform manner on a perfectly rough horizontal plane so that the plane of the wheel makes a constant angle a with the vertical. Round the circumference there is a uniform smooth canal of very small section, and a heavy particle which just fits the canal can slide freely along it under the action of gravity. If m be the particle, B the point where the wheel touches the plane, and $\theta = \angle BCm$, and if n be the angular rate at which B describes the circular trace on the horizontal plane, prove that $\left(\dfrac{d\theta}{dt}\right)^2 = \dfrac{2g}{a}\cos a \cos \theta - n^2 \cos^2 a \cos^2 \theta + \text{const.}$, where a is the radius of the wheel. *Annales de Gergonne*, Tome XIX.

Ex. 8. A regular homogeneous prism, whose normal section is a regular polygon of n sides, the radius of the circumscribing circle being a, rolls down a perfectly rough inclined plane whose inclination to the horizon is a. If ω_n be the angular velocity just before the nth edge becomes the instantaneous axis, then

$$\omega_n{}^2 - \frac{g \sin a}{a \sin \dfrac{\pi}{n}} \frac{8 + \cos \dfrac{2\pi}{n}}{5 + 4\cos \dfrac{2\pi}{n}} = \left(\frac{2 + 7\cos \dfrac{2\pi}{n}}{8 + \cos \dfrac{2\pi}{n}}\right)^2 \left(\omega^2{}_{n-1} - \frac{g \sin a}{a \sin \dfrac{\pi}{n}} \frac{8 + \cos \dfrac{2\pi}{n}}{5 + 4\cos \dfrac{2\pi}{n}}\right).$$

The Principle of Similitude.

367. What are the conditions necessary that two systems of particles which are initially geometrically similar should also be mechanically similar, *i.e.* that the relative positions of the particles in one system after a time t should always be similar to the relative positions in the other system after another time t', such that t' bears to t a constant ratio?

In other words, a model is made of a machine, and is found to work satisfactorily, what are the conditions that a machine made according to the model should work as satisfactorily?

The principle of similitude was first enunciated by Newton in Prop. 32, Sect. VII. of the second book of the *Principia*. But the demonstration has been very much improved by M. Bertrand in *Cahier xxxii.* of the *Journal de l'école Polytechnique*. He derives the theorem from the principle of virtual work so as to avoid that necessity of considering the unknown reactions which enters into some other modes of proof. Since all the equations of motion may be deduced from the general principle of virtual work, that principle seems to afford the simplest method of investigating any general theorem in Dynamics.

368. Let (x, y, z) be the co-ordinates of any particle of mass m in one system referred to any rectangular axes fixed in space, and let (X, Y, Z) be the resolved part of the impressed *moving*

forces on that particle. Let accented letters refer to corresponding quantities in the other system.

Then the principle of virtual work supplies the two following equations:

$$\Sigma \{(X - m\ddot{x}) \, \delta x + \&c.\} = 0,$$

$$\Sigma \{(X' - m'\ddot{x}') \, \delta x' + \&c.\} = 0.$$

It is evident that one of these equations will be changed into the other if we put $X' = FX$, $Y' = FY$, &c., $x' = lx$, $y' = ly$, &c., $m' = \mu m$, &c., $t' = \tau t$, &c., where F, l, μ, τ are all constants, provided that $\mu l = F\tau^2$. In two geometrically similar systems we have but one ratio of similarity, viz. that of the linear dimensions, but in two mechanically similar systems we have three other ratios, viz. that of the masses of the particles, that of the forces which act on them, and that of the times at which the systems are to be compared. It is clear that, if the relation just established hold between these four ratios of similitude, the motions of the two systems will be similar.

Suppose then that the two systems are initially geometrically similar, that the masses of corresponding particles are proportional each to each, and that they begin to move in parallel directions with like motions and in proportional times, then they will continue to move with like motions and in proportional times provided the external moving forces in either system are proportional to $\dfrac{\text{mass} \times \text{linear dimensions}}{(\text{time})^2}$. Since the resolved velocities of any particle are $\dfrac{dx}{dt}$, &c., it is clear that in two similar systems the velocities of corresponding points at corresponding times are proportional to $\dfrac{\text{linear dimensions}}{\text{time}}$. If we eliminate the time between these two relations, we may state, briefly, that the condition of similitude between two systems is that the moving forces must be proportional to $\dfrac{\text{mass} \times (\text{velocity})^2}{\text{linear dimensions}}$.

369. **On Models.** M. Bertrand remarks that, in comparing the working of a model with that of a large machine, we must take care that all the forces bear their proper ratios. The weights of the several parts will vary as their masses. Hence we infer that the velocity of working the model must be made to be proportional to the square root of its linear dimensions. The times of describing corresponding arcs will also be in the same ratio.

When the speeds of working the model and the large machine are thus related it is convenient to apply to them the terms "*corresponding velocities.*"

If there be any forces besides gravity which act on the model, these must bear the same ratio to the corresponding forces in the machine, if the model is to be similar to the machine. If the model be made of the same material as the machine, the weights of the several parts will vary as the cubes of the linear dimensions. Hence the impressed forces must be made to vary as the cubes of the linear dimensions. For example, in the case of a model of a steam-engine, the pressure of the steam on the piston varies as the product of the area of the piston into the elastic force. Hence, the elastic force of the steam used must be proportional to the linear dimensions of the model.

Supposing the impressed forces in the two systems to have, each to each, the proper ratio, the mutual reactions between the parts of the systems will, of themselves, assume the same ratio. For if, by giving proper displacements according to the principle of virtual work, we form equations of motion to find these reactions, it is easy to see that they will be, each to each, in the same ratio as the forces. Since sliding friction varies as the normal pressure, and is independent of the areas in contact, these frictions will bear their proper ratio in the model and machine. This, however, is not the case with rolling friction. Recurring to Art. 164, we see that the rolling friction varies inversely as the diameter of the wheel, and therefore bears a greater ratio to the other forces in the model than it does in the machine. If the resistance of the air be proportional to the product of the area exposed and the square of the velocity, the resistances will bear the proper ratio in the model and the machine.

370. **Examples.** As an example, let us apply the principle to the case of a rigid body oscillating about a fixed axis under the action of gravity. That the motions of two pendulums may be similar they must describe equal angles, corresponding times are therefore proportional to the times of oscillation. Since the forces vary as the mass into gravity, we see that when a pendulum oscillates through a given angle, the square of the time of oscillation must vary as the ratio of the linear dimensions to gravity.

As a second example consider the case of a particle describing an orbit round a centre of attraction whose force is equal to the product of the inverse square of the distance and some constant μ. The principle at once shows that the square of the periodic time must vary as the cube of the distance directly, and as μ inversely. This is Kepler's third law.

In Mr Froude's experiments to determine the resistance to ships he employed small models. The following rule used by him will be a third example. If the linear dimensions of a ship be n times those of the model, and if at a speed V the measured resistance to the model be R, then at the corresponding speed, viz., $n^{\frac{1}{2}}V$, the resistance to the ship will be n^3R.

Ex. Experiments are to be made on the deflection of a bridge 50 feet long and weighing 100 tons, when an engine weighing 20 tons passes with a velocity of 40 miles per hour, by means of a model bridge 5 feet long and weighing 100 oz.

Find the weight of the model engine, and if the model bridge be of such stiffness that its statical central deflection under the model engine be one-tenth of the statical central deflection of the bridge due to the engine, show that the velocity of the model engine must be 18·55 feet per second. [Coll. Exam.]

371. **Savart's Theorem.** In the twenty-ninth volume of the *Annales de Chimie* (Paris, 1825) Savart describes numerous experiments which he made on the notes sounded by similar vessels containing air. He says that if we construct cubical boxes and set the air in motion, as is ordinarily done in organ pipes, we find that the number of vibrations in a given time is proportional to the reciprocals of the linear dimensions of the masses of air. This law was verified between extreme limits, and its truth tested over many notes. He says that he frequently used the law during his researches, and never once found that it led him wrong. This result having been obtained for cubes, it was natural to examine whether the same law held for prismatic tubes on square bases. After numerous experiments he found the same law to be true.

He then tested the law with conical pipes in which the opening was always of the same solid angle, then with cylindrical pipes, then with pipes whose bases were equilateral triangles. These he made to sound in different ways, putting the mouth-piece for instance at different points of the length of the tube. In all cases the same law was found to hold, for tubes whose diameters were very small compared with their lengths as well as for those whose diameters were very great. This law he again found applicable to masses of air set in motion by communication from other vibrating bodies. Hence he inferred the following general law, which he enunciated as an experimental fact.

When masses of air are contained in two similar vessels, the number of vibrations in a given time [*i.e.* the pitch of the note sounded] is inversely proportional to the linear dimensions of the vessel.

This theorem of Savart's follows at once from the principle of similarity. Divide the similar vessels into corresponding elements, then the motions of these elements will be similar each to each if the forces vary as $\dfrac{\text{mass} \times \text{lin. dim.}}{(\text{time})^2}$. But by Marriotte's law the force between two elements varies as the product of the area of contact into the density. Hence the times of oscillation of corresponding particles of air must vary as the linear dimensions of the vessel.

372. The first person who gave a theoretical explanation of Savart's law was Cauchy, who showed, in a *Memoire* presented to the Academy of Sciences in 1829, that it followed from the linearity of the equations of motion. He refers to the general equations of motion of an elastic body whose particles are but slightly displaced even though the elasticity is different in different directions. These equations, which serve to determine the displacements (ξ, η, ζ) of a particle in terms of the time t and the co-ordinates (x, y, z), are of two kinds. One applies to all points of the interior of the elastic body and the other to all points on its surface. These are to be found in all treatises on elasticity. An inspection of the equations shows that they will continue to exist if we replace $\xi, \eta, \zeta, x, y, z, t$ by $\kappa\xi, \kappa\eta, \kappa\zeta, \kappa x, \kappa y, \kappa z, \kappa t$, where κ is any constant, provided that we alter the accelerating forces in the ratio κ to 1. Hence if the accelerating forces are zero, it is sufficient to increase the dimensions of the elastic body and the initial values of the displacements in the ratio 1 to κ, in order that the general values of ξ, η, ζ and the durations of the vibrations may vary in the same ratio. Hence we deduce Cauchy's extension of

Savart's law, viz., if we measure the pitch of the note given by a body, a plate or an elastic rod, by the number of vibrations produced in a unit of time, the pitch will vary inversely as the linear dimensions of the body, plate, or rod, supposing all its dimensions altered in a given ratio.

373. **Theory of Dimensions.** These results may be also deduced from the theory of dimensions. Following the notation of Art. 332, a force F is measured by $m\dfrac{d^2x}{dt^2}$. We may then state the general principle, that all dynamical equations must be such that the dimensions of terms added together are the same in space, time and mass, the dimensions of force being taken to be $\dfrac{\text{mass . space}}{(\text{time})^2}$.

Let us apply this to the case of a simple pendulum of length l, oscillating through a given angle α, under the action of gravity. Let m be the mass of the particle, F the moving force of gravity, then the time τ of oscillation can be a function of F, l, m and α only. Let this function be expanded in a series of powers of F, l and m. Thus

$$\tau = \Sigma A\, F^p l^q m^r,$$

where A, being a function of α only, is a number. Since τ is of no dimensions in space, we have $p + q = 0$. Also τ is of one dimension in time ; $\therefore - 2p = 1$. Finally τ is of no dimensions in mass ; $\therefore p + r = 0$. Hence $p = -\frac{1}{2}$, $q = r = \frac{1}{2}$, and since p, q, r have each only one value, there is but one term in the series. We infer that in any simple pendulum $\tau = A\sqrt{\dfrac{ml}{F}}$ where A is an undetermined number. See also Art. 370.

374. Ex. 1. A particle moves from rest towards a centre of force, whose attraction varies as the distance, in a medium resisting as the velocity, show by the theory of dimensions that the time of reaching the centre of force is independent of the initial position of the particle.

Ex. 2. A particle moves from rest in vacuo towards a centre of force whose attraction varies inversely as the nth power of the distance, show that the time of reaching the centre of force varies as the $\dfrac{n+1}{2}$th power of the initial distance of the particle.

Clausius' theory of stationary motion.

375. *To determine the mean vis viva of a system of material points in stationary motion.* Clausius, *Phil. Mag., August,* 1870.

By stationary motion is meant any motion in which the points do not continually move further and further from their original position, and the velocities do not alter continuously in the same direction, but the points move within a limited space and the velocities only fluctuate within certain limits. Of this nature are all periodic motions, such as those of the planets about the sun, and the vibrations of elastic bodies, and further, such irregular motions as are attributed to the atoms and molecules of a body in order to explain its heat.

Let x, y, z be the co-ordinates of any particle in the system and let its mass be m. Let X, Y, Z be the components of the forces on this particle. Then $m \dfrac{d^2 x}{dt^2} = X$. We have by simple differentiation,

$$\frac{d^2 (x^2)}{dt^2} = 2 \frac{d}{dt} \left(x \frac{dx}{dt} \right) = 2 \left(\frac{dx}{dt} \right)^2 + 2x \frac{d^2 x}{dt^2},$$

and therefore

$$\frac{m}{2} \left(\frac{dx}{dt} \right)^2 = -\frac{1}{2} xX + \frac{m}{4} \frac{d^2 (x^2)}{dt^2}.$$

Let this equation be integrated with regard to the time from 0 to t and let the integral be divided by t, we thereby obtain

$$\frac{m}{2t} \int_0^t \left(\frac{dx}{dt} \right)^2 dt = -\frac{1}{2t} \int_0^t xX dt + \frac{m}{4t} \left[\frac{d (x^2)}{dt} - \left(\frac{d (x^2)}{dt} \right)_0 \right],$$

in which the application of the suffix zero to any quantity implies that the initial value of that quantity is to be taken.

The left-hand side of this equation and the first term on the right-hand side are evidently the mean values of $\dfrac{m}{2} \left(\dfrac{dx}{dt} \right)^2$ and $-\dfrac{1}{2} xX$ during the time t. For a periodic motion the duration of a period may be taken for the time t; but for irregular motions (and if we please for periodic ones also) we have only to consider that the time t, in proportion to the times during which the point moves in the same direction in respect of any one of the directions of co-ordinates, is very great, so that in the course of the time t many changes of motion have taken place, and the above expressions of the mean values have become sufficiently constant. The last term of the equation, which has its factor included in square brackets, becomes, when the time is periodic, equal to zero at the end of each period. When the motion is not periodic, but irregularly varying, the factor in brackets does not so regularly become zero, yet its value cannot continually increase with the time, but can only fluctuate within certain limits; and the divisor t, by which the term is affected, must accordingly cause the term to become vanishingly small for very great values of t. The same reasoning will apply to the motions parallel to the other co-ordinates. Hence adding together our results for each particle, we have, if v be the velocity of the particle m,

$$\text{mean } \frac{1}{2} \Sigma m v^2 = - \text{ mean } \frac{1}{2} \Sigma (Xx + Yy + Zz).$$

The mean value of the expression $-\dfrac{1}{2} \Sigma (Xx + Yy + Zz)$ has been called by Clausius the *virial* of the system. His theorem may therefore be stated thus, *the mean semi-vis viva of the system is equal to its virial.*

376. To apply this theorem to the kinetic theory of heat we premise that every body is to be regarded as a system of particles in motion. So far as this proposition is concerned, the particles may describe paths of any kind, and any particle may pass as close as we please to another. But, as no account of impacts has *here* been considered, we must either suppose the particles to be restrained from actual contact by strong repulsive forces at close quarters, or (which amounts to the same

thing) suppose the particles to be perfectly elastic, so that the total vis viva is unaltered by the impacts.

The forces which act on the system consist in general of two parts. In the first place, the elements of the body exert on each other attractive or repulsive forces, and, secondly, forces may act on the system from without. The virial will therefore consist of two parts, which are called the *internal* and *external virials*. It has just been shown that *the mean semi-vis viva is equal to the sum of these two parts*.

If $\phi(r)$ be the law of repulsion between two particles whose masses are m and m', we have $Xx + X'x' = -\phi(r)\dfrac{x'-x}{r}x - \phi(r)\dfrac{x-x'}{r}x' = \phi(r)\dfrac{(x'-x)^2}{r}$. And, since for the two other co-ordinates corresponding equations may be formed, we have for the internal virial $-\frac{1}{2}\Sigma(Xx + Yy + Zz) = -\Sigma\frac{1}{2}r\phi(r)$.

Let the volume be increased, the system remaining similar to itself. Every r is now increased so that $dr = \beta r$, where β is an infinitely small quantity. If W be the work of the internal repulsions, we have $dW = \Sigma\phi(r)\beta r$. If V be the volume of the body, $dV = 3\beta V$. Hence $-\Sigma\frac{1}{2}r\phi(r) = -\frac{3}{2}V\dfrac{dW}{dV}$. This supplies another expression for the internal virial, if we understand W to represent the *mean* work.

As to the external forces, in the case most frequently to be considered the body is acted on by a uniform pressure normal to the surface. If p be this pressure, $d\sigma$ an element of the surface, l the cosine of the angle the normal makes with the axis of x, $-\dfrac{1}{2}\Sigma Xx = \dfrac{1}{2}\int xp\,l\,d\sigma = \dfrac{p}{2}\iint x\,dy\,dz$. If V be the volume of the body this is $\dfrac{1}{2}pV$, and therefore the whole external virial is $\dfrac{3}{2}pV$.

Let us suppose that a gas is composed of particles (such as those here described) each in motion, but not acting on each other, and equally distributed throughout the containing vessel. It follows from this proposition that $\frac{1}{2}\Sigma mv^2 = \frac{3}{2}pV$. Hence the resulting continuous pressure p produced by their impacts on the containing surface, when referred to a unit of area, is equal to one-third of the vis viva of the particles which occupy any unit of volume.

The reader who is interested in these matters is referred to *Applications of dynamics to physics and chemistry* by Prof. J. J. Thomson, 1888.

Ex. 1. Show that the virial of a system of forces is independent of the origin and the directions of the axes, supposed rectangular.

The first result is clear, since in stationary motion $\Sigma X = 0$, &c. The second follows from the equality $Xx + Yy + Zz = R\rho$, where R is the resultant of X, Y, Z, and ρ is the projection of the radius vector on the direction of R.

Ex. 2. For a free system of particles moving under the influence of their mutual attractions only, where the force function U is a homogeneous function of degree n, prove the equation

$$\frac{d^2}{dt^2}(\Sigma mR^2) = 2(n+2)U + \text{constant},$$

where $R_1, R_2, \ldots$ are the distances of the particles $m_1, m_2, \ldots$ from the common centre of gravity. If the law of attraction be the inverse cube of the distance, prove that

$$\Sigma mR^2 = A + Bt + Ct^2. \qquad\qquad \text{[June Exam.]}$$

General Theorems on Impulses.

377. General equation of virtual work. Let (x, y, z) be the co-ordinates of any particle m, and (X, Y, Z) the resolved parts in the directions of the axes of the impulses which act on that particle. Let (u, v, w), (u', v', w') be the resolved parts of the velocity of the particle in the same directions just before and just after the impulse.

The momenta $m(u' - u)$, $m(v' - v)$, $m(w' - w)$ being reversed for every particle, will be in equilibrium with the impulsive forces. Hence by the principle of virtual work we have

$$\Sigma m \{(u' - u)\,\delta x + (v' - v)\,\delta y + (w' - w)\,\delta z\} = \Sigma (X\delta x + Y\delta y + Z\delta z),$$

where δx, δy, δz are any small arbitrary displacements of the particle m consistent with the geometrical conditions of the system.

This is the general equation of virtual work, and it will be seen further on that the subsequent motion of the system may be deduced from it. At present we are only concerned with such general properties of the motion as may be deduced from this equation by a proper choice of the arbitrary displacement.

378. Carnot's first theorem. Let us first suppose that the only impulsive forces are those produced by the actions and reactions of the bodies forming the system. (For example, two bodies may impinge on each other, or two points may be suddenly connected together by an inelastic string.) Then these mutual actions and reactions are in equilibrium, and the sum of their virtual works is zero for all displacements which do not alter the distance apart of the particles acting on each other. Suppose the bodies impinging to be inelastic, then *just after the impact* the points of the two bodies which impinge have no velocity of separation normal to the common surface of the bodies. If therefore we take as our arbitrary displacement the actual displacement of the system during the time dt just after the impact, the sum of the virtual works of the impulses will be zero. Hence, writing $\delta x = u'\delta t$, $\delta y = v'\delta t$, $\delta z = w'\delta t$, we have

$$\Sigma m \{(u' - u)\,u' + (v' - v)\,v' + (w' - w)\,w'\} = 0.$$

This gives us

$$\Sigma m (u'^2 + v'^2 + w'^2) = \Sigma m (uu' + vv' + ww')$$

which may be put into the form

$$\Sigma m (u'^2 + v'^2 + w'^2) - \Sigma m (u^2 + v^2 + w^2)$$
$$= -\Sigma m \{(u' - u)^2 + (v' - v)^2 + (w' - w)^2\}.$$

Therefore *in the impact of inelastic bodies vis viva is always lost.* This is the first part of Carnot's general Theorem.

379. **Generalization of Carnot's theorem.** It should be noticed that Carnot's demonstration applies not exclusively to collisions but to all impulses which do not appear in the equation of virtual work as applied to the subsequent displacement. Let a system be moving in any way, and let us suddenly introduce some new restraints by which some of the particles are compelled to take new courses. The impulses which produce this change of motion are of the nature of reactions, and are such that in the subsequent motion their virtual works are zero. It therefore follows that *vis viva is lost and that the amount of vis viva lost is equal to the vis viva of the relative motion.* This is sometimes called Bertrand's Theorem.

380. **Carnot's second theorem.** Let us next suppose that an explosion takes place in any body of the system. Then, *just before the impulse,* any two particles about to separate are moving so that the virtual works of their mutual actions are equal and opposite, but just after the explosion this may not be the case. Hence we now put $\delta x = u\delta t$, $\delta y = v\delta t$, $\delta z = w\delta t$ and we have from the equation of virtual moments

$$\Sigma m \left\{ (u' - u)\, u + (v' - v)\, v + (w' - w)\, w \right\} = 0.$$

This may be put into the form

$$\Sigma m \left(u'^2 + v'^2 + w'^2 \right) - \Sigma m \left(u^2 + v^2 + w^2 \right)$$
$$= \Sigma m \left\{ (u' - u)^2 + (v' - v)^2 + (w' - w)^2 \right\}.$$

Therefore *in cases of explosion vis viva is always gained.* This is the second part of Carnot's Theorem.

381. Thirdly, let the bodies of the system be perfectly elastic. If two elastic bodies impinge, the whole action consists of two parts, a force of compression as if the bodies were inelastic, and a force of restitution of the nature of an explosion. The circumstances of these two forces are equal and opposite to each other. Hence the vis viva lost in compression is exactly balanced by the vis viva gained in the restitution. This is the last part of Carnot's Theorem.

382. **Three forms of the equation of virtual work.** Let us now resume the general equation of virtual work for a system in motion acted on by any impulses. We have already seen that there are two displacements, either of which we may with advantage choose as our arbitrary displacement. One of these coincides with the motion just before, and the other with the motion just after, the action of the impulses. These equations may be written

$$\Sigma m \left\{ (u' - u)\, u + (v' - v)\, v + (w' - w)\, w \right\} = \Sigma \left(Xu + Yv + Zw \right)$$
$$\Sigma m \left\{ (u' - u)\, u' + (v' - v)\, v' + (w' - w)\, w' \right\} = \Sigma \left(Xu' + Yv' + Zw' \right).$$

But besides these there is a great variety of motions which are geometrically possible. Let (u'', v'', w'') be the components of the velocity of the typical particle m for any one of these possible motions. Then we may write $\delta x = u'' \delta t$, $\delta y = v'' \delta t$, $\delta z = w'' \delta t$, and we obtain

$$\Sigma m \left\{ (u' - u) u'' + (v' - v) v'' + (w' - w) w'' \right\} = \Sigma (X u'' + Y v'' + Z w'').$$

This equation of course includes the two former as special cases.

This possible motion might have been produced from the initial state by the application of proper impulses. Let these be represented by X', Y', Z'. Then with these forces the state (u'', v'', w'') becomes the actual subsequent motion, and our former subsequent motion becomes a mere variation from this. Thus we may write down three more equations, obtained from these by interchanging (u', v', w') with (u'', v'', w'') and (X, Y, Z) with (X', Y', Z').

By comparing these equations we may deduce several general theorems. In order to avoid a great deal of analytical reasoning we shall adopt a simple notation.

383. Vis Viva of the Relative Motion. Let $2T$ be the initial vis viva of the system. Let $2T'$ be the vis viva after the application of a set of impulses which we shall designate as the set A, and let the resulting motion be called the motion A. Let $2T''$ be the vis viva of any possible variation of this motion which we shall call the motion B, and let the forces which produce it be called the forces B. We shall want to use also the vis viva of the relative motion of any two of these. Thus, taking the two first and expressing the vis viva of the relative motion by $2R_{01}$, we have

$$2R_{01} = \Sigma m \left\{ (u' - u)^2 + (v' - v)^2 + (w' - w)^2 \right\}$$
$$= 2T' + 2T - 2\Sigma m (uu' + vv' + ww'),$$
$$\therefore \quad \Sigma m (uu' + vv' + ww') = T + T' - R_{01}.$$

Similarly if we call the vires vivæ of the other relative motions R_{02} and R_{12} we have

$$\Sigma m (uu'' + vv'' + ww'') = T + T'' - R_{02},$$
$$\Sigma m (u'u'' + v'v'' + w'w'') = T' + T'' - R_{12}.$$

Thus the accents of the T's on the right-hand side and the suffixes of the R's correspond in all three equations to the accents on the left-hand side.

The three equations deduced from the principle of virtual work in Art. 382 may therefore be written

$$T' - T - R_{01} = \text{vir. wk. of forces } A \text{ in initial motion,}$$
$$T' - T + R_{01} = \text{vir. wk. of forces } A \text{ in motion } A,$$
$$T' - T - R_{12} + R_{02} = \text{vir. wk. of forces } A \text{ in motion } B,$$

where the divisor dt on the right-hand side has been dropped for the sake of brevity. Or we may say that the right-hand sides express the rates at which the forces A are doing work in the respective motions. Or again, the right-hand sides express the sums of the products obtained by multiplying each force by the velocity of its point of application resolved in the direction of the force, for the particular motion concerned.

384. **Change of vis viva due to impulses.** If we add the first two equations together we see that R_{01} disappears, and thus we are led to the theorem; *If any impulses act on a system in motion, the change in the semi vis viva is equal to the sum of the products obtained by multiplying each impulse by the mean of the velocities of its point of application just before and just after the impulse, both velocities being resolved in the direction of the impulse.*

This theorem and the next may be simply proved by respectively adding and subtracting the first two equations of Art. 382. This differs from the proof just given only in not using so many symbols. A different proof for the case of a *single body* has been given in Art. 346.

385. **Vis Viva of the Relative Motion.** If we subtract one equation from the other we are led to the theorem; *If any impulses act on a system in motion, the semi vis viva of the relative motion is equal to the sum of the products obtained by multiplying each impulse by half the excess of the resolved velocity of its point of application just after, over that just before, the impulse, both velocities being resolved in the direction of the impulse.*

386. Let us now consider the third equation in Art. 383, and let us choose the hypothetical motion B so that the virtual work of the forces A may in it be the same as in the actual motion. Then we have

$$R_{02} = R_{01} + R_{12}.$$

Therefore, *if any impulsive forces act on a system in motion, the vis viva of the relative motion is less than if the particles took any other motion for which the virtual work of these impulses was the same.* Of course this hypothetical motion must be consistent with the geometrical conditions of the system.

387. **Sir W. Thomson's Theorem.** If the system start from rest we have $T = 0$, $R_{01} = T'$, $R_{02} = T''$, and thus we obtain $T'' = T' + R_{12}$. This gives us Sir W. Thomson's theorem. *Suppose a system to be at rest and to be set in motion by jerks or impulses at given points so that the motions of these points are prescribed, then the vis viva of the subsequent motion is less than that of any other hypothetical motion the system can take in which these points*

have the prescribed motion. By prescribing the motion of the points of application of the impulses (*i.e.* the forces called A in the fundamental equations of Art. 383) we secure the fact that their virtual work is the same for all hypothetical displacements of the system. *Natural Philosophy,* by Thomson and Tait, Art. 312.

By the use of this proposition the actual motion may be found by the application of the ordinary processes for maxima and minima.

388. **Bertrand's Theorem.** We may write down the equations for the motion B corresponding to those given in Art. 383 for the motion A. They are

$$T'' - T - R_{02} = \text{vir. wk. of forces } B \text{ in initial motion,}$$

$$T'' - T + R_{02} = \text{vir. wk. of forces } B \text{ in motion } B,$$

$$T'' - T - R_{12} + R_{01} = \text{vir. wk. of forces } B \text{ in motion } A,$$

where the divisor dt has been omitted as before.

Comparing the second of these with the last of the three given in Art. 383 we see that, if we choose the hypothetical motion B so that the right-hand sides of the two equations are the same, we have

$$T'' + R_{12} = T'.$$

In order that the right-hand sides may be equal we may suppose the motion B to differ from the motion A only by the introduction of some constraints, so that the forces B differ from the forces A only by some reactions whose virtual works are zero in the motion B.

We thus arrive at a theorem of Lagrange generalized first by Delaunay (Liouville's *Journal,* vol. v.) and afterwards by Bertrand in his notes to the *Mécanique Analytique. Suppose a system in motion to be acted on by any impulses, then the vis viva of the subsequent motion is greater than if the system were subjected to any constraints and acted on by the same impulses.* See Art. 379.

Comparing Sir W. Thomson's and Bertrand's theorems we perceive that, when the motions of the points of application of the impulses are given, the subsequent motion may be found by making the vis viva a *minimum,* but, when the impulses are given, the subsequent motion may be found by introducing some constraints and making the vis viva a *maximum.*

389. **Imperfectly elastic and rough bodies.** When two bodies of an imperfectly elastic and rough system impinge on each other, we may deduce from the equations of Art. 382 some extensions of Carnot's theorems.

Let $(uvw)\,(u'v'w')\,(u''v''w'')$ be the resolved velocities of a particle m just before the impact begins, at the moment of greatest compression, and just after the conclusion of the impact. Let the vis viva of the system at any one of these epochs be represented by one of the symbols $2T$, $2T'$, $2T''$. Let the vis viva of the relative motion at any two of these epochs be represented by R_{01}, R_{12}, R_{02}.

If the bodies impinging are perfectly smooth we have by the same reasoning as in Art. 378 and 380

$$\Sigma m \left\{ (u' - u) \, u' + \&\text{c.} \right\} = 0 \quad \dots\dots\dots\dots\dots\dots\dots\dots(1),$$

$$\Sigma m \left\{ (u'' - u) \, u' + \&\text{c.} \right\} = 0 \quad \dots\dots\dots\dots\dots\dots\dots (2).$$

Since the whole impulse between the two bodies bears to the impulse up to the moment of greatest compression the ratio $1 + e : 1$ we may deduce from Art. 382 the two following equations

$$\Sigma m \left\{ (u'' - u) \, u + \&\text{c.} \right\} = (1 + e) \, \Sigma m \left\{ (u' - u) \, u + \&\text{c.} \right\} \quad \dots\dots\dots\dots(3),$$

$$\Sigma m \left\{ (u'' - u) \, u'' + \&\text{c.} \right\} = (1 + e) \, \Sigma m \left\{ (u' - u) \, u'' + \&\text{c.} \right\} \dots\dots\dots\dots(4).$$

The left-hand side of either of these equations, after multiplication by dt, is equal to the virtual work of the whole impulse, and the summation on the right-hand side, after multiplication by dt, is equal to the virtual work of the impulse of compression. These are taken for the same displacement and are therefore in the ratio $1 + e : 1$. In the first equation the displacement chosen is the actual displacement just before impact. In the second equation the displacement chosen is that just after impact. These are both consistent with the geometrical conditions.

The above four equations may be conveniently expressed in the forms

$$T' - T = - R_{01} \dots\dots\dots\dots\dots\dots\dots\dots\dots\dots\dots(5),$$

$$T'' - T' = R_{12} \dots\dots\dots\dots\dots\dots\dots\dots\dots\dots\dots(6),$$

$$T'' - T' (1 + e) + eT = R_{02} - (1 + e) \, R_{01} \quad \dots\dots\dots\dots\dots(7),$$

$$T'' - T' (1 + e) + eT = eR_{02} - (1 + e) \, R_{12} \quad \dots\dots\dots\dots\dots(8).$$

If we eliminate the R's from these equations, we find

$$T'' - T' = - e^2 \, (T' - T) \quad \dots\dots\dots\dots\dots\dots\dots\dots\dots(9),$$

thus the *gain of vis viva due to restitution or explosion is* e^2 *into the loss of vis viva due to compression.*

If we eliminate the T's, we find

$$R_{01} = \frac{R_{02}}{(1 + e)^2} = \frac{R_{12}}{e^2} \dots\dots\dots\dots\dots\dots\dots\dots\dots\dots(10).$$

If we eliminate T', R_{01}, R_{12}, we find

$$T'' - T = - \frac{1 - e}{1 + e} R_{02} \dots\dots\dots\dots\dots\dots\dots\dots\dots(11),$$

which may be regarded as an extension of Carnot's third theorem in Art. 381.

Suppose next that the bodies impinging are rough, and slide on each other during the whole impact, the friction acting always in the same direction. The friction now bears a constant ratio to the normal pressure throughout the impact. The equations (3) and (4) hold as before. The separate equations (1) and (2) no longer hold, but instead we may form the single equation

$$\Sigma m \left\{ (u'' - u) \, u' + \&\text{c.} \right\} = (1 + e) \, \Sigma m \left\{ (u' - u) \, u' + \&\text{c.} \right\} \dots\dots\dots\dots (12),$$

by the same reasoning as in equations (3) and (4). The equation (12) may be expressed in the form

$$T'' - T' (1 + e) + eT = R_{12} + eR_{01} \dots\dots\dots\dots\dots\dots\dots\dots(13).$$

Joining (13) to (7) and (8) we have three equations connecting the six quantities T, T', T'', R_{01}, R_{02}, R_{12}. We easily find

$$R_{01} = \frac{R_{02}}{(1+e)^2} = \frac{R_{12}}{e^2} = \frac{T'' - T'(1+e) + eT}{e(1+e)} \quad \ldots\ldots\ldots\ldots\ldots (14).$$

We may deduce from these equations the following theorem. *When one body of a system impinges on another, the three states of motion (viz. that just before, that just after, and that at the moment of greatest compression) are so related that the vis viva of the relative motion of any two bears to the vis viva of the relative motion of any other two a ratio which depends only on the coefficient of elasticity.*

Let us suppose a system to be acted on by an impulsive force whose direction in space remains unchanged during its time of action. A theorem similar to that just enunciated applies to any three epochs in the time of action of this impulse, provided these epochs are such that the whole impulse exerted in the interval from the first epoch to the second bears a known ratio (say $1 : e$) to the whole impulse exerted in the interval from the second to the third.

Representing the vires vivæ of the system at the three epochs by $2T$, $2T'$, $2T''$ as before, and the vires vivæ of the relative motions by $2R_{01}$, $2R_{02}$, $2R_{12}$, we notice that the equations (3), (4) and (12) apply to the motions of the system at the three epochs. The equation (14) will therefore give the same relations as before between the six quantities T, T', T'', R_{01}, R_{02}, R_{12}.

We may obtain an easy proof of this theorem by combining the results of Arts. 385, 386 with Art. 313. Let X be an impulse, and let the axis of x be taken parallel to its direction. By Art. 385 the vis viva of the relative motion before and after the impulse is proportional to $X(u' - u)$. But, by Art. 313, $u' - u$ is a linear function of X, and vanishes with X. It is therefore proportional to X. The vis viva of the relative motion is therefore proportional to X^2. It immediately follows that R_{01}, R_{02}, R_{12} are proportional to 1, $(1+e)^2$, e^2.

The remaining part of the theorem follows from Art 386. Letting X now represent the impulse from the first to the second epoch, we have

$$T' - T = \tfrac{1}{2} X (u' + u), \qquad T'' - T' = \tfrac{1}{2} Xe (u'' + u').$$

It easily follows that

$$T'' - T' - e (T' - T) = \tfrac{1}{2} Xe (u'' - u).$$

Since the right-hand side of this equation is $R_{02}e/(1+e)$, by Art. 385, the remaining part of equation (14) has been proved.

When two elastic systems impinge on each other, the theorems contained in equation (14) are true for the impulse on each system. They therefore follow by simple addition for the two impinging systems regarded as one.

Examples. To understand these two principles properly we should examine their application to some simple cases of motion.

Ex. 1. *A body at rest having one point O fixed is struck by a given impulse, find the resulting motion.* See Art. 308 and Art. 310.

Let L, M, N be the given components of the impulse about the principal axes at O. Then, if the body begin to turn about an axis *fixed in space* whose direction cosines are (l, m, n), the angular velocity ω is found by Art. 89 from

$$(Al^2 + Bm^2 + Cn^2)\, \omega = Ll + Mm + Nn.$$

To find the axis about which the body begins to turn *when free*, we must by Lagrange's Theorem make the vis viva a maximum. That is, we have

$$(Al^2 + Bm^2 + Cn^2)\,\omega^2 = \text{maximum}.$$

We have also the condition $l^2 + m^2 + n^2 = 1$.

Treating these three equations in the usual manner indicated in the differential calculus, we find

$$\frac{Al}{L} = \frac{Bm}{M} = \frac{Cn}{N}\,.$$

These equations determine the direction cosines of the axis about which the body begins to turn.

Ex. 2. *A body is at rest with one point O fixed in space. Suddenly a straight line OC fixed in the body begins to move round O in a known manner, find the motion of the body.* See Art. 293.

Take the instantaneous position of OC as the axis of z, and let O be the origin. Let the motion of OC be given by the angular velocities θ, ϕ about the axes Ox, Oy, and let ω be the required angular velocity of the body about Oz. Then, by Sir W. Thomson's theorem, we make the vis viva of the body a minimum. We have therefore

$$A\theta^2 + B\phi^2 + C\omega^2 - 2D\phi\omega - 2E\theta\omega - 2F\theta\phi = \text{min.},$$

where A, B, &c. are the moments and products of inertia at O. Differentiating we have

$$C\omega - D\phi - E\theta = 0.$$

Thus ω has been found. This last equation expresses the fact that the angular momentum about the axis of z is unaltered by the blow.

Ex. 3. A rod AB at rest is acted on by an impulse F perpendicular to its length at the extremity A, and that extremity begins to move with a velocity f. Find the point O in AB about which the rod will begin to turn (1) when F is given and (2) when f is given. If $AO = x$, show that both Sir W. Thomson's theorem and Lagrange's or Bertrand's theorem require the same function of x to be made a minimum.

Ex. 4. A system is moving in any manner. A blow is given at any point perpendicular to the direction of motion of that point. Prove that the vis viva is increased.

This follows from the first of the equations in Art. 383; for the virtual work of this force (there called A) vanishes in the initial motion. Hence $T' = T + R_{01}$.

Ex. 5. A system at rest, if acted on by two different sets of impulses called A and B, will take two different motions. Prove that the sum of the virtual works of the forces A for displacements represented by the velocities in the motion B is equal to the sum of virtual works of the forces B for displacements represented by the velocities in the motion A.

Since $T = 0$, $T' = R_{01}$, and $T'' = R_{02}$, the result follows by comparing the third equations in Art. 383 and Art. 388.

390. **Gauss' measure of the "constraint."** The expression, called R in the previous articles, which represents the vis viva of the relative motion, has been interpreted by Gauss in another manner. Let the particles m_1, m_2, &c. of a system just before the action of any impulses occupy positions which we shall call p_1, p_2, &c. Let us suppose that the particles *if free* would under the action

of these impulses and their previous momenta acquire such velocities that in the time dt subsequent to the impulses they would describe the small spaces $p_1 q_1$, $p_2 q_2$, &c. But if the particles were *constrained in any manner* consistent with the geometrical conditions which hold just before the action of the impulses, let us suppose that they would under the same impulses and their previous momenta describe in the time dt subsequent to the impulses the small spaces $p_1 r_1$, $p_2 r_2$, &c. Then the spaces $q_1 r_1$, $q_2 r_2$, &c. may be called the deviations from free motion due to the constraints. The sum $\Sigma m\,(qr)^2$ is called the "constraint."

391. We may also measure the constraint by the ratio of this sum to $(dt)^2$. We then take $p_1 q_1$, &c. $p_1 r_1$, &c.' to represent, not the displacements in the time dt, but the velocities of the particles just after the action of the forces in the two cases in which the particles are free or constrained. Referring to D'Alembert's principle in Art. 67, we see that pq represents the resultant of the previous velocity and of the velocity generated by the impressed force on the typical particle m, while qr represents the velocity generated by the molecular forces*.

If we suppose that the lengths pq, qr, &c. represent velocities and not displacements, let (u, v, w) be the components of pq in any motion, and (u', v', w') the components of pr in any other motion; then

$$\Sigma m\,(qr)^2 = \Sigma m\,\{(u'-u)^2 + (v'-v)^2 + (w'-w)^2\}$$

measures the "constraint" from one motion to the other. This is precisely what we have represented by the symbol $2R$, with suffixes to define the two motions compared.

392. **Gauss' principle of least constraint.** Suppose a system of particles in motion and constrained in any given manner to be acted on by any given set of impulses. Let $2T'$ be the vis viva of the subsequent motion. This is the actual motion taken by the system. Let us now suppose that the particles were forced to take some hypothetical motion consistent with the geometrical conditions by introducing some further constraints. Let $2T''$ be the subsequent vis viva in this hypothetical motion. Thirdly, let us suppose that all constraints were removed so that the particles were acted on solely by the given set of impulses. Let $2T'''$ be the subsequent vis viva in this free motion. Let $2T$ be the initial vis viva common

---

* Gauss' proof of the principle is nearly as follows. By D'Alembert's principle the particles m_1, m_2, &c., if placed in the positions r_1, r_2, &c., would be in equilibrium under the action of these molecular forces alone. Let us apply the principle of virtual work, and displace the system so that the typical particle m describes a space $r\rho$, making an angle ϕ with the direction rq of the molecular force on m. Then since the product $m\,(rq)$ measures the molecular force on m, we have

$$\Sigma m\,(rq)\,(r\rho \cos \phi) = 0.$$

But $\qquad\qquad\qquad q\rho^2 = qr^2 + r\rho^2 - 2qr\,.\,r\rho \cos \phi.$

Hence we easily find $\qquad \Sigma m\,(q\rho)^2 = \Sigma m\,(qr)^2 + \Sigma m\,(r\rho)^2.$

In the actual motion the particles move from p_1, &c. to r_1, &c. and the "constraint" is $\Sigma m\,(qr)^2$. If the particles had been forced to take any other hypothetical courses, by which they were brought into the positions ρ_1, &c., the "constraint" would be $\Sigma m\,(q\rho)^2$. Gauss' Principle asserts that the former is always less than the latter.

to all the motions. Let $2R_{12}$, $2R_{13}$, $2R_{23}$ be the vires vivæ of the relative motions of the first, second and third subsequent motions as denoted by the suffixes.

By Bertrand's theorem, since the hypothetical motion is more constrained than the actual motion, we have
$$T' = T'' + R_{12}.$$
Also, since each of these is more constrained than the free motion,
$$T''' = T' + R_{13},$$
$$T''' = T'' + R_{23}.$$
Hence we have $\qquad R_{23} = R_{13} + R_{12}.$

Therefore R_{23} is always greater than R_{13}. It follows that the motion which the system actually takes when subject to any impulses is such that the "constraint" from the free motion is less than if the system took any other motion consistent with the geometrical conditions. This result is true whichever way the "constraint" is measured.

393. If we suppose the system to be acted on by a series of indefinitely small impulses, these impulses may be regarded as finite forces. We therefore infer the following theorem, which is usually called *Gauss' principle of least constraint.*

The motion of a system of material points connected by any geometrical relations is always as nearly as possible in accordance with free motion; i.e. if the constraint during any time dt *is measured by the sum of the products of the mass of each particle into the square of its distance at the end of that time from the position it would have taken if it had been free, then the actual motion during the time* dt *is such that the constraint is less than if the particles had taken any other position.*

Gauss remarks that the free motions of the particles when they are incompatible with the geometrical conditions of the system are modified in exactly the same way as geometers modify results which have been obtained by observation, i.e. by applying the method of Least Squares so as to render them compatible with the geometrical conditions of the question.

394. **Ex.** *Any number of particles* m_1, m_2, &c. *are acted on by any forces whose components are* m_1X_1, m_1Y_1, m_1Z_1, &c. *Their co-ordinates* x_1, y_1, z_1; x_2, y_2, z_2; &c. *are connected together by some relation such as* $\phi(x_1, \&c.) = 0$. *(For instance the particles may be beads slung on a string of given length whose extremities are tied together.) It is required to form the equations of motion.*

Let U, V, W be the resolved velocities of the typical particle m at the time t; u, v, w its resolved velocities just after the action of the impulse whose resolved parts are $mXdt$, $mYdt$, $mZdt$, on the supposition that the particle is perfectly free. But as the typical particle is not perfectly free, let u', v', w' be its actual resolved velocities at the same instant. Then to find u', v', w' we make
$$R_{13} = \Sigma m \left[(u' - u)^2 + (v' - v)^2 + (w' - w)^2 \right] = \text{minimum},$$
where the Σ implies summation for all the particles. This quantity is to be a minimum for all variations of u', v', w' subject to the condition
$$\Sigma (\phi_x u' + \phi_y v' + \phi_z w') = 0,$$
where the Σ here also implies summation for all suffixes.

To make R_{13} a minimum we take the total differential of each of these quantities with regard to all the accented letters, multiply the second by some indeterminate multiplier λ, and add the results together. Equating to zero the coefficients of du', &c. we obtain the three typical equations

$$m\,(u'-u)+\lambda\phi_x=0 \qquad m\,(v'-v)+\lambda\phi_y=0 \qquad m\,(w'-w)+\lambda\phi_z=0.$$

Putting suffixes we have equations sufficient to find λ and the (u', v', w') of every particle.

We may write these equations in another form. Since U and u' are two successive values at an interval dt of the same quantity in the continuous motion which we are considering, we write $u' - U = \dfrac{dU}{dt}\,dt$. Since u is the resolved velocity after the impulse when the particle is free, we have $u - U = Xdt$. The equations therefore become

$$m\left(\frac{dU}{dt}-X\right)+\mu\phi_x=0,\ \&c.,$$

where μdt has been written for λ.

The equations in this form might have been derived directly from the principle of virtual work. By that principle we have

$$\Sigma m\left[\left(\frac{du}{dt}-X\right)\delta x+\&c.\right]=0$$

with the condition　　　　　$\Sigma\left[\phi_x\delta x+\&c.\right]=0.$

Multiplying the second by an indeterminate multiplier μ, adding the results together, and equating to zero the coefficients of δx, &c. we obtain the same results as before.

EXAMPLES*.

1. A screw of Archimedes is capable of turning freely about its axis, which is fixed in a vertical position: a heavy particle is placed at the top of the tube and runs down through it; determine the whole angular velocity communicated to the screw.

Result. Let n be the ratio of the mass of the screw to that of the particle, a the angle which the tangent to the screw makes with the horizon, h the height descended by the particle. If ω be the angular velocity generated, prove that

$$\omega^2 a^2\,(n+1)\,(n+\sin^2 a)=2gh\cos^2 a.$$

2. A fine circular tube, carrying within it a heavy particle, is set revolving about a vertical diameter. Show that the difference of the squares of the absolute velocities of the particle at any two given points of the tube equidistant from the axis is the same for all initial velocities of the particle and tube.

3. A circular wire ring, carrying a small bead, lies on a smooth horizontal table; an elastic thread, the natural length of which is less than the diameter of the ring, has one end attached to the bead and the other to a point in the wire; the bead is placed initially so that the thread coincides very nearly with a diameter of the ring; find the vis viva of the system when the string has contracted to its original length.

---

* These examples, except the last two, are taken from the Examination Papers which have been set in the University and in the Colleges.

4. A straight tube of given length is capable of turning freely in a horizontal plane about one extremity, two equal particles are placed at different points within it at rest; an angular velocity being given to the system, determine the velocity of each particle on leaving the tube.

5. A smooth circular tube of mass M has placed within it two equal particles of mass m, which are connected by an elastic string whose natural length is $\frac{2}{3}$ of the circumference. The string is stretched until the particles are in contact, when the tube is placed flat on a smooth horizontal table and left to itself. Show that, when the string arrives at its natural length, the actual energy of the two particles is to the work done in stretching the string as $2\,(M^2 + Mm + m^2) : (M + 2m)\,(2M + m)$.

6. An endless flexible and inextensible chain, in which the mass per unit of length is μ through one continuous half, and μ' through the other half, is stretched over two equal perfectly rough uniform circular discs (radius a, mass M) which can turn freely about their centres at a distance b in the same vertical line. Prove that the time of a small oscillation of the chain under the action of gravity is

$$2\pi \sqrt{\frac{M + (\pi a + b)\,(\mu + \mu')}{2\,(\mu - \mu')\,g}}\,.$$

7. Two particles of masses m, m' are connected by an inelastic string of length a. The former is placed in a smooth straight groove, and the latter is projected in a direction perpendicular to the groove with a velocity V. Prove that the particle m will oscillate through a space $\dfrac{2am'}{m + m'}$, and that, if m be large compared with m', the time of oscillation is nearly $\dfrac{2\pi a}{V}\left(1 - \dfrac{m'}{4m}\right).$

8. A rough plane rotates with uniform angular velocity n about a horizontal axis which is parallel to it but not in it. A heavy sphere of radius a, being placed on the plane when in a horizontal position, rolls down it under the action of gravity. If the centre of the sphere be originally in the plane containing the moving axis and perpendicular to the moving plane, and if x be its distance from this plane at a subsequent time t, before the sphere leaves the plane, then

$$x = \frac{1}{24\sqrt{35}}\left(\frac{35g}{n^2} - 84a - 60c\right)\left(e^{\sqrt{\frac{5}{7}}nt} - e^{-\sqrt{\frac{5}{7}}nt}\right) - \frac{5}{12}\frac{g}{n^2}\sin nt,$$

c being the distance from the axis to the plane measured upwards.

9. The extremities of a uniform heavy beam of length $2a$ slide on a smooth wire in the form of the curve whose equation is $r = a\,(1 - \cos\theta)$, the prime radius being vertical and the vertex of the curve downwards. Prove that, if the beam be placed in a vertical position and displaced with a velocity just sufficient to bring it into a horizontal position, $\tan\theta = \dfrac{1}{2}\left\{e^{\sqrt{\frac{3g}{2a}}t} - e^{-\sqrt{\frac{3g}{2a}}t}\right\}$, where θ is the angle through which the rod has turned during a time t.

10. A rigid body, whose radius of gyration about G the centre of gravity is k, is attached to a fixed point C by a string fastened to a point A on its surface. $CA\,(=b)$ and $AG\,(=a)$ are initially in one line, and to G is given a velocity V at right angles to that line. No impressed forces are supposed to act, and the string is attached so as always to remain in one right line. If θ be the angle between AG and AC

at time t, show that $\left(\dfrac{d\theta}{dt}\right)^2 = \dfrac{V^2}{b^2} \dfrac{k^2 - 4ab \sin^2 \frac{\theta}{2}}{k^2 + a^2 \sin^2 \theta}$, and if the amplitude of θ, i.e.

$2 \sin^{-1} \dfrac{k}{2 \sqrt{ab}}$, be very small, the period is $\dfrac{2\pi bk}{V \sqrt{a(a+b)}}$.

11. A fine weightless string having a particle at one extremity is partially coiled round a hoop, which is placed on a smooth horizontal plane, and is capable of motion about a fixed vertical axis through its centre. If the hoop be initially at rest and the particle be projected in a direction perpendicular to the length of the string, and if s be the portion of the string unwound at any time t, then

$$s^2 - b^2 = \frac{\mu}{m+\mu} V^2 t^2 + 2Vat,$$

where b is the initial value of s, m and μ the masses of the hoop and particle, a the radius of the hoop and V the velocity of projection.

12. A square, formed of four similar uniform rods jointed freely at their extremities, is laid upon a smooth horizontal table, one of its angular points being fixed: if angular velocities ω, ω' in the plane of the table be communicated to the two sides containing this angle, show that the greatest value of the angle ($2a$) between them is given by the equation $\cos 2a = -\dfrac{5}{6} \dfrac{(\omega - \omega')^2}{\omega^2 + \omega'^2}$.

13. Two particles of masses m, m' lying on a smooth horizontal table are connected by an inelastic string extended to its full length and passing through a small ring on the table. The particles are at distances a, a' from the ring and are projected with velocities v, v' at right angles to the string. Prove that, if $mv^2a^2 = m'v'^2a'^2$, their second apsidal distances from the ring will be a', a respectively.

14. If a uniform thin rod PQ move, in consequence of a primitive impulse, between two smooth curves in the same plane, prove that the square of the angular velocity varies inversely as the difference between the sum of the squares of the normals OP, OQ to the curves at the extremities of the rods and $\frac{5}{12}$ of the square of the whole length of the rod.

15. Assuming that the muscular power or moving force of an animal varies as the sectional area of its limbs, and that its weight varies as its volume, prove that two animals of similar forms, but of different dimensions, can make jumps of exactly the same height, the height being measured by the vertical distance described by the centre of gravity after the animal has left the ground.

16. The extremities of a uniform beam of length $2a$, slide on two slender rods without inertia, the plane of the rods being vertical, their point of intersection fixed, and the rods inclined at angles $\dfrac{\pi}{4}$ and $-\dfrac{\pi}{4}$ to the horizon. The system is set rotating about the vertical line through the point of intersection of the rods with an angular velocity ω, prove that if θ be the inclination of the beam to the vertical at the time t and a the initial value of θ

$$4\left(\frac{d\theta}{dt}\right)^2 + \frac{(3\cos^2 a + \sin^2 a)^2}{3\cos^2 \theta + \sin^2 \theta} \omega^2 = (3\cos^2 a + \sin^2 a) \omega^2 + \frac{6g}{a}(\sin a - \sin \theta).$$

17. A perfectly rough sphere of radius a is placed close to the intersection of the highest generating lines of two fixed equal horizontal cylinders of radius c, the axes being inclined at an angle $2a$ to each other, and is allowed to roll down be-

tween them.　Prove that the vertical velocity of its centre in any position will be

$$\sin \alpha \cos \phi \left\{ \frac{10g\,(a+c)\,(1-\sin \phi)}{7-5\cos^2 \phi \cos^2 \alpha} \right\}^{\frac{1}{2}}, \text{ where } \phi \text{ is the inclination to the horizon of the}$$

radius to either point of contact.

18.　Let a complete integral of the equation $\dfrac{d^2x}{dt^2} = \dfrac{dT}{dx}$, in which T is a function of x, be $x = X$, X being a known function of a and b, two arbitrary constants, and t. Then the solution of $\dfrac{d^2x}{dt^2} = \dfrac{dT}{dx} + \dfrac{dR}{dx}$, R being a function of x, may also be represented by $x = X$ provided that a and b are variable quantities determined by the equations $\dfrac{da}{dt} = k\,\dfrac{dR}{db}$, $\dfrac{db}{dt} = -k\,\dfrac{dR}{da}$, where k is a function of a and b which does not contain the time explicitly.

19.　A satellite, considered as a particle, revolves about its primary with an angular velocity Ω, and the primary rotates about an axis which is perpendicular to the plane of the satellite's orbit with an angular velocity n.　Show that the angular momentum h of the system about its centre of gravity and the energy E are given by

$$h = Cn + D\Omega^{-\frac{1}{3}}, \qquad 2E = Cn^2 - D\Omega^{\frac{2}{3}},$$

where C is the moment of inertia of the primary about the axis of rotation and D is a quantity depending on the masses of the bodies.

Trace the curves whose ordinates are h and E and abscissa is $x = D\Omega^{-\frac{1}{3}}$.　Show that the latter curve belongs to one or other of two species according as a maximum and a minimum ordinate do or do not exist, i.e., according as the biquadratic

$$h = x + CD^3x^{-3}$$

has two real roots or none.　Show also that the real roots correspond to the case in which the primary always turns the same face to the satellite.

20.　Assuming the results of the last example, determine the effect on the motion of a continual loss of energy (due to tidal friction or any other cause), the angular momentum h being constant.　Show that, when the circumstances of the system are such that the energy curve is of the second species, the satellite must ultimately fall into the planet.　If the energy curve is of the first species, show that, according to the initial value of Ω, the satellite will either fall into the planet or will approach the planet until it reaches a certain distance, when the two will revolve as a rigid body.

To obtain these results imagine two points to be placed with the same abscissa, one on the momentum line and the other on the energy curve, and suppose the one on the energy curve to guide that on the momentum line.　Since the energy decreases, it is clear that, however the two points are set initially, the point on the energy curve must always slide down a slope, carrying with it the other point.　The final positions of the points will thus depend on the existence or absence of a minimum ordinate in the energy curve.　See a paper by G. H. Darwin on the secular effects of tidal friction in *the Proceedings of the Royal Society*, June 1879, or Thomson and Tait's *Treatise on Natural Philosophy*, Vol. I, Part II. App. Gb.

CHAPTER VIII.

LAGRANGE'S EQUATIONS.

395. **Two advantages of Lagrange's equations.** Our object in this section is to form the general equations of motion of a dynamical system freed from all the unknown reactions and expressed, so far as is possible, in terms of any kind of co-ordinates which may be convenient in the problem under consideration.

In order to eliminate the reactions we shall use the principle of virtual work. This principle has already been applied to obtain the equation of vis viva, by giving the system that particular displacement which it would have taken if it had been left to itself. But since every dynamical problem can, by D'Alembert's principle, be reduced to one in statics, it is clear that, by giving the system proper displacements, we must be able to deduce, as in Art. 357, not the vis viva equation only, but all the equations of motion.

396. Let the co-ordinates of any particle m of the system referred to any fixed rectangular axes be (x, y, z). These are not independent of each other, being connected by the geometrical relations of the system. But they may be expressed in terms of a certain number of independent variables whose values will determine the position of the system at any time. Extending the definition given in Art. 73, we shall call these the co-ordinates of the system. Let them be called θ, ϕ, ψ, &c. Then x, y, z, &c. are functions of θ, ϕ, &c. Let

$$x = f(t, \theta, \phi, \&c.) \dots\dots\dots\dots\dots\dots(1),$$

with similar equations for y and z. It should be noticed that these equations are not to contain $\dfrac{d\theta}{dt}$, $\dfrac{d\phi}{dt}$, &c. *The independent variables in terms of which the motion is to be found may be any we please, with this restriction, that the co-ordinates of every particle of the body can, if required, be expressed in terms of them by means of equations which do not contain any differential coefficients with regard to the time.*

The number of independent co-ordinates to which the position of a system is reduced by its geometrical relations is sometimes spoken of as the *number of degrees of freedom of that system.* Sometimes it is referred to as being the *number of independent motions* of which the system admits.

In this chapter *total* differential coefficients with regard to t will in general be denoted by *accents.* Occasionally dots will be used as before, and sometimes the differential coefficients will be written at length. Thus $\dfrac{dx}{dt}$ and $\dfrac{d^2x}{dt^2}$ will in general be written x' and x''.

If $2T$ be the vis viva of the system, we have

$$2T = \Sigma m \, (x'^2 + y'^2 + z'^2) \ldots\ldots\ldots\ldots\ldots\ldots(2);$$

we also have, since the geometrical equations do not contain θ', ϕ', &c.,

$$x' = \frac{dx}{dt} + \frac{dx}{d\theta} \, \theta' + \frac{dx}{d\phi} \, \phi' + \&c. \ \ldots\ldots\ldots\ldots(3),$$

with similar equations for y' and z'. In these the differential coefficients $\dfrac{dx}{dt}$, $\dfrac{dx}{d\theta}$, &c. are all partial. Substituting in the expressions for $2T$, we find

$$2T = F \, (t, \, \theta, \, \phi, \, \&c. \ \theta', \, \phi', \, \&c.).$$

When the system of bodies is given, the form of F is known. It will appear presently that it is only through the form of F that the effective forces depend on the nature of the bodies considered; so that two dynamical systems which have the same F are dynamically equivalent.

It should be noticed that no powers of θ', ϕ', &c. above the second enter into this function, and that, when the geometrical equations do not contain the time explicitly, it is a homogeneous function of θ', ϕ', &c. of the second order.

397. Virtual work of the effective forces. *To find the virtual moment of the momenta of a system, and also that of the effective forces, corresponding to a displacement produced by varying one co-ordinate only.*

Let this co-ordinate be θ, and let us follow the notation already explained. Let all differential coefficients be partial, unless it be otherwise stated, excepting those denoted by accents. Since x', y', z' are the components of the velocity, the virtual moment of the momenta is $\Sigma m \, (x' \delta x + y' \delta y + z' \delta z)$, where δx, δy, δz are the small changes produced in the co-ordinates of the particle m by a variation $\delta\theta$ of θ. This is the same as

$$\Sigma m \left(x' \, \frac{dx}{d\theta} + y' \, \frac{dx}{d\theta} + z' \, \frac{dz}{d\theta} \right) \delta\theta.$$

If $2T$ be the vis viva given by (2) of the last article

$$\frac{dT}{d\theta'} = \Sigma m \left(x' \frac{dx'}{d\theta'} + \&c. \right).$$

But, differentiating (3) partially with regard to θ', we see that $\dfrac{dx'}{d\theta'} = \dfrac{dx}{d\theta}$. Hence the virtual moment of the momenta is equal to $\dfrac{dT}{d\theta'} \delta\theta$.

398. The virtual work of the effective forces is

$$\Sigma m \left(x'' \frac{dx}{d\theta} + y'' \frac{dy}{d\theta} + z'' \frac{dz}{d\theta} \right) \delta\theta.$$

Omitting the factor $\delta\theta$ for the moment, this may be written in the form

$$\frac{d}{dt} \Sigma m \left(x' \frac{dx}{d\theta} + \&c. \right) - \Sigma m \left(x' \frac{d}{dt} \frac{dx}{d\theta} + \&c. \right),$$

where the $\dfrac{d}{dt}$ represents a total differential coefficient with regard to t. We have already proved that the first of these terms is $\dfrac{d}{dt} \dfrac{dT}{d\theta'}$. It remains to express the second term also as a differential coefficient of T. Differentiating the expression for $2T$ partially with regard to θ,

$$\frac{dT}{d\theta} = \Sigma m \left(x' \frac{dx'}{d\theta} + \&c. \right).$$

But, differentiating the expression for x' with regard to θ,

$$\frac{dx'}{d\theta} = \frac{d^2x}{d\theta\,dt} + \frac{d^2x}{d\theta^2}\,\theta' + \frac{d^2x}{d\theta\,d\phi}\,\phi' + \&c.,$$

and this is the same as $\dfrac{d}{dt} \dfrac{dx}{d\theta}$. Hence the second term may be written $\dfrac{dT}{d\theta}$, and the virtual work of the effective forces is therefore $\left(\dfrac{d}{dt} \dfrac{dT}{d\theta'} - \dfrac{dT}{d\theta} \right) \delta\theta.$

The following explanation will make the argument clearer. The virtual work of the effective forces is clearly the ratio to dt of the difference between the virtual moments of the momenta of the particles of the system at the times $t+dt$ and t, the displacements being the same at each time. The virtual moment of the momenta at the time t is first shown to be $\dfrac{dT}{d\theta'} \delta\theta$. Hence $\left(\dfrac{dT}{d\theta'} + \dfrac{d}{dt}\dfrac{dT}{d\theta'}\,dt \right) \delta\theta$ is the virtual moment of the momenta at the time $t+dt$ corresponding to a displacement $\delta\theta$ consistent with the positions of the particles at that time. To make the displacements the same, we must subtract from this the virtual moment of the momenta for a displacement which is the difference between the two displacements at the times t and $t + dt$. Since $\delta x = \dfrac{dx}{d\theta}\,\delta\theta$, this difference for the variable x is $\dfrac{d}{dt}\left(\dfrac{dx}{d\theta}\right) dt\,\delta\theta$. We therefore subtract on the whole $\Sigma m\left\{x'\dfrac{d}{dt}\left(\dfrac{dx}{d\theta}\right) dt + \&\text{c.}\right\}\delta\theta$, and this is shown to be $\dfrac{dT}{d\theta}\,dt\,\delta\theta$.

399. **Lagrange's equations for finite forces.** *To deduce the general equations of motion referred to any co-ordinates.*

Let U be the force-function, then U is a function of θ, ϕ, &c. and t. The virtual work of the impressed forces corresponding to a displacement produced by varying θ only is $\dfrac{dU}{d\theta}\,\delta\theta$. But by D'Alembert's principle this must be the same as the virtual work of the effective forces. Hence

$$\frac{d}{dt}\frac{dT}{d\theta'} - \frac{dT}{d\theta} = \frac{dU}{d\theta}.$$

Similarly we have $\dfrac{d}{dt}\dfrac{dT}{d\phi'} - \dfrac{dT}{d\phi} = \dfrac{dU}{d\phi}$, &c. = &c.

It may be remarked that if V be the potential energy we must write $-V$ for U. We then have

$$\frac{d}{dt}\frac{dT}{d\theta'} - \frac{dT}{d\theta} + \frac{dV}{d\theta} = 0,$$

with similar equations for ϕ, ψ, &c.

In using these equations, it should be remembered that all the differential coefficients are partial except that with regard to t.

Let us write $L = T + U$, so that L is the difference of the kinetic and potential energies. Then, since U is not a function of θ', ϕ', &c., the Lagrangian equations of motion may be written in the typical form

$$\frac{d}{dt}\frac{dL}{d\theta'} - \frac{dL}{d\theta} = 0.$$

Thus it appears that, when the one function L is known, *all* the differential equations of motion may be deduced by simple partial differentiations. The function L is called the *Lagrangian function*.

These are called Lagrange's general equations of motion. Lagrange only considers the case in which the geometrical equations do not contain the time explicitly, but it has been shown by Vieille, in *Liouville's Journal*, 1849, that the equations are still true when this restriction is removed. In the proof given above we have included Vieille's extension, and adopted in part Sir W. Hamilton's mode of proof, *Phil. Trans.*, 1834. It differs from Lagrange's in two respects; firstly, he makes the arbitrary displacement such that only one co-ordinate varies at a time, and secondly, he operates directly on T instead of $\Sigma m x'^2$.

Ex. 1. If we change the co-ordinates in Lagrange's equation from θ, ϕ, &c. to any others x, y, which are connected with θ, ϕ, &c. by equations which do not contain differential coefficients with regard to the time, show by an *analytical transformation* that the form of Lagrange's equations is not altered, i.e. that the transformed equations are the same as the original ones with x, y, &c. written for θ, ϕ, &c. This is of course evident by dynamics.

Ex. 2. If two sides b, c and the included angle A of any triangle be taken as the co-ordinates θ, ϕ, ψ, prove that the Lagrangian equations are satisfied by $L = B'$.

This easily follows from the last example by a change of co-ordinates.

400. **Indeterminate Multipliers.** In order to use these equations it is necessary to express the Lagrangian function L in terms of the independent co-ordinates of the system. If the geometrical conditions are somewhat complex it may be very troublesome to do this. It is sometimes convenient to express L as a function of more than the necessary number of co-ordinates and to have geometrical relations connecting them. Suppose that we have L expressed as a function of the co-ordinates θ, ϕ, ψ, &c., θ', ϕ', ψ', &c., and that there are two geometrical equations connecting these co-ordinates, viz.

$$f(\theta, \phi, \&c.) = 0, \quad F(\theta, \phi, \&c.) = 0 \ldots\ldots\ldots\ldots(1).$$

To simplify the explanation, we suppose that there are only *two* such geometrical equations, but it will be seen that the process is quite general and will apply to any number of conditions.

By the principle of virtual work we have

$$\left(\frac{d}{dt}\frac{dL}{d\theta'} - \frac{dL}{d\theta}\right)\delta\theta + \left(\frac{d}{dt}\frac{dL}{d\phi'} - \frac{dL}{d\phi}\right)\delta\phi + \&c. = 0 \ldots\ldots(2).$$

Also

$$\frac{df}{d\theta}\delta\theta + \frac{df}{d\phi}\delta\phi + \&c. = 0 \ldots\ldots(3),$$

and

$$\frac{dF}{d\theta}\delta\theta + \frac{dF}{d\phi}\delta\phi + \&c. = 0 \ldots\ldots(4).$$

Since the co-ordinates θ, ϕ, &c. are connected by two geometrical equations, two of them are dependent variables; let these be θ, ϕ. Following the argument explained in the differential calculus, we multiply (3) and (4) by two arbitrary quantities λ and μ, and add the products to (1). We now choose λ and μ so that the coefficients of $\delta\theta$, $\delta\phi$ may be zero. The remaining co-ordinates ψ, &c., being independent, the coefficients of $\delta\psi$, &c., must also vanish. We thus have

$$\left.\begin{array}{l}
\dfrac{d}{dt}\dfrac{dL}{d\theta'} - \dfrac{dL}{d\theta} + \lambda\,\dfrac{df}{d\theta} + \mu\,\dfrac{dF}{d\theta} = 0 \\[2mm]
\dfrac{d}{dt}\dfrac{dL}{d\phi'} - \dfrac{dL}{d\phi} + \lambda\,\dfrac{df}{d\phi} + \mu\,\dfrac{dF}{d\phi} = 0 \\[2mm]
\qquad\qquad\&c. = 0
\end{array}\right\} \ldots\ldots\ldots\ldots(5).$$

There are here as many equations as co-ordinates. Joining these to the equations (1) we have sufficient equations to find all the co-ordinates and the two multipliers λ and μ.

These equations may be put into a simpler form. We notice that the geometrical functions f and F do not contain θ', ϕ', &c. (see also Art. 396). Let us then write

$$L_1 = L + \lambda f + \mu F \ldots\ldots\ldots\ldots\ldots\ldots(6),$$

and treat L_1 as if it were the Lagrangian function. If we substitute this value of L_1 in the *typical equation*

$$\frac{d}{dt}\frac{dL_1}{d\theta'} - \frac{dL_1}{d\theta} = 0 \ldots\ldots\ldots\ldots\ldots\ldots(7),$$

where θ stands for any one of the co-ordinates, and simplify the results by remembering that $f = 0$, $F = 0$, we obtain in turn all the equations (5). The same process will also supply the geometrical equations (1), if we include λ and μ among the co-ordinates. Thus, since L_1 contains no λ, we have $dL_1/d\lambda' = 0$; hence, writing λ for θ, the equation (7) gives $f = 0$.

401. Lagrange's equations for impulsive forces. *To deduce the general equations of motion for impulsive forces.*

Let δU_1 be the virtual moment of the impulsive forces produced by a general displacement of the system. Then from the geometry of the system, we can express δU_1 in the form

$$\delta U_1 = P\delta\theta + Q\delta\phi + \ldots.$$

The virtual moment of the momenta imparted to the particles of the system is

$$\Sigma m \left\{ (x_1' - x_0')\,\delta x + (y_1' - y_0')\,\delta y + (z_1' - z_0')\,\delta z \right\},$$

where (x_0', y_0', z_0'), (x_1', y_1', z_1') are the values of (x', y', z') just before and just after the action of the impulsive forces.

Let θ_0', ϕ_0', &c., θ_1', ϕ_1', &c. be the values of θ', ϕ', &c. just before and just after the impulses, and let T_0, T_1 be the values of T when these are substituted for θ', ϕ', &c. Then, as in Art. 397, the virtual moment of the momenta is $\left(\dfrac{dT_1}{d\theta_1'} - \dfrac{dT_0}{d\theta_0'}\right)\delta\theta$. The Lagrangian equations of impulses may therefore be written

$$\frac{dT_1}{d\theta_1'} - \frac{dT_0}{d\theta_0'} = P,$$

with similar equations for ϕ, ψ, &c.

402. These equations are sometimes written in the convenient forms

$$\left(\frac{dT}{d\theta'}\right)^1_0 = P, \quad \left(\frac{dT}{d\phi'}\right)^1_0 = Q, \text{ &c.,}$$

where the brackets enclosing any quantity imply that that quantity is to be taken between the limits mentioned. Sometimes when no mistake can arise as to the particular limits meant, these are omitted, and only the brackets, with perhaps some distinguishing marks, retained.

When the quantity in brackets (as in our case) is a linear function of the variables θ', ϕ', &c. of the first order, another meaning can be given to the expressions. The brackets may then be said to indicate that $\theta_1' - \theta_0'$, $\phi_1' - \phi_0'$, &c. are to be written for θ', ϕ', &c. after all other operations indicated within the brackets have been performed.

403. If we interpret our equations by the general principles of Art. 283, viz., that the momenta of the particles just after an impulse compounded with the reversed momenta just before are equivalent to the impulse, we see that it will be convenient to call $\frac{dT}{d\theta'}$ the generalized component of the momenta with regard to θ, a name suggested in Thomson and Tait's *Natural Philosophy* More briefly we may say that the θ-component of the momentum is $\frac{dT}{d\theta'}$. In the same way we may define the θ-component of the effective forces to be $\frac{d}{dt}\frac{dT}{d\theta'} - \frac{dT}{d\theta}$.

Suppose for example that a variation $\delta\theta$ of any co-ordinate has the effect of turning the system as a whole about some straight line through an angle $\delta\theta$, then $\frac{dT}{d\theta'}$ is equal to the angular momentum about that straight line. But, if the variation $\delta\theta$ move the system as a whole parallel to some straight line through a space $\delta\theta$, then $\frac{dT}{d\theta'}$ is the linear momentum parallel to that straight line. See Arts. 306, 308.

These results also follow immediately from the general expression

$$\frac{dT}{d\theta'} = \Sigma m \left(x'\frac{dx'}{d\theta'} + y'\frac{dy'}{d\theta'} + z'\frac{dz'}{d\theta'} \right),$$

given in Art. 397. Let the given straight line be the axis of z. In the first case $x' = -y\theta'$, $y' = x\theta'$, $z' = 0$, hence the expression reduces to $\Sigma m(-x'y + y'x)$, which is the angular momentum. In the second case $x' = 0$, $y' = 0$, $z' = \theta'$, hence the expression becomes $\Sigma mz'$, which is the linear momentum.

404. The equations for impulsive forces were not given by Lagrange. They seem to have been first deduced by Prof. Niven from the Lagrangian equation

$$\frac{d}{dt}\frac{dT}{d\theta'} - \frac{dT}{d\theta} = \frac{dU}{d\theta}.$$

We may regard an impulse as the limit of a very large force acting for a very short time. Let t_0, t_1 be the times at which the force begins and ceases to act. Let us integrate this equation between the limits $t = t_0$ and $t = t_1$. The integral of the first term is $\left[\frac{dT}{d\theta'}\right]_{t_0}^{t_1}$ which is the difference between the initial and final values of $\frac{dT}{d\theta'}$. The integral of the second term is zero. For $\frac{dT}{d\theta}$ is a function of θ, ϕ, &c., θ', ϕ', &c. which, though variable, remains finite during the time $t_1 - t_0$. If A be its greatest value during this time, the integral is less than $A(t_1 - t_0)$, which ultimately vanishes. Hence the Lagrangian equation becomes $\left[\frac{dT}{d\theta'}\right]_{t_0}^{t_1} = \frac{dU_1}{d\theta}$. See a paper in the *Mathematical Messenger* for May, 1867.

405. **Examples of Lagrange's equations.** Before proceeding to discuss the properties of Lagrange's equations, we may illustrate their use by the following problems.

A body, two of whose principal moments at the centre of gravity are equal, turns under the action of gravity about a fixed point O, situated in the axis of unequal moment. To determine the conditions that there may be a simple equivalent pendulum.

Def. If a body be suspended from a fixed point O under the action of gravity, and if the angular motion of the straight line joining O to the centre of gravity be the same as that of a string of length l to the extremity of which a heavy particle is attached, then l is called the *length of the simple equivalent pendulum*. This is an extension of the definition in Art. 92.

Let OC be the axis of unequal moment, A, A, C the principal moments at the fixed point, and let the rest of the notation be the same as in Art. 365, Ex. 1. Then

$$2T = A\,(\theta'^2 + \sin^2\theta\,\psi'^2) + C\,(\phi' + \psi'\cos\theta)^2, \,-$$
$$U = Mgh\cos\theta + \text{constant},$$

where h is the distance of the centre of gravity from the fixed point, and gravity is supposed to act in the positive direction of the axis of z. Lagrange's equations will be found to become

$$\frac{d}{dt}\,(A\theta') - A\sin\theta\cos\theta\,\psi'^2 + C\psi'\,(\phi' + \psi'\cos\theta)\sin\theta = -Mgh\sin\theta,$$

$$\frac{d}{dt}\,\{C\,(\phi' + \psi'\cos\theta)\} = 0,$$

$$\frac{d}{dt}\,\{C\,(\phi' + \psi'\cos\theta)\cos\theta + A\sin^2\theta\,\psi'\} = 0.$$

Integrating the second of Lagrange's equations, we have

$$\phi' + \psi'\cos\theta = n,$$

where n is a constant expressing the angular velocity about the axis of unequal moment. (See Art. 256.) Integrating the third we have

$$Cn\cos\theta + A\sin^2\theta\,\psi' = a,$$

where a is another constant expressing the moment of the momentum about the vertical through O. (See Arts. 264 and 265, also Art. 403.)

There are errors, sometimes made in using Lagrange's equations, which we should here guard against. If ω_3 be the angular velocity about OC, we know by Euler's equations, Art. 251, that ω_3 is constant. If n be this constant, the vis viva of the body may be correctly written in the form

$$2T = A\left(\theta'^2 + \sin^2\theta\,\psi'^2\right) + Cn^2.$$

But, if this value of T were substituted in Lagrange's equations, we should obtain results altogether erroneous. The reason is, that, in Lagrange's equations, all the differential coefficients except those with regard to t are partial. Though ω_3 is constant, and therefore its *total* differential coefficient with regard to t is zero, yet its *partial* differential coefficients with regard to θ, ϕ, &c. are not zero. In writing down the value of T, preparatory to using it in Lagrange's equation, no properties of the motion are to be assumed which involve differential coefficients of the co-ordinates. This has been already indicated in Art. 396. But we must introduce into the expression any geometrical relations which exist between the co-ordinates and which therefore reduce the number of independent variables.

Instead of the first equation, we may use the equation of vis viva, which gives

$$A\left(\sin^2\theta\,\psi'^2 + \theta'^2\right) = \beta + 2Mgh\cos\theta.$$

To find the arbitrary constants a and β we must have recourse to the initial values of θ and ψ. Let θ_0, ψ_0, $\dfrac{d\theta_0}{dt}$, $\dfrac{d\psi_0}{dt}$ be the initial values of θ, ψ, $\dfrac{d\theta}{dt}$, $\dfrac{d\psi}{dt}$; then the above equations become

$$\left.\begin{aligned}
\sin^2\theta\,\frac{d\psi}{dt} + \frac{Cn}{A}\cos\theta &= \sin^2\theta_0\,\frac{d\psi_0}{dt} + \frac{Cn}{A}\cos\theta_0 \\[2mm]
\sin^2\theta\left(\frac{d\psi}{dt}\right)^2 + \left(\frac{d\theta}{dt}\right)^2 &= \sin^2\theta_0\left(\frac{d\psi_0}{dt}\right)^2 + \left(\frac{d\theta_0}{dt}\right)^2 + 2\frac{Mgh}{A}\left(\cos\theta - \cos\theta_0\right)
\end{aligned}\right\}\;\ldots\ldots(1).$$

These equations, when solved, give θ and ψ in terms of t, and thus determine the motion of the line OG. The corresponding equations for the motion of the simple equivalent pendulum OL are found by making $C = 0$, $A = Ml^2$, and $h = l$, where l is the length of the pendulum. These changes give

$$\left.\begin{aligned}
\sin^2\theta\,\frac{d\psi}{dt} &= \sin^2\theta_0\,\frac{d\psi_0}{dt} \\[2mm]
\sin^2\theta\left(\frac{d\psi}{dt}\right)^2 + \left(\frac{d\theta}{dt}\right)^2 &= \sin^2\theta_0\left(\frac{d\psi_0}{dt}\right)^2 + \left(\frac{d\theta_0}{dt}\right)^2 + 2\frac{g}{l}\left(\cos\theta - \cos\theta_0\right)
\end{aligned}\right\}\;\ldots\ldots(2).$$

In order that the motions of the two lines OG and OL may be the same, the two equations (1) and (2) must be the same. This will be the case if either $Cn = 0$, or $\theta = \theta_0$. In the first case, we must have $n = 0$, or $C = 0$, so that either the body must have no rotation about OG, or the body must be a rod. In the second case, we must have throughout the motion θ and ψ' constant, so that the body must be moving in steady motion making a constant angle with the vertical. In either case, the two sets of equations are identical if $Mhl = A$. This is the same formula as that obtained in Art. 92.

406. **Ex. 1.** *Show how to deduce Euler's equations, Art. 251, from Lagrange's equations.*

Taking as axes of reference the principal axes at the fixed point,

$$2T = A\omega_1^2 + B\omega_2^2 + C\omega_3^2.$$

We cannot take $(\omega_1, \omega_2, \omega_3)$ as the independent variables, because the co-ordinates of every particle of the body cannot be expressed in terms of them without introducing differential coefficients into the geometrical equations. (See Art. 396.) Let us therefore express $\omega_1, \omega_2, \omega_3$ in terms of θ, ϕ, ψ. By Art. 256, we have

$$\left. \begin{aligned} \omega_1 &= \theta' \sin\phi - \psi' \sin\theta \cos\phi \\ \omega_2 &= \theta' \cos\phi + \psi' \sin\theta \sin\phi \\ \omega_3 &= \phi' + \psi' \cos\theta \end{aligned} \right\}.$$

As it is only necessary to establish one of Euler's equations, the others following by symmetry, we need only use that one of Lagrange's equations which gives the simplest result. Since ϕ' does not enter into the expressions for ω_1, ω_2, it is most convenient to use the equation

$$\frac{d}{dt}\frac{dT}{d\phi'} - \frac{dT}{d\phi} = \frac{dU}{d\phi}.$$

Now $\dfrac{dT}{d\phi'} = C\omega_3 \dfrac{d\omega_3}{d\phi'} = C\omega_3$, and $\dfrac{dT}{d\phi} = A\omega_1\dfrac{d\omega_1}{d\phi} + B\omega_2\dfrac{d\omega_2}{d\phi} = A\omega_1\omega_2 - B\omega_2\omega_1$; as may be seen by differentiating the expressions for ω_1, ω_2. Also, by Art. 340, if N be the moment of the forces about the axis of C, $\dfrac{dU}{d\phi} = N$.

Substituting we have $\dfrac{d}{dt}(C\omega_3) - (A - B)\,\omega_1\omega_2 = N$, which is a typical form of Euler's equations.

Ex. 2. A body turns about a fixed point and its vis viva is given by

$$2T = A\omega_1^2 + B\omega_2^2 + C\omega_3^2 - 2D\omega_2\omega_3 - 2E\omega_3\omega_1 - 2F\omega_1\omega_2.$$

Show that, if the axes are fixed in the body, but are not necessarily principal axes, Euler's equations of motion may be written in the form

$$\frac{d}{dt}\frac{dT}{d\omega_1} - \frac{dT}{d\omega_2}\omega_3 + \frac{dT}{d\omega_3}\omega_2 = L,$$

with two similar equations. This result is given by Lagrange.

407. Ex. *Deduce the equation of vis viva from Lagrange's equations.*

If the geometrical equations do not contain the time explicitly, T is a homogeneous function of θ', ϕ', &c. of the second degree. Hence $2T = \dfrac{dT}{d\theta'}\theta' + \dfrac{dT}{d\phi'}\phi' + \ldots$

Differentiating this totally, we have $2\dfrac{dT}{dt} = \theta'\dfrac{d}{dt}\left(\dfrac{dT}{d\theta'}\right) + \dfrac{dT}{d\theta'}\theta'' + \&c.$,

where the &c. implies similar expressions for ϕ, ψ, &c. If we now substitute on the right-hand side from Lagrange's equations, we have

$$2\frac{dT}{dt} = \frac{dT}{d\theta}\theta' + \frac{dT}{d\theta'}\theta'' + \frac{dU}{d\theta}\theta' + \&c.$$

But, since T is a function of $\theta, \theta', \phi, \phi'$, &c., $\dfrac{dT}{dt} = \dfrac{dT}{d\theta}\theta' + \dfrac{dT}{d\theta'}\theta'' + \&c.$,

subtracting this from the last expression we have $\dfrac{dT}{dt} = \dfrac{dU}{d\theta}\theta' + \dfrac{dU}{d\phi}\phi' + \&c.$

Integrating, we have the equation of vis viva, $T - U = h$, where h is an arbitrary constant, sometimes called the constant of vis viva.

408. **Ex.** As an illustration of the application of Lagrange's equations to impulsive forces, let us consider the example already discussed in Art. 176.

Let x be the altitude of the centre of gravity of the rhombus at any time, then x and a may be taken as the independent variables.

We have
$$T = 2 \left\{ x'^2 + (k^2 + a^2) \, a'^2 \right\}.$$

Let P be the impulsive action between the rhombus and the plane, then the virtual moment of the impulsive forces is

$$\delta U = P \delta \, (x - 2a \cos a) = P \delta x + 2aP \sin a \delta a.$$

The Lagrangian equations are therefore, by Art. 401,

$$\left. \begin{array}{l} 4 \, (x_1' - x_0') = P \\ 4 \, (k^2 + a^2) \, (a_1' - a_0') = 2aP \sin a \end{array} \right\}.$$

Now the initial and final values of x' are $x_0' = -V$, $x_1' = -2a\omega \sin a$; those of a' are $a_0' = 0$, $a_1' = \omega$. Hence eliminating P we have

$$\omega' = \frac{3}{2} \frac{V}{a} \frac{\sin a}{1 + 3 \sin^2 a},$$

which is the same result as in Art. 176.

If we wish to avoid introducing the impulse into the Lagrangian equations we must choose such co-ordinates that the variation of one, while the other is constant, does not alter the point of application of the blow. This will be the case if we take as co-ordinates a and the ordinate y of the point A which strikes the plane. When the co-ordinates chosen are x and a a variation of either *alone* alters the position of the point of application A of the blow. Hence the virtual moment of the blow enters into each of the dynamical equations formed by varying x and a. But, when the co-ordinates chosen are y and a, a variation of a alone does not alter the position of A, so that the virtual moment of any force acting at A does not enter into the equation thus formed. In the same way if the magnitude of the blow at A were wanted we should use an equation formed by the variation of some co-ordinate, such as y, which would alter the position in space of A.

Taking as co-ordinates y and a, we find

$$T = 2 \left\{ y'^2 - 4ay'a' \sin a + (k^2 + a^2 + 4a^2 \sin^2 a) \, a'^2 \right\}.$$

The single equation required is now $\left(\dfrac{dT}{da'} \right)_0^1 = 0$ so that it is unnecessary to calculate U. The limits of y' are $y_0' = -V$, $y_1' = 0$; those of a' are the same as before. The value of ω' follows without difficulty.

Ex. Six equal uniform rods form a regular hexagon loosely jointed at the angular points; a blow is given perpendicularly to one of them at its middle point, show that the opposite rod begins to move with one-tenth of the velocity of the rod struck.　　　　　　　　　　　　　　　　　　　　　[Math. Tripos, 1882.]

409. Sir W. R. Hamilton has put the general equations of motion into another form, which is sometimes more convenient for investigating the general properties of a dynamical system. This transformation may be made to depend on the lemma given in the following article.

In what follows we confine ourselves to the elementary properties of reciprocation. The subject will be resumed and treated more fully in the second volume. Sir W. Hamilton's demonstration of his equations requires that T should be a homogeneous quadratic function of the velocities, and this is generally true in dynamics. The extension to the case in which the geometrical equations contain the time explicitly is due to Donkin, *Phil. Trans.* 1854.

410. The Reciprocal Function*. *Let* T_1 *be a function of any quantities which it will be presently found convenient to call* θ', ϕ', &c. *Let*

$$\frac{dT_1}{d\theta'} = u, \quad \frac{dT_1}{d\phi'} = v, \ \&c.,$$

then θ', ϕ', &c. *may be found in terms of* u, v, &c., *from these equations. Let*

$$T_2 = - T_1 + u\theta' + v\phi' + \&c.,$$

and let T_2 *be expressed in terms of* u, v, &c., *the quantities* θ', ϕ', &c. *being eliminated. Then will*

$$\frac{dT_2}{du} = \theta', \quad \frac{dT_2}{dv} = \phi', \ \&c.$$

It may be that T_1 *is a function of some other quantities, which it will presently be found convenient to designate by the unaccented letters* θ, ϕ, &c. *Then* T_2 *will also be a function of these, and we shall have*

$$\frac{dT_2}{d\theta} = - \frac{dT_1}{d\theta}, \quad \frac{dT_2}{d\phi} = - \frac{dT_1}{d\phi}, \ \&c.$$

---

* We may deduce from this lemma the *method of solving partial differential equations by reciprocation*, sometimes called Legendre's method and sometimes De Morgan's method. Let the partial differential equation be $\phi(x, y, z_1, p, q) = 0$, where p and q are the partial differential coefficients of z_1 with regard to x and y. If we write $z_2 = - z_1 + px + qy$, we have by the lemma $x = \frac{dz_2}{dp}$. $y = \frac{dz_2}{dq}$. Hence this rule; substitute for x, y, z_1 from the auxiliary equations

$$x = \frac{dz_2}{dp}, \quad y = \frac{dz_2}{dq}, \quad z_1 = - z_2 + p\frac{dz_2}{dp} + q\frac{dz_2}{dq},$$

and treat p, q as the independent variables. Thus we have a new differential equation which it may be more easy to solve than the former. Let the solution be $z_2 = f(p, q)$, then, by the auxiliary equations, x, y and z_2 have all been found in terms of two auxiliary quantities p and q, and further these quantities have a geometrical meaning. This method may be extended to any number of variables and orders. Also as in Art. 418 we may if we please *modify* the equation for some only of these variables.

Ex. If the differential equation be $xp^2 + yq^2 = z_1$, show that

$$z_2 = \frac{p}{1-p} F\left(\frac{p(1-q)}{q(1-p)}\right),$$

whence x, y, z can be found in terms of the auxiliary quantities by differentiation.

To prove this, let us take the total differential of T_2, we have

$$dT_2 = -\frac{dT_1}{d\theta}\,d\theta + \left(-\frac{dT_1}{d\theta'} + u\right)d\theta' + \theta'\,du + \&c.$$

By the conditions of the lemma the quantity in brackets vanishes. Now if T_2 be expressed as a function of θ, u, ϕ, v, &c. only, and not θ', ϕ', &c., we have

$$dT_2 = \frac{dT_2}{d\theta}\,d\theta + \frac{dT_2}{du}\,du + \&c.$$

Comparing these two expressions for dT_2 we have

$$\frac{dT_2}{d\theta} = -\frac{dT_1}{d\theta} \text{ and } \frac{dT_2}{du} = \theta'.$$

Thus we have a reciprocal relation between the functions T_1 and T_2. We find T_2 from T_1 by eliminating θ', ϕ', &c. by the help of certain equations, we now see that we could deduce T_1 from T_2 by eliminating u, v, &c. by the help of similar equations. We shall therefore call T_2 the *reciprocal function* of T_1 with regard to the accented letters θ', ϕ', &c.

411. It should be noticed that, if T_1 be a homogeneous quadratic function of the accented letters θ', ϕ', &c , then $u\theta' + v\phi' + \&c. = 2T_1$, and therefore $T_2 = T_1$, but is *differently expressed*. Thus T_1 is a function of θ', ϕ', &c. and not of u, v, &c., while T_2 is a function of u, v, &c. and not of θ', ϕ', &c. We notice that in this case T_2 is a homogeneous quadratic function of u, v, &c.

412. If T_1 be the semi vis viva of a dynamical system, this process is really equivalent to changing from the use of component velocities to the use of the corresponding component momenta. Either may be used to determine the motion of the system, sometimes the one set being the more convenient and sometimes the other.

413. **Examples on the Reciprocal Function. Ex. 1.** The position in space of a body of mass M is given by x, y, z, the rectangular co-ordinates of its centre of gravity, and θ, ϕ, ψ the angular co-ordinates of its principal axes at the centre of gravity, as used in Chap. v. Art. 256. If two of the principal moments of inertia are equal, and if ξ, η, ζ, u, v, w, be the components of momentum corresponding respectively to x, y, z, θ, ϕ, ψ, the vis viva $2T_1$ is given in Art. 365, Ex. 1. Show that the reciprocal function is

$$2T_2 = \frac{\xi^2 + \eta^2 + \zeta^2}{M} + \frac{u^2}{A} + \frac{v^2}{C} + \frac{(w - v\cos\theta)^2}{A\sin^2\theta}.$$

Ex. 2. If the vis viva $2T_1$ be given by the general expression

$$2T_1 = A_{11}\theta'^2 + 2A_{12}\theta'\phi' + \ldots$$

show that the reciprocal function of T_1 may be written in the form

$$T_2 = -\frac{1}{2\Delta} \begin{vmatrix} 0 & u & v & \dots \\ u & A_{11} & A_{12} & \dots \\ v & A_{12} & A_{22} & \dots \\ \dots\dots\dots\dots\dots\dots \end{vmatrix}$$

where Δ is the discriminant of T_1. Thus the coefficients of u^2, v^2, $2uv$, &c. in T_2 are the minors after division by 2Δ of the corresponding terms in T_1. See also Chap. I. Art. 28, Ex. 3.

Ex. 3. If ξ, η, &c. be partial differential coefficients of a function P of x, y, &c. with regard to those variables respectively, prove that x, y, &c. are also partial differential coefficients of a function Q of ξ, η, &c. with regard to these variables respectively. If P be homogeneous and of n dimensions prove also that $Q = (n-1)P$. For instance P may be the potential function in Attractions, or the velocity potential in Hydrodynamics.

Ex. 4. Regarding T_1 as a function of θ', ϕ', &c., let Δ be the Hessian of T_1, i.e. the Jacobian of its first differential coefficients with regard to θ', ϕ', &c. Then will $\dfrac{d^2T_2}{du^2}$, $\dfrac{d^2T_2}{du\,dv}$, &c. be equal to the minors of the corresponding constituents of the determinant Δ, each minor having its proper sign and being divided by Δ.

To prove this, we take the total differential of the two sets of equations, $u = \dfrac{dT_1}{d\theta'}$, &c., $\theta' = \dfrac{dT_2}{du}$, &c. From the first set we find $d\theta'$, $d\phi'$, &c. in terms of du, dv, &c. Substituting in the second set the theorem follows at once.

414. The Hamiltonian Transformation. Let us put $L = T + U$, so that L is the *difference* between the kinetic and the potential energies. Then, as explained in Art. 399, L is called the *Lagrangian function* and the Lagrangian equations may be written in the typical form

$$\frac{d}{dt}\frac{dL}{d\theta'} = \frac{dL}{d\theta},$$

there being corresponding equations for all the co-ordinates.

Let H be the reciprocal function of L, then H is called the *Hamiltonian function.* The equations of transformation are

$$u = \frac{dL}{d\theta'} = \frac{dT}{d\theta'},$$

with similar equations for all the co-ordinates. We have by the reciprocal property $\theta' = \dfrac{dH}{du}$; and by Lagrange's equation we have $u' = \dfrac{dL}{d\theta} = -\dfrac{dH}{d\theta}$, with similar equations for all the co-ordinates. Thus the single typical Lagrangian equation written down above is transformed into the two Hamiltonian equations

$$\theta' = \frac{dH}{du}, \qquad u' = -\frac{dH}{d\theta}.$$

There are of course similar equations for all the co-ordinates.

When the geometrical equations do not contain the time explicitly, T is a homogeneous quadratic function of $(\theta', \phi', \&c.)$, and therefore $u\theta' + v\phi' + \&c. = 2T$. Hence

$$H = -L + u\theta' + v\phi' + \&c. = T - U.$$

Thus H is the *sum* of the kinetic and potential energies, and is therefore the whole energy of the system.

415.   *To express the Lagrangian equations of impulses in the Hamiltonian form.*

Referring to Art. 402, we see that the Lagrangian equations of motion may be written in the typical form

$$\left(\frac{dT}{d\theta'}\right)_0^1 = P.$$

Let H be the reciprocal function of T, and let us replace u, v, &c. by P, Q, &c. Then these equations take the typical form

$$\theta_1' - \theta_0' = \frac{dH}{dP}.$$

416. Examples on the Hamiltonian Equations. Ex. 1. To deduce the equation of Vis Viva from the Hamiltonian equations.

Since H is a function of $(\theta, \phi, \&c.)$, $(u, v, \&c.)$ we have, if accents denote total differential coefficients with regard to the time,

$$H' = \frac{dH}{dt} + \frac{dH}{d\theta}\theta' + \frac{dH}{du}u' + \&c. = \frac{dH}{dt},$$

so that the *total* differential coefficient of H with regard to t is always equal to the *partial* differential coefficient. If the geometrical equations do not contain the time explicitly, this latter vanishes and therefore we have $H = h$, where h is a constant.

Ex. 2. To deduce Euler's equations of motion from the Hamiltonian equations.

Taking the same notation as in the corresponding proposition for Lagrange's equations, Art. 406, we have

$$u = \frac{dT}{d\theta'} = A\omega_1 \sin\phi + B\omega_2 \cos\phi, \quad v = \frac{dT}{d\phi'} = C\omega_3,$$

$$w = \frac{dT}{d\psi'} = (-A\omega_1 \cos\phi + B\omega_2 \sin\phi)\sin\theta + C\omega_3 \cos\theta.$$

Before we can use the Hamiltonian equations we must by Art. 411 express T in terms of (u, v, w). To do this we solve these equations to find ω_1, ω_2, ω_3 in terms of u, v, w. We find

$$A\omega_1 = u \sin\phi + (v \cos\theta - w)\frac{\cos\phi}{\sin\theta},$$

$$B\omega_2 = u \cos\phi - (v \cos\theta - w)\frac{\sin\phi}{\sin\theta}.$$

Also by Art. 414

$$H = \tfrac{1}{2}(A\omega_1^2 + B\omega_2^2 + C\omega_3^2) - U.$$

As we only require one of Euler's equations, let us use $\dfrac{dH}{d\phi} = -v'$, $\dfrac{dH}{dv} = \phi'$.

The former of these gives $A\omega_1\dfrac{d\omega_1}{d\phi}+B\omega_2\dfrac{d\omega_2}{d\phi}-\dfrac{dU}{d\phi}=-C\dfrac{d\omega_3}{dt}$,

which is the same as $\qquad A\omega_1\dfrac{B\omega_2}{A}-B\omega_2\dfrac{A\omega_1}{B}-\dfrac{dU}{d\phi}=-C\dfrac{d\omega_3}{dt}$,

and this leads at once to the third Euler's equation in Art. 251. The latter of the two Hamiltonian equations leads to one of the geometrical equations of Art. 256. Thus the six Hamiltonian equations are equivalent to all the three dynamical and the three geometrical Eulerian equations.

Ex. 3. A sphere rolls down a rough inclined plane as described in Art. 144. We have $T=\tfrac{7}{10}ma^2\theta'^2$ and $U=mga\theta\sin a$. Is it correct to equate H to the difference of these functions? Verify the answer by obtaining the equations of motion given in Art. 144.

Ex. 4. A system being referred to co-ordinates θ, ϕ, &c., and the corresponding momenta u, v, &c., in the Hamiltonian manner, it is desired to change the co-ordinates to x, y, &c., where θ, ϕ, &c. are given functions of x, y, &c. Show that if ξ, η, &c. be the corresponding momenta, then

$$\xi=u\theta_x+v\phi_x+\ldots,\qquad \eta=u\theta_y+v\phi_y+\ldots,\qquad \&c.=\&c.$$

where the suffixes as usual denote partial differentiations. Show also by a purely analytical transformation that the Hamiltonian equations with θ, u, &c. change into the corresponding ones with x, ξ, &c.

Ex. 5. The Lagrangian function is a function of θ, ϕ, &c. and θ', ϕ', &c. In what precedes we have taken the reciprocal function with regard to θ', ϕ', &c., but we might also have taken the reciprocal function with regard to θ, ϕ, &c. The following example will illustrate this.

Let T_1, or L, be the Lagrangian function, and in order to keep the notation as nearly the same as possible, let $U=\dfrac{dT_1}{d\theta}$, $V=\dfrac{dT_1}{d\phi}$, &c. Then if T_3 be the reciprocal function of T_1, the transformation corresponding to Sir W. Hamilton's leads to the typical equations $\qquad \theta=\dfrac{dT_3}{dU}$, $\quad U=-\dfrac{d}{dt}\dfrac{dT_3}{d\theta'}$.

To show this, it is sufficient to notice that $T_3=-T_1+U\theta+V\phi+\ldots$ Then by the lemma in Art. 410 we have $\dfrac{dT_3}{d\theta'}=-\dfrac{dT_1}{d\theta'}$, and $\dfrac{dT_3}{dU}=\theta$, whence the results follow at once by Lagrange's equations.

417. The analogy to reciprocation in Geometry. The Hamiltonian transformation of Lagrange's equations bears a remarkable analogy to the transformation by reciprocation in Geometry. Thus suppose the system to have three co-ordinates θ, ϕ, ψ, and let the semi vis viva T_1 be a homogeneous quadratic function of the velocities θ', ϕ', ψ'. We may regard θ', ϕ', ψ' as the Cartesian co-ordinates of a representative point P, the position and path of which will exhibit to the eye the instantaneous motion of the system. These co-ordinates of P may be found from Lagrange's equations. In the same way we may regard the Hamiltonian variables u, v, w as the Cartesian co-ordinates of another point Q whose position and path will also exhibit the instantaneous motion of the system.

Taking any instantaneous values of θ', ϕ', ψ' the point P will lie somewhere on the quadric $T_1=U$ where U is the instantaneous value of the force function. Then since $u = \dfrac{dT_1}{d\theta'}$, $v = \dfrac{dT_1}{d\phi'}$, $w = \dfrac{dT_1}{d\psi'}$, we see that Q will also lie on a quadric, which is the polar reciprocal of the quadric T_1 with regard to a sphere whose centre is at the origin, and whose radius is equal to $\sqrt{2U}$.

Let this reciprocal quadric be $T_2 = U$. Then, since these quadrics possess reciprocal properties, we see that $\theta' = \dfrac{dT_2}{du}$, $\phi' = \dfrac{dT_2}{dv}$, $\psi' = \dfrac{dT_2}{dw}$.

Ex. 1. If the coefficients of the two quadrics T_1 and T_2 be functions of any quantity θ, show *geometrically* that $\dfrac{dT_1}{d\theta} = -\dfrac{dT_2}{d\theta}$. Thence deduce the remaining three of the Hamiltonian equations, viz. $-u' = \dfrac{dH}{d\theta}$, $-v' = \dfrac{dH}{d\phi}$, $-w' = \dfrac{dH}{d\psi}$, where $H = T_2 - U$. See the author's essay on "Stability of Motion," p. 62.

Ex. 2. Show that the form of T_2 as used in Geometry is the same as that given in Art. 413, Ex. 2.

418. The Modified Lagrangian Function.

Sir W. Hamilton transforms *all* the accented letters θ', ϕ', &c. into the corresponding letters u, v, &c. But we may also apply the Lemma to change *some only* of the Lagrangian co-ordinates into the corresponding Hamiltonian co-ordinates, leaving the others unchanged. We may thus use a mixture of the two kinds of equations. With one and the same function we can use Lagrange's equations for those co-ordinates for which they are best adapted, and the Hamiltonian equations with the remaining co-ordinates, if we think their forms preferable.

The substance of this theory, as given in Arts. 418 to 425, is taken from the author's essay on "Stability of Motion," 1876.

419. To explain this more clearly let us consider a system depending on four co-ordinates, θ, ϕ, ξ, η. Let L_1 be the Lagrangian function. Let us now suppose that we wish to use Lagrange's equations for the co-ordinates ξ, η and the Hamiltonian equations for the co-ordinates θ, ϕ. To do this we use the *two* formulæ of transformation $\dfrac{dL_1}{d\theta'} = u$, $\dfrac{dL_1}{d\phi'} = v$, and we put

$$L_2 = -L_1 + u\theta' + v\phi'.$$

We have in consequence the two sets of Hamiltonian equations,

$$\theta' = \frac{dL_2}{du}, \qquad u' = -\frac{dL_2}{d\theta},$$

$$\phi' = \frac{dL_2}{dv}, \qquad v' = -\frac{dL_2}{d\phi}.$$

We must now include ξ', η' among the unaccented letters spoken of in the Lemma of Art. 410, so that we have

$$\frac{dL_2}{d\xi'} = -\frac{dL_1}{d\xi'}, \qquad \frac{dL_2}{d\xi} = -\frac{dL_1}{d\xi},$$

with two similar equations for η. Thus the two Lagrangian equations for ξ, η are still true if we replace L_1 by L_2; so that we have the two sets of Lagrangian equations,

$$\frac{d}{dt}\frac{dL_2}{d\xi'} = \frac{dL_2}{d\xi}, \qquad \frac{d}{dt}\frac{dL_2}{d\eta'} = \frac{dL_2}{d\eta}.$$

420. The function L_2 might be called the *modified function*, but it is more convenient to give this name to the function with its sign changed. The definition may be repeated thus:—

If the Lagrangian function L be a function of θ, θ', ϕ, ϕ', &c., then the function modified for (say) the two co-ordinates θ, ϕ will be

$$L' = L - u\theta' - v\phi',$$

where $u = \dfrac{dL}{d\theta'}$, $v = \dfrac{dL}{d\phi'}$, and we suppose θ', ϕ' eliminated from the function L'. Thus L is a function of θ, ϕ, θ', ϕ' and all the other letters, L' is a function of θ, ϕ, u, v and all the other letters.

These two functions L, L' possess the property (by Art. 410) that their partial differential coefficients are the same with respect to all letters except θ', ϕ', u, v. As regards these four we have

$$\frac{dL}{d\theta'} = u, \quad \frac{dL}{d\phi'} = v, \text{ and } \frac{dL'}{du} = -\theta', \quad \frac{dL'}{dv} = -\phi'.$$

We may form the dynamical equations, for the co-ordinates with regard to which the function has been modified by the Hamiltonian rule, as if $L_2 = -L'$ were the Hamiltonian function, and for the remaining co-ordinates by the Lagrangian rule, as if either L_2 or L' were the Lagrangian function.

The function L_2 may be also called the reciprocal function of the Lagrangian function L_1 with regard to the co-ordinates θ, ϕ, &c., because it is obtained from L_1 just as T_2 is obtained from T_1 in Art. 410, except that we operate only on such of the co-ordinates as we please. It is however convenient to use the two words in slightly different senses. We shall use the word Reciprocation when we change *all* the co-ordinates, and Modification when we change only some.

421. *To find a general expression for the modified Lagrangian function after the necessary eliminations have been performed.*

Let the vis viva $2T$ be given by the homogeneous quadratic expression

$$T = T_{\theta\theta}\frac{\theta'^2}{2} + T_{\theta\phi}\theta'\phi' + \ldots + T_{\xi\xi}\frac{\xi'^2}{2} + T_{\theta\xi}\theta'\xi' + \ldots,$$

so that the Lagrangian function is $L = T + U$, where U is a function of the co-ordinates θ, ϕ, ξ, &c. We intend to modify L with regard to θ, ϕ, &c., leaving ξ, η, &c. to be operated on by Lagrange's rule. We therefore have according to

Art. 420 to eliminate θ', ϕ', &c. by help of the equations

$$\left.\begin{array}{l} T_{\theta\theta}\theta' + T_{\theta\phi}\phi' + \ldots = u - T_{\theta\xi}\xi' - T_{\theta\eta}\eta' - \ldots \\ T_{\theta\phi}\theta' + T_{\phi\phi}\phi' + \ldots = v - T_{\phi\xi}\xi' - T_{\phi\eta}\eta' + \ldots \\ \quad\&\text{c.} = \&\text{c.} \end{array}\right\} \ldots\ldots\ldots\ldots(1).$$

For the sake of brevity let us call the right-hand members of these equations $u - X$, $v - Y$, &c. Since T is a homogeneous function, we have

$$\left.\begin{array}{l} T = T_{\xi\xi}\dfrac{\xi'^2}{2} + T_{\xi\eta}\xi'\eta' + \ldots \\ \quad + \tfrac{1}{2}\theta'(u + X) + \tfrac{1}{2}\phi'(v + Y) + \&\text{c.} \end{array}\right\}\ldots\ldots\ldots\ldots\ldots(2).$$

But by definition the modified function $L' = -L_2$ is

$$L' = L - u\theta' - v\phi' - \ldots$$

$$\left.\begin{array}{l} = T_{\xi\xi}\dfrac{\xi'^2}{2} + T_{\xi\eta}\xi'\eta' + \ldots + U \\ \quad - \tfrac{1}{2}\theta'(u - X) - \tfrac{1}{2}\phi'(v - Y) - \&\text{c.} \end{array}\right\}\ldots\ldots\ldots\ldots(3).$$

Solving equations (1) we find θ', ϕ', &c. in terms of ξ', η', &c. by the help of determinants. Substituting their values in the expression (3), we find

$$L' = T_{\xi\xi}\frac{\xi'^2}{2} + T_{\xi\eta}\xi'\eta' + \&\text{c.} + U + \frac{1}{2\Delta}\begin{vmatrix} 0, & u - X, & v - Y, & \ldots \\ u - X, & T_{\theta\theta}, & T_{\theta\phi}, & \ldots \\ v - Y, & T_{\theta\phi}, & T_{\phi\phi}, & \ldots \\ \ldots & \ldots & \ldots & \ldots \end{vmatrix},$$

where Δ is the discriminant of the terms in T which contain only θ', ϕ', &c. It may also be derived from the determinant just written down by omitting the first row and the first column.

We may expand this determinant, and write the modified function in the form

$$L' = T_{\xi\xi}\frac{\xi'^2}{2} + T_{\xi\eta}\xi'\eta' + \&\text{c.} + U$$

$$+ \frac{1}{2\Delta}\begin{vmatrix} 0, & u, & v, & \ldots \\ u, & T_{\theta\theta}, & T_{\theta\phi}, & \ldots \\ v, & T_{\theta\phi}, & T_{\phi\phi}, & \ldots \\ \ldots & \ldots & & \end{vmatrix} + \frac{1}{2\Delta}\begin{vmatrix} 0, & X, & Y, & \ldots \\ X, & T_{\theta\theta}, & T_{\theta\phi}, & \ldots \\ Y, & T_{\theta\phi}, & T_{\phi\phi}, & \ldots \\ \ldots & \ldots & & \end{vmatrix} - \frac{1}{\Delta}\begin{vmatrix} 0, & X, & Y, & \ldots \\ u, & T_{\theta\theta}, & T_{\theta\phi}, & \ldots \\ v, & T_{\theta\phi}, & T_{\phi\phi}, & \ldots \\ \ldots & \ldots & \ldots & \ldots \end{vmatrix},$$

where u, v, &c. as usual stand for $\dfrac{dT}{d\theta'}$, $\dfrac{dT}{d\phi'}$, &c., and X, Y, &c. are given by

$$X = T_{\theta\xi}\xi' + T_{\theta\eta}\eta' + \ldots, \qquad Y = T_{\phi\xi}\xi' + T_{\phi\eta}\eta' + \ldots, \qquad \&\text{c.} = \&\text{c.},$$

so that X, Y, &c. may be obtained from u, v, &c. by omitting the terms which contain θ', ϕ', &c., i.e. the co-ordinates to which we intend to apply the Hamiltonian equations.

It should be noticed that the first of the three determinants in the expression for L' contains only the momenta u, v, &c. and the co-ordinates. The second does not contain u, v, &c. but is a quadratic function of ξ', η', &c. The third contains terms of the first degree in ξ', η', &c. multiplied by the momenta u, v, &c.

422. Case of absent co-ordinates. In many cases of small oscillations about a state of steady motion, and in some other problems, the Lagrangian function L does not contain some of the co-ordinates as θ, ϕ, &c., though it is a function of their differential coefficients θ', ϕ', &c.; at the same time it may contain the other co-ordinates ξ, η, &c., as well as their differential coefficients ξ', η', &c. When this occurs, the Lagrangian equations for θ, ϕ, &c. become $\dfrac{d}{dt}\dfrac{dL}{d\theta'} = 0$, &c. Integrating, we have

$$\frac{dL}{d\theta'} = u, \qquad \frac{dL}{d\phi'} = v, \quad \&c.$$

where u, v, &c. are absolute constants whose values are known from the initial conditions. By the help of these equations we may find θ', ϕ', &c. in terms of ξ', η', &c., so that the problem is really reduced to that of finding ξ, η, &c.

The names *kinosthenic* and *speed co-ordinates* have both been suggested by Prof. J. J. Thomson for co-ordinates which enter into the Lagrangian function only through their differential coefficients, (*Phil. Trans.* 1885, and *Applications of dynamics to physics and chemistry*, 1888).

We may now simplify the process of finding these remaining co-ordinates ξ, η, &c. by modifying the Lagrangian function so as to eliminate the variables θ', ϕ', &c., and introducing in their place the constant quantities u, v, &c. *We write*

$$L' = L - u\theta' - v\phi' \ldots,$$

and eliminate θ', ϕ', &c. by help of the integrals just found. The equations to find ξ, η, &c. may be deduced by treating $\pm L'$ as the Lagrangian function.

423. *When the system starts from rest the modified function takes a simple form.* Suppose the Lagrangian function L to be a homogeneous quadratic function of θ', ϕ', &c. Then, referring to the first integrals found above, and remembering that the initial values of θ', ϕ', &c. are all zero, we have

$$u = 0, \qquad v = 0, \qquad \&c. = 0.$$

Thus the modified function L' *is equal to the original function, but is differently expressed.* The function L is a function of θ', ϕ', &c.; the function L' is the value of L after we have eliminated the differential coefficients θ', ϕ', &c. by help of the first integrals.

The result of the elimination can be deduced from **Art. 421**. The first and third determinants are here zero. We have therefore

$$L' = T_{\xi\xi}\frac{\xi'^2}{2} + T_{\xi\eta}\xi'\eta' + \&c. + U + \frac{1}{2\Delta}\begin{vmatrix} 0, & X, & Y, & \ldots \\ X, & T_{\theta\theta}, & T_{\theta\phi}, & \ldots \\ Y, & T_{\theta\phi}, & T_{\phi\phi}, & \ldots \\ \ldots & \ldots & \ldots & \ldots \end{vmatrix}.$$

We may deduce this expression from the Lagrangian function L by a simple rule, viz., *omit all the terms which contain the differential coefficients* θ', ϕ', *&c. to be eliminated, and add the determinantal term written down above.*

424. Example of the Solar System. As an example let us consider the case of three particles whose masses are m_1, m_2, m_3 mutually attracting each other according to the Newtonian law and moving in any manner in one plane. Referring these to any rectangular axes, their vis viva and force-function will be functions of the six Cartesian co-ordinates and their differential coefficients. But we may move the origin and turn the axes round the origin without altering the vis viva or the force-function. It follows that each of these functions is independent of three of the co-ordinates, though it may depend on their differential coefficients with regard to the time. We may therefore *modify* the Lagrangian function and make it depend only on the three other co-ordinates.

The vis viva of the system is equal to the vis viva of the whole mass collected at the centre of gravity together with the vis viva relative to the centre of gravity. The former is easily written down and is in our case a constant; let us turn our attention to the latter.

Let G be the centre of gravity, draw $G\alpha$, $G\beta$, $G\gamma$ to represent in direction and magnitude the velocities of the three particles, *i.e.* let α, β, γ trace out their hodographs. Then the sides of the triangle α, β, γ represent the relative velocities of the particles, and the vis viva of the system is represented by $m_1 G\alpha^2 + m_2 G\beta^2 + m_3 G\gamma^2$. Since the momentum of the system relative to its centre of gravity resolved in any direction is zero, it follows that G is the centre of gravity of three particles m_1, m_2, m_3 placed at α, β, γ. By a well-known property of the centre of gravity we have

$$m_1 m_2 (\alpha\beta)^2 + \ldots\ldots = \mu \{ m_1 (G\alpha)^2 + \ldots \},$$

where μ is the sum of the masses. It immediately follows that the

vis viva of any system relative to its centre of gravity $= \dfrac{\Sigma m_1 m_2 v_{12}{}^2}{\Sigma m}$,

where v_{12} is the relative velocity of the particles m_1, m_2. This formula for the relative vis viva is evidently true for any number of particles. It was obtained by Sir R. Ball by an analytical demonstration in the Astronomical Notices for March, 1877.

Let a, b, c, A, B, C be, as usual, the sides and angles of the triangle formed by joining the particles. Let θ be the angle made by the side c with any straight line fixed in space. Let accents as usual denote differential coefficients with regard to the time. Then we have

$$\Sigma m_1 m_2 v_{12}{}^2 = m_1 m_2 \{ c'^2 + c^2 \theta'^2 \} + m_1 m_3 \{ b'^2 + b^2 (\theta' + A')^2 \} + m_2 m_3 \{ a'^2 + a^2 (\theta' - B')^2 \}.$$

Thus, if $2T$ be the vis viva relative to the centre of gravity, we have

$$2T = P\theta'^2 + 2Q\theta' + R,$$

where P, Q, R are functions only of the triangle, and not of θ. We have

$$\mu P = m_1 m_2 c^2 + m_1 m_3 b^2 + m_2 m_3 a^2,$$
$$\mu Q = \quad m_3 (m_1 b^2 A' - m_2 a^2 B'),$$
$$\mu R = m_1 m_2 c'^2 + m_1 m_3 (b'^2 + b^2 A'^2) + m_2 m_3 (a'^2 + a^2 B'^2).$$

How we shall express these must depend on the co-ordinates we wish to use. Thus we may choose any three parts of the triangle, except the three angles, as co-ordinates.

Ex. Supposing it to be convenient to choose the distances b and c of two of the particles from the third, and the angle A subtended by those two at that third particle, as the co-ordinates of the triangle, show that P, Q, R may be expressed in terms solely of b, c, A and their differential coefficients by the help of the following results $a^2 = b^2 + c^2 - 2bc \cos A$,

$$\frac{d}{dt} (bc \sin A) = b^2 A' + a^2 B' + 2bc' \sin A,$$

$$a'^2 + a^2 B'^2 = b'^2 + c'^2 - 2b'c' \cos A + b^2 A'^2 + 2bA'c' \sin A.$$

These admit of easy geometrical demonstrations.

425. We may also modify the Lagrangian function with regard to θ. To do this we put $u = \dfrac{dT}{d\theta'} = P\theta' + Q$. We notice that, since the force-function U is not a function of θ, u is by Art. 422 an absolute constant. We now form the modified function

$$L' = L - u\theta' = \frac{PR - Q^2 + 2uQ - u^2}{2P} + U.$$

This function may now be used as if it were the Lagrangian function to find any changes in the triangle joining the three particles.

We may also notice that the angular velocity in space of the side of the triangle joining m_1, m_2 is given by the equation

$$P\theta' + Q = u,$$

where θ' is the angular velocity required and u is a constant.

Ex. 1. Show that P is equal to the moment of inertia of the three particles about the centre of gravity.

Ex. 2. Show that $\mu^2 (PR - Q^2)$ may be written in the symmetrical form

$$\{m_1 m_2 c^2 + m_2 m_3 b^2 + m_3 m_1 a^2\} \, \{m_1 m_2 c'^2 + m_2 m_3 b'^2 + m_3 m_1 a'^2\}$$
$$+ m_1 m_2 m_3 \{m_1 (bcA')^2 + m_2 (caB')^2 + m_3 (abC')^2\}.$$

Ex. 3. Show that the quantity u is equal to the angular momentum of the system about the centre of gravity. See Arts. 397 and 403.

Ex. 4. Show that we may take for μQ either of the forms $m_1 (m_2 c^2 B' - m_3 b^2 C')$, or $m_2 (m_3 a^2 C' - m_1 c^2 A')$, the effect of the change being to add to the Lagrangian function L' a quantity equal to B' or C' respectively. See Art. 400.

426. **Non-Conservative Forces.** *To explain how Lagrange's equations are to be used when some of the forces are non-conservative.*

Lagrange's equations in the form given in Art. 399 can be used only when the forces which act on the system have a force-function. If however $P\delta\theta$ be the virtual work of the impressed forces obtained by varying θ only, $Q\delta\phi$ the virtual work obtained by varying ϕ only, and so on, it is clear that Lagrange's equations may be written in the typical form $\dfrac{d}{dt} \dfrac{dT}{d\theta'} - \dfrac{dT}{d\theta} = P$.

427. It is often convenient to separate the forces which act on the system into two sets. *Firstly* those which are conservative. The parts of P, Q, &c. due to these forces may be found by differentiating the force-function with regard to θ, ϕ, &c. *Secondly* those which are non-conservative, such as friction, some kinds of resistances, &c. The parts of P, Q, &c. due to these must be found by the usual methods given in statics for writing down virtual work.

Though the non-conservative forces do not admit of a force-function, yet sometimes their virtual works may be represented by a differential coefficient of another kind. Thus suppose some of the forces acting on a particle of a body to be such that their resolved parts parallel to three rectangular axes fixed in space are proportional to the velocities of the particle in those directions. The virtual work of these forces is

$$\Sigma\,(\mu_1 x'\delta x + \mu_2 y'\delta y + \mu_3 z'\delta z),$$

where μ_1, μ_2, μ_3 are three constants which are negative if the forces are resistances. For example, if the particles be moving in a medium whose resistance is equal to the velocity multiplied by a constant κ, then μ_1, μ_2, μ_3 are each equal to $-\kappa$. Put

$$-F = \frac{1}{2}\,\Sigma\,(\mu_1 x'^2 + \mu_2 y'^2 + \mu_3 z'^2).$$

Since (x, y, z) are functions of θ, ϕ, &c. given by the geometry of the system we have, as in Art. 396,

$$x' = \frac{dx}{dt} + \frac{dx}{d\theta}\,\theta' + \dots$$

with similar expressions for the other co-ordinates. Substituting we have F expressed as a function of θ, ϕ, &c., θ', ϕ', &c. We also notice that, as in Art. 397, $\dfrac{dx'}{d\theta'} = \dfrac{dx}{d\theta}$. Differentiating F partially we have

$$-\frac{dF}{d\theta'} = \Sigma\left(\mu_1 x'\frac{dx'}{d\theta'} + \&c.\right) = \Sigma\left(\mu_1 x'\frac{dx}{d\theta} + \&c.\right).$$

$$\therefore\ -\frac{dF}{d\theta'}\,\delta\theta - \frac{dF}{d\phi'}\,\delta\phi - \&c. = \Sigma\left\{\mu_1 x'\left(\frac{dx}{d\theta}\,\delta\theta + \frac{dx}{d\phi}\,\delta\phi + \dots\right) + \&c.\right\}$$

$$= \Sigma\,(\mu_1 x'\delta x + \&c.).$$

In this case, therefore, if U be the force-function of the conservative forces, F the function just defined, $\Theta\delta\theta$, $\Phi\delta\phi$, &c. the virtual works of the remaining forces, Lagrange's equations may be written

$$\frac{d}{dt}\frac{dT}{d\theta'} - \frac{dT}{d\theta} = \frac{dU}{d\theta} - \frac{dF}{d\theta'} + \Theta,$$

with similar equations for ϕ, ψ, &c.

We may notice that, if the geometrical equations do not contain the time explicitly, the function F is a quadratic homogeneous function of θ', ϕ', &c.

If the forces whose effects are included in F be *resistances*, then μ_1, μ_2, μ_3, &c. are all negative. In this case F is essentially a positive function of the velocities, and in this respect it resembles the function T representing half the vis viva.

If we treat the equations written down above exactly as Lagrange's equations are treated in Art. 407 to obtain the principle of vis viva we find

$$\frac{d}{dt}\,(T - U) = \theta'\Theta + \&c. - \frac{dF}{d\theta'}\,\theta' - \&c.,$$

but in this case F also is a homogeneous function of θ', &c. Hence we find

$$\frac{d}{dt}\,(T - U) = \theta'\Theta + \&c. - 2F.$$

We therefore conclude that, if the geometrical equations do not contain the time explicitly, and if there be no forces present but those which may be included in the potential function U and in the function F, then F represents half the rate at which energy is leaving the system, *i.e.* is dissipated.

The use of this function was suggested by Lord Rayleigh in the *Proceedings of the London Mathematical Society*, June, 1873. The function F has been called by him the Dissipation function.

428. Ex. 1. If any two particles of a dynamical system act and react on each other with a force whose resolved parts in three fixed directions at right angles are proportional to the relative velocities of the particles in those directions, show that these may be included in the dissipation function F. If V_x, V_y, V_z be the components of the velocities, $\mu_1 V_x$, $\mu_2 V_y$, $\mu_3 V_z$ the components of the force of repulsion, the part of F due to these is $-\frac{1}{2} \Sigma (\mu_1 V_x^2 + \mu_2 V_y^2 + \mu_3 V_z^2)$. This example is taken from the paper just referred to.

Ex. 2. A solid body moves in a medium which acts on *every* element of the surface with resisting forces partly frictional and partly normal to the surface. Each of these when referred to a unit of area is equal to the velocity resolved in its own direction multiplied by the same constant κ. Show that these resistances may be included in a dissipation function F, where

$$F = \frac{\kappa}{2} \{ \sigma (u^2 + v^2 + w^2) + A\omega_x^2 + B\omega_y^2 + C\omega_z^2 - 2D\omega_y\omega_z - 2E\omega_z\omega_x - 2F\omega_x\omega_y \},$$

where σ is the area, A, B, &c. the moments and products of inertia of the *surface* of the body, and (u, v, w) the resolved velocities of the centre of gravity of σ.

429. Indeterminate Multipliers, &c. *To explain how Lagrange's equations can be used in some cases when the geometrical equations contain differential coefficients with regard to the time.*

It has been pointed out in Art. 396, that the independent variables θ, ϕ, &c. used in Lagrange's equations must be so chosen that all the co-ordinates of the bodies in the system can be expressed in terms of them without introducing θ', ϕ', &c. But when we have to discuss a motion like that of a body rolling on a perfectly rough surface, the condition that the relative velocity of the points in contact is zero may sometimes be expressed by an equation which, like that given in Art. 137, necessarily involves differential coefficients of the co-ordinates. In some cases the equation expressing this condition is integrable. For example: when a sphere rolls on a rough plane, as in Art. 144, the condition is $x' - a\theta' = 0$, which by integration becomes $x - a\theta = b$, where b is some constant. In such cases we may use the condition as one of the geometrical relations of the motion, thus reducing by one the number of independent variables.

But when the conditions cannot easily be cleared of differential coefficients, it is often convenient to introduce the reactions and frictions into the equations among the non-conservative forces in the manner explained in Art. 427. Each reaction has an accompanying equation of condition, and thus we always have sufficient equations to eliminate the reactions and determine the co-ordinates of the system.

The elimination of the reactions may generally be most easily effected by recurring to the general equation of virtual work and giving only such displacements to the system as make the virtual work of these forces disappear. Suppose, to fix our ideas, that a body is rolling on a perfectly rough surface. Let θ, ϕ, &c. be the six co-ordinates of the body, then by Art. 137, there will be three equations of the form

$$L_1 = A_1\theta' + B_1\phi' + \ldots = 0 \ldots\ldots\ldots\ldots\ldots(1),$$

the other two being derived from this by writing 2 and 3 for the suffix. These three equations express the fact that the resolved velocities in three directions of the point of contact are zero. The equation of virtual work may be written (Art. 398)

$$\left(\frac{d}{dt}\frac{dT}{d\theta'} - \frac{dT}{d\theta}\right)\delta\theta + \&c. = \frac{dU}{d\theta}\delta\theta + \&c. \ldots\ldots\ldots\ldots(2),$$

where U is the force-function of the impressed forces. Since the virtual work of the reactions at the point of contact have been omitted, this equation is not true for all variations of θ, ϕ, &c., but only for such as make the body roll on the rough surface. But the geometrical equations L_1, L_2, L_3 express the fact that the body rolls in some manner, hence $\delta\theta$, $\delta\phi$, &c. are connected by three equations of the form

$$A_1\delta\theta + B_1\delta\phi + \ldots = 0\ldots\ldots\ldots\ldots\ldots(3).$$

If we use the method of indeterminate multipliers (see Art. 400), the equations of virtual work are transformed in the usual manner into

$$\frac{d}{dt}\frac{dT}{d\theta'} - \frac{dT}{d\theta} = \frac{dU}{d\theta} + \lambda\frac{dL_1}{d\theta'} + \mu\frac{dL_2}{d\theta'} + \nu\frac{dL_3}{d\theta'} \ldots\ldots\ldots(4),$$

with similar equations for the other co-ordinates ϕ, ψ, &c. These joined to the three equations L_1, L_2, L_3 are sufficient to determine the co-ordinates of the body and λ, μ, ν.

This process will be very much simplified, if we prepare the geometrical equations L_1, L_2, L_3 by elimination, so that one differential coefficient, as θ', is absent from all but the first equation, another, as ϕ', absent from all but the second, and so on. When this has been done, the equation for θ becomes

$$\frac{d}{dt}\frac{dT}{d\theta'} - \frac{dT}{d\theta} = \frac{dU}{d\theta} + \lambda\frac{dL_1}{d\theta'} \ldots\ldots\ldots\ldots\ldots(5).$$

Thus λ is found at once. The values of μ and ν may be found from the corresponding equations for ϕ, ψ. We may then substitute their values in the remaining equations.

430. The method of indeterminate multipliers is really an introduction of the unknown reactions into Lagrange's equations. Thus let R_1, R_2, R_3 be the resolved parts of the reaction at the point of contact in the directions of the three straight lines

used in forming the equations L_1, L_2, L_3. Then L_1, L_2, L_3 are proportional to the resolved relative velocities of the points of contact. Let these velocities be $\kappa_1 L_1$, $\kappa_2 L_2$, $\kappa_3 L_3$. Then if θ only be varied the virtual work of R_1 is $\kappa_1 A_1 \delta\theta$, which may be written $\kappa_1 \dfrac{dL_1}{d\theta'} \delta\theta$. Similarly the virtual works of R_2 and R_3 are $\kappa_2 \dfrac{dL_2}{d\theta'} \delta\theta$ and $\kappa_3 \dfrac{dL_3}{d\theta'} \delta\theta$. Hence, by Art. 426, Lagrange's equations are of the form

$$\frac{d}{dt}\frac{dT}{d\theta'} - \frac{dT}{d\theta} = \frac{dU}{d\theta} + \kappa_1 R_1 \frac{dL_1}{d\theta'} + \kappa_2 R_2 \frac{dL_2}{d\theta'} + \kappa_3 R_3 \frac{dL_3}{d\theta'}.$$

Comparing this with the equations obtained by the method of indeterminate multipliers we see that λ, μ, ν are proportional to the resolved parts of the reactions. The advantage of using the method of indeterminate multipliers is that the reactions are introduced with the least amount of algebraic calculation, and in just that manner which is most convenient for the solution of the problem.

431. Ex. *Form by Lagrange's method the equations of motion of a homogeneous sphere rolling on an inclined plane under the action of gravity.*

Let the axis of x be taken down the plane along the line of greatest slope, and let the axis of y be horizontal and that of z normal to the plane. Let (x, y, a) be the co-ordinates of the centre of gravity of the sphere. θ, ϕ, ψ the angular co-ordinates of three diameters at right angles fixed in the sphere in the manner explained in Art. 256. Then, if the mass be taken as unity, the vis viva is by Art. 365, Ex. 1,

$$2T = x'^2 + y'^2 + k^2\{(\phi' + \psi'\cos\theta)^2 + \theta'^2 + \sin^2\theta\psi'^2\}.$$

The resolved velocities parallel to the axes of x and y of the point of the sphere in contact with the plane are to be zero. These give the conditions $x' - a\omega_y = 0$, $y' + a\omega_x = 0$. By Art. 257 these conditions will be found to lead to the equations

$$L_1 = x' - a\theta'\cos\psi - a\phi'\sin\theta\sin\psi = 0$$
$$L_2 = y' - a\theta'\sin\psi + a\phi'\sin\theta\cos\psi = 0.$$

Also, if g be the resolved part of gravity along the plane, and C any constant, we have $U = gx + C$.

The general equation of motion is

$$\frac{d}{dt}\frac{dT}{dq'} - \frac{dT}{dq} = \frac{dU}{dq} + \lambda\frac{dL_1}{dq'} + \mu\frac{dL_2}{dq'},$$

where q stands for any one of the five quantities x, y, θ, ϕ, ψ. Taking these in turn, we have

$$\left.\begin{array}{c} x'' = g + \lambda, \qquad y'' = \mu, \\ k^2(\theta'' + \phi'\psi'\sin\theta) = -\lambda a\cos\psi - \mu a\sin\psi, \\ k^2\dfrac{d}{dt}(\phi' + \psi'\cos\theta) = -\lambda a\sin\theta\sin\psi + \mu a\sin\theta\cos\psi, \\ k^2\dfrac{d}{dt}(\phi'\cos\theta + \psi') = 0, \end{array}\right\}.$$

The last equation shows that $\phi' \cos\theta + \psi'$ is constant. From this we infer, by Art. 257, that the angular velocity ω_z of the sphere about a normal to the plane is constant throughout the motion. Eliminating μ from the two preceding equations, and substituting for ψ'' from the last we find

$$-\frac{\lambda a}{k^2} = \theta'' \cos\psi + \phi'' \sin\theta \sin\psi + \phi'\psi' \sin\theta \cos\psi - \theta'\psi' \sin\psi + \theta'\phi' \cos\theta \sin\psi.$$

But this is $\dfrac{x''}{a}$. In the same way we find $-\dfrac{\mu a}{k^2} = \dfrac{y''}{a}$. Substituting these values of λ and μ in the first two of Lagrange's equations we have

$$x''(a^2 + k^2) = a^2 g \qquad y''(a^2 + k^2) = 0.$$

These are the equations of motion of a particle acted on by a constant force parallel to the axis of x. The centre of gravity of the sphere therefore describes a parabola, as if it were under a constant acceleration, equal to $\frac{5}{7}g$, tending along the line of greatest slope.

This solution is rather complicated, but the problem has been selected to show how we may use Lagrange's equations as specially illustrating the remarks made in Art. 429. So far as this particular problem is concerned a very simple and short solution may be obtained by the ordinary processes of resolving and taking moments. For this we refer the reader to Art. 269 and also to the chapter *on the motion of a body under any forces* in the second part of this work.

EXAMPLES *.

1. Two weights of masses m and $2m$ respectively are connected by a string which passes over a smooth pulley of mass m. This pulley is suspended by a string passing over a smooth fixed pulley, and carrying a mass $4m$ at the other end. Prove that the mass $4m$ moves with an acceleration which is one twenty-third part of gravity.

2. A uniform rod of mass $3m$ and length $2l$ has its middle point fixed, and a mass m attached at one extremity. The rod when in a horizontal position is set rotating about a vertical axis through its centre, with an angular velocity equal to $\sqrt{\dfrac{2ng}{l}}$. Show that the heavy end of the rod will fall till the inclination of the rod to the vertical is $\cos^{-1}(\sqrt{n^2+1} - n)$, and will then rise again.

3. A rod of length $2l$ is constrained to move on the surface of a hyperboloid of revolution of one sheet with its axis of symmetry vertical, so that the rod always lies along a generator. If the rod start from rest, show that

$$r'^2 - 2ar'\theta' \sin\alpha + a^2\theta'^2 + \sin^2\alpha\,(r^2 + \tfrac{1}{3}l^2)\,\theta'^2 + 2g\cos\alpha\,(r - r_0) = 0,$$
$$\{a^2 + \sin^2\alpha\,(r^2 + \tfrac{1}{3}l^2)\}\,\theta' - ar'\sin\alpha = 0,$$

where r is the distance measured along a generator from the centre of gravity to the principal circular section, θ is the excentric angle of the point in which the generator meets this circular section, a is the radius of the circular section, and α is the inclination of the rod to the vertical.

* These examples are taken from the Examination Papers which have been set in the University and in the Colleges.

4. A ring of mass m and radius b rolls inside a perfectly rough ring of mass M and radius a, which is moveable about its centre in a vertical plane. If θ, ϕ be the angles turned through by the rings from their position of equilibrium, prove that

$$a\theta + b\phi = (a - b)\,\psi, \qquad Ma\theta'' = mb\phi'', \qquad (2M + m)\,(a - b)\,\psi'' = - (M + m)\,g\sin\psi.$$

5. If l, m, n be the direction-cosines with respect to fixed axes of a rod moving in any manner in space, and if V be the potential energy, prove that

$$\frac{1}{l}\left(I\frac{d^2 l}{dt^2} + \frac{dV}{dl} \right) = \frac{1}{m}\left(I\frac{d^2 m}{dt^2} + \frac{dV}{dm} \right) = \frac{1}{n}\left(I\frac{d^2 n}{dt^2} + \frac{dV}{dn} \right),$$

where I is the moment of inertia of the rod about an axis through its centre perpendicular to its length. See Art. 400.

6. A particle of mass m moves in one plane, and its motion is referred to areal co-ordinates x, y, z. If $2T$ be the vis viva, and V the potential energy expressed as a homogeneous function of the areal co-ordinates, prove that

$$2T = - m\,(a^2 y'z' + b^2 z'x' + c^2 x'y'),$$

$$m\,(b^2 z'' + c^2 y'') - 2\frac{dV}{dx} = m\,(c^2 x'' + a^2 z'') - 2\frac{dV}{dy} = m\,(a^2 y'' + b^2 x'') - 2\frac{dV}{dz}.$$

7. A heavy rod, whose length is $2a$, slips down with its extremities in contact with a smooth horizontal floor and a smooth vertical wall; the rod not being initially in a plane perpendicular to the wall. If θ be the inclination of the rod to the vertical, and ψ the inclination of the horizontal projection of the rod to the intersection of the planes, prove that

$$4\frac{d^2}{dt^2}\,(\cos\theta) = \cot\theta\,\sec\psi\,\frac{d^2}{dt^2}\,(\sin\theta\cos\psi) - \frac{3g}{a},$$

$$4\frac{d^2}{dt^2}\,(\sin\theta\sin\psi) = \tan\psi\,\frac{d^2}{dt^2}\,(\sin\theta\cos\psi).$$

8. A particle moves under the action of two centres of repulsive force F and G tending from two fixed points, at a distance $2c$ from each other. Show that the Lagrangian equations of motion may be written in the form

$$\frac{d}{dt}\frac{dT}{d\lambda'} - \frac{dT}{d\lambda} = F + G, \qquad \frac{d}{dt}\frac{dT}{d\mu'} - \frac{dT}{d\mu} = F \sim G,$$

where λ and μ are the elliptic co-ordinates of the particle referred to the fixed points as foci, and

$$\frac{2T}{\lambda^2 - \mu^2} = \frac{\lambda'^2}{\lambda^2 - c^2} + \frac{\mu'^2}{c^2 - \mu^2}.$$

9. If r, θ be the polar co-ordinates of a particle of mass m which describes an orbit under the action of a central force F tending to the pole, and u, v be the corresponding momenta, prove that the Hamiltonian function is $H = \dfrac{u^2}{2m} + \dfrac{v^2}{2mr^2} + \int F\,dr$. Thence deduce the Hamiltonian equations of motion $u = mr'$, $v = mr^2\theta'$, $mr^3\,(u' + F) = v^2$, $v' = 0$.

CHAPTER IX.

Small Oscillations.

Oscillations with One Degree of Freedom.

432. When a system of bodies admits of only one independent motion and is making small oscillations about some mean position, or some mean state of motion, it is in general our object to reduce the equation of motion to the form

$$\frac{d^2x}{dt^2} + 2a\frac{dx}{dt} + bx = c,$$

where x is some small quantity which determines the position of the system at the time t. This reduction is effected by neglecting the square of the small quantity x.

433. **Meaning of the Terms.** We suppose the equation to be obtained by writing down the equations of motion of all the particles, and then eliminating the reactions. Let us consider the case in which the system is displaced from a position of equilibrium. We represent the amount of displacement by some letter x such that, x being known, the position of every particle can be deduced from the geometrical conditions of the system. The displacement ξ of any particle m is therefore some function of x, and since the square of x is to be neglected in a small oscillation we have by Maclaurin's theorem $\xi = G + Hx$, where G and H are some constants depending on the position of the particle in the system. The effective forces on m are (1) $Hm\ddot{x}$ along the tangent to its arc of oscillation, and (2) a centrifugal force which has $m\dot{x}^2$ in the numerator, and may therefore be neglected. The effective forces therefore contribute terms of the form $\ddot{x}$ to the differential equation.

Next let us consider the impressed forces on the system. These are of three kinds.

(1) The system being displaced the forces of the system tend to bring it back to its position of equilibrium, if this position be stable. These forces are all functions of x, and since the square of x is neglected, they contribute terms of the form $c - bx$ to the equation. The terms $c - bx$ therefore represent the *natural forces of restitution*.

(2) There may be some *forces of resistance* acting at special points of the system which depend on the velocities of the particles. The velocity of any such

particle m will be some function of $\dot{\xi}$, which, as before, may be taken equal to $H\dot{x}$. These resistances will therefore contribute terms of the form $a\dot{x}$ to the equation.

(3) We may have some small *external forces* which are functions of the time. We may, when they exist, represent them by a term $f(t)$ on the right hand side of the equation.

We see that the effective forces and the three kinds of impressed forces contribute different kinds of terms to the equation, and, since the products of these terms are to be neglected, each term comes exclusively from the source mentioned.

We propose in the first instance to omit the external forces, and to consider the motion of a system acted on only by the forces of restitution and the forces of resistance. The oscillation produced by these two together is called the *natural or free vibration*. The oscillations produced by the external forces are sometimes called *forced vibrations*, and will be considered under that heading.

434. **Solution of the Equation.** It generally happens that a, b, c are all constants, and in this case we can completely determine the oscillation. By putting $x = \dfrac{c}{b} + \xi e^{-at}$, when b is not zero, we reduce the equation to the well-known form

$$\frac{d^2\xi}{dt^2} + (b - a^2)\,\xi = 0.$$

When $b - a^2$ is positive, let us, for the sake of brevity, put $b - a^2 = n^2$. We then have

$$x = \frac{c}{b} + A e^{-at} \sin(nt + B),$$

where A and B are two undetermined constants which depend on the initial conditions of the motion. The physical interpretation of this equation is not difficult. It represents an oscillatory motion. The central position about which the system oscillates is determined by $x = \dfrac{c}{b}$. The system passes through this central position whenever $nt + B$ is a multiple of π. We therefore infer that the time of a complete oscillation is $\dfrac{2\pi}{n}$. To find the times at which the system comes momentarily to rest we put $\dfrac{dx}{dt} = 0$. This gives $\tan(nt + B) = \dfrac{n}{a}$. The extent of the oscillations on each side of the central position may be found by substituting the values of t given by this equation in the expression for $x - \dfrac{c}{b}$. Since these must occur at a constant interval equal to $\dfrac{\pi}{n}$, we see that the extent of the oscillation continually decreases, and that the successive arcs on each side of the position of equilibrium form a geometrical progression whose common ratio is $e^{-\frac{a\pi}{n}}$.

The quantity n is called the *frequency* of the oscillation. This very useful term has been introduced by Lord Rayleigh in his *Theory of Sound*.

When $b - a^2$ is negative, we put $b - a^2 = -\nu^2$. In this case the *sine* in the solution must be replaced by its exponential value, and the integral becomes,

$$x = \frac{c}{b} + Ce^{(-a+\nu)t} + De^{(-a-\nu)t},$$

where C and D are two undetermined constants. The motion is now no longer oscillatory. If a and b are both positive, ν is less than a, and in this case, whatever the initial conditions may be, x ultimately becomes equal to $\frac{c}{b}$, and the system continually approaches the position determined by this value of x. The same thing occurs if ν be greater than a, provided that the initial conditions are such that the coefficient of the exponential which has a positive index is zero.

If $b - a^2 = 0$, the integral takes a different form, and we have

$$x = \frac{c}{b} + (Et + F)\, e^{-at},$$

where E and F are two undetermined constants. If a be positive, the system continually approaches the position determined by $bx = c$.

435. When the value of x as given by these equations becomes large, the terms depending on x^2 which have been neglected in forming the equation may also become great. It is possible that these terms may alter the whole character of the motion. In such cases the equilibrium, or the undisturbed motion of the system as the case may be, is called unstable, and these equations can represent only the nature of the motion with which the system *begins* to move from its undisturbed state.

436. Ex. 1. The initial conditions of the system are such that

$$\frac{dx}{dt} = -(a+\nu)\left(x - \frac{c}{b}\right),$$

find the ultimate value of x.

Ex. 2. Show that the complete integral of $\dfrac{d^2x}{dt} + 2a\dfrac{dx}{dt} + bx = f(t)$ is

$$x = e^{-at}\left\{\dot{x}_0\,\frac{\sin nt}{n} + x_0\left(\cos nt + \frac{a}{n}\sin nt\right)\right\} + \frac{1}{n}\int_0^t e^{-a(t-t')}\sin n\,(t-t')\,f(t')\,dt',$$

where x_0, $\dot{x}_0$ are the values of x and $\dot{x}$ when $t = 0$. [Math. Tripos, 1876.]

437. It will be often found advantageous to trace the motion of the system by a figure. Let equal increments of the abscissa

of a point P represent on any scale equal increments of the time, and let the ordinate represent the deviation of the co-ordinate x from its mean value. Then the curve traced out by the representative point P will exhibit to the eye the whole motion of the system. In the case in which a and $b - a^2$ are both positive the curve takes the form here represented.

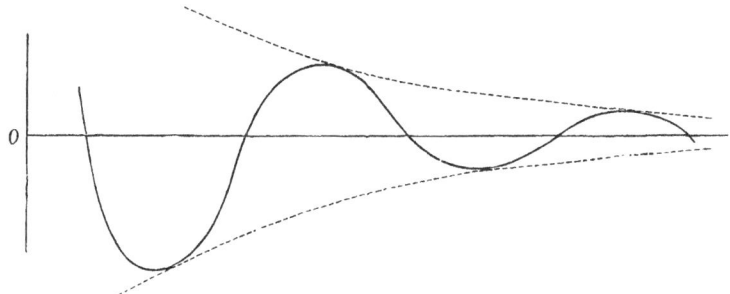

The dotted lines correspond to the ordinate $\pm Ae^{-at}$. The representative point P oscillates between these, and its path alternately touches each of them. In just the same way we may trace the representative curve for other values of a and b.

The most important case in dynamics is that in which $a = 0$. The motion is then given by

$$x - \frac{c}{b} = A \sin (\sqrt{b}\,t + B).$$

The representative curve is then the curve of sines. In this case the oscillation is usually called *harmonic*.

438. Ex. 1. A system oscillates about a mean position, and its deviation is measured by x. If x_0 and $\dot{x}_0$ be the initial values of x and $\dot{x}$, show that the system will never deviate from its mean position by so much as $\left\{\dfrac{\dot{x}_0{}^2 + 2ax_0\dot{x}_0 + bx_0{}^2}{b - a^2}\right\}^{\frac{1}{2}}$ if b be greater than a^2.

Ex. 2. A system oscillates about a position of equilibrium. It is required to find by observations on its motion the numerical values of a, b, c.

Any three determinations of the co-ordinate x at three different times will generally supply sufficient equations to find a, b, c, but some measurements can be made more easily than others. For example, the values of x when the system comes momentarily to rest can be conveniently observed, because the system is then moving slowly, and a measurement at a time slightly wrong will cause an error only of the second order, while the values of t at such times cannot be conveniently observed, because, owing to the slowness of the motion, it is difficult to determine the precise moment at which $\dot{x}$ vanishes.

If three successive values of x thus found be x_1, x_2, x_3, the ratio of the two successive arcs $x_2 - x_1$ and $x_3 - x_2$ is a known function of a and b, and one equation can thus be formed to find the constants. If the position of equilibrium is unknown, we may form a second equation from the fact that the three arcs $x_1 - \dfrac{c}{b}$, $x_2 - \dfrac{c}{b}$, $x_3 - \dfrac{c}{b}$ also form a geometrical progression. In this way we find $\dfrac{c}{b}$, which is the value of x corresponding to the position of equilibrium.

The position of equilibrium being known, the interval between two successive passages of the system through it is also a known function of a and b, and thus a third equation may be formed.

Ex. 3. A body performs rectilinear vibrations in a medium whose resistance is proportional to the velocity, under the action of an attractive force tending towards a fixed centre and proportional to the distance therefrom. If the observed period of vibration is T, and the co-ordinates of the extremities of three consecutive semi-vibrations are p, q, r, prove that the co-ordinate of the position of equilibrium, and the time of vibration if there were no resistance are respectively

$$\frac{pr - q^2}{p + r - 2q} \text{ and } T \left\{ 1 + \frac{1}{\pi^2} \left(\log \frac{p - q}{r - q} \right)^2 \right\}^{-\frac{1}{2}}. \quad \text{[Math. Tripos, 1870.]}$$

First Method of forming the Equations of Motion.

439. When the system under consideration is a single body there is a simple method of forming the equation of motion which is sometimes of great use.

Let the motion be in two dimensions.

It has been shown in Art. 205, that if we neglect the squares of small quantities we may take moments about the instantaneous centre as a fixed centre. Usually the unknown reactions will be such that their lines of action will pass through this point, their moments will then be zero, and thus we shall have an equation containing only known quantities.

Since the body is supposed to be turning about the instantaneous centre as a point fixed for the moment, the direction of motion of any point of the body is perpendicular to the straight line joining it to the centre. Conversely, when the directions of motion of two points of the body are known, the position of the instantaneous centre can be found. For if we draw perpendiculars at these points to their directions of motion, the perpendiculars must meet in the instantaneous centre of rotation.

The equation may, in general, be reduced to the form

$$Mk^2 \frac{d^2\theta}{dt^2} = \begin{pmatrix} \text{moment of impressed forces about} \\ \text{the instantaneous centre} \end{pmatrix},$$

where θ is the angle some straight line fixed in the body makes with a fixed line in space. In this formula Mk^2 is the moment of inertia of the body about the instantaneous centre, and since the left-hand side of the equation contains the small factor $\dfrac{d^2\theta}{dt^2}$

we may here suppose the instantaneous centre to have its mean or undisturbed position.   On the right-hand side there is no small factor, and we must therefore be careful either to take the moment of the forces about the *instantaneous centre in its disturbed position*, or to include the moment of any unknown reaction which passes through the instantaneous centre.

Ex. If a body with only one independent motion can be in equilibrium in the same position under two different systems of forces, and if L_1, L_2 are the lengths of the simple equivalent pendulums for these systems acting separately, then the length L of the equivalent pendulum when they act together is given by

$$\frac{1}{L} = \frac{1}{L_1} + \frac{1}{L_2}.$$

440. Ex. 1. *A homogeneous hemisphere performs small oscillations on a perfectly rough horizontal plane : find the motion.*

Let C be the centre, G the centre of gravity of the hemisphere, N the point of contact with the rough plane.   Let the radius $= a$, $CG = c$, $\theta = \angle NCG$.

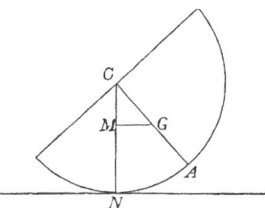

Here the point N is the centre of instantaneous rotation, because, the plane being perfectly rough, sufficient friction is called into play to keep N at rest. Hence taking moments about N

$$(k^2 + GN^2)\,\ddot\theta = -gc \,.\, \sin\theta.$$

Since we can put $GN = a - c$ in the small terms, this reduces to

$$\{k^2 + (a-c)^2\}\,\ddot\theta + gc \,.\, \theta = 0.$$

Therefore the time of a small oscillation is $2\pi \sqrt{\dfrac{k^2 + (a-c)^2}{cg}}$.

It is clear that $k^2 + c^2 =$ sq. of rad. of gyration about $C = \dfrac{2}{5} a^2$, and that $c = \dfrac{3}{8} a$.

If the plane had been smooth, M would have been on the instantaneous axis, GM being the perpendicular on CN. For the motion of N is in a horizontal direction, because the sphere remains in contact with the plane, and the motion of G is vertical by Art. 79. Hence the two perpendiculars GM, NM meet on the instantaneous axis. By reasoning similar to the above the time is found to be $2\pi \sqrt{\dfrac{k^2}{cg}}$.

Ex. 2. Two circular rings, each of radius a, are firmly jointed together at one point so that their planes make an angle $2a$ with one another, and are placed on a perfectly rough horizontal plane. Shew that the length of the simple equivalent pendulum is $\frac{1}{2}a\,(1 + 3\cos^2 a)\cos a \csc^2 a$.   [Math. Tripos.]

441. Oscillations of Cylinders. *A cylindrical surface of any form rests in stable equilibrium under gravity on another perfectly rough cylindrical surface, the axes of the cylinders being horizontal and parallel. A small disturbance being given to the upper surface, find the time of a small oscillation.*

Let BAP, $B'A'P$ be the sections of the cylinders perpendicular to their axes. Let OA, CA' be normals at those points A, A'

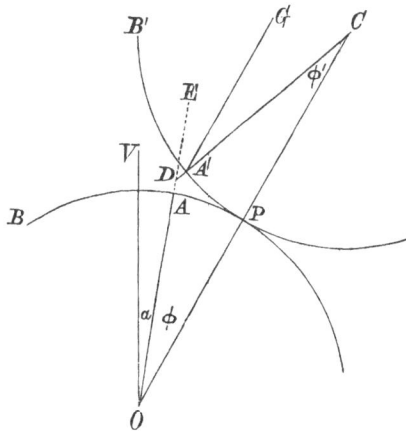

which before disturbance were in contact, and let α be the angle made by AO with the vertical. Let OPC be the common normal at the time t. Let G be the centre of gravity of the moving body, then before disturbance $A'G$ was vertical. Let $A'G = r$.

Now we have only to determine the time of oscillation when the motion decreases without limit. Hence the arcs AP, $A'P$ will be ultimately zero, and therefore C and O may be taken as the centres of curvature of AP, $A'P$. Let $\rho = OA$, $\rho' = CA'$, and let the angles AOP, $A'CP$ be denoted by ϕ, ϕ' respectively.

Let θ be the angle turned round by the body in moving from the position of equilibrium into the position $B'A'P$. Then, since before disturbance $A'C$ and AO were in the same straight line, we have $\theta = \angle CDE = \phi + \phi'$, where CA' meets OAE in D. Also, since one body rolls on the other, the arc $AP = $ arc $A'P$, $\therefore \rho\phi = \rho'\phi'$,

$$\therefore \phi = \frac{\rho'}{\rho + \rho'}\theta.$$

Again, in order to take moments about P, we require the horizontal distance of G from P; this may be found by projecting the broken line $PA' + A'G$ on the horizontal. The projection of $PA' = PA'\cos(\alpha + \theta) = \rho\phi\cos\alpha$ when we neglect the squares of

23—2

small quantities. The projection of $A'G$ is $r\theta$. Thus the horizontal distance required is $\left(\dfrac{\rho\rho'}{\rho+\rho'}\cos\alpha - r\right)\theta$.

If k be the radius of gyration about the centre of gravity, the equation of motion is

$$(k^2 + GA^2)\frac{d^2\theta}{dt^2} = -g\theta\left(\frac{\rho\rho'}{\rho+\rho'}\cos\alpha - r\right).$$

If L be the length of the simple equivalent pendulum, we have

$$\frac{k^2 + r^2}{L} = \frac{\rho\rho'}{\rho+\rho'}\cos\alpha - r.$$

442. Circle of Stability. Along the common normal at the point of contact A of the two cylindrical surfaces measure a length $AS = s$, where $\dfrac{1}{s} = \dfrac{1}{\rho} + \dfrac{1}{\rho'}$, and describe a circle on AS as diameter. Let AG, produced if necessary, cut this circle in N.

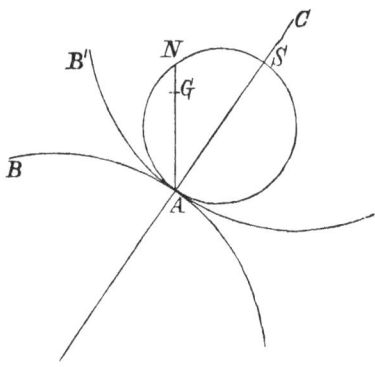

Then $GN = s\cos\alpha - r$, the positive direction being from N towards A. The length L of the simple equivalent pendulum is given by the formula

$$L \cdot GN = \text{sq. of rad. of gyration about } A.$$

It is clear from this formula, that if G* lie without the circle

---

* Let R be the radius of curvature of the path traced out by G as the one cylinder rolls on the other, then we know that $R = -\dfrac{AG^2}{NG}$, so that all points without the circle described on AS as diameter are describing curves whose concavity is turned towards A, while those within the circle are describing curves whose convexity is turned towards A. It is then clear that the equilibrium is stable, unstable, or neutral, according as the centre of gravity lies within, without, or on the circumference of the circle.

and above the tangent at A, L is negative and the equilibrium is unstable, if within, L is positive and the equilibrium is stable. This circle is called the *circle of stability*.

This rule will be found very convenient to determine not only the condition of stability of a heavy cylinder resting in equilibrium on one side of a rough fixed cylinder, but also to determine the time of oscillation when the equilibrium is disturbed. An extension of the rule to cases of rough cones and other surfaces will be given further on.

443. It may be noticed that the preceding result is perfectly general and may be used in all cases in which the locus of the instantaneous axis is known. Thus ρ' is the radius of curvature of the locus in the body, ρ that of the locus in space, and α the inclination of its tangent to the horizon.

If dx be the horizontal displacement of the instantaneous centre produced by a rotation $d\theta$ of the body, the equation to find the length of the simple equivalent pendulum of a body oscillating under gravity may be written

$$\frac{k^2 + r^2}{L} = \frac{dx}{d\theta} - r.$$

This follows at once from the reasoning in Art. 441. It may also be easily seen that the diameter of the circle of stability is equal to the ratio of the velocity in space of the instantaneous axis to the angular velocity of the body.

Ex. 1. A homogeneous sphere makes small oscillations inside a fixed sphere so that its centre moves in a vertical plane. If the roughness be sufficient to prevent all sliding, prove that the length of the equivalent pendulum is seven-fifths of the difference of the radii. If the spheres were smooth the length of the equivalent pendulum would be equal to the difference of the radii.

Ex. 2. A homogeneous hemisphere being placed on a rough fixed plane, which is inclined to the horizon at an angle $\sin^{-1}\frac{1}{4}\sqrt{2}$, makes small oscillations in a vertical plane. Show that, if a is the radius of the hemisphere, the length of the equivalent pendulum is $\left(\frac{46}{5} - \frac{\sqrt{56}}{4}\right)a$.

444. If the body be acted on by any force which passes through the centre of gravity, the results must be slightly modified. Just as before, the force in equilibrium must act along the straight line joining the centre of gravity G to the instantaneous centre A. When the body is displaced, the force cuts its former line of action in some point F, which we shall assume to be known. Let $AF=f$, taking f positive when G and F are on opposite sides of the locus of the instantaneous centre. Then it may be shown by similar reasoning, that the length L of the simple equivalent pendulum under this force, supposed constant and equal to gravity, is given by $\dfrac{k^2 + r^2}{L} = \dfrac{\rho\rho'}{\rho + \rho'}\cos a - \dfrac{fr}{f + r}$, where a is the angle the direction of the force makes with the normal to the path of the instantaneous centre.

If we measure along the line AG a length AG' so that $\dfrac{1}{AG'} = \dfrac{1}{AG} + \dfrac{1}{AF}$, then the expression for L takes the form $\dfrac{k^2 + r^2}{L} = G'N$. The equilibrium is therefore stable or unstable according as G' lies within or without the circle of stability.

445. **Oscillations of a body resting on two curves.** *Two points* A, B *of a body are constrained to describe given curves, and the body is in equilibrium under the action of gravity. A small disturbance being given, find the time of an oscillation.*

Let C, D be the centres of curvature of the given curves at the two points A, B. Let AC, BD meet in O. Let G be the centre of gravity of the body, GE a perpendicular on AB. Then in the position of equilibrium OG is vertical. Let i, j be the angles which CA, BD make with the vertical, and let a be the angle AOB. Let A', B', G', E' denote the positions into which A, B, G, E are moved when the body is turned through an angle θ, and let O' be the point of intersection of the normals at A', B'. Let $ACA' = \phi$, $BDB' = \phi'$. Since the body may be brought from the position AB into the position $A'B'$ by turning it about O through an angle θ, we have $\dfrac{CA \cdot \phi}{OA} = \dfrac{BD \cdot \phi'}{OB} = \theta$. Also GG' is ultimately perpendicular to OG, and we have $GG' = OG \cdot \theta$. Also let x, y be the projections of OO' on the horizontal and vertical through O. Then by projections

$$x \cos j + y \sin j = \text{distance of } O' \text{ from } OD = OD \cdot \phi',$$

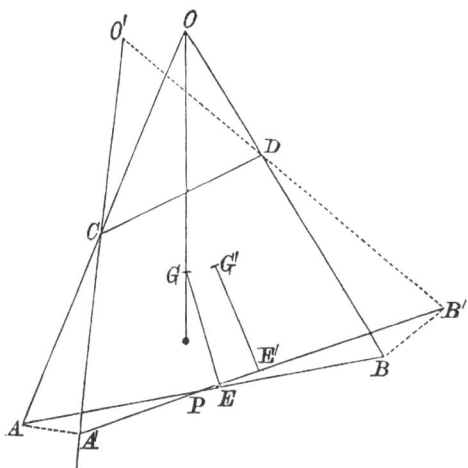

$$x \cos i - y \sin i = \text{distance of } O' \text{ from } OC = OC \cdot \phi;$$
$$\therefore x = \frac{OD \cdot \sin i \cdot \phi' + OC \cdot \sin j \cdot \phi}{\sin a}.$$

Now, taking moments about O' as the centre of instantaneous rotation, we have

$$(k^2 + OG^2)\frac{d^2\theta}{dt^2} = -g \cdot (GG' + x)$$

$$= -g\theta\left(OG + \frac{OD \cdot OB}{BD}\frac{\sin i}{\sin a} + \frac{OC \cdot OA}{CA}\frac{\sin j}{\sin a}\right),$$

where k is the radius of gyration about the centre of gravity.

Hence, if L be the length of the simple equivalent pendulum, we have

$$\frac{k^2 + OG^2}{L} = OG + \frac{OD \cdot OB}{BD} \frac{\sin i}{\sin \alpha} + \frac{OC \cdot OA}{AC} \frac{\sin j}{\sin \alpha}.$$

If the given curves, on which the points A, B are constrained to move, be straight lines, the centres of curvature C and D are at infinity. In this case, we may put $\frac{OD}{BD} = -1$, $\frac{OC}{AC} = -1$, and the expression becomes

$$\frac{k^2 + OG^2}{L} = OG - OB \cdot \frac{\sin i}{\sin \alpha} - OA \cdot \frac{\sin j}{\sin \alpha}.$$

If OA and OB be at right angles, this takes the simple form

$$\frac{k^2 + OG^2}{L} = OG - 2OF,$$

where F is the projection on OG of the middle point of AB.

Ex. 1. A heavy rod ACB rests in equilibrium in a horizontal position within a surface of revolution whose axis is vertical. Let $2a$ be the length of the rod, ρ the radius of curvature of the generating curve at either extremity of the rod, i the inclination of this radius of curvature to the vertical. Prove that, if the rod be slightly disturbed, so that it makes small oscillations in a vertical plane, the length of the equivalent pendulum is $\dfrac{a\rho \sin^2 i \cos i \, (1 + 3 \cot^2 i)}{3 \, (a - \rho \sin^3 i)}$.

Ex. 2. The extremities of a uniform heavy rod of length $2c$ slide on a smooth wire in the form of a parabola, whose axis is vertical, and whose latus rectum is equal to $4a$. If the rod be slightly displaced from its position of stable equilibrium, prove that the length of the equivalent pendulum is $\dfrac{2ac}{3 \, (c - 2a)}$, or $\dfrac{2a}{3} \dfrac{c^2 + 12a^2}{4a^2 - c^2}$, according as the length of the rod is greater or less than the latus rectum of the parabola.

In the first case the rod in its stable position of equilibrium passes through the focus and is inclined to the horizon. In the second case the rod is horizontal. When the length of the rod is equal to the latus rectum the oscillation is not tautochronous, see Art. 450. If the rod start from rest at a small inclination α to the horizon, it will become horizontal after a time $\left(\dfrac{4c}{8g}\right)^{\frac{1}{2}} \displaystyle\int_0^1 (1 - \phi^4)^{-\frac{1}{2}} d\phi$. The first case of this question was set in a Caius Coll. paper.

Ex. 3. The extremities of a rod of length $2a$ slide upon two smooth wires, which form the upper sides of a square whose diagonal is vertical, prove that the length of the equivalent pendulum is $\frac{4}{3}a$.   [Math. Tripos.]

446. **Oscillation when path of centre of gravity is known.** *A body oscillates about a position of equilibrium under the action of gravity, the radius of curvature of the path of the centre of gravity being known, find the time of an oscillation.*

Let A be the position of the centre of gravity of the body when it is in its position of equilibrium, G the position of the centre of gravity at the time t. Then since in equilibrium the altitude of the centre of gravity is a maximum or minimum, the tangent at A to the curve AG is horizontal. Let the normal GC to the curve at G meet the normal at A in C. Then, when the oscillation becomes indefinitely small, C is the centre of curvature of the curve at A. Let $AG = s$, the angle $ACG = \psi$, and let R be the radius of curvature of the curve at A.

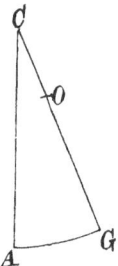

Let θ be the angle turned round by the body in moving from the position of equilibrium into the position in which the centre of gravity is at G; then $\dfrac{d\theta}{dt}$ is the angular velocity of the body. Since G is moving along the tangent at G, the centre of instantaneous rotation lies in the normal GC, at such a point O that

$$OG \, \frac{d\theta}{dt} = \text{vel. of } G = \frac{ds}{dt}, \quad \therefore \; GO = \frac{ds}{d\theta}.$$

Let Mk^2 be the moment of inertia of the body about its centre of gravity, then taking moments about O, we have

$$(k^2 + OG^2) \frac{d^2\theta}{dt^2} = -g \cdot OG \sin \psi.$$

Now ultimately when the angle θ is indefinitely small $\dfrac{\psi}{\theta} = \dfrac{d\psi}{d\theta} = \dfrac{OG}{R}$; $\therefore$ the equation of motion becomes

$$(k^2 + OG^2) \frac{d^2\theta}{dt^2} = -g \, \frac{OG^2}{R} \cdot \theta.$$

Hence if L be the length of the simple equivalent pendulum we have

$$L = \left(1 + \frac{k^2}{OG^2}\right) R.$$

447. **Oscillations found by Vis Viva.** When the system of bodies in motion admits of only one independent motion, the time of a small oscillation may frequently be deduced from the equation of vis viva. This equation is one of the second order of small quantities, and in forming the equation it is thus necessary to take into account small quantities of that order. This sometimes involves rather troublesome considerations. On the other hand, the equation is free from all the unknown reactions, and we thus frequently save much elimination.

The method of proceeding will be made clear by the following example, by which a comparison may be made with the method of the last article.

The motion of a body in space of two dimensions is given by the co-ordinates x, y *of its centre of gravity, and the angle θ which any fixed line in the body makes with a line fixed in space. The body being in equilibrium under the action of gravity, it is required to find the time of a small oscillation.*

Since the body is capable of only one independent motion, we may express (x, y) as functions of θ, thus

$$x = F(\theta), \quad y = f(\theta).$$

Let Mk^2 be the moment of inertia of the body about an axis through its centre of gravity, then the equation of vis viva becomes $\dot{x}^2 + \dot{y}^2 + k^2\dot{\theta}^2 = C - 2gy$, where C is an arbitrary constant.

Let a be the value of θ when the body is in the position of equilibrium, and suppose that, at the time t, $\theta = a + \phi$. Then, by Maclaurin's theorem,

$$y = y_0 + y_0'\phi + y_0''\frac{\phi^2}{2} + \dots,$$

where y_0', y_0'' are the values of $\dfrac{dy}{d\theta}$, $\dfrac{d^2y}{d\theta^2}$ when $\theta = a$. But in the position of equilibrium y is a maximum or minimum; $\therefore y_0' = 0$. Hence the equation of vis viva becomes $(x_0'^2 + k^2)\dot{\phi}^2 = C - gy_0''\phi^2$, where x_0' is the value of $\dfrac{dx}{d\theta}$ when $\theta = a$; differentiating we get $(x_0'^2 + k^2)\ddot{\phi} = -gy_0''\phi$.

If L be the length of the simple equivalent pendulum, we have

$$L\frac{d^2y}{d\theta^2} = k^2 + \left(\frac{dx}{d\theta}\right)^2,$$

where for θ we are to write its value a after the differentiations have been effected. It is not difficult to see that the geometrical meaning of this result is the same as that given in the last article.

This analytical result was given by Mr Holditch, in the eighth volume of the *Cambridge Transactions*. It is a convenient formula to use when the motion of the oscillating body is known with reference to its centre of gravity.

Ex. 1. The lower extremity of a heavy uniform beam of length a slides on a weightless inextensible string of length $2a$, whose extremities are attached to two fixed points in the same horizontal line, and the upper extremity slides on a vertical rod which bisects the line joining the two fixed points. Prove that the only position of equilibrium is vertical, and that the time of a small oscillation about this position is $\dfrac{2\pi a \sqrt{(2)}}{\sqrt{\{3g(2b-a)\}}}$, where $2\sqrt{(a^2 - b^2)}$ is the distance between the two fixed points.

[Math. Tripos.]

The lower extremity of the rod may be regarded as moving in a circle of radius a^2/b. Express the co-ordinates (x, y) of the middle point in terms of the angle θ which the rod makes with the vertical. The result follows by the principle of vis viva.

Ex. 2. The extremities of a rod slide on the circumference of a three-cusped hypocycloid whose plane is vertical. The radius of the circumscribing circle is $3a$, and one of the cusps is at the highest point of the circle. Prove that the length of the equivalent pendulum is $\frac{4}{3}a$.   [Math. T. 1872.]

First prove that in this hypocycloid the rod as it slides with its two ends on the side branches BE, DE always touches the lowest branch BD. Its middle point R describes a circle with centre O, and radius a. If $BOR = \phi$, the angle which the rod makes with the tangent at the cusp B is $\frac{1}{2}\phi$. The result then follows by using the principle of vis viva.

448. Moments about the Instantaneous Axis. When a body moves in space with one independent motion there is not in general an instantaneous axis. It has, however, been proved in

Art. 225 that the moment may always be reduced to a rotation about some central axis and a translation along that axis.

Let I be the moment of inertia of the body about the instantaneous central axis, Ω the angular velocity about it, V the velocity of translation along it, M the mass of the body, then by the principle of vis viva $\frac{1}{2} I\Omega^2 + \frac{1}{2} MV^2 = U + C$, where U is the force-function, and C some constant. Differentiating we get

$$I \frac{d\Omega}{dt} + \frac{1}{2} \Omega \frac{dI}{dt} + M \frac{V}{\Omega} \frac{dV}{dt} = \frac{dU}{\Omega dt}.$$

Let L be the moment of the impressed forces about the instantaneous central axis, then $L = \frac{dU}{\Omega dt}$ by Art. 340.

Let p be the pitch of the screw-motion of the body, then $V = p\Omega$. The equation of motion therefore becomes

$$(I + Mp^2) \frac{d\Omega}{dt} + \frac{1}{2} \Omega \frac{dI}{dt} + Mp\Omega \frac{dp}{dt} = L.$$

If the body be performing small oscillations about a position of equilibrium, we may reject the second and third terms, and the equation becomes

$$(I + Mp^2) \frac{d\Omega}{dt} = L.$$

If there be an instantaneous axis, $p = 0$, and we see that we may take moments about the instantaneous axis exactly as if it were fixed in space and in the body.

Second Method of forming the Equations of Motion.

449. Let the general equations of motion of all the bodies be formed. If the position about which the system oscillates be known, some of the quantities involved will be small. The squares and higher powers of these may be neglected, and all the equations will become linear. If the unknown reactions be then eliminated the resulting equations may be easily solved.

If the position about which the system oscillates be unknown, *it is not necessary to solve the statical problem first*. We may by one process determine the positions of rest, ascertain whether they are stable or not, and find the time of oscillation. The method of proceeding will be best explained by an example.

450. Ex. *The ends of a uniform heavy rod* AB *of length* 2l *are constrained to move, the one along a horizontal line* Ox, *and the other along a vertical line* Oy. *If the whole system turn round* Oy

with a uniform angular velocity ω, it is required to find the positions of equilibrium and the time of a small oscillation.

Let x, y be the co-ordinates of G the middle point of the rod, θ the angle OAB which the rod makes with Ox. Let R, R' be the reactions at A and B resolved in the plane xOy. Let the mass of a unit of length be taken as the unit of mass.

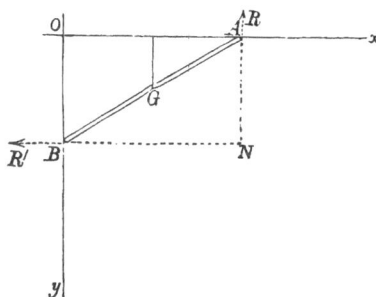

The accelerations of any element dr of the rod whose co-ordinates are (ξ, η) are $\dfrac{d^2\xi}{dt^2} - \omega^2\xi$ parallel to Ox, $\dfrac{1}{\xi}\dfrac{d}{dt}(\xi^2\omega)$ perpendicular to the plane xOy, and $\dfrac{d^2\eta}{dt^2}$ parallel to Oy.

As it will not be necessary to take moments about Ox, Oy, or to resolve perpendicular to the plane xOy, the second acceleration will not be required. The resultants of the effective forces $\ddot{\xi}dr$ and $\ddot{\eta}dr$, taken throughout the body, are $2l\ddot{x}$ and $2l\ddot{y}$ acting at G, and a couple $2lk^2\ddot{\theta}$ tending to turn the body round G. The resultants of the effective forces $\omega^2\xi dr$ taken throughout the body are a single force acting at $G = \displaystyle\int_{-l}^{+l} \omega^2(x + r\cos\theta)\,dr = \omega^2x\,.\,2l$, and a couple* round $G = \displaystyle\int_{-l}^{+l} \omega^2(x + r\cos\theta)\,r\sin\theta\,dr = \omega^2\,.\,2l\,.\,\dfrac{l^2}{3}\sin\theta\cos\theta$, the distance r being measured from G towards A.

Then we have, by resolving along Ox, Oy, and by taking moments about G, the *dynamical equations*

---

* If a body in one plane be turning about an axis in its own plane with an angular velocity ω, a general expression can be found for the resultants of the centrifugal forces on all the elements of the body. Take the centre of gravity G as origin and the axis of y parallel to the fixed axis. Let c be the distance of G from the axis of rotation. Then all the centrifugal forces are equivalent to a single resultant force at G

$$= \int \omega^2(c + x)\,dm = \omega^2\,.\,Mc, \text{ since } \bar{x} = 0,$$

and a single resultant couple

$$= \int \omega^2(c + x)\,y\,dm = \omega^2 \int xy\,dm, \text{ since } \bar{y} = 0.$$

$$2l\ddot{x} = -R' + \omega^2 x \cdot 2l$$
$$2l\ddot{y} = -R + g \cdot 2l$$
$$2lk^2\ddot{\theta} = Rx - R'y - \omega^2 \cdot 2l \cdot \frac{l^2}{3} \sin \theta \cos \theta$$

$$\left.\rule{0pt}{40pt}\right\} \quad \text{......(1)}.$$

We have also the *geometrical equations*

$$x = l \cos \theta, \quad y = l \sin \theta \quad \text{..................(2)}.$$

Eliminating R, R', from the equations (1), we get

$$x\ddot{y} - y\ddot{x} + k^2\ddot{\theta} = gx - \omega^2 xy - \omega^2 \frac{l^2}{3} \sin \theta \cos \theta \text{......(3)}.$$

To find the position of rest. We observe that if the rod were placed at rest in that position it would always remain there, and that therefore $\ddot{x} = 0$, $\ddot{y} = 0$, $\ddot{\theta} = 0$. These give

$$f(x, y, \theta) = gx - \omega^2 xy - \omega^2 \frac{l^2}{3} \sin \theta \cos \theta = 0 \text{.........(4)}.$$

Joining this with equations (2), we get $\theta = \dfrac{\pi}{2}$, or $\sin \theta = \dfrac{3g}{4\omega^2 l}$, and thus the positions of equilibrium are found. Let any one of these positions be represented by $\theta = \alpha$, $x = a$, $y = b$.

To find the motion of oscillation. Let $x = a + x'$, $y = b + y'$, $\theta = \alpha + \theta'$, where x', y', θ' are all small quantities, then we must substitute these values in equation (3). On the left-hand side, since $\ddot{x}$, $\ddot{y}$, $\ddot{\theta}$, are all small, we have simply to write a, b, α, for x, y, θ. On the right-hand side the substitution should be made by Taylor's Theorem, thus

$$f(a + x', b + y', \alpha + \theta') = \frac{df}{da} x' + \frac{df}{db} y' + \frac{df}{d\alpha} \theta'.$$

We know that the first term $f(a, b, \alpha)$ will be zero, because this is the very equation (4) from which a, b, α were found. We therefore get

$$a\ddot{y}' - b\ddot{x}' + k^2\ddot{\theta}' = (g - \omega^2 b) x' - \omega^2 a y' - \omega^2 \frac{l^2}{3} \cos 2\alpha \cdot \theta'.$$

But, by putting $\theta = \alpha + \theta'$ in equations (2), we get by Taylor's Theorem $x' = -l \sin \alpha \cdot \theta'$, $y' = l \cos \alpha \cdot \theta'$.

Hence the equation to determine the motion is

$$(l^2 + k^2) \frac{d^2\theta'}{dt^2} + \left(gl \sin \alpha + \frac{4}{3} \omega^2 l^2 \cos 2\alpha\right) \theta' = 0.$$

Now, if $gl \sin \alpha + \dfrac{4}{3} \omega^2 l^2 \cos 2\alpha = n$ be positive when either of the two values of α is substituted, the corresponding position of equilibrium is *stable*, and the time of a small oscillation is

$$2\pi \sqrt{\frac{l^2 + k^2}{n}}.$$

If n be negative the equilibrium is *unstable*, and there can be no oscillation.

If $\omega^2 > \dfrac{3g}{4l}$, there are two positions of equilibrium of the rod. It will be found by substitution that the position in which the rod is inclined to the vertical is stable, and the other position unstable. If $\omega^2 < \dfrac{3g}{4l}$ the only position in which the rod can rest is vertical, and this position is stable.

If $n = 0$, the body is in a position of neutral equilibrium. To determine the small oscillations we must retain terms of an order higher than the first. By a known transformation we have

$$x\ddot{y} - y\ddot{x} = \frac{d}{dt}(l^2 \dot{\theta}).$$

Hence the left-hand side of equation (3) becomes $(l^2 + k^2)\ddot{\theta}$. The right-hand side becomes by Taylor's Theorem

$$\frac{d^2}{d\alpha^2}\left(gl\cos\alpha - \frac{2}{3}\omega^2 l^2 \sin 2\alpha\right)\frac{\theta'^2}{1 \cdot 2} + \&c.$$

When $n = 0$, we have $\alpha = \dfrac{\pi}{2}$ and $\omega^2 = \dfrac{3g}{4l}$. Making the necessary substitutions the equation of motion becomes

$$(l^2 + k^2)\frac{d^2\theta'}{dt^2} = -\frac{gl}{2}\theta'^3.$$

Since the lowest power of θ' on the right-hand side is odd, and its coefficient negative, the equilibrium is stable for a displacement on either side of the position of equilibrium. Let α be the initial value of θ', then the time T of reaching the position of equilibrium is

$$T = \sqrt{\frac{4(l^2 + k^2)}{gl}} \int_0^\alpha \frac{d\theta'}{\sqrt{\alpha^4 - \theta'^4}};$$

put $\theta' = \alpha\phi$, then

$$T = \sqrt{\frac{4(l^2 + k^2)}{gl}} \cdot \int_0^1 \frac{d\phi}{\sqrt{1 - \phi^4}} \cdot \frac{1}{\alpha}.$$

Hence the time of reaching the position of equilibrium varies inversely as the arc. When the initial displacement is indefinitely small, the time becomes infinite.

This definite integral may be otherwise expressed in terms of the Gamma function. It may be easily shown that $\int_0^1 \dfrac{d\phi}{\sqrt{1-\phi^4}} = \dfrac{\{\,\Gamma\,(\frac{1}{4})\,\}^2}{4\sqrt{2\pi}}$.

451. This problem might have been easily solved by the first method. For, if the two perpendiculars to Ox, Oy at A and B meet in N, N is the instantaneous axis. Taking moments about N, we have the equation

$$(l^2 + k^2)\ddot\theta = gl\cos\theta - \int_{-l}^{+l} \omega^2 (l+r)^2 \sin\theta\cos\theta\, \frac{dr}{2l}$$

$$= gl\cos\theta - \frac{4l^2}{3}\,\omega^2 \sin\theta\cos\theta.$$

If we represent the right-hand side of this equation by $f(\theta)$, the position of equilibrium can be found from the equation $f(\alpha) = 0$ and the time of oscillation from the equation

$$(l^2 + k^2)\,\ddot\theta = \frac{df(\alpha)}{d\alpha}\,\theta'.$$

452. Ex. 1. If the mass of the rod AB is M, show that the magnitude of the couple which constrains the system to turn round Oy with uniform angular velocity is $M\,\dfrac{4l^2}{3}\,\omega\,\dfrac{d\theta}{dt}\,\sin 2\theta$.

Would the magnitude of this couple be altered if Ox or Oy had any mass?

Ex. 2. The upper extremity of a uniform beam of length $2l$ is constrained to slide on a smooth horizontal rod without inertia, and the lower along a smooth vertical rod, through the upper extremity of which the horizontal rod passes; the system rotates freely about the vertical rod, prove that if α be the inclination of the beam to the vertical when in a position of relative equilibrium, the angular velocity of the system will be $\left(\dfrac{3g}{4l\cos\alpha}\right)^{\frac{1}{2}}$, and, if the beam be slightly displaced from this position, show that it will make a small oscillation in the time

$$\frac{4\pi}{\left\{\dfrac{3g}{l}\,(\sec\alpha + 3\cos\alpha)\right\}^{\frac{1}{2}}}.$$ [Coll. Exam.]

In the example in the text the system is constrained to turn round the vertical with uniform angular velocity, but in this example the system rotates freely. The angular velocity about the vertical is therefore not constant, and its small variations must be found by the principle of angular momentum.

Lagrange's Method of forming the Equations of Motion.

453. **Advantages of the Method.** We now propose to state Lagrange's method of forming the equations of motion. This method has several advantages. It gives us the equations of motion free from all reactions, and is therefore specially useful when we have to consider the motions of several bodies connected

together. It also gives us a larger choice of quantities which we may take as co-ordinates. Again, as soon as we have written down the Lagrangian function we may deduce from this one function all the equations of motion, instead of deriving each from a separate principle. On the other hand, this function must be calculated so as to include the squares of the small quantities. Now in small oscillations we generally retain only the first powers of the small quantities, so that, when only a few equations are wanted, it is often more convenient to obtain these by resolving and taking moments.

It will be seen, therefore, that the method is best adapted to oscillations which have more than one degree of freedom. For this reason we shall here only state the general mode of forming the equations of motion, so that we may be able to apply the method to the solution of problems. But we shall postpone the general discussion of Lagrange's determinant to the second part of this work.

454. The object of Lagrange's method is to determine the *oscillations of a system about a position of equilibrium.* It does not apply to oscillations about a state of steady motion. For example, if a heavy particle were suspended by a string from a fixed point, the string is vertical when the system is in equilibrium, and the oscillations about this position could be found by Lagrange's method. If however the particle were made to describe a horizontal circle, as in the conical pendulum, the oscillations about the circular steady motion could not be found by this method. In the same way when a hoop rolls on the ground in a vertical plane, it may make small oscillations from one side to the other of the plane. These oscillations cannot be found by Lagrange's method. A method of investigating the oscillations of a system about a state of steady motion will be given in the next volume.

We shall assume, for the present, that the forces which act on the system have a force function. We shall also assume that the geometrical equations do not contain the time explicitly, and do not contain any differential coefficient with regard to the time.

In Lagrange's method it is essential that the co-ordinates chosen should be such small quantities that we may reject all powers of them except the lowest which occur. They should generally be so chosen that they vanish in the position of equilibrium. But with this restriction they may be any whatever. Let us represent them by the letters θ, ϕ, &c. Then if the system oscillate about the position of equilibrium, *these quantities will be small throughout the motion.* Let n be the number of these co-ordinates.

As before, let accents denote differential coefficients with regard to the time.

Let $2T$ be the vis viva of the system when disturbed from its position of equilibrium, then as in Art. 396 we may express T as a homogeneous quadratic function of θ', ϕ', &c. of the form

$$2T = A_{11}\theta'^2 + 2A_{12}\theta'\phi' + A_{22}\phi'^2 + \&c. \dots\dots\dots(1).$$

Here the coefficients A_{11} &c. are all functions of θ, ϕ, &c. and we may suppose them expanded in a series of some powers of these co-ordinates. If the oscillations are so small that we may reject all powers of the small quantities except the lowest which occur, we may reject all except the constant terms of these series. We shall therefore regard the coefficients A_{11} &c. as constants.

Let U be the force function of the system when disturbed from the position of equilibrium. Then we may also expand U in a series of powers of θ, ϕ, &c.

Let this expansion be

$$2U = 2U_0 + 2B_1\theta + 2B_2\phi + \&c. + B_{11}\theta^2 + 2B_{12}\theta\phi + \&c.\dots(2).$$

Here U_0 is a constant, which is evidently the value of U when θ, ϕ, &c. are all zero. It is necessary for the success of Lagrange's method that both these expansions should be possible.

In the position of equilibrium, we must have, by the principle of virtual work, $\dfrac{dU}{d\theta} = 0$, $\dfrac{dU}{d\phi} = 0$, &c. $= 0$ (see also Art. 340). If the co-ordinates chosen are such that they vanish in the position of equilibrium, it immediately follows that $B_1 = 0$, $B_2 = 0$, &c. $= 0$. If the co-ordinates have not been so chosen they must yet vanish for some position of the system close to the position of equilibrium. The differential coefficients of U, i.e. B_1, B_2, &c., are therefore necessarily *small*. The terms $B_1\theta$, $B_2\phi$, &c. are thus of the second order of small quantities and the quadratic terms of U cannot be neglected in comparison with them.

We may also notice that the *equilibrium values* of θ, ϕ, &c. may be found beforehand by equating to zero the several first differential coefficients of U. But this is generally unnecessary, as these values of θ, ϕ, &c. will appear in the sequel (see also Art. 449).

We have now to substitute the expanded values of T and U in the n Lagrange's equations

$$\frac{d}{dt}\frac{dT}{d\theta'} - \frac{dT}{d\theta} = \frac{dU}{d\theta} \dots\dots\dots\dots\dots (3),$$

with similar equations for ϕ, ψ, &c. Since the expression for T does not contain θ, ϕ, &c., we have

$$\frac{dT}{d\theta} = 0, \frac{dT}{d\phi} = 0, \&c.$$

The n equations (3) therefore become

$$\left. \begin{aligned} A_{11}\theta'' + A_{12}\phi'' + \ldots &= B_1 + B_{11}\theta + B_{12}\phi + \ldots \\ A_{12}\theta'' + A_{22}\phi'' + \ldots &= B_2 + B_{12}\theta + B_{22}\phi + \ldots \\ \&\text{c.} &= \&\text{c.} \end{aligned} \right\} \ldots\ldots(4).$$

These are Lagrange's equations to determine the small oscillations of any system about a position of equilibrium.

455. **Method of Solution.** We have now to solve these equations. We notice that they are all linear, and that therefore θ, ϕ, &c. are properly represented by a series of exponentials of the form Me^{qt}. But, as we are seeking an oscillatory motion, it is more convenient to replace these exponentials by the corresponding trigonometrical expressions. Since the equations do not contain any differential coefficients of the first order, it will be found possible, on making the trial, to satisfy them by means of the following assumption.

$$\left. \begin{aligned} \theta &= \alpha + M_1 \sin(p_1 t + \epsilon_1) + M_2 \sin(p_2 t + \epsilon_2) + \&\text{c.} \\ \phi &= \beta + N_1 \sin(p_1 t + \epsilon_1) + N_2 \sin(p_2 t + \epsilon_2) + \&\text{c.} \\ \&\text{c.} &= \&\text{c.} \end{aligned} \right\} \ldots (5).$$

Taking the trigonometrical terms separately, they may be written in the typical form

$$\theta = M \sin(pt + \epsilon), \quad \phi = N \sin(pt + \epsilon), \quad \&\text{c.} = \&\text{c.}$$

If we now substitute these in equations (4) we have

$$\left. \begin{aligned} (A_{11}p^2 + B_{11}) M + (A_{12}p^2 + B_{12}) N + \&\text{c.} &= 0 \\ (A_{12}p^2 + B_{12}) M + (A_{22}p^2 + B_{22}) N + \&\text{c.} &= 0 \\ \&\text{c.} \qquad\qquad \&\text{c.} \qquad\quad &= 0 \end{aligned} \right\} \ldots\ldots(6).$$

Eliminating M, N, &c. we have the determinantal equation

$$\begin{vmatrix} A_{11}p^2 + B_{11}, & A_{12}p^2 + B_{12}, & \&\text{c.} \\ A_{12}p^2 + B_{12}, & A_{22}p^2 + B_{22}, & \&\text{c.} \\ \&\text{c.} & \&\text{c.} & \&\text{c.} \end{vmatrix} = 0 \ldots\ldots\ldots\ldots(7).$$

This determinant, it will be observed, is symmetrical about the leading diagonal. If there be n co-ordinates, it is an equation of the nth degree to find p^2. It will be shown in the second part of this work that all the values of p^2 are real.

Taking any root positive or negative, the equations (6) determine the ratios of N, P, &c. to M, and we notice that these ratios also are all real. If all the roots of the determinantal equation are positive, the equations (5) give the whole motion, with $2n$ arbitrary constants, viz. $M_1, M_2, M_3 \ldots M_n$ and $\epsilon_1, \epsilon_2, \ldots \epsilon_n$. These have to be determined by the initial values of θ, ϕ, &c., θ', ϕ', &c. If any root of the determinantal equation is negative, the corresponding *sine* will resume its exponential form, the coefficient being rationalized by giving the coefficient M an imaginary form. In this case there is no oscillation about the position of equilibrium. The position is then said to be unstable.

It may be noticed that for every positive value of p^2 given by the equation (7) there are two equal values of p with opposite signs. No attention however should be here given to the negative values of p. To prove this, we notice that the solution of the linear differential equations is properly represented by a series of exponentials. Now each sine is the sum of two exponentials with indices of opposite signs. Both the values of p have therefore been included in the trigonometrical expressions assumed for θ, ϕ, &c.

The constants α, β, &c. in the trial solution (5) are evidently the co-ordinates of the central position about which the system oscillates. Substituting these values of θ, ϕ, &c. in the equations (4) we have

$$\left.\begin{array}{l} 0 = B_1 + B_{11}\alpha + B_{12}\beta + \&c. \\ 0 = B_2 + B_{12}\alpha + B_{22}\beta + \&c. \\ 0 = \&c. \end{array}\right\} \quad \dots\dots\dots\dots(8).$$

These equations determine the values of α, β, &c. Since the equations of motion are satisfied by these constant values of the co-ordinates without any terms containing the time, it follows that α, β, &c. *are the co-ordinates of the equilibrium position of the system.* That this is so, follows also from the rules given in statics to find the position of equilibrium of a system when the function U is known. According to these rules, we find the equilibrium values of the co-ordinates θ, ϕ, &c. by equating to zero the first differential coefficients of U with regard to θ, ϕ, &c. But the equations thus obtained are evidently the same as the equations (8).

456. **Periods of Oscillation.** We see from (5) that each of the n co-ordinates θ, ϕ, &c. is expressed in a series of as many sines as there are separate values of p^2. Thus, when there are several independent ways in which the system can move, there are as many periods of oscillation. These are clearly equal to $\dfrac{2\pi}{p_1}$, $\dfrac{2\pi}{p_2}$, &c. Generally we want only these periods of oscillation and not the particular position occupied by the system at any instant. In such a case we may in any problem omit all the steps of the argument and write down the determinantal equation at once. We then use the following rule. *Expand the force function* U *and the semi vis viva* T *in ascending powers of the co-ordinates θ, ϕ, &c., and their differential coefficients θ', ϕ', &c., all powers above the second being rejected. Then, omitting the accents or dots*

in the expression for T *and retaining only the quadratic term in* U, *equate to zero the discriminant of* $p^2 T + U$. *The roots of the equation thus formed will give the required values of* p.

The mode of using this rule in conjunction with the method of indeterminate multipliers will be given in the second part of this treatise.

457. **Position of the system.** If it be also required to find the position of the system at any time, we must determine the values of the constants. Referring to equations (6) we see that the ratios of M, N, P, &c. for any particular trigonometrical term in the solution (5) are the same as the ratios of the minors of the constituents of any line we please in the Lagrangian determinant (7). In these minors we of course substitute the value of p^2 which belongs to the particular trigonometrical term we are considering. In this manner the coefficients of all the trigonometrical terms are found in terms of those which occur in the series for any one co-ordinate. As already explained, these remaining n coefficients, and the n constants from $\epsilon_1, \ldots \epsilon_n$ must be found from the given initial values of the n co-ordinates θ, ϕ, &c., and their n initial velocities θ', ϕ', &c. The constants α, β, &c. have been already found from equations (8). Thus all the undetermined constants in equations (5) have been found, and these equations thus give the position of the system at any time t.

Ex. Show that, when the determinant (7) is zero, the ratios of the minors of the constituents of any one line are equal to the ratios of the minors of the constituents of any other line.

458. **Examples of Lagrange's Method.** The following examples will show how we may use Lagrange's method to find the small oscillations of a system. When only the periods are required, the process may be summed up thus:—*Form the terms of* T *and* U *which depend on the squares of small quantities, and equate to zero the discriminant of* $p^2 T + U$.

Ex. 1. A rod AB, whose length is $2a$ and mass is m, is suspended from a fixed point O by a string OA, the length of which is l. The rod oscillates under gravity in a vertical plane, find the periods of the small oscillations.

Let θ, ϕ be the angles which the string and rod make with the vertical. Proceeding as in Art. 147 we find that, when powers of θ and ϕ higher than the second are neglected,

$$T = \tfrac{1}{2} m \left\{ l^2 \theta'^2 + 2al\theta'\phi' + (k^2 + a^2)\, \phi'^2 \right\},$$

$$U = U_0 - \tfrac{1}{2} mg \left(l\theta^2 + a\phi^2 \right).$$

Forming the discriminant of $p^2 T + U$, and dividing by the common factor m, we have

$$\begin{vmatrix} p^2 l^2 - gl & alp^2 \\ alp^2 & p^2(k^2 + a^2) - ag \end{vmatrix} = 0.$$

24—2

This quadratic gives two values of p^2. If these be $p_1{}^2$ and $p_2{}^2$, we have

$$\theta = M_1 \sin (p_1 t + \epsilon_1) + M_2 \sin (p_2 t + \epsilon_2),$$

$$\phi = - M_1 \frac{p_1{}^2 l^2 - gl}{a l p_1{}^2} \sin (p_1 t + \epsilon_1) - M_2 \frac{p_2{}^2 l^2 - gl}{a l p_2{}^2} \sin (p_2 t + \epsilon_2).$$

Writing $3k^2 = a^2$, show that the ratio of the periods of the two oscillations cannot lie between $2 \pm \surd 3$.

Ex. 2.   Two heavy particles, masses M and m, are tied to a string and suspended from a fixed point O, the lengths OM, Mm of the string being respectively a and b. If the particles make small transverse oscillations find the two periods of oscillation, and show that they cannot be equal. Show also that one period is double the other if $4 (M+m) (a+b)^2 = 2Mab$.

Ex. 3.   A smooth thin shell of mass M and radius a rests on a smooth inclined plane by means of an elastic string, which is attached to the sphere, and to a peg at the same distance from the plane as the centre of the sphere, while a particle of mass m rests on the inner surface of the shell. In the position of equilibrium the string is parallel to the plane, find the times of oscillation of the system when it is slightly displaced in a vertical plane, and prove that the arc traversed by the particle and the distance traversed by the centre of the shell from their positions of equilibrium can always be equal if $(M+m+m \cos a) gl = Ea (1 + \cos a)$, where E is the coefficient of elasticity of the string, l its natural length, and a the inclination of the plane to the horizon.   Caius Coll.

Ex. 4.   A three-legged table is made by supporting a heavy triangular lamina on three equal legs, the points of support being the angular points of the lamina; if the legs be equally compressible and their weights be neglected, then the system of co-existent oscillations of the top consist of one vertical oscillation and two angular oscillations about two axes at right angles in its plane, and the periods of the latter are equal and double that of the former.   St John's Coll.

Ex. 5.   A bar AB of mass m and length $2a$ is hung by two equal elastic cords AC, BD, which have no sensible mass, and have unstretched lengths l_0. C and D are fixed points in the same horizontal line, and $CD = 2a$. Investigate the small oscillation of the bar when it is displaced from its position of equilibrium in the vertical plane through CD, and show that the periodic times of the horizontal and vertical oscillations of the centre of gravity of the bar, and of the rotational oscillation, are those of pendulums of lengths l, $l - l_0$, $\tfrac{1}{3} (l - l_0)$ respectively, where l is the length of either cord when the system is in equilibrium.   Math. Tripos.

Ex. 6.   Three equal particles mutually attracting each other according to the Newtonian law are constrained to move like beads along the smooth sides of an equilateral triangle. In equilibrium they occupy the middle points of the sides. Prove that the equilibrium is *unstable* unless the initial displacements and the initial velocities are equal, and in this latter case find the time of a small oscillation.

Ex. 7.   A heavy body whose centre of gravity is H is suspended from a fixed point O. A second body whose centre of gravity is G is attached to the first at some point A situated in OH produced. The system oscillates freely in a vertical plane, prove that the quadratic giving the periods is

$$\{(MK^2 + ma^2) p^2 - (Mh + ma) g\} \{k^2 p^2 - bg\} = ma^2 b^2 p^4,$$

where MK^2 and mk^2 are the moments of inertia of the two bodies about O and A respectively. Also $OH = h$, $OA = a$, $AG = b$. What do these periods become when

(1) the upper body, and (2) the lower, is reduced to a short pendulum of slight mass ? The first case occurs when the attachment of a pendulum to its point of support is not quite rigid, so that the pendulum may be regarded as supported by a short string. The second case occurs when a small part of the mass of a pendulum is loose and swings to and fro at each oscillation.

459. **Principal Co-ordinates.** *To explain what is meant by the principal co-ordinates of a dynamical system.*

When we have two homogeneous quadratic functions of any number of variables, one of which is essentially positive for all values of the variables, it is known that by a real linear transformation of the variables we may clear both expressions of the terms containing the products of the variables, and also make the coefficients of the squares in the positive function each equal to unity. If the co-ordinates θ, ϕ, &c. be changed into ξ, η, &c. by the equations

$$\left.\begin{aligned}\theta &= \lambda_1\xi + \lambda_2\eta + \&c.\\ \phi &= \mu_1\xi + \mu_2\eta + \&c.\\ \&c. &= \&c.\end{aligned}\right\} \quad\dots\dots\dots\dots\dots(9),$$

we observe that θ', ϕ', &c. are changed into ξ', η', &c. by the same transformation. Also the vis viva is essentially positive. Hence we infer that by a proper choice of new co-ordinates, we may express the vis viva and the force function in the forms

$$\left.\begin{aligned}2T &= \xi'^2 + \eta'^2 + \zeta'^2 + \dots\\ 2(U - U_0) &= 2b_1\xi + 2b_2\eta + \&c. + b_{11}\xi^2 + b_{22}\eta^2 + \dots\dots\end{aligned}\right\}.$$

These new co-ordinates ξ, η, &c. are called *principal co-ordinates* of the dynamical system. A great variety of other names has been given to these co-ordinates; such as *harmonic, simple* and *normal* co-ordinates.

It is usually understood (when not otherwise stated) that principal co-ordinates are so chosen that they vanish in the position of equilibrium. We then have $b_1 = 0$, $b_2 = 0$, &c. $= 0$.

460. When a dynamical system is referred to principal co-ordinates which do not necessarily vanish in the position of equilibrium, Lagrange's equations take the form

$$\xi'' - b_{11}\xi = b_1, \quad \eta'' - b_{22}\eta = b_2, \quad \&c. = \&c.$$

so that the whole motion is given by

$$\xi = a + E\sin(p_1 t + \epsilon_1), \quad \eta = b + F\sin(p_2 t + \epsilon_2), \quad \&c.,$$

where E, F, &c., ϵ_1, ϵ_2, &c. are arbitrary constants to be determined by the initial conditions, and $p_1^2 = -b_{11}$, $p_2^2 = -b_{22}$, &c. and a, b, &c. are the values of ξ, η, &c. in equilibrium.

If we substitute the trigonometrical values of ξ, η, &c. in the formulæ of transformation given above, we obviously reproduce the equations (5) of Art. 455, where the general co-ordinates θ, ϕ, &c. are expressed as trigonometrical functions of t. We may therefore obtain one set of principal co-ordinates, viz. ξ_1, η_1, &c., which vanish in the position of equilibrium, by writing

$$\left.\begin{aligned} \theta &= \alpha + M_1\xi_1 + M_2\eta_1 + \dots \\ \phi &= \beta + N_1\xi_1 + N_2\eta_1 + \dots \\ \&c. &= \&c. \end{aligned}\right\} \dots\dots\dots\dots\dots(10),$$

where the values of α, β, &c., M_1, M_2, &c., N_1, N_2, &c. may be found by the methods explained in Art. 455. All other sets of principal co-ordinates may be found from these by taking

$$\xi = a + E\xi_1, \quad \eta = b + F\eta_1, \quad \&c.$$

When the initial conditions are such that throughout the motion all the principal co-ordinates are constant except one, the system is said to be performing a *principal* or *harmonic* oscillation. It performs a *compound oscillation* when any two or more are variable. We may therefore say that any possible oscillation of the system about a position of equilibrium is *analysed* by Lagrange's method into its simple or component oscillations.

From this reasoning we infer the important theorem that *if the equilibrium of a system is stable for the principal oscillations it is stable for all oscillations.*

It is therefore important to determine the peculiarities of a principal oscillation by which it can be recognized apart from all mathematical symbols.

461. The physical peculiarities of a principal oscillation are:

1. The motion recurs at constant intervals, i.e. after one of these intervals the system occupies the same position in space as before, and is moving in exactly the same way.

2. The system passes through the position of equilibrium, twice in each complete oscillation. For, taking ξ as the variable co-ordinate, we see that $\xi - a$ vanishes twice while $p_1 t$ increases by 2π.

3. The velocity of every particle of the system becomes zero at the same instant, and this occurs twice in every complete oscillation. For $\dfrac{d\xi}{dt}$ vanishes twice while $p_1 t$ increases by 2π. The positions of rest may be called the *extreme positions* of the oscillation.

4. Let the system be referred to any co-ordinates θ, ϕ, &c. whose equilibrium values are (as before) α, β, &c. When the system is performing a principal oscillation these are all variable, but the ratios of $\theta - \alpha$, $\phi - \beta$, &c. to each other are constant

throughout the motion*. For, referring to the formulæ of transformation (10), we see that, when η_1, ζ_1, &c. are all zero and only ξ_1 is variable,

$$\frac{\theta - \alpha}{M_1} = \frac{\phi - \beta}{N_1} = \&c. = \xi_1.$$

This theorem may be expressed by saying that every point of the system is in the same *phase* of motion.

The periods of oscillation may all have a least common multiple. If this be so, no matter what initial small disturbance is given to the system, the initial state will be repeated over and over again at intervals equal to the least common multiple. If on the other hand no two of the periods of oscillation are commensurable, the initial state can recur only when the system is performing a principal oscillation. If the periods of some of the principal oscillations are commensurable, the initial state will recur whenever the initial conditions are such that the coefficients E, F, &c. of the other principal oscillations are either zero or so small that the corresponding displacements are imperceptible. The interval is of course equal to the least common multiple of the periods of the existing oscillations.

As an illustration of the method of finding the principal oscillation let us refer to the example (1), already solved in Art. 458. There are two principal oscillations, which are given respectively by

$$\theta_1 = M_1 \sin (p_1 t + \epsilon_1) \qquad\qquad \theta_2 = M_2 \sin (p_2 t + \epsilon_2)$$
$$\phi_1 = - M_1 \frac{p_1^2 l^2 - gl}{alp_1^2} \sin (p_1 t + \epsilon_1) \quad \phi_2 = - M_2 \frac{p_2^2 l^2 - gl}{alp_2^2} \sin (p_2 t + \epsilon_2) \Bigg\}.$$

Thus in each principal oscillation both the string and the rod oscillate. If we saw the rod performing either of these oscillations we should recognize the fact by observing that both the string and the rod become vertical at the same instant, that both reach their extreme positions at the same moment, and so on.

Remembering that p_1^2, p_2^2 are the roots of the quadratic equation given in Art. 458, we easily deduce that $l\theta_1\theta_2 + a\phi_1\phi_2 = 0$. This relation connecting the principal oscillations is a case of a general theorem which will be established in the second volume. See the Method of Multipliers.

Ex. Two equal particles are attached to a string OAB at the points A and B, and suspended from a fixed point O, where $OA = a$, $AB = b$. If these make small transversal oscillations under gravity, find the two principal oscillations. If in one of these the strings make angles θ_1, ϕ_1, with the vertical and in the other make angles θ_2, ϕ_2, for the same value of t, prove that $2a\theta_1\theta_2 + b\phi_1\phi_2 = 0$, whatever the initial conditions may be.

462. Equal Roots in Lagrange's Determinant.

When some of the roots of the equation giving p^2 are equal, we know by the theory of linear differential equations that either (1) terms of

---

* This property is mentioned by Lagrange, who on several occasions uses principal co-ordinates, though not by name.

the form $(At + B) \sin pt$ enter into the values of θ, ϕ, &c., or (2) there must be an indeterminateness in the coefficients M, N, &c. given by Art. 455. Referring the system to principal co-ordinates, which vanish in the position of equilibrium, we see by Art. 460, that the first alternative is in general excluded. If two values of p^2 are equal, say b_{11} and b_{22}, the trigonometrical expressions for ξ and η have equal periods, but terms which contain t as a factor do not make their appearance. The physical peculiarity of this case is that the system has more than one set of principal or harmonic oscillations. For it is clear that, without introducing any terms containing the products of the co-ordinates into the expressions for T or U, we may change ξ, η into any other co-ordinates ξ_1, η_1, which make $\xi^2 + \eta^2 = \xi_1^2 + \eta_1^2$, the other co-ordinates ζ, &c. remaining unchanged. For example we may put $\xi = \xi_1 \cos \alpha - \eta_1 \sin \alpha$ and $\eta = \xi_1 \sin \alpha + \eta_1 \cos \alpha$, where α has any value we please. These new quantities ξ_1, η_1, ζ, &c., are evidently principal co-ordinates, according to the definition of Art. 459.

One important exception must however be noticed, viz., when one or more of the values of p are zero. If, for example, $b_{11} = 0$, we have $\xi = At + B$, where A and B are two undetermined constants. The physical peculiarity of this case is that the position of equilibrium from which the system is disturbed is not solitary. To show this, we remark that the equations giving the position of equilibrium are $\dfrac{dU}{d\xi} = 0$, $\dfrac{dU}{d\eta} = 0$, &c., where U has the value

$$2 (U - U_0) = b_{11}\xi^2 + b_{22}\eta^2 + \dots$$

These in general require that ξ, η, &c. should all vanish, but if $b_{11} = 0$ they are satisfied whatever ξ may be, provided that η, ζ, &c. are zero. In any case however ξ must be very small, because the cubes of ξ, η, &c. have been rejected. It follows therefore that there are other positions of equilibrium in the immediate neighbourhood of the given position. Unless the initial conditions of disturbance are such as to make the terms of the form $At + B$ zero, it may be necessary to examine the terms of higher orders to obtain an approximation to the motion.

Ex. 1. A heavy particle of mass m rests in equilibrium within a right circular smooth fixed cylinder whose generating lines are horizontal. If the particle be disturbed, form Lagrange's equations of motion, and show that in their solution there may be terms of the form $At + B$.

Ex. 2. A rough thin cylinder of mass m and radius b is free to roll inside another thin cylinder of mass M and radius a. The whole system is placed in equilibrium on a smooth horizontal plane. A small disturbance being given, show that the three values of p^2 are $p^2 = 0$, $p^2 = 0$ and $p^2 = \dfrac{M+m}{2M} \dfrac{g}{a-b}$. Interpret this result. If x be the space rolled over, ϕ the angle turned through by the outer cylinder, and θ the inclination to the vertical of the plane containing the axes, show that all three co-ordinates have a common periodic term, while x and ϕ each have additional independent terms of the form $At + B$.

How would the results be altered if the horizontal plane were perfectly rough?

463. Initial Motions. We may also use Lagrange's method to find the initial motion of any system as it starts from a position of rest. See Art. 199. As before we must choose for our co-ordinates some quantities whose higher powers can be rejected. It is generally convenient to choose them so that they vanish in the initial position. As in Art. 454 we have

$$2T = A_{11}\theta'^2 + 2A_{12}\theta'\phi' + A_{22}\phi'^2 + \&c.,$$

where A_{11}, &c. are functions of θ, ϕ, &c. Since the system starts *from rest, θ, ϕ,* &c. are in the beginning of the motion all small quantities. If we reject all powers of θ, ϕ, &c. except the lowest which occur, we may regard A_{11} &c. as constants whose values are found by substituting for θ, ϕ, &c. their initial values.

We require also the expansion of U given in the same article, viz.,

$$2\,(U - U_0) = 2B_1\theta + 2B_2\phi + \&c.$$

Since the initial position of the system is not close to a position of equilibrium, the first differential coefficients of U with regard to θ, ϕ, &c. are not small. The terms $B_1\theta$, $B_2\phi$, &c. are not now small quantities of the second order and hence it is unnecessary to retain the quadratic terms of U. Proceeding exactly as in Art. 454 the equations of motion are

$$\left.\begin{array}{l} A_{11}\theta'' + A_{12}\phi'' + \ldots = B_1 \\ A_{12}\theta'' + A_{22}\phi'' + \ldots = B_2 \\ \quad\&c. \qquad\qquad = \&c. \end{array}\right\}\ldots\ldots\ldots\ldots\ldots\ldots(1).$$

From these equations we may determine the initial values of θ'', ϕ'', &c.

If x, y, z be the Cartesian co-ordinates of any point P of the system, we may, by the geometry of the question, express these as functions of θ, ϕ, &c., Art. 396. Thus suppose that $x = f(\theta, \phi, \&c.)$, then we have initially, since θ', ϕ' are zero,

$$x'' = \frac{df}{d\theta}\,\theta'' + \frac{df}{d\phi}\,\phi'' + \&c.,$$

with similar expressions for y and z. The quantities x'', y'', z'' are evidently proportional to the direction cosines of the initial direction of motion of the point P. In this way the initial direction of motion of every point of the system may be found.

464. Initial Radius of Curvature. As explained in Art. 200, we sometimes want more than the initial direction of motion of any point P of the system. Suppose that we also want the initial radius of curvature of the path of P. We must find the values of x'', x''', &c., and then substitute in any of the formulæ given in Art. 200. If, as before, $x = f(\theta, \phi, \&c.)$ we find by differentiation that *initially*

$$x'' = f_\theta \theta'' + f_\phi \phi'' + ...,$$
$$x''' = f_\theta \theta''' + f_\phi \phi''' + ...,$$
$$x'''' = 3\left(f_{\theta\theta}\theta''^2 + 2f_{\theta\phi}\theta''\phi'' + ...\right) + f_\theta\theta'''' + f_\phi\phi'''' + ...,$$

where suffixes as usual indicate partial differential coefficients with respect to θ, ϕ, &c. If $y = F(\theta, \phi, \&c.)$ there are of course similar expressions for y'', &c., and in three dimensions for z'', &c.

If the point P be so situated that for every possible motion of the system it can begin to move only in some one direction, we take the axis of x perpendicular to that direction. We then have $x'' = 0$ for all initial variations of θ, ϕ, &c. It follows that $f_\theta = 0$, $f_\phi = 0$, &c. $= 0$. Hence $x''' = 0$, and the value of x'''' depends only on θ'', ϕ'', &c., and not on θ'''', ϕ'''', &c. It is therefore unnecessary to differentiate the dynamical equations (1) to find these higher differential coefficients. The axis of y being parallel to the initial direction of the motion of P, the value of y'' is finite. Hence, taking the formula at the end of Art. 200, we find that the initial radius of curvature ρ of the path of P is given by

$$\rho = \frac{(F_\theta \theta'' + F_\phi \phi'' + ...)^2}{f_{\theta\theta}\theta''^2 + 2f_{\theta\phi}\theta''\phi'' + ...}.$$

465. In order to find the higher differential coefficients of θ, ϕ, &c. when they are required, it may be necessary to form the equations of motion (1) to a higher degree of approximation. There can of course be no difficulty in retaining the first few powers of θ, ϕ, &c. which occur on either side of the equation. After differentiation we put zero for each of the quantities θ, ϕ, &c., θ', ϕ', &c.

But it is often more convenient to use Leibnitz' theorem. We have to substitute

$$2T = A_{11}\theta'^2 + 2A_{12}\theta'\phi' + ...$$

in the Lagrangian equations, to differentiate the results, and to put $\theta = 0$, $\theta' = 0$, &c. after the differentiations have been performed. Taking the first differential coefficient with regard to t we find $\theta''' = 0$, $\phi''' = 0$, &c. Taking the second differential coefficient of the θ-equation we find

$$\left. \begin{array}{l} A_{11}\theta'''' + A_{12}\phi'''' + ... \\[1mm] + 3\left(\theta''\dfrac{d^2 A_{11}}{dt^2} + \phi''\dfrac{d^2 A_{12}}{dt^2} + ...\right) \\[3mm] - \left(\theta''^2\dfrac{dA_{11}}{d\theta} + 2\theta''\phi''\dfrac{dA_{12}}{d\theta} + ...\right) \end{array} \right\} = B_{11}\theta'' + B_{12}\phi'' + ...$$

where for the sake of brevity we have written

$$\frac{d^2 A}{dt^2} = \frac{dA}{d\theta}\theta'' + \frac{dA}{d\phi}\phi'' + ...$$

which is evidently true *initially*. The other equations are formed in the same way.

466. **Examples of Initial Motion.** Ex. 1. A smooth plane of mass M is freely moveable about a horizontal axis lying within it and passing through its centre of gravity, the radius of gyration of the plane about the axis being k. The plane being inclined at an angle a to the horizon, a sphere of mass m is placed gently on it. If initially the centre of the sphere be in a vertical through the axis of the plane, and

if h be its initial height above that axis, show that the angle ϕ which the initial direction of motion of the centre makes with the vertical is given by
$$(Mk^2 + mh^2) \tan \phi = Mk^2 \cot a. \qquad \text{[Math. Tripos, 1879.]}$$

Ex: 2. n rods of lengths a_1, $a_2 \ldots a_n$ are jointed together in one straight line and have initial angular accelerations ω_1, $\omega_2 \ldots \omega_n$ in one plane. If one end be fixed, prove that the initial radius of curvature of the path of the free end is
$$\frac{(\Sigma a \omega)^2}{\Sigma a \omega^2}. \qquad\qquad \text{[St John's Coll.]}$$

Ex. 3. BC is a diameter of a sphere, and rods AB, CD are jointed at B and C each equal in length to BC. A being fixed, the system is held so that $ABCD$ is a horizontal straight line, and then let fall. If the mass of each rod be equal to that of the sphere, the initial radius of curvature of the path of D is $\frac{511}{314}AB$.

[St John's Coll.]

The Energy test of Stability.

467. Stability of equilibrium. The principle of the Conservation of Energy may be conveniently used in some cases to determine whether a system of bodies *at rest* is in stable or unstable equilibrium.

Let the system be in equilibrium in any position, and let V_0 be the potential energy of the forces in this position. Let the system be displaced into any initial position very near the position of equilibrium and be started with any very small initial kinetic energy T_1, and let V_1 be the potential energy of the forces in this position. At any subsequent time let T and V be the kinetic and potential energies. Then by the principle of energy
$$T + V = T_1 + V_1 \dots\dots\dots\dots\dots\dots(1).$$

Let V be an absolute minimum in the position of equilibrium, so that V is greater than V_0 for all neighbouring positions. The initial disturbed position being included amongst these, it follows that $V_1 - V_0$ is a small positive quantity. Now the kinetic energy T is necessarily a positive quantity, and since V is $> V_0$, the equation (1) shows that T is $< T_1 + V_1 - V_0$. Thus throughout the subsequent motion the vis viva lies between zero and a small positive quantity, and therefore the motion of the system can never be great.

Also, since T is necessarily positive, the system can never deviate so far from the position of equilibrium as to make V greater than $T_1 + V_1$. These two results may be stated thus:—

If a system be in equilibrium in a position in which the potential energy of the forces is a minimum, or the work a maximum, for all displacements, then the system if slightly displaced will never acquire any large amount of vis viva, and will never deviate far from the position of equilibrium. The equilibrium is then said to be stable.

It will be shown that this reasoning may in certain cases be extended to determine whether a *given state of motion* as well as a given state of *equilibrium* is stable. See also the *Treatise on the Stability of Motion*, Chap. VI., 1877.

468. If the potential energy be an absolute maximum in the position of equilibrium, V is less than V_0 for all neighbouring positions. By the same reasoning we see that T is always greater than $T_1 + V_1 - V_0$, and the system cannot approach so near the position of equilibrium as to make V greater than $T_1 + V_1$. So far therefore as the equation of vis viva is concerned, there is nothing to prevent the system from departing widely from the position of equilibrium. To determine this point we must examine the other equations of motion*.

If any principal oscillation can exist, let the system be placed at rest in an extreme position of that oscillation, then the system will describe the complete oscillation and will therefore pass through the position of equilibrium. But, if T_1 be zero, V can never exceed V_1, and can therefore never become equal to V_0. Hence the system cannot pass through the position of equilibrium.

It is unnecessary to pursue this line of reasoning further, for the argument will be made clearer in the next article.

469. We may also *deduce the test of stability from the equations which determine the small oscillations of a system about a position of equilibrium.* Let the system be referred to its principal co-ordinates, and let these be θ, ϕ, &c. Then we have

$$2T = \theta'^2 + \phi'^2 + \dots\dots$$
$$2(U - U_0) = b_{11}\theta^2 + b_{22}\phi^2 + \dots\dots$$

where b_{11}, b_{22}, &c. are constants, and U_0 is the value of U in the position of equilibrium. Taking as a type any one of Lagrange's equations

$$\frac{d}{dt}\frac{dT}{d\theta'} - \frac{dT}{d\theta} = \frac{dU}{d\theta},$$

we have $\qquad \theta'' - b_{11}\theta = 0$,

with similar equations for ϕ, ψ, &c. If b_{11} is positive, this equation gives θ in terms of real exponentials, and the equilibrium is unstable for all disturbances which affect θ, except such as make the coefficient of the term containing the positive exponent

---

* This demonstration is twice given by Lagrange in his *Mécanique Analytique*. In the form in which it appears in the first part of that work, V is expanded in powers of the co-ordinates, which are supposed very small; but in Section VI. of the second part this expansion is no longer used, and the proof appears almost exactly as it is given in this treatise up to the asterisk. The demonstration in the next article is simplified from that of Lagrange by the use of principal co-ordinates.

vanish. If b_{11} is negative, θ is expressed by a trigonometrical term, and the equilibrium is stable for all disturbances which affect θ only. In this demonstration the values of b_{11}, b_{22}, &c. are supposed not to be zero.

If in the position of equilibrium U is a maximum for all possible displacements of the system, we must have b_{11}, b_{22}, &c. all negative. Whatever disturbance is given to the system, it will oscillate about the position of equilibrium, and that position is then stable. If U is a maximum for some displacements and a minimum for others, some of the coefficients b_{11}, b_{22}, &c. will be negative and some positive. In this case if the system be disturbed in some directions, it will oscillate about the position of equilibrium; if disturbed in other directions, it may deviate more and more from the position of equilibrium. The equilibrium is therefore stable for all disturbances in certain directions, and unstable for disturbances in other directions. If U is a minimum in the position of equilibrium for all displacements, the coefficients b_{11}, b_{22}, &c. are all positive, and the equilibrium is then unstable for displacements in all directions. Briefly, we may sum up the results thus:

The system will oscillate about the position of equilibrium for all disturbances if the potential energy is a minimum for all displacements. It will oscillate for some disturbances and not for others if the potential energy, though stationary, is neither a maximum nor a minimum. It will not oscillate for any disturbance if the potential energy is a maximum for all displacements.

It appears from this theorem that the stability or instability of a position of equilibrium depends, not on the inertia of the system, but only on the force function. The rule is, give the system a sufficient number of small arbitrary displacements, so that all possible displacements may be compounded of these. By examining the work done by the forces in these displacements we can determine whether the potential energy is a maximum or minimum or neither.

Ex. 1. A perfectly free particle is in equilibrium under the attraction of any number of fixed bodies. Show that, if the law of attraction be the inverse square, the equilibrium is unstable. [*Earnshaw's Theorem.*]

Let O be the position of equilibrium, Ox, Oy, Oz any three rectangular axes, then if V be the potential of the bodies, $b_{11} = \dfrac{d^2 V}{dx^2}$, $b_{22} = \dfrac{d^2 V}{dy^2}$, $b_{33} = \dfrac{d^2 V}{dz^2}$. But, since the sum of these is zero, b_{11}, b_{22}, b_{33} cannot all have the same sign.

Ex. 2. Hence show that, if any number of particles mutually repelling each other be contained in a vessel, and be in equilibrium, the equilibrium will be unstable unless they all lie on the containing surface. [Sir W. Thomson, *Camb. Math. Journal*, 1845.]

The Cavendish Experiment.

470. As an example of the mode in which the theory of small
oscillations may be used as a means of discovery we have selected
the Cavendish Experiment. The object of this experiment is to
compare the mass of the earth with that of some given body. The
plan of effecting this by means of a torsion-rod was first suggested
by the Rev. John Michell. As he died before he had time to
enter on the experiments, his plan was taken up by Mr Cavendish,
who published the result of his labours in the *Phil. Trans.* for
1798. His experiments being few in number, it was thought
proper to have a new determination. Accordingly, in 1837 a
grant of £500 was obtained from the Government to defray the
expenses of the experiments. The theory and the analytical
formulæ were supplied by Sir G. Airy, while the arrangement of
the plan of operation and the task of making the experiments
were undertaken by Mr Baily. Mr Baily made upwards of two
thousand experiments with balls of different weights and sizes,
and suspended in a variety of ways, a full account of which is
given in the *Memoirs of the Astronomical Society*, Vol. XIV.
The experiments were, in general, conducted in the following
manner.

471. Two small equal balls are attached to the extremities
of a fine rod called the torsion-rod, and the rod itself is suspended by a string fixed to its middle point C. Two large
spherical masses A, B are fastened on the ends of a plank
which can turn freely about its middle point O. The point O is

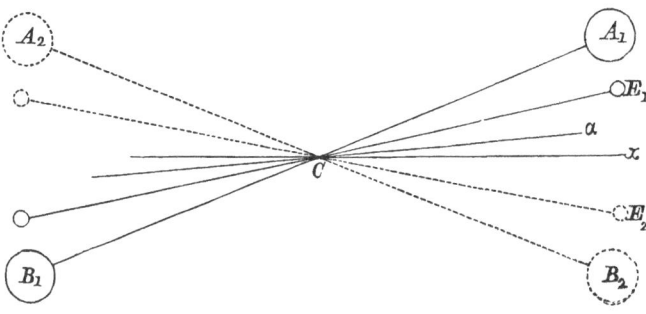

vertically under C and so placed that the four centres of gravity
of the four balls are in one horizontal plane.

Firstly, suppose the plank to be placed at right angles to the
torsion-rod, then the rod will take up some position of equilibrium
called the neutral position, in which the string has no torsion.

Let this be represented in the figure by $C\alpha$. Now let the masses A and B be moved round O into some position B_1A_1, making a not very large angle with the neutral position of the torsion-rod. The attractions of the masses A and B on the balls will draw the torsion-rod out of its neutral position into a new position of equilibrium, in which the attraction is balanced by the torsion of the string. Let this be represented in the figure by CE_1. The angle of deviation $E_1C\alpha$, and the time of oscillation of the rod about this position of equilibrium are observed.

Secondly, replace the plank AB at right angles to the neutral position of the rod, and move it in the opposite direction until the masses A and B come into some position A_2B_2 near the rod but on the side opposite to B_1A_1. Then the torsion-rod will perform oscillations about another position of equilibrium CE_2 under the influence of the attraction of the masses and the torsion of the string. As before, the time of oscillation and the deviation $E_2C\alpha$ are observed.

In order to eliminate the errors of observation, this process is repeated over and over again, and the mean results are taken. The positions B_1A_1 and A_2B_2, into which the masses are alternately put, are as nearly as possible the same throughout all the experiments. The neutral position $C\alpha$ of the rod very nearly bisects the angle between B_1A_1 and A_2B_2, but as this neutral position, possibly owing to changes in the torsion of the string, is found to undergo slight changes of position, it is not to be considered in any one experiment coincident with the bisector of the angle A_1CB_2.

Let Cx be any line fixed in space from which the angles may be measured. Let b be the angle $xC\alpha$, which the neutral position of the rod makes with Cx; A and B the angles which the alternate positions, B_1A_1 and A_2B_2, of the straight line joining the centres of the masses, make with Cx; and let $a = \frac{1}{2}(A + B)$. Also let x be the angle which the torsion-rod makes with Cx at the time t.

Supposing the masses to be in the position A_1, B_1, the moment about CO of their attractions on the two balls and on the rod will be a function only of the angle between the rod and the line A_1B_1; let this moment be represented by $\phi(A - x)$. The whole apparatus is enclosed in a wooden casing to protect it from any currents of air. The attraction of this casing cannot be neglected. As it may be different in different positions of the rod, let the moment of its attraction about CO be $\psi(x)$. Also the torsion of the string is very nearly proportional to the angle through which it has been twisted. Let its moment about CO be $E(x - b)$.

If then I be the moment of inertia of the balls and rod about the axis CO, the equation of motion is

$$I\frac{d^2x}{dt^2} = \phi(A - x) + \psi(x) - E(x - b).$$

Now $a - x$ is a small quantity, let it be represented by ξ. Substituting for x and expanding by Taylor's theorem in powers of ξ, we get

$$-I\frac{d^2\xi}{dt^2} = \phi(A - a) + \psi(a) - E(a - b) + \{\phi'(A - a) - \psi'(a) + E\}\xi.$$

Let
$$n^2 = \frac{\phi'(A - a) - \psi'(a) + E}{I},$$

and
$$e = a + \frac{\phi(A - a) + \psi(a) - E(a - b)}{In^2}.$$

Then
$$x = e + L\sin(nt + L'),$$

where L and L' are two arbitrary constants. We see therefore that in the position of equilibrium the angle made by the torsion-rod with the axis of x is e, and the time of oscillation about the position of equilibrium is $\dfrac{2\pi}{n}$.

Let us now suppose the masses to be moved into their alternate position A_2B_2; the moment of their attraction on the balls and rod is now $-\phi(x - B)$. The equation of motion is therefore

$$I\frac{d^2x}{dt^2} = -\phi(x - B) + \psi(x) - E(x - b).$$

Let $a = x - \xi$, then, substituting for B its value $2a - A$, we find by the same reasoning as before

$$x = e' + N\sin(nt + N'),$$

where n has the same value as before, and

$$e' = a + \frac{-\phi(A - a) + \psi(a) - E(a - b)}{In^2}.$$

In these expressions, the attraction $\psi(a)$ of the casing, the coefficient of torsion E and the angle b are all unknown. But they all disappear together, if we take the difference between e and e'. We then find

$$\frac{\phi(A - a)}{I} = \frac{e - e'}{2} \cdot \left(\frac{2\pi}{T}\right)^2 \quad\ldots\ldots\ldots\ldots(A),$$

where T is the time of a complete oscillation of the torsion-rod about either of the *disturbed* positions of equilibrium. Thus the attraction $\phi(A - a)$ can be found if the angle $e - e'$ between the two positions of equilibrium and also the time of oscillation about either can be observed.

472. It is sometimes wrongly objected to the Cavendish Experiment that the attractions of the balls A and B are supposed to be great enough to be measured, while the much greater attractions of surrounding objects, such as the house, &c., are neglected. But this is not the case. The attractions of all *fixed bodies* are included in that of the casing. These are therefore not neglected but *eliminated* from the result. It is to effect this elimination that we have to observe both $e' - e$ and the time of oscillation. We thus really form two equations, and from these we eliminate those attractions which we do not want to find.

473. The function $\phi (A - a)$ is the moment of the attractions of the masses and the plank on the balls and rod, when the rod has been placed in a position Cf, bisecting the angle $A_1 C B_2$ between the alternate positions of the masses. Let M be the mass of either of the bodies A and B, m that of one of the small balls, m' that of the rod. Let the attraction of M on m be represented by $\mu \dfrac{Mm}{D^2}$, where D is the distance between their centres. If (p, q) be the co-ordinates of the centre of A_1 referred to Cf as the axis of x, the moment about C of the attraction of both the masses on both the balls is

$$2\mu Mm \left\{ \frac{cq}{\{(p - c)^2 + q^2\}^{\frac{3}{2}}} - \frac{cq}{\{(p + c)^2 + q^2\}^{\frac{3}{2}}} \right\},$$

where c is the distance of the centre of either small ball from the centre C of motion. Let this be represented by μMmP. The moment of the attractions of the masses on the rod may by integration be found to be $\mu Mm'Q$, where Q is a known function of the linear dimensions of the apparatus. The attraction of the plank may also be taken account of. Thus we find

$$\phi (A - a) = \mu M (mP + m'Q).$$

If r be the radius of either ball, we have

$$I = 2m \left\{ c^2 + \frac{2}{5} r^2 \right\} + m' \frac{(c - r)^2}{3},$$

which may be represented by $I = mP' + m'Q'$, where P' and Q' are known functions of the linear dimensions of the rod and balls. Hence we find by substituting in equation (A)

$$\mu M \cdot \frac{mP + m'Q}{mP' + m'Q'} = \frac{\overset{\bullet}{e - e'}}{2} \cdot \left(\frac{2\pi}{T} \right)^2 .$$

Let E be the mass of the earth, R its radius and g the force of gravity, then $g = \mu \dfrac{E^*}{R^2}$. Substituting for μ, we find

$$\frac{M}{E} = \frac{e - e'}{2} \cdot \left(\frac{2\pi}{T}\right)^2 \cdot \frac{1}{gR^2} \cdot \frac{\dfrac{m}{m'}P' + Q'}{\dfrac{m}{m'}P + Q}.$$

The ratio $\dfrac{m}{m'}$ was taken equal to the ratio of the weights of the ball and rod *weighed in vacuo*, but it would clearly have been more accurate to have taken it equal to their ratio when weighed in air. For, since the masses attract the air as well as the balls, the pressure of the air on the side of a ball nearest the attracting mass is greater than that on the furthest side. The difference of these pressures is equal to the attraction of the mass on the air displaced by the ball.

474. By this theory the discovery of the mass of the earth has been reduced to the determination of two elements, (1) the time of oscillation of the torsion-rod, and (2) the angle $e - e'$ between its two positions of equilibrium when under the influence of the masses in their alternate positions. To observe these, a small mirror was attached to the rod at C, with its plane nearly perpendicular to the rod. A scale was engraved on a vertical plate at a distance of 108 inches from the mirror, and the image of the scale formed by reflection on the mirror was viewed in a telescope placed just over the scale. The telescope was furnished with three vertical wires in its focus. As the torsion-rod turned on its axis, the image of the scale was seen in the telescope to move horizontally across the wires, and at any instant the number of the scale coincident with the middle wire constituted the reading. The scale was divided by vertical lines one-thirteenth of an inch apart and numbered from 20 to 180 to avoid negative readings. The angle turned through by the rod when the image of the scale moved through a space corresponding to the interval of two divisions was therefore $\dfrac{1}{13} \cdot \dfrac{1}{108} \cdot \dfrac{1}{2} = 73''\!\cdot\!46$. But the division lines were cut diagonally and subdivided decimally by horizontal lines; so that not only could the tenth of a division be clearly distinguished, but, after some little practice, the fractional parts of these tenths. The arc of oscillation of the torsion-rod was so small that the square of its circular measure could be

---

* In Baily's experiment, a more accurate value of g was used. If ϵ be the ellipticity of the earth, m the ratio of centrifugal force at the equator to equatoreal gravity, and λ the latitude of the place, we have $g = \mu \dfrac{E}{R^2} \left\{ 1 - 2\epsilon + \left(\dfrac{5}{2}\, m - \epsilon\right) \cos^2 \lambda \right\}$.

neglected; but as it extended over several divisions it is clear that it could be observed with accuracy. A minute description of the mode in which the observations were made would not find a fit place in a treatise on Dynamics, we must therefore refer the reader to Baily's Memoir.

In this investigation no notice has been taken of the effect of the resistance of the air on the arc of vibration. This was, to some extent at least, eliminated by a peculiar mode of taking the means of the observations. In this way also some allowance was made for the motion of the neutral position of the torsion-rod.

We have also not considered what relative dimensions should be given to the different parts of the instrument, consistent with its proper support, so as to obtain the most accurate result. Such considerations are hardly suited to a general treatise on dynamics. In the original experiments the attracting masses A and B were large, and brought near the small balls m and m. As a rapid oscillation of the rod was inadmissible, the moment of inertia I of the rod and balls was large and the torsion of the string was small. The size of the instrument was not handy. A plan of using a *quartz fibre* as the supporting string has been proposed by C. V. Boys, by which the whole apparatus can be made on so small a scale that the two difficulties of keeping the temperature uniform and of dealing with large balls as the attracting masses are very much reduced. See the *Proceedings of the Royal Society*, May, 1889.

475. The density of water in which the weight of a cubic inch is 252·725 grains (7000 grains being equal to one pound avoirdupois) was taken as the unit of density. The final result of all the experiments was to determine for the mean density of the earth the value 5·6747.

The most important experiments after Baily which were conducted on this plan were those of Cornu and Baille. See *Comptes Rendus*, Tome LXXVI., 1873 and Tome LXXXVI., 1878. They made several improvements in the apparatus which we cannot here describe. They made the mean density to be 5·56. They considered that they had found an error in Baily's method of taking his means. If this were corrected Baily's result would become 5·55.

476. Two other methods of finding the mean density have been employed. In 1772 Dr Maskelyne, then Astronomer Royal, suggested that the mass of the earth might be compared with that of a mountain by observing the deviation produced in a plumb-line by the attraction of the latter. The mountain chosen was Schehallien, and the density of the earth was found to be a little less than five times that of water. See *Phil. Trans.* 1788 and 1811. From some observations near Arthur's Seat, the mean density of the earth was given by Lieut.-Col. James of the Ordnance Survey, as 5·316. See *Phil. Trans.* 1856.

The other method, used by Sir G. Airy, is to compare the force of gravity at the bottom of a mine with that at the surface, by observing the times of vibration of a pendulum. In this way the mean density of the earth was found to be 6·566. See *Phil. Trans.* 1856.

Within the last ten years the density of the earth has been found by observing how a very delicate balance is disturbed by the near approach of large attracting masses. The experiments were conducted by Jolly in Munich and Poynting in Manchester. The result was 5·69.

EXAMPLES*.

1. A uniform rod of length $2c$ rests in stable equilibrium with its lower end at the vertex of a cycloid whose plane is vertical and vertex downwards, and passes through a small smooth fixed ring situated on the axis at a distance b from the vertex. Show that, if the equilibrium be slightly disturbed, the rod will perform small oscillations with its lower end on the arc of the cycloid in the time

$$4\pi \sqrt{\frac{a\{c^2 + 3(b-c)^2\}}{3g(b^2 - 4ac)}},$$ where $2a$ is the length of the axis of the cycloid.

2. A small smooth ring slides on a circular wire of radius a which is constrained to revolve about a vertical axis in its own plane, at a distance c from the centre of the wire, with a uniform angular velocity $\sqrt{\dfrac{g\sqrt{2}}{c\sqrt{2}+a}}$; show that the ring will be in a position of stable relative equilibrium when the radius of the circular wire passing through it is inclined at an angle 45° to the horizon ; show also that, if the ring be slightly displaced, it will perform a small oscillation in the time

$$2\pi \left\{ \frac{a\sqrt{2}}{g} \cdot \frac{c\sqrt{2}+a}{c\sqrt{8}+a} \right\}^{\frac{1}{2}}.$$

3. A uniform bar of length $2a$, suspended by two equal parallel strings each of length b from two points in the same horizontal line, is turned through a small angle about the vertical line through the middle point, show that the time of a small oscillation is $2\pi \sqrt{\dfrac{bk^2}{ga^2}}$.

4. Two equal heavy rods connected by a hinge which allows them to move in a vertical plane rotate about a vertical axis through the hinge, and a string whose length is twice that of either rod is fastened to their extremities and bears a weight at its middle point. If M, M' be the masses of a rod and the particle, and $2a$ the length of a rod, prove that the angular velocity about the vertical axis when the rods and string form a square is $\sqrt{\dfrac{3g}{2a\sqrt{2}} \cdot \dfrac{M+2M'}{M}}$; prove also that, if the weight be slightly depressed in a vertical direction and the system left to itself, the time of a small oscillation is $2\pi\sqrt{\dfrac{4a\sqrt{2}}{15g} \cdot \dfrac{M+3M'}{M+2M'}}$.

5. A ring of weight W which slides on a rod inclined to the vertical at an angle a is attached by means of an elastic string to a point in the plane of the rod, so

---

* These examples are taken from the Examination Papers which have been set in the University and in the Colleges.

situated that its least distance from the rod is equal to the natural length of the string. Prove that, if θ be the inclination of the string to the rod when in equilibrium, $\cot\theta - \cos\theta = \dfrac{W}{w}\cos\alpha$, where w is the modulus of elasticity of the string. Also if the ring be slightly displaced the time of a small oscillation will be $2\pi \sqrt{\dfrac{Wl}{wg}\dfrac{1}{1-\sin^3\theta}}$, where l is the natural length of the string.

6. A circular tube of radius a contains an elastic string fastened at its highest point equal in length to $\dfrac{1}{8}$ of its circumference, and having attached to its other extremity a heavy particle which hanging vertically would double its length. The system revolves about the vertical diameter with an angular velocity $\sqrt{\dfrac{g}{a}}$. Find the position of relative equilibrium, and prove that, if the particle be slightly disturbed, the time of a small oscillation is $\dfrac{2\pi\sqrt{\pi}}{\sqrt{\pi+4}}\sqrt{\dfrac{a}{g}}$.

7. A heavy uniform rod AB has its lower extremity A fixed to a vertical axis, and an elastic string connects B to another point C in the axis such that $AC = \dfrac{AB}{\sqrt{2}} = a$; the whole is made to revolve round AC with such angular velocity that the string is double its natural length and horizontal when the system is in relative equilibrium, and then left to itself. If the rod be slightly disturbed in a vertical plane, prove that the time of a small oscillation is $2\pi\sqrt{\dfrac{4a}{21g}}$, the weight of the rod being sufficient to stretch the string to twice its length.

8. Three equal elastic strings AB, BC, CA surround a circular arc, the ends being fixed at A. At B and C two equal particles of mass m are fastened. If l be the natural length of each string supposed always stretched, and λ the modulus of elasticity, show that if the equilibrium be disturbed the particles will be at equal distances from A after intervals $\pi\sqrt{\dfrac{ml}{\lambda}}$.

9. A particle of mass M is placed near the centre of a smooth circular horizontal table of radius a, strings are attached to the particle and pass over n smooth pullies which are placed at equal intervals round the circumference of the circle; to the other end of each of these strings a particle of mass M is attached; show that the time of a small oscillation of the system is $2\pi\left(\dfrac{2+n}{n}\dfrac{a}{g}\right)^{\frac{1}{2}}$.

10. Two discs slide in a circular tube of uniform bore containing air, exactly fitting the tube. The two discs are placed initially so that the line joining their centres passes through the centre of the tube, and the air in the tube is initially of its natural density. One disc is projected so that the initial velocity of its centre is a small quantity. If the inertia of the air be neglected, prove that the point on the axis of the tube equidistant from the centres of the discs moves uniformly and that the time of an oscillation of each disc is $2\pi\sqrt{\dfrac{Ma\pi}{4P}}$, where M is the mass of each disc, a the radius of the axis of tube, and P the pressure of air on the disc in its natural state.

11. A uniform beam of mass M and length $2a$ can turn round a fixed horizontal axis at one end; to the other end of the beam a string of length l is attached and at the other end of the string a particle of mass m. If, during a small oscillation of the system, the inclination of the string to the vertical is always twice that of the beam, then $M(3l - a) = 6m(l + a)$.

12. A conical surface of semivertical angle a is fixed with its axis inclined at an angle θ to the vertical, and a smooth right cone of semivertical angle β is placed within it so that the vertices coincide. Show that time of a small oscillation $= 2\pi \sqrt{\dfrac{a \sin(a - \beta)}{g \sin \theta}}$, where a is the distance of the centre of oscillation of the cone from the vertex.

13. A number of bodies, the particles of which attract each other with forces varying as the distance, are capable of motion on certain curves and surfaces. Prove that, if A, B, C be the moments of inertia of the system about three axes mutually at right angles through its centre of gravity, the positions of stable equilibrium will be found by making $A + B + C$ a minimum.

14. A particle is in motion within a triangle ABC, and is attracted perpendicularly to the sides with forces each equal to μ times the perpendicular distance. Show that the motion is expressed by two periodic terms of the form
$$P \sin\{t \sqrt{(\lambda\mu)} + a\},$$
where $(\lambda - 1)(\lambda - 2) + 2 \cos A \cos B \cos C = 0$.

Shew that the roots of this quadratic are real and positive.

Examine the case of an equilateral triangle, and in that case verify the above result independently.

15. The force between two small masses attracting according to the law of the inverse square of the distance is equal, at distance a, to a very small fraction $\dfrac{1}{n}$ of the weight of either. They are suspended by two strings of length l from two points situated in a horizontal plane, at a distance apart equal to a, and are set to perform small vibrations in the same vertical plane; prove that the motion of each is compounded of two harmonic motions whose periods are very nearly as
$$1 : 1 + \frac{2l}{na}.$$

CHAPTER X.

Oscillations of a Rocking Body in three dimensions.

477. *A heavy body oscillates in three dimensions with one degree of freedom on a fixed rough surface of any form in such a manner that there is no rotation about the common normal. Find the motion.*

478. The Relative Indicatrix. Let O be the point of contact when the heavy body is in equilibrium. Let the common normal be the axis of z, and let the other two axes be at right angles in the common tangent plane. The equations to the portions of the surfaces in the neighbourhood of O may be written in the forms

$$z = \tfrac{1}{2} (ax^2 + 2bxy + cy^2) + \&c.$$
$$z' = \tfrac{1}{2} (a'x^2 + 2b'xy + c'y^2) + \&c.$$

Let an ordinate move round the origin so that the portion $z - z'$ between the surfaces is constant and equal to any indefinitely small quantity Δ. This ordinate traces out an evanescent conic on the plane of xy whose equation is

$$(a - a')\, x^2 + 2(b - b')\, xy + (c - c')\, y^2 = 2\Delta.$$

Any conic similar and similarly situated to this, lying in the tangent plane and having its centre at O, is called the *Relative Indicatrix* of the two surfaces.

Let OR be any radius vector of this indicatrix, then the difference of the curvatures of the two sections made by a normal plane zOR (or their sum, if they are measured in opposite directions) varies inversely as the square of OR. This of course follows from the definition of the conic by a well-known argument in solid geometry. Thus, let $(r, z)\, (r, z')$ be the co-ordinates of two points on the two circles of curvature at the same distance from the axis of z. We have ultimately $2\rho z = r^2$ and $2\rho' z' = r^2$. Also $z - z' = \Delta$, hence, eliminating z and z', we see that the difference of the curvatures varies inversely as r^2.

Let OR be a tangent to the arc of rolling determined by the geometrical conditions of the question. Let ρ, ρ' be the radii of curvature of the normal sections through OR, taken positively when the curvatures are in *opposite directions*, and let $\dfrac{1}{s} = \dfrac{1}{\rho} + \dfrac{1}{\rho'}$. Then s may be called the *radius of relative curvature*.

We have the three following propositions which are of use in Dynamics.

479.   PROP.   **The Instantaneous Axis.**   Let OI and Oy be two conjugate diameters of the relative indicatrix, then, if Oy be a tangent to the arc of rolling, OI is the instantaneous axis, and, if θ be the indefinitely small angle turned round the instantaneous axis, the arc σ of rolling is given by $\sigma = \theta s \sin yOI$.

To prove this, measure in the plane yz along the surfaces two lengths OP and OP' each equal to σ. Then in the limit $P'P$ is parallel to the normal Oz. Let it cut the plane of xy in M. Draw another ordinate $Q'QN$ indefinitely near to $P'PM$ so that $PP' = QQ'$, then MN is an elementary arc of that relative indicatrix which passes through M, and is therefore parallel to OI the conjugate diameter of OM. Also $PQP'Q'$ is a parallelogram.

The planes OPQ, $OP'Q'$ are ultimately tangent planes at P and P', and must intersect in a straight line OJ parallel to PQ or $P'Q'$. If then we turn the body round OJ, the tangent planes at P and P' will be brought into coincidence and the one body will roll on the other. Thus OJ is the instantaneous axis.

Now, since MN is the projection of PQ or $P'Q'$ on the plane of xy, it follows that OI, a parallel to MN, is the projection of OJ, a parallel to PQ or $P'Q'$. Also the parallels PQ and $P'Q'$, being tangents to the surfaces, make indefinitely small angles with the plane of xy, hence OJ makes an equal indefinitely small angle with OI. If ϕ be this small angle and θ the angle of rotation about OJ, the motion of the body is represented by rotations $\theta \sin \phi$ about Oz and $\theta \cos \phi$ about OI. Since θ is indefinitely small, the former is of the second order and is to be neglected. The latter reduces to θ.

To prove the last part of the proposition, we may again resolve this latter rotation into a rotation $\theta \cos yOI$ about Oy and a rotation $\theta \sin yOI$ about Ox. The former does not affect the arc of rolling along Oy, the latter obviously gives $\sigma = s\theta \sin yOI$.

480.   PROP.   **The Cylinder of Stability.**   Measure a length $s \sin^2 yOI$ along the common normal Oz and describe a circular cylinder having this length as a diameter of the base, the axis being parallel to OI. If the centre of gravity of the body be inside this cylinder, the equilibrium is stable: if outside and above the plane of xy, the equilibrium is unstable. The cylinder may therefore be called *the cylinder of stability*.

These results follow from the second expression for the moment of gravity about OI found in the next proposition.

481. PROP. **The time of Oscillation.** Let G be the centre of gravity and K the radius of gyration of the body about OI, then the length L of the simple equivalent pendulum is given by

$$\frac{K^2}{L} = s \cos GOz \cdot \sin^2 yOI - OG \cdot \sin^2 GOI.$$

If OG produced cut the cylinder of stability in V, then

$$\frac{K^2}{L} = GV \cdot \sin^2 GOI.$$

We deduce from this that the time of oscillation of the body is the same as if the fixed surface were plane, and the curvatures of the upper body at the point of contact were altered so that the Relative Indicatrix remained the same as before.

482. These results may be obtained by taking moments about the instantaneous axis, see Art. 448. The general course of reasoning may be indicated as follows. In equilibrium O is the point of contact and OG is vertical; as the body rolls on the surface, say in the direction $y'P$, let P be the point of contact at the time t and let O', G' be the positions in space occupied by the points O and G of the body. These points are not marked in the figure, but O and O' will obviously lie indefinitely close to each other between y' and P, so that OO' is perpendicular to Py', while G' will move from G a little to the right, as seen from any point in PI'. Draw PW vertical, and PF parallel and equal to $O'G'$. If PI' be the instantaneous axis at the time t, θ is the angle between the planes WPI' and FPI'.

To find the moment of the weight about PI' we resolve gravity parallel and perpendicular to PI'. The former component has no moment about PI', the latter is $g \sin WPI'$. Let this latter act parallel to some straight line KP. The moment required is the product of resolved gravity into the shortest distance between the line of action of this force and the straight line PI'. This shortest distance is equal to the sum of the projections (with their proper signs) of PO', $O'G'$ on a straight line perpendicular to both KP and PI'. Let this straight line be PH. To find these projections we shall use a little spherical trigonometry. Let the

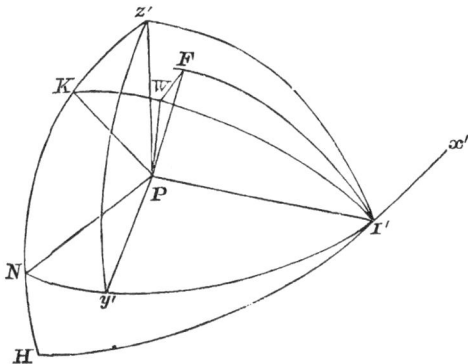

figure represent the spherical triangles formed by the arcs on a sphere subtending the various angles at the centre P. Also Py' is a tangent to PO' the arc of rolling, and Pz' is normal to the surface at P. The projection of PO' on PH is $\sigma \cos y'PH$ $= \sigma \cos y'PN \cos NPH = \sigma \sin y'PI' \cos KPz'$. The projection of $O'G'$ is the same as the projection of $PF = PF \cos HPF = -PF \sin WPF = -OG \cdot \theta \sin WPI'$.

The differential equation is therefore

$$K^2\ddot{\theta} = -\theta g \{s \cdot \sin^2 y'PI' \cdot \sin WPI \cdot \cos KPz' - OG \cdot \sin^2 WPI'\}.$$

We now replace $\sin WPI' \cdot \cos KPz'$ by its equivalent $\cos WPz'$. In the small terms containing the factor θ we may remove the accents, and replace P and W by O and G. We immediately obtain one of the results.

To obtain the other, we write the equation of moments in the form

$$K^2\ddot{\theta} = -\theta g \sin^2 WPI' \left\{ s \sin^2 y'PI' \frac{\cos KPz'}{\cos KPV} - OG \right\}.$$

But, if D be the diameter of the cylinder of stability drawn with its axis parallel to PI', and if PW cut the cylinder in V, we have $PV \cdot \cos KPW = D \cos KPz'$. Substituting in the equation, the expression in brackets takes the form $PV - OG$, which is ultimately equal to GV. We thus obtain the second result.

We might also find the periods by the method of vis viva.

Oscillations of Cones in three dimensions.

483. Oscillations of Cones to the first order. *A heavy cone of any form oscillates on a fixed rough conical surface, the vertices being coincident. It is required to find the time of a small oscillation.*

The motion of a cone about its vertex regarded as a fixed point is conveniently discussed by the help of spherical trigonometry.

Let O be the common vertex, G the centre of gravity of the moving cone, $OG = h$. With centre O, and radius equal to OG, describe a sphere; it is on this sphere that we shall suppose our spherical triangles to be constructed. Let OI be the instantaneous axis of the moving cone, *i.e.* the common generator along which the two cones touch, and let it cut the sphere in I. Let OW be a vertical drawn upwards to cut the same sphere in W. Let the arcs $WI = z$, $GI = r$. In the position of equilibrium the three straight lines OW, OG, OI are in the same vertical plane, and they are so represented in the figure.

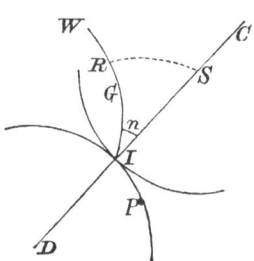

Let n be the inclination of the vertical plane GOI to the normal plane to the two cones along OI. Let ρ, ρ' be the semi-angles of the two right circular osculating cones of contact along $0, I$, taken positively when the curvatures are in opposite directions. In the figure their axes cut the sphere in C and D.

If K be the radius of gyration of the moving cone about OI, the length L of the simple equivalent pendulum is given by

$$\frac{K^2}{hL} = \sin(z - r) \cos n \frac{\sin \rho \sin \rho'}{\sin(\rho + \rho')} - \sin r \sin z.$$

The dynamical principle used in obtaining this result is that of taking moments about the instantaneous axis, Art. 448. If G' be the position of the centre of gravity at the time t, and θ the angle between the planes GOI, $G'OI$, we have

$$K^2\ddot{\theta} = M \quad\dots\dots\dots\dots\dots\dots\dots\dots\dots(1),$$

where M is the moment of g acting at G' about the instantaneous axis at the time t.

If OP be a neighbouring generator of the fixed cone and the angle POI be σ, the moment M' about OP of g acting at G' is a function of θ and σ. We therefore have to the first order of small quantities

$$M' = A\sigma + B\theta \quad\dots\dots\dots\dots\dots\dots\dots\dots(2),$$

where A and B are two expressions which depend on the form of the cone.

Finally, if OP be the instantaneous axis at the time t, we have $M' = M$ and

$$\sigma \sin(\rho + \rho') = \theta \sin \rho \sin \rho' \quad\dots\dots\dots\dots\dots(3).$$

Eliminating either σ or θ from these equations the time of oscillation can be deduced.

The relations (2) and (3) are established in an elementary manner in Arts. 484 and 485. The steps in the investigation correspond to those used in the oscillation of cylinders (Art. 441), the chief difference being that the straight lines used in the figure for cylinders are here replaced by spherical arcs. The proof of the relation (3) presents no difficulty, but in the general case when both the rolling and the fixed cone are of any forms the figure required to obtain the relation (2) is rather complicated. In particular cases, such as when the fixed surface is plane or the rolling cone is one of revolution, there is considerable simplification, the extent of which is pointed out in some of the examples in Art. 486. In these the proof, as adapted to the special case under consideration, is again briefly sketched.

By considering the parts of M' due to θ and σ separately, we may arrive at their values without requiring any figure more

complicated than that already drawn in this Article.   The proof
is as follows.

Suppose (1) that $\sigma = 0$, then M' is the moment round OI of g acting at G'
parallel to the vertical WO.   Since the body is turned round OI through an angle
θ, the arc $GG' = h\theta \sin GI$.   Resolving g parallel and perpendicular to OI, the
latter component is $g \sin WI$ and its moment round OI is $g \sin WI . GG'$.   Substituting for the spherical arcs WI and GI their values z and r, the moment
becomes $-gh\theta \sin r . \sin z$.

Suppose (2) that $\theta = 0$, then M' is the moment round the neighbouring generator
OP of g acting at G parallel to WO.   Resolving g along and perpendicular to GO,
the latter component is $g \sin WG$, and acts at G along a tangent to the spherical
arc GI.   To find its moment round OP we resolve it perpendicular to the plane
OGP and multiply the component by $h \sin GP$.   The required moment is therefore
the product of $g \sin WG$, $\sin IGP$ and $h \sin GP$.   Since $\sigma \cos n$ and $IGP . \sin GP$
both express the perpendicular distance of P from the arc GI, the required moment
becomes $gh\sigma \sin (z - r) \cos n$, where $z - r$ has been written for WG.

The complete value of M is therefore

$$M = gh \{\sigma \cos n \sin (z - r) - \theta \sin r \sin z\}.$$

484.   As the heavy cone rolls on the surface, the point on the sphere which is at
I in equilibrium takes the position I', and P is the new point of contact.   Let the
arc IG assume the position $I'G'$, and let the centre C of the osculating cone move
to C'.   Let $\sigma = IP$ be the arc rolled over, and let θ be the angle turned round by

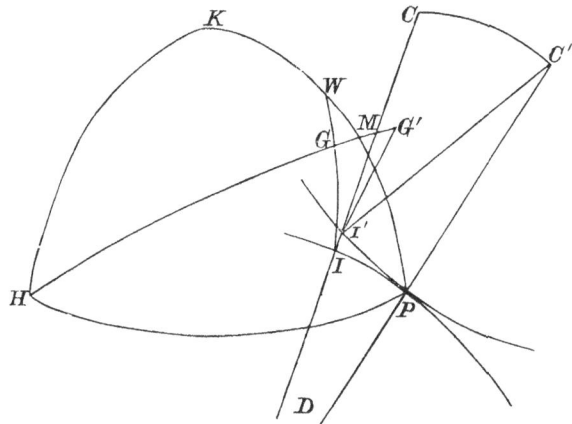

the cone.   Since this angle is ultimately the same as CPC', we have $CC' = \theta \sin \rho$.
Also $CC' \operatorname{cosec} (\rho + \rho')$ and $\sigma \operatorname{cosec} \rho'$ are each equal to the angle IDP.   We thus
find $\sigma = \theta \dfrac{\sin \rho \sin \rho'}{\sin (\rho + \rho')}$.

485.   The vertical OW cuts the sphere in W.   To find the moment of the weight
about OP we must resolve gravity parallel and perpendicular to OP.   The former
component has no moment, and the latter is $g \sin WP$.   Let this latter act parallel

to some straight line KO. The moment required is the product of resolved gravity into the projection of OG' on a straight line OH, which is perpendicular to both OK and OP. Thus the spherical triangle HKP has all its sides right angles. In equilibrium G lies in the vertical plane WOI, and as the cone rolls G moves to G', so that the arc GG' is perpendicular to WI, and equal to $\theta \sin r$. Let this arc cut WP in M. The projection required is $h \cos HG' = -h \cdot MG'$ since HM is a right angle. Since PI makes with PH an angle which is ultimately equal to n, we have $\dfrac{GM}{\sigma \cos n} = \dfrac{\sin WG}{\sin WI} = \dfrac{\sin (z-r)}{\sin z}$ ultimately. The moment required, urging the cone back to its position of equilibrium, is $gh \sin z (GM - GG')$, which on substitution becomes

$$M = gh \{ \sigma \cos n \sin (z-r) - \theta \sin r \sin z \}.$$

Equating this moment with the sign changed to $K^2 \ddot{\theta}$, the result to be proved follows immediately.

We may obtain this equation by the analytical method given in Art. 509. We there replace the geometry here used by a process of differentiation, which may be extended to any higher degree of approximation.

486. **Examples. Ex. 1.** If the upper body be a right cone of semi angle ρ, and if it be on the top of any conical surface, we have $n = 0$ and $r = \rho$. The preceding expression then takes the form

$$\frac{K^2}{hL} = -\frac{\sin (z + \rho') \sin^2 \rho}{\sin (\rho + \rho')}.$$

Ex. 2. A right cone of angle 2ρ and altitude a, suspended by its vertex from a fixed point in a rough vertical wall, makes small oscillations, prove that the length of the equivalent pendulum is $\dfrac{a}{5} \dfrac{1 + 5 \cos^2 \rho}{\cos \rho}$.

Let the cone when in equilibrium touch the plane along the vertical Oz. At the time t, let the generator ON be the line of contact, where $zON = \sigma$. Let OA be the axis. Resolving gravity along and perpendicular to the line ON, and taking moments about the instantaneous axis ON, we have

$$K^2 \ddot{\theta} = -g \sin \sigma \cdot \tfrac{3}{4} a \sin \rho.$$

Now, if the cone turn round ON through an angle θdt, the centre A of the base advances a space $a \sin \rho \cdot \theta dt$, hence, if AH be a perpendicular on ON, H advances an equal space. But it does advance a space $OH \cdot d\sigma$ i.e. $a \cos \rho d\sigma$. We therefore have $\dot{\theta} \tan \rho = \dot{\sigma}$. Substituting this value of θ in the above equation and quoting the value of K^2 from Art. 18, Ex. 7, the length of the equivalent pendulum is found without difficulty.

Ex. 3. A right cone of angle 2ρ and altitude a oscillates on a perfectly rough plane inclined to the vertical at an angle z, the length of the equivalent pendulum is

$$\frac{a (1 + 5 \cos^2 \rho)}{5 \cos \rho \cos z}.$$

Resolve gravity into $g \cos z$ acting down the plane and a perpendicular component which can be neglected. Then proceed as in the last question.

Ex. 4. A right cone of angle 2ρ and altitude a is divided by a plane through the axis. One of the halves rests in equilibrium with its axis along a generator of a fixed right cone of angle $2\rho'$, the vertices being coincident, prove that the length L of the equivalent pendulum is given by

$$\{9\pi^2 + 16\tan^2\rho\}^{\frac{1}{2}}\frac{2a\tan^2\rho}{5L} = 3\pi\sin z\tan\rho' - 4\tan\rho\frac{\sin(\rho'+z)}{\cos\rho'},$$

where z is the inclination of the line of contact to the vertical measured upwards.

487. Condition of Stability of Cones to the first order.

To determine the condition of stability when a heavy cone rests in equilibrium on a perfectly rough cone fixed in space.

It is evident that we must have the length L of the equivalent pendulum, found in Art. 483, equal to a positive quantity. This leads to the following construction, which is represented in the figure of Art. 483. Measure along the common normal CI to the cones a length $IS = s$, such that $\cot s = \cot\rho + \cot\rho'$. From S draw an arc SR perpendicular to IGW, then

$$\cos n = \cot s \cdot \tan IR.$$

Then L is positive and the equilibrium is stable if the centre of gravity of the moving cone be either below the common generator of the two cones, or above the generator at an angle r such that

$$\cot r > \cot z + \cot IR.$$

When the vertex O is very distant the cones become cylinders. In this case, if the arc z become a quadrant, the condition of stability is reduced to $r < IR$. This agrees with the condition given in Art. 442.

Large Tautochronous Motions.

488. When the oscillations of a system are not small, the equation of motion cannot always be reduced to a linear form, and no general rule can be given for the solution. But the oscillation may still be tautochronous, and it is sometimes important to ascertain whether this is the case. Various rules to determine this question are given in the following Articles.

489. *Show that, if the equation of motion be*

$$\frac{d^2x}{dt^2} = a \text{ homogeneous function of the first degree of } \frac{dx}{dt} \text{ and } x,$$

then, in whatever position the system is placed at rest, the time of arriving at the position determined by $x = 0$ *is the same.*

Let the homogeneous function be written $xf\left(\frac{1}{x}\frac{dx}{dt}\right)$. Let x and ξ be the co-ordinates of two systems starting from rest in two different positions, and let $x = a$, $\xi = \kappa a$ initially. It is easy to see that the differential equation of one system is changed into that of the other by writing $\xi = \kappa x$. If therefore the motion of one system is given by $x = \phi\,(t, A, B)$, that of the other is given by $\xi = \kappa\phi\,(t, A', B')$. To determine the arbitrary constants A, B and A', B', we have exactly the same conditions, viz. that, when $t = 0$, $\phi = a$ and $\dfrac{d\phi}{dt} = 0$. Since only one motion can follow from a single set of initial conditions, we have $A' = A$, and $B' = B$. Hence throughout the motion $\xi = \kappa x$, and therefore x and ξ vanish together. It follows that the motions of the two systems are perfectly similar.

This result may be obtained also by integrating the differential equation. If we put $\dfrac{1}{x}\dfrac{dx}{dt} = p$, we find $x = A\phi(t + B)$. When $t = 0$, $\dfrac{dx}{dt} = 0$, and therefore $\phi'(B) = 0$. Thus B is known and x vanishes when $\phi\,(t + B) = 0$, whatever be the value of A.

490. It must be noticed that if the force be a homogeneous function of the velocity and x, the motion is tautochronous only in a certain sense. It may happen that the system arrives at the position determined by $x = 0$ only after an infinite time, or the time of arrival may be imaginary. Thus, suppose the homogeneous function to be $m^2 x$, where m^2 is positive, then the system starting from rest moves continually away from the position $x = 0$. The value of x is evidently represented by an exponential function of x which never ceases to increase with the time. It is therefore necessary in applying the rule to ascertain whether the time given by the equation $\phi\,(t + B) = 0$ is real or not.

We may in general determine this from the known circumstances of each particular case. The two following general conditions will guide us in our decision. If the time before arrival at the position $x = 0$ is to be real and finite, and the same from all initial positions, it is clear that the position $x = 0$ *must be one of equilibrium.* For, if not, place the system at rest indefinitely close to that position, then the time of arrival will be zero, unless the acceleration be also zero. Further, the position of arrival must be a position of *stable equilibrium* for all displacements; or at least for all displacements on that side of the position of equilibrium on which the motion is to take place.

491. **Lagrange's rule.** *If the equation of motion of the system be*

$$\frac{d^2x}{dt^2} = \left(\frac{dx}{dt}\right)^2 \frac{f'(x)}{f(x)} + F\left\{\frac{dx}{dt},\ f(x)\right\},$$

where F *is a homogeneous function of the first degree, and* $f(x)$ *is any function of* x, *show that, in whatever position the system is*

placed, the time of arriving at the position determined by $f(x) = 0$
is the same.

This is Lagrange's general expression for a force which causes a tautochronous
motion. The formula was given by him in the Berlin Memoirs for 1765 and 1770,
and in other places. Another very complicated demonstration was given by D'Alembert, requiring variations as well as differentiations. Lagrange seems to have
believed that his expression for a tautochronous force was both necessary and
sufficient. But it has been pointed out by M. Fontaine and M. Bertrand that
though sufficient it is not necessary. At the same time the latter reduced the
demonstration to a few simple principles. A more general expression than
Lagrange's has been lately given by Brioschi, but it does not appear to contain any
cases of tautochronous motion not already given by Lagrange's formula.

Lagrange's result may be arrived at by the following reasoning.
The motion from rest is tautochronous with regard to the point
$x = 0$, if the equation of motion be $\dfrac{d^2x}{dt^2} = xF\left(\dfrac{1}{x}\dfrac{dx}{dt}\right)$. Put $x = \phi(y)$
we easily find

$$\phi'\frac{d^2y}{dt^2} + \phi''\left(\frac{dy}{dt}\right)^2 = \phi F\left(\frac{\phi'}{\phi}\frac{dy}{dt}\right),$$

where ϕ stands for $\phi(y)$ and accents as usual denote differential
coefficients. Let $\dfrac{\phi}{\phi'} = f(y)$, substituting we have

$$\frac{d^2y}{dt^2} = \frac{f'}{f}\left(\frac{dy}{dt}\right)^2 - \frac{1}{f}\left(\frac{dy}{dt}\right)^2 + fF\left(\frac{1}{f}\frac{dy}{dt}\right),$$

where f has been written for $f(y)$. The last two terms of this
expression form a homogeneous function of f and $\dfrac{dy}{dt}$ of the first
degree, and therefore Lagrange's formula has been proved. This
demonstration is due to Bertrand.

The motion begins from rest with any initial value of x and
ends when $x = 0$. Hence, writing $x = \phi(y)$, we see that in the
second equation the motion begins with $\dfrac{dy}{dt} = 0$ and with any
initial value of y, and terminates when $\phi(y) = 0$. Now $\dfrac{dx}{dt}$ does
not in general vanish when $x = 0$, since the system arrives with
some velocity at the position of equilibrium. But $\dfrac{dx}{dt} = \phi'(y)\dfrac{dy}{dt}$,
hence $\phi'(y)$ does not vanish when $x = 0$. It follows therefore, since
$\phi = \phi' \cdot f(y)$, that the motion terminates when $f(y) = 0$.

When the particle is constrained to move on a rough curve
under the action of a tangential force, Lagrange's rule may be put
under another form. If F be the tangential force tending to

diminish the arc s, the equation of motion takes the form (Art. 495)

$$\frac{d^2s}{dt^2} = \left(\frac{ds}{dt}\right)^2 \frac{\mu}{\rho} - F.$$

It is proved in Arts. 495 and 496 that the motion will be tautochronous if

$$m^2\rho = \frac{dF}{d\psi} - F,$$

where m is some constant and ψ is the angle the tangent makes with a straight line fixed in space. The tautochronous motion terminates at the point determined by $F = 0$ and the constant interval is $\dfrac{\pi}{2m}$.

492. Effect of a resisting medium. If the motion of any system is tautochronous according to Lagrange's formula in vacuo, it will also be tautochronous in a resisting medium, if the effect of the resistance is to add on to the differential equation of motion a term proportional to the velocity. This theorem is due to Lagrange.

The proof of this is easy, for to introduce the resistance of such a medium into the equation of motion is merely to increase the homogeneous function F by a term of the form $\kappa \dfrac{dx}{dt}$. This is permissible, since the only restriction on the form of F is that it must be a homogeneous function of the first degree.

493. Motion on a rough cycloid. *A heavy particle slides from rest on a rough cycloid placed with its axis vertical, show that the motion is tautochronous.*

Let O be the lowest point of the cycloid, P the particle, $OP = s$, so that the arc is measured from O in the direction opposite to that of the motion. Let the normal at P make an angle ψ with the vertical, let ρ be the radius of curvature, and a the diameter of the generating circle. Then, by known properties of the cycloid, $s = 2a \sin \psi$, $\rho = 2a \cos \psi$. Let μ be the coefficient of friction, g the accelerating force of gravity, and let the mass be unity. Then, if R be the pressure on the particle measured inwards when positive, and v the velocity, we have

$$\left.\begin{aligned}\frac{v^2}{\rho} &= R - g\cos\psi \\ \frac{d^2s}{dt^2} &= \mu R - g\sin\psi\end{aligned}\right\} \quad \dots\dots\dots\dots\dots\dots(1).$$

Eliminating R the equation of motion becomes

$$\frac{d^2s}{dt^2} = \frac{\mu}{\rho}\left(\frac{ds}{dt}\right)^2 - g(\sin\psi - \mu\cos\psi) \quad \ldots\ldots\ldots\ldots(2).$$

Substituting for ρ and s their values in terms of ψ, this becomes

$$-\cos\psi\,\frac{d^2\psi}{dt^2} + (\sin\psi + \mu\cos\psi)\left(\frac{d\psi}{dt}\right)^2 = \frac{g}{2a}(\sin\psi - \mu\cos\psi).$$

Writing $\mu = \tan\epsilon$ this is identical with

$$\frac{d^2}{dt^2}\{\epsilon^{-\mu\psi}\sin(\psi-\epsilon)\} + \frac{g}{2a\cos^2\epsilon}\{\epsilon^{-\mu\psi}\sin(\psi-\epsilon)\} = 0.$$

Since $\dfrac{d\psi}{dt}$ is initially zero, the solution of this equation is

$$\epsilon^{-\mu\psi}\sin(\psi-\epsilon) = A\cos\left(\sqrt{\frac{g}{2a}}\frac{t}{\cos\epsilon}\right),$$

where A is a constant depending on the initial value of ψ.

The motion is therefore tautochronous. At whatever point of the cycloid the particle is placed at rest, it arrives at a point A determined by $\epsilon^{-\mu\psi}\sin(\psi-\epsilon) = 0$ in the same time, and this time is $\dfrac{\pi}{2}\cos\epsilon\sqrt{\dfrac{2a}{g}}$. The point A at which the tautochronous motion terminates is clearly an extreme position of equilibrium in which the limiting friction just balances gravity.

We have here given an independent proof of this result, but it might have been deduced at once from Lagrange's rule. We may write his theorem in the form

$$\frac{d^2s}{dt^2} = \left(\frac{ds}{dt}\right)^2\frac{f'(s)}{f(s)} + A\left(\frac{ds}{dt}\right)^2\frac{1}{f(s)} + Bf(s),$$

since the last two terms on the right-hand side constitute a homogeneous function of $f(s)$ and ds/dt of the first order. Comparing this form with the equation (2) we have

$$\frac{f'(s) + A}{f(s)} = \frac{\mu}{\rho}, \qquad f(s) = \sin\psi - \mu\cos\psi.$$

Substituting in the first of these equalities the value of $f(s)$ given by the second, we find that the former is identically satisfied by choosing $1 + 2aA = -\mu^2$. It follows that the equation (2) is merely an example of Lagrange's rule. The motion is therefore tautochronous for arcs terminating at the point determined by $\tan\psi = \mu$.

The same result follows immediately from the general theorem proved in Art. 495. We have $F = g(\sin\psi - \mu\cos\psi)$. The equation of condition $m^2\rho = \dfrac{dF}{d\psi} - \mu F$ is therefore satisfied identically if $m = \sec \epsilon \cdot \sqrt{g/2a}$. The tautochronous arcs terminate at the point given by $F = 0$ and the period is $\pi/2m$.

494.   That cycloidal oscillations in a medium in which the resistance varies as the velocity are tautochronous has been proved by Newton in the second book of the *Principia*, Prop. XXVI. That the oscillations are tautochronous when the cycloid is rough has been deduced by M. Bertrand from Lagrange's formula, given in Art. 491, see *Liouville's Journal*, Vol. XIII. M. Bertrand ascribes the proposition to M. Necker, who published it in the fourth volume of the *Mémoires présentés à l'Academie des Sciences par des savants étrangers.* It follows of course from Lagrange's proposition (Art. 492) that the cycloid is tautochronous when the medium resists as the velocity, and at the same time the cycloid is rough.

495.   **Motion on any rough curve.** *A particle starts from rest and is constrained to move along a rough curve under the action of any forces, find the conditions of tautochronous motion.*

Let A be the point at which the tautochronous motion terminates, P the position of the particle at any time t, $AP = s$, so that s is measured from A in the direction opposite to that of motion. Let the tangent at P make an angle ψ with the axis of x, and let ψ and s increase together. Let the tangential and normal components of the force on P be G and H; the tangential component G acting on P to urge the particle *towards* A, and the normal component H acting outwards, *i.e.* opposite to the direction in which ρ is measured. Let the letters R, v, μ have the same meaning as before. We shall suppose ρ to be positive throughout the arc.

The equations of motion are therefore

$$\frac{v^2}{\rho} = R - H, \qquad v\frac{dv}{ds} = \mu R - G \dots\dots\dots\dots\dots\dots\dots(1).$$

Since the particle starts from rest, we see that R and H are initially equal and thus have the same sign. We shall suppose that H is positive throughout the motion, so that the impressed force urges the particle outwards. It follows that R also is positive throughout the motion. The friction continues therefore to be represented by μR, without any discontinuous changes in the sign of μ, such as would happen if R were to change sign without a corresponding change in the direction of the friction. (See Art. 159.) Eliminating R we find

$$v\frac{dv}{ds} = \mu\frac{v^2}{\rho} - (G - \mu H) \dots\dots\dots\dots\dots\dots\dots(2).$$

Let $F = G - \mu H$, so that F is the whole impressed force urging the particle along the tangent towards the point A. We may prove that F must be positive throughout the motion until the particle reaches A. If F were zero at any point B, then, placing the particle at rest at B, it will remain there in equilibrium, and therefore the times of reaching A from all points will not be the same. We see also by the same reasoning that the point A must be one at which F is zero. (See Art. 490.) Writing $\frac{ds}{d\psi}$ for ρ, equation (2) becomes

$$\frac{dv^2}{d\psi} - 2\mu v^2 = -2\rho F \dots\dots\dots\dots\dots\dots\dots\dots(3),$$

therefore
$$v^2 e^{-2\mu\psi} = c^2 - 2\int_\alpha^\psi \rho F e^{-2\mu\psi} d\psi \dots\dots\dots\dots\dots\dots(4),$$

where α is the angle made by the tangent at A with the axis of x. As ψ is greater than α throughout the motion, the constant of integration, viz. c^2, must be positive. If we put

$$2\int_\alpha^\psi \rho F e^{-2\mu\psi} d\psi = z^2, \quad e^{-\mu\psi} ds = \phi(z)\, dz,$$

this equation reduces to the form

$$-\int_c^0 \frac{\phi(z)\, dz}{\sqrt{c^2 - z^2}} = t \dots\dots\dots\dots\dots\dots\dots\dots(5).$$

The lower limit is determined by the value of z at the point where the motion begins. Referring to equation (4) we see that, since $v = 0$, we have $z = c$. The upper limit is determined by the value of z at the point where all the tautochronous motions are to end. This has been defined by $\psi = \alpha$, and therefore by $z = 0$.

If the force F be such that $\phi(z)$ is constant, the integration of (5) presents no difficulty. Writing $\dfrac{1}{m}$ for $\phi(z)$ we then have $t = \dfrac{\pi}{2m}$. Since this result is independent of c, the motion is tautochronous. Wherever the particle be placed at rest on the curve, it will reach the position A in the same time, and this time will be $\dfrac{\pi}{2m}$.

The supposition we have made regarding the value of z gives

$$\left\{ m\int_\alpha^\psi e^{-\mu\psi} ds \right\}^2 = 2\int_\alpha^\psi \rho F e^{-2\mu\psi} d\psi,$$

differentiating and reducing we find

$$F = m^2 e^{\mu\psi} \int_\alpha^\psi e^{-\mu\psi} \rho\, d\psi \dots\dots\dots\dots\dots\dots\dots(6).$$

This expression gives the tangential force required to make the motion on a given curve tautochronous. When the force is given and the curve is required, we may write the equation in the form

$$m^2\rho = \frac{dF}{d\psi} - \mu F \dots\dots\dots\dots\dots\dots\dots\dots(7).$$

496. *We shall now show that unless $\psi(z)$ be constant the motion will not be tautochronous.* To prove this we must find what forms of $\phi(z)$ will make the integral (5) independent of c. Put $z = c\xi$, and let $\phi(z)$ be expanded in a series of some powers of z not necessarily integral, say let $\phi(z) = \Sigma N z^n$. Then we have

$$\Sigma N c^n \int_0^1 \frac{\xi^n d\xi}{\sqrt{1 - \xi^2}} = t.$$

Now since ξ is less than unity, the integrals in all the terms are less than the integral in the term defined by $n = 0$, and therefore they are finite. Since every element in each integral is positive, none of the integrals can vanish. Hence this series of powers of c cannot be independent of c unless it contain only the one term determined by $n = 0$. But this makes $\phi(z)$ a constant and leads to the solution we have already discussed.

497. Ex. Show that this law of force coincides with that given by Lagrange's formula.

We have here to determine when equation (2) coincides with Lagrange's formula. We therefore take as his form

$$\frac{d^2s}{dt^2} = \frac{f'(s)}{f(s)} v^2 + A \frac{v^2}{f(s)} + f(s).$$

Comparing this with (2), term for term, we find $\dfrac{dF}{ds} - A = \mu F \dfrac{d\psi}{ds}$, which leads to the required form for F.

498. **Motion on a rough epicycloid.** *A particle acted on by a repulsive force, varying as the distance and tending from a fixed point, is constrained to move along a rough curve, find the curve that the motion may be tautochronous.*

Let r be the radius vector of the particle, λr the repulsive force acting along it. Let p be the perpendicular on the tangent from the origin, then the projection of the radius vector on the tangent is known to be $\dfrac{dp}{d\psi}$. Let A be the point on the curve at which the tautochronous motion is to terminate. Resolving the radial force along the tangent towards A and along the normal outwards, the components are respectively $G = -\lambda \dfrac{dp}{d\psi}$ and $H = \lambda p$, Art. 495. Since $F = G - \mu H$, we have $F = -\lambda \dfrac{dp}{d\psi} - \mu\lambda p$. Substituting in equation (7) of Art. 495 we find

$$-\frac{m^2}{\lambda} \rho = \frac{d^2p}{d\psi^2} - \mu^2 p = \rho - (\mu^2 + 1) p,$$

therefore
$$\rho = \frac{\mu^2 + 1}{\dfrac{m^2}{\lambda} + 1} p \quad \dots\dots\dots\dots\dots\dots\dots\dots\dots\dots\dots(1).$$

The equation to an epicycloid generated by the rolling of a circle whose radius is b on a fixed circle whose radius is a is known to be

$$p^2 = c^2 \frac{r^2 - a^2}{c^2 - a^2}, \quad \therefore \ \rho = \frac{c^2 - a^2}{c^2} p \dots\dots\dots\dots\dots\dots\dots\dots(2),$$

where r is the radius vector measured from the centre O of the fixed circle as origin and $c = a + 2b$. We find therefore that the epicycloid is a tautochronous curve for a central repulsive force varying as the distance. The time of arriving at the position of equilibrium is $\dfrac{\pi}{2m}$, where m is given by $\dfrac{m^2}{\lambda} = \dfrac{\mu^2 c^2 + a^2}{c^2 - a^2}$.

If ϕ be the angle made by the tangent with the radius vector, the position A of equilibrium is given by $\cot\phi = \mu$. Since $p = r\sin\phi$, we have by (2) $r^2 = \dfrac{a^2 c^2 (1 + \mu^2)}{a^2 + \mu^2 c^2}$. Let this value of r be represented by r_0. Since a is less than c, it is easy to see that a circle described with centre O and radius r_0 intersects the epicycloid. If A be any one of these points of intersection and C the nearest cusp, a particle starting from rest at any point D between A and C will describe the arc DA in a time which is independent of the position of D.

Ex. 1. Show that the equiangular spiral is not included in the formula (1).

For if we write $\rho \sin^2 a = p$, we have $(\mu^2 + 1)\sin^2 a$ greater than unity, which may be shown to be impossible if the particle is to move at all. Since the angle made by the tangent with the radius vector is the same at all points of the curve, we also notice that there is no point of equilibrium at which a tautochronous motion could properly terminate.

Ex. 2. If the central force be attractive and vary as the distance, show that a hypocycloid is a tautochronous curve.

499. Effect of a Resisting Medium on the time. We know by Lagrange's theorem that, if the motion on a rough curve be tautochronous in a vacuum, it is also tautochronous when the motion occurs in a medium resisting as the velocity. *Let us now determine how the time of arrival at the position of equilibrium is affected by the presence of such a resisting medium.*

Referring to Art. 495, the equation of motion there marked (2) now takes the form

$$\frac{d^2s}{dt^2} = \frac{\mu}{\rho}\left(\frac{ds}{dt}\right)^2 - 2\kappa\frac{ds}{dt} - F,$$

where 2κ is the coefficient of the resistance due to the medium. This equation may be put into the form

$$\frac{d}{dt}\left(e^{-\mu\psi}\frac{ds}{dt}\right) + 2\kappa\left(e^{-\mu\psi}\frac{ds}{dt}\right) = -Fe^{-\mu\psi}.$$

Let us put $e^{-\mu\psi}ds = dw$ and suppose that w vanishes when $\psi = a$. Substituting for F its value given by equation (6) of Art. 495 we find

$$\frac{d^2w}{dt^2} + 2\kappa\frac{dw}{dt} + m^2w = 0.$$

This solution of this by Art. 434 is

$$\int_a^\psi e^{-\mu\psi}\rho\, d\psi = w = Ae^{-\kappa t}\cos\left(\sqrt{m^2-\kappa^2}\,t + B\right).$$

To find the constants of integration we notice that $\frac{d\psi}{dt}$, and therefore $\frac{dw}{dt}$, is zero when $t = 0$. This gives $\tan B = \dfrac{-\kappa}{\sqrt{m^2-\kappa^2}}$. To find the time of arriving at the position of equilibrium we put $\psi = a$, this gives $\sqrt{m^2-\kappa^2}\,t + B = \dfrac{\pi}{2}$. The time t of the tautochronous motion is therefore given by the equation

$$\sqrt{m^2-\kappa^2}\,t = \frac{\pi}{2} + \tan^{-1}\frac{\kappa}{\sqrt{m^2-\kappa^2}}.$$

We notice that the time depends only on m and κ and not on the form of the curve.

Ex. If the resistance of the medium be so great that κ is equal to or greater than m, the solution by Art. 434 takes another form. Show that in both these cases the time of arriving at the position of rest is made infinite by the resistance of the medium.

Oscillations of Cylinders and Cones to the second order.

500. Condition of Stability of Cylinders to the higher orders. When a heavy cylinder rests in equilibrium on one side of a fixed rough cylinder as in Art. 441, the condition of stability is that the centre of gravity should lie within a certain circle called the circle of stability. If the centre of gravity lie on the boundary of this circle the equilibrium is called *neutral*, but it is generally either stable or unstable, a higher degree of approximation however, being required to distinguish the two. We may reach any degree of approximation by the following easy process, which amounts to the continued differentiation of a certain quantity until we arrive at a result which is not zero. The sign of this result distinguishes between the stability and instability of the cylinder. The magnitude of the result, joined to some other elements, enables us to form the equation of motion.

501. In equilibrium the centre of gravity is in the vertical through the point of contact. Let the body be turned round through any angle θ, so that G in the figure is the position of the centre of gravity, and I the point of contact. Let IV

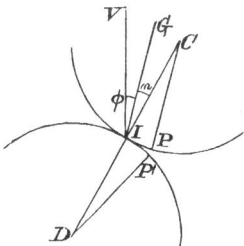

be vertical. Let CID be the common normal to the two cylinders, C and D being the centres of curvature of their transverse sections. Let $\rho = CI$, $\rho' = DI$, and let $\frac{1}{z} = \frac{1}{\rho} + \frac{1}{\rho'}$, so that z is the radius of relative curvature.

Let $IG = r$, the angles $GIC = n$, $GIV = \phi$, and let $IP = ds$. Then we have the four following subsidiary equations

$$\frac{dr}{ds} = \sin n, \qquad \frac{dn}{ds} = \frac{\cos n}{r} - \frac{1}{\rho}$$

$$\frac{d\phi}{ds} = \frac{1}{z} - \frac{\cos n}{r}, \qquad \frac{ds}{d\theta} = z.$$

Since GI is the radius vector of the upper curve referred to an origin G fixed relatively to it, and $\frac{\pi}{2} - n$ is the angle made by this radius vector with the tangent at I, the first of these subsidiary equations is evident. To obtain the second we notice that C is the centre of curvature, so that the distance GC is constant as well as the radius of curvature, when I moves a short distance ds along the arc. Now

$$GC^2 = r^2 + \rho^2 - 2\rho r \cos n,$$

therefore $0 = (r - \rho \cos n)\, dr + \rho r \sin n\, dn.$

Substituting for dr its value from the first subsidary equation, this immediately gives the second. To obtain the third equation we notice that $\phi + n$ is the angle made by the normal DI to the lower curve, which is fixed in space, with a straight line also fixed in space. Hence $\frac{d\phi}{ds} + \frac{dn}{ds} = \frac{1}{\rho'}$, whence the third equation follows from the second. The fourth equation has been proved in Art. 441; the proof may be summed up as follows. If CP, DP' be the two normals which are in a straight line when the body has turned through an angle $d\theta$, then $d\theta = PCI + P'DI$, which gives $ds\left(\frac{1}{\rho} + \frac{1}{\rho'}\right) = \frac{ds}{z}$.

502. In equilibrium the centre of gravity of the body must be vertically over the point of support. Hence $\phi = 0$. In any other position of the body the value of ϕ is given by the series

$$\frac{d\phi}{ds} s + \frac{d^2\phi}{ds^2} \frac{s^2}{1 \cdot 2} + \&c.$$

If in this series the first coefficient which does not vanish be positive and of an odd order, it is clear that the line IG moves to the same side of the vertical as that to which the body is moved. The equilibrium is therefore unstable for displacements on either side of the position of equilibrium. If the coefficient be negative the equilibrium is stable. On the other hand if the term be of an even order, it does not change sign with s, the equilibrium is therefore stable for a displacement on one side and unstable for one on the other side.

The first differential coefficient is given by the third subsidiary equation. The second differential coefficient is found by differentiating this subsidiary equation and substituting for $\frac{dn}{ds}$ and $\frac{dr}{ds}$ from the others. The third differential coefficient may be found by repeating the process. In this way we may find any differential coefficient which may be required.

503. If the first differential coefficient $\frac{d\phi}{ds}$ be not zero, the equilibrium is stable or unstable according as its sign is negative or positive. This leads to the condition that r must be respectively less or greater than $z \cos n$, which agrees with the rule given in Art. 441.

If $\frac{d\phi}{ds} = 0$, we have $r = z \cos n$, so that the centre of gravity lies on the circumference of the circle of stability. Differentiating we have

$$\frac{d^2\phi}{ds^2} = \frac{d}{ds}\left(\frac{1}{z}\right) + \frac{2 \sin n \cos n}{r^2} - \frac{\sin n}{r\rho} \quad\ldots\ldots\ldots\ldots\ldots\ldots(1).$$

Substituting for r and, z we have

$$\frac{d^2\phi}{ds^2} = \frac{d}{ds}\left(\frac{1}{\rho} + \frac{1}{\rho'}\right) + \tan n \left(\frac{1}{\rho} + \frac{1}{\rho'}\right)\left(\frac{1}{\rho} + \frac{2}{\rho'}\right).$$

If this be not zero, the equilibrium is stable for displacements on one side of the position of equilibrum and unstable for displacements on the other.

If $\frac{d^2\phi}{ds^2} = 0$ also, we differentiate (1) again. After some reduction we find

$$\frac{d^2\phi}{ds^2} = \frac{d^2}{ds^2}\left(\frac{1}{z}\right) + \frac{1}{z\rho'}\left(\frac{1}{\rho} + \frac{2}{\rho'}\right) - \frac{\tan n}{z}\frac{d}{ds}\left(\frac{1}{\rho}\right) - \frac{3 \tan^2 n}{z^2}\left(\frac{1}{\rho} + \frac{2}{\rho'}\right).$$

The equilibrium is stable or unstable according as this expression is negative or positive.

If the tranverse section be a circle or a straight line these expressions admit of great simplification.

504. Ex. 1. A heavy body rests in neutral equilibrium on a rough plane inclined to the horizon at an angle n. Show that, unless $\frac{d\rho}{ds} = \tan n$, the equilibrium is stable for displacements on the one side and unstable for displacements on the other. But, if this equality hold, the equilibrium is stable or unstable according as $\frac{d^2\rho}{ds^2}$ is positive or negative. Here ds is measured along the arc in the direction down the plane.

Show also that these conditions imply that the equilibrium is stable or unstable according as the centre of the conic of closest contact to the upper body is without or within the circle of stability.

Ex. 2. If a convex spherical surface rest on the summit of a fixed convex spherical surface in neutral equilibrium, the equilibrium is really unstable. But if the lower surface have its concavity upwards the equilibrium is stable or unstable according as its radius is greater or less than twice that of the upper surface, and is really neutral if its radius is equal to twice that of the upper surface.

The moveable spherical surface in this example must of course be weighted so that its centre of gravity is at such an altitude above the point of support that the equilibrium is neutral to a first approximation. Thus, when the radius of the lower surface is twice its radius, its centre of gravity lies on its surface, i.e. at a distance twice its radius from the point of contact. The centre of gravity is outside or within the surface according as the radius of the lower surface is less or greater than twice its radius, and when the lower surface is plane the centre of gravity lies at the centre. In this last case also the equilibrium is really neutral.

505. **Oscillations of Cylinders to the higher orders.** *To form to any degree of approximation the general equation of motion of a cylinder oscillating about a position of equilibrium.*

Following the same notation as before and taking the figure of Art. 501, the equation of vis viva is

$$(k^2 + r^2)\, \dot{\theta}^2 = C + 2U,$$

where U is the force function and k the radius of gyration of the body about its centre of gravity. Differentiating this with regard to θ, as in Art. 448, we have

$$(k^2 + r^2)\, \ddot{\theta} + r\, \frac{dr}{d\theta}\, \dot{\theta}^2 = \frac{dU}{d\theta}.$$

The right-hand side of this equation is by Art. 340 the moment of the forces about the instantaneous axis, and is therefore in our case equal to $gr \sin \phi$. Substituting for $\frac{dr}{d\theta}$ from the subsidiary equations of Art. 501, the equation of motion is therefore

$$(k^2 + r^2)\, \ddot{\theta} + rz \sin n\, \dot{\theta}^2 = gr \sin \phi.$$

The method of proceeding is the same as that in Art. 502. We expand each coefficient by Taylor's theorem in powers of θ, which is to be so chosen as to vanish in the position of equilibrium. To do this we require the successive differentials of these coefficients to any order expressed in terms of the *initial* values only of ϕ, n, and r. The first differentials are given in the subsidiary equations of Art. 501. To find the others we continually differentiate these subsidiary equations, until we have obtained as many differential coefficients as we require.

506. *To form the equation to the first order.* Let the initial or equilibrium values of n and r be a and h. The equation is therefore

$$(h^2 + k^2)\, \ddot{\theta} = gr \sin \phi.$$

We have to find $r \sin \phi$ to the first power of θ. Now

$$\frac{d}{d\theta}\, (r \sin \phi) = \frac{dr}{d\theta} \sin \phi + r\, \frac{d\phi}{d\theta} \cos \phi = z \sin n \sin \phi + rz \cos \phi \left(\frac{1}{z} - \frac{\cos n}{r} \right),$$

by substituting from the subsidiary equations. This by reduction becomes

$$\frac{d}{d\theta}\, (r \sin \phi) = r \cos \phi - z \cos (\phi - n).$$

R. D.   27

In equilibrium G lies in the vertical through the point of contact, hence the initial value of ϕ is zero. The equation of motion is therefore

$$(h^2 + k^2)\,\ddot{\theta} = (h - z \cos a)\,g\theta,$$

which is the same as that given in Art. 441.

507.　*To form the equation to the second order.*

We have already found the first differential coefficient of $r \sin \phi$, we must differentiate this again and retain only the terms which do not vanish when $\phi = 0$. We have

$$\frac{d^2}{d\theta^2}(r \sin \phi) = z^2 \left\{ z \cos a \, \frac{d}{ds} \frac{1}{z} + \frac{\sin 2a}{h} - \frac{\sin a}{\rho} \right\}.$$

The equation of motion to the second order is therefore

$$(k^2 + h^2 + 2hz \sin a \cdot \theta)\,\ddot{\theta} + hz \sin a\,\dot{\theta}^2$$
$$= -(z \cos a - h)\,g\theta + gz^2 \left\{ z \cos a \, \frac{d}{ds} \frac{1}{z} + \frac{\sin 2a}{h} - \frac{\sin a}{\rho} \right\} \frac{\theta^2}{2}.$$

This may be reduced to the form

$$\ddot{\theta} + a^2\theta = -b^2\dot{\theta}^2 + c\theta^2,$$

where

$$a^2 = \frac{z \cos a - h}{k^2 + h^2}\,g, \qquad b^2 = \frac{hz \sin a}{k^2 + h^2},$$

$$c = 2a^2b^2 + \tfrac{1}{2}g\,\frac{z^2}{k^2 + h^2}\left\{ z \cos a \, \frac{d}{ds} \frac{1}{z} + \frac{\sin 2a}{h} - \frac{\sin a}{\rho} \right\}.$$

Supposing a not to be zero, we find as the solution

$$\theta = A \sin(at + B) + \frac{c - a^2b^2}{2a^2}\,A^2 + \frac{c + a^2b^2}{6a^2}\,A^2 \cos 2\,(at + B),$$

where A and B are two undetermined constants, and the first term represents the first approximation. Thus it appears that the first approximation is substantially correct unless a be small, that is, unless the equilibrium is nearly neutral. The effect of the small terms is to make the extent of the oscillation on the lower side of the position of equilibrium slightly greater than that on the upper side.

508.　**Oscillations of Cones to the higher orders.** *To form the general equation of motion of a heavy cone rolling on a perfectly rough fixed cone.*

Let us follow the same line of argument with the same notation as in Art. 483. We have however one point of difference. Since the moving cone is not in equilibrium its centre of gravity is not in the vertical plane WOI. As before let the arcs $IG = r$, $IW = z$, and the angles $GIC = n$, $WIC = \psi$.

Let Ω be the angular velocity of the moving cone about its instantaneous axis OI. Then, by Art. 448,

$$K^2 \frac{d\Omega}{dt} + \frac{1}{2}\,\Omega\,\frac{dK^2}{dt} = L \quad\text{...............................(1)},$$

where L is the moment of gravity about OI.

As the cone rolls, the point I moves along the intersection of the fixed cone with the sphere. Let $IP = ds$ be the arc described in a time dt. It will be convenient to take s as the co-ordinate by which the position of the cone is determined.

By the same reasoning as in Art. 484 we find

$$\Omega = \frac{ds}{dt}\,\frac{\sin(\rho + \rho')}{\sin \rho \cdot \sin \rho'}, \quad\text{.................................. (2).}$$

We have now to find the moment of gravity about OI. We again use the same argument as in Art. 485. Resolving gravity along and perpendicular to OI, the

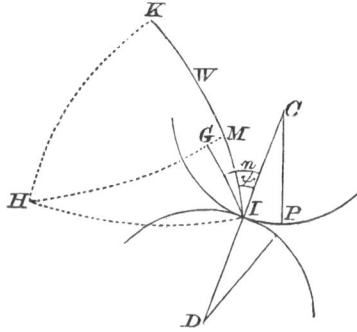

former component has no moment, and the latter is $g \sin z$. Let this latter component act parallel to some straight line KO, then KWI is an arc in a vertical plane. The moment required is then the product of resolved gravity into the projection of OG on OH, where H is the pole of the arc KWI. Thus the moment is $gh \sin z \cos HG$. To find $\cos HG$ produce HG to cut KWI in M. Then, in the right-angled triangle GIM, we have $\sin GM = \sin GI \sin GIM$. The moment L is therefore

$$L = -gh \sin r \sin z \sin (n - \psi) \quad \dots\dots\dots\dots\dots\dots\dots(3).$$

When the forms of the cones are known, we can express K, r, z, n and ψ in terms of s or any other co-ordinate we may choose. The equation of motion will then be known. This process may be effected by the help of the four following subsidiary equations

$$\left.\begin{aligned}
\frac{dr}{ds} &= \sin n, \quad \frac{dz}{ds} = \sin \psi \\
\frac{dn}{ds} &= \cot r \cos n - \cot \rho \\
\frac{d\psi}{ds} &= \cot z \cos \psi + \cot \rho'
\end{aligned}\right\} \quad \dots\dots\dots\dots\dots\dots\dots(4).$$

The proof of these is left to the reader. They may be obtained by the same reasoning as in the case of the cylinder, with only such modifications as are made necessary by using spherical instead of plane triangles.

509. *To find to any degree of approximation the equation of motion of a right cone oscillating about a position of equilibrium.*

Since the cone is a right cone, we have K^2 constant. The equation of motion is therefore $K^2 \dfrac{d\Omega}{dt} = L$, where Ω and L have the values given in equations (2) and (3) of Art. 508.

We notice that $L = 0$, and therefore $n = \psi$ in the position of equilibrium. Let the co-ordinate s be so chosen that it also vanishes in this position. We have therefore now to expand Ω and L in powers of s. To effect this we use Taylor's theorem, thus

$$L = \left(\frac{dL}{ds}\right) s + \left(\frac{d^2L}{ds^2}\right) \frac{s^2}{1 \cdot 2} + \dots,$$

where the bracket implies that s is to be made equal to zero after the differentiations have been performed. Now these differentiations may all be performed without any difficulty, by using the expression for L given in (3) and continually substituting for $\dfrac{dr}{ds}$, $\dfrac{dz}{ds}$, &c. their values given in the subsidiary equations (4). We may treat Ω in the same way.

The formation of the equation of motion is thus reduced to the differentiation of a known expression and the substitution of known functions.

We may use this method to obtain the equation of motion to the first power. Thus we have

$$K^2 \frac{d\Omega}{dt} = -gh \frac{d}{ds} \left\{ \sin r \sin z \sin (n - \psi) \right\} s.$$

Substituting for Ω and retaining on the right-hand side only those terms which do not vanish when $\psi = n$ we obtain

$$\frac{K^2}{gh} \frac{d^2 s}{dt^2} = - \left\{ \sin (z - r) \cos n \frac{\sin \rho \sin \rho'}{\sin (\rho + \rho')} - \sin r \sin z \right\} s,$$

which gives the same result as in Art. 483.

If the cone is not a right cone, we may express K^2 in terms of r and n and proceed in the same way.

510.   **Ex.**   A heavy right cone rests in neutral equilibrium on another right cone which is fixed in space, the vertices being coincident. Show that the equation of motion, including the squares of small quantities, is

$$\frac{K^2}{gh} \frac{d^2 s}{dt^2} = - \frac{\sin \rho \sin \rho'}{\sin (\rho + \rho')} \sin (r - z) \sin n \left\{ (\cot z + 2 \cot r) \cos n - \cot \rho \right\} \frac{s^2}{2}.$$

END OF THE FIRST VOLUME.

CAMBRIDGE: PRINTED BY C. J. CLAY, M.A. AND SONS, AT THE UNIVERSITY PRESS.